HANDBOOK OF COMPOSITES
VOLUME 3

HANDBOOK OF COMPOSITES

VOLUME 3

Series Editors

A. Kelly

Vice-Chancellor's Office
University of Surrey
Guildford
Surrey GU2 5XH
England

Yu.N. Rabotnov

Academy of Sciences of the USSR
Department of Mechanics and Mathematics
Moscow State University
117234 Moscow
USSR

NORTH-HOLLAND – AMSTERDAM • NEW YORK • OXFORD

FAILURE MECHANICS OF COMPOSITES

Volume Editors

G.C. Sih

Institute of Fracture and Solid Mechanics
Lehigh University
Bethlehem, PA 18105
USA

A.M. Skudra

Department of Structural Mechanics
Riga Polytechnic Institute
Riga 226355
USSR

NORTH-HOLLAND – AMSTERDAM • NEW YORK • OXFORD

ISBN: 0 444 86879 8

First edition: 1985
Second printing: 1986

Publishers:
ELSEVIER SCIENCE PUBLISHERS B.V.
P.O. Box 1991
1000 BZ Amsterdam
The Netherlands

Sole distributors for the U.S.A. and Canada:
ELSEVIER SCIENCE PUBLISING COMPANY, INC.
52 Vanderbilt Avenue
New York, N.Y. 10017
U.S.A.

Library of Congress Cataloging in Publication Data
Main entry under title:

Failure mechanics of composites.

(Handbook of composites; v. 3)
Includes index.

1. Composite materials--Fracture. 2. Composite materials--Testing.
I. Sih, G.C. (George C.) II. Skudra, Albert Martynovich. III. Series
TA418.9C6F33 1984 620.118 84-1628
ISBN 0-444-86879-8 (U.S.)

PRINTED IN THE NETHERLANDS

Preface to Series

During 1978 Dr. W.H. WIMMERS, at that time Managing Director of the North-Holland Publishing Company, made arrangements with the Publishers Atomizdat and his own Company, the North-Holland Publishing Company, for the publication simultaneously within the Soviet Union and in the West of Handbooks concerning technical subjects of importance. The idea was that Handbooks consisting of a number of volumes should be edited jointly by an Editor from the Soviet Union and an Editor from the West. These Editors would agree a format for each particular volume and would encourage Authors of individual chapters to write these, the Authors to be chosen from the Soviet Union and from the Western World.

It was hoped, by this initiative, that scientists and technologists in the West and within the Soviet Union would learn something more of one another than they presently glean from the currently published literature.

It would, of course, have been nice to include within this idealistic framework Authors from the so-called Third World to which, of course, the People's Republic of China belongs but a journey of a thousand miles starts with a single step and the first arrangements were possible with the publishing house Atomizdat and with North-Holland.

These volumes are offered by us as a first step along the road towards the idea of publication throughout the world of tracts of importance. There have been a number of difficulties to overcome in reconciling the different cultural systems and the different ways of operating of the scientific communities of the Soviet Union, those in Western Europe and in the United States. At times the interaction has been slow and uncertain. However, we are grateful to all the Authors for their patience and their ready acceptance of Editorial comment.

We are very glad to see this volume appear in print and we hope there will be many others like it.

YU.N. RABOTNOV

A. KELLY

Preface to Volume 3

Fiber-reinforced composites continue to generate interest in their application to high performance structures. They are being used increasingly more as load bearing structural components in both commercial and military aircrafts. This is mainly because they are superior in strength, light in weight and tolerable to damage. Composites can be more readily designed to fail in a non-catastrophic manner such that yielding and/or cracking are distributed more uniformly rather than concentrated locally. High local energy release should be avoided or minimized as it is conducive to sudden fracture. Although much research has been done since the early 1960's, predictive capability in composite behavior is still lacking. A better understanding of specimen testing and structure application is needed before composites can be successfully applied in design and manufacture. This calls for expanding knowledge in several areas of discipline. The influence of fiber and matrix constituent properties and bonding process on composite behavior is not always clear. Improved methods of testing for determining the constituent and composite properties are needed so that reliable data can be generated to better predict composite structures long-term integrity under service conditions such as creep, impact, etc. Imperfections and defects introduced during processing are essential as they can alter composite response. It is with these objectives in mind that this volume on "Failure Mechanics of Composites" is edited. Contributors from the U.S. and U.S.S.R. represent a joint effort towards the further advancement of composite technology.

Chapter 1 deals with the micromechanics of failure of reinforced plastics. The integration of constituents involving fiber orientation, volume fraction, spacing and matrix into composite materials and their characterization are considered for different fiber and matrix type. Practical methods for determining the strength of reinforced plastics subjected to uniaxial and biaxial stress states are discussed. Presented are strength criteria that can be applied to optimize the matrix, fiber and bonding such that the simultaneous failure of constituents can be achieved. An extensive review of conventional composite failure criteria is given in Chapter 2. Predicted results are compared with those obtained by tests. Limited attention is focused on identifying failure loads with failure modes while the majority of the work is concerned with predicting only the onset of global failure. Various utilized and proposed biaxial specimens and testing techniques are reviewed, and their advantages and disadvantages noted. Relative agreement between predicted and experimentally observed results is demonstrated for those theories whose stress-strain information is generated numerically to failure. Emphases are placed on the correlation between prediction and observation. Dynamic analyses of nonhomogeneous materials such as composites are not only complex but difficult. An attempt is made in Chapter 3 to give an overview of the type of dynamic composite systems with cracks that can be analyzed analytically. Both stationary and moving cracks are considered. Because of the lack of load, geometry and material symmetry, failure by non-self-similar crack growth is the rule rather than the exception. The effect of local material nonhomogeneity on crack growth is shown to have a significant influence on the global behavior of the composite system. Assumed are the existence of unique threshold levels of the strain energy density function associated with the different failure modes such as fiber breaking, matrix cracking, etc. Because of the fracture toughness information of the constituents and their precise failure modes, the load carrying capacity of the composite system can be predicted. Only in this way can predictive capability be developed to design composite structures. Any parameters whether they represent strength or toughness are of limited use if their values tend to change when the relative orientation of load and composite microstructure is altered. Chapter 4 emphasizes the need to continuously develop new testing methods in order to keep pace with composite technology advancement. Test methods for unidirectional composites subjected to tensile, compressive, shear and bending loads are presented. Composites reinforced by fibers oriented at 0°, 90° and ±45° in different layers bonded together are known as laminates. The loading scheme for testing laminate plates, rings and cylinders is presented. Results are discussed in connection with estimating the strength and stiffness of composite systems. The main objective of Chapter 5 is to investigate the long-time or creep failure of composites in terms of the constituent properties. To this end, a phenomenological model by identifying the conditions under which composites rupture under creep. Attention is focused on the time-dependent strength that changes with different failure modes caused by alteration in loading rates for a given composite. The results are represented by the $\sigma \sim \log \dot{\sigma}$ curve containing three distinct stages. The first corresponds to strength increasing with loading rate attributed to damage accumulation; the second to decreasing strength with increasing loading rate caused by the growth of defects or cracks; and the third to instantaneous failure due to breaking of the weakest element. Chapter 6 concentrates on the application of experimental methods for analyzing the stress and failure of composites. Described are several techniques including photoelasticity, strain gages, Moiré, holography and interferometric technique and nondestructive evaluation methods such as ultrasonics, acoustic emission, X-ray and thermography. Examples on specific applications are also given. Defects and imperfections are known to have a marked influence on composite behavior. This, as pointed out in Chapter 7, can arise in processing. Imperfections can be classified in terms of fiber discontinuity, debonding between fiber and matrix, cracks in matrix, nonuniform distribution of fibers, delaminations, etc. These individual defects may be accounted through different coefficients and related to the degradation of composite strength when compared with data for the defect-free specimens. Studied are also the interaction of fiber diameter and length with defects in terms of the load carrying capacity of the composite. The final chapter of this volume is concerned with the mathematical formulation of fiber reinforced composite in which the influence of interface cracks on the global behavior of the composite is accounted for. The size, orientation and concentration of these cracks can affect the failure modes depending on the nature of loading. Results are discussed in example problems.

In closing, it should be reiterated that the basic composition, characteristics and processing technology of composites should be integrated with the mechanics of composite behavior and failure and fabrication and testing so that high performance composite structures can be designed with confidence. The multifaceted nature of composite material technology cannot be overemphasized. The editors wish to take this opportunity to thank the authors of this volume. It is their cooperative efforts that has led to the completion of this work.

March, 1983

G.C. Sih
Bethlehem, Pennsylvania, U.S.A.

A.M. Skudra
Riga, U.S.S.R.

Contents

List of Contributors

I.M. Daniel, *Illinois Institute of Technology, Chicago, IL* (Chapter VI)
G.M. Gunyaev, *All-Union Institute of Aviation Materials, Moscow* (Chapter VII)
T. Kincis, *Academy of Sciences of the Latvian S.S.R., Riga* (Chapter V)
R.E. Rowlands, *University of Wisconsin, Madison, WI* (Chapter II)
G.C. Sih, *Lehigh University, Bethleham, PA* (Chapter III)
A.M. Skudra, *Riga Polytechnic Institute, Riga* (Chapter I)
J.V. Suvorova, *Academy of Sciences of the U.S.S.R., Moscow* (Chapter IV)
Yu. M. Tarnopol'skii, *Academy of Sciences of the Latvian S.S.R., Riga* (Chapter V)
G.A. Vanin, *Academy of Sciences of the Ukrainian S.S.R., Kiev* (Chapter VIII)

CHAPTER I

Micromechanics of Failure of Reinforced Plastics

A.M. Skudra

Department of Structural Mechanics (Strength Micromechanics)
Riga Polytechnic Institute
Riga 226355
U.S.S.R.

Contents

HANDBOOK OF COMPOSITES, VOL. 3 – Failure Mechanics of Composites
Edited by G.C. SIH and A.M. SKUDRA

List of Symbols

x_1, x_2, x_3 Elastic symmetry axes of a laminate

$\|, \perp, \perp\!\!\perp$ Elastic symmetry axes of a unidirectionally reinforced laminate;

r, z Elastic symmetry axes of transversally isotropic fibres

$\sigma_{ij}, i, j=1, 2, 3$

Components of stress tensor

$\sigma_i, i=1\cdots6$ Components of stress matrixes for a homogeneous material

$\langle\langle\sigma_i\rangle\rangle, \langle\langle e_i\rangle\rangle$ Mean stresses and strains of an orthotropic laminate in elastic symmetry axes

$\langle\sigma_\|\rangle, \langle\sigma_\perp\rangle$ Mean normal stresses in the unidirectionally reinforced lamina parallel and perpendicularly to the reinforcement

$\langle\tau_{\|\perp}\rangle$ Mean stresses of longitudinal shear in a unidirectionally reinforced lamina

r, θ Polar coordinates of a point

$\sigma_{rr}, \sigma_{\theta\theta}, \sigma_{zz}$ Normal stresses in polar coordinates

$\sigma_{r\theta}, \sigma_{rz}$ Shear stresses in polar coordinates

$\bar{\sigma}_r, \bar{\sigma}_\theta, \bar{\sigma}_z, \bar{\sigma}_{r\theta}, \bar{\sigma}_{rz}$

Dimensionless structural parametres of a composite characterizing the stress concentration

E_1, E_2, E_3 Elastic moduli of a laminate in the elastic symmetry directions X_1, X_2 and X_3

G_{12}, G_{13}, G_{23} Shear moduli of a laminate

ν_{ij} Poisson's ratio determining the transverse stain in direction j with the load acting in i-direction

E_{fz}, E_{fr} Elastic moduli of anisotropic fibres in axial and radial directions

G_{frz} Shear modulus of anisotropic fibres

$\nu_{fr\theta}, \nu_{frz}$ Poisson's ratios for anisotropic fibres

E_f, G_f, ν_f Elastic characteristics of isotropic fibres

r_f Fibre radius

p, l Fibre packaging parametres

E_m, G_m, ν_m Elastic characteristics of the matrix

v_f Relative volume content of fibres

$E_\|, E_\perp, G_{\|\perp}, \nu_{\|\perp}, \nu_{\perp\|}, \nu_{\perp\perp\!\!\perp}$

Elastic characteristics of a unidirectionally reinforced material

m_1 The ratio between the volume reinforced in the direction X_1 and the total volume of the cross-ply composite (fiber embedding ratio)

Q_{ij} Stiffness matrix of a unidirectionally reinforced material

$\bar{Q}_{ij}$ Stiffness matrix of a unidirectionally reinforced plastic in the axes inclined at angle β

A_{ij} Stiffness matrix of a laminate
$\bar{A}_{ij}$ The relative stiffness matrix of a laminate
$\sigma_m^+, \sigma_m^-, \tau_m^u$ Tensile, compressive and shear strengths of the matrix
σ_b^u, τ_b^u Tensile and shear strengths of the bond
e_{fu}^+, e_{fu}^- Ultimate strains of fibres in tension and compression
$\sigma_\parallel^+, \sigma_\perp^+$ Tensile strengths of a unidirectionally reinforced plastic in the elastic symmetry directions
$\sigma_\parallel^-, \sigma_\perp^-$ Compressive strengths of a unidirectionally reinforced plastic in the elastic symmetry directions
$\tau_{\parallel\perp}^u$ Strength of a unidirectionally reinforced plastic in longitudinal shear
$\sigma_1^+, \sigma_2^+, \sigma_1^-, \sigma_2^-$
Tensile and compressive strengths of a laminate in the elastic symmetry directions X_1 and X_2
τ_{12}^u Shear strength of a laminate
β Angle determining fibre orientation
φ Angle determining orientation of failure plane

1. Introduction – The character of fracture

Reinforced plastics consist of a polymer matrix and fibres. The combined performance of the matrix and the fibres is guaranteed by the bond between them. Fracture of reinforced plastics may result from flaws in fibres or matrix as well as may be caused by the failure of the bond. Therefore the strength of such a composite is governed by the strength characteristics of the three constituent elements, i.e., fibres, matrix and bond.

The peculiarity of fracture is that the three constituents of reinforced plastics do not fail simultaneously. This is explained by their differing ultimate characteristics. Initial hypothesis applied to develop strength criteria, should take into account precisely which of the constituents will fail first.

The strength criteria given by GOLDENBLAT and KOPNOV (1968) and TSAI and WU (1971) may be used for evaluating the strength of a reinforced plastic under combined loading. These criteria include reinforced plastic strength under simple loading (tension, compression and shear) which depend largely upon the structure of the material and the volume content of the fibres.

The application of these criteria is limited by the fact, that they are valid only for one particular type of material (i.e., under definite structure conditions). The peculiarity of the criteria is, that they are based on mean stresses which means they are valid for homogeneous anisotropic materials. It should be remarked that in the case of reinforced plastics the term 'mean stress' may be regarded from two aspects, depending on the degree to which we ignore the structure of the material. Further the term 'mean stress' will be used to denote those stresses which are calculated without any consideration of the composite structure; the stresses in the polymer matrix and the fibres, the microstructure of which is not considered in the calculation, will be called "structural stresses in the components." In works by other authors the two types of stresses are often referred to as macro- and micro-stresses respectively.

In the case of non-homogeneous anisotropic materials like reinforced plastics the actual stresses in the components differ essentially from the mean stresses. There are both quantitative and qualitative differences. Thus, for instance, the strength criteria for homogeneous anisotropic materials cannot embrace the stresses within the individual layers of a composite, the stress concentrations, the interlaminar shear stresses, initial stresses in the components etc. Besides, at uniaxial loading tension or compression the reinforced plastics as to their mean stresses, will be in a linear (uniaxial) stress state. Actually the components like the polymer matrix will be under plane or volume state of stresses even in this simple case of loading. The strength determination of each component on which the total strength of the composite depends will require the application of such criteria where the actual state of stress has been considered. Consequently, a promising method for solving strength problems with due attention to the actual behaviour of the reinforced plastic, is the predetermination of the composite strength on the basis of the existing states of stress or strains in the constituents – the matrix, the fibres and their interface. The mathematical method which makes this problem solving possible will further be called micromechanics of failure of composite materials.

2. Strength criteria for the constituents

2.1. *Strength of the polymer matrix*

The strength and the ultimate strain in the polymer matrix depend largely upon the mode of loading and temperature. It has been established by experiments (SKUDRA et al. (1971)) that under constant loading the ultimate strain of polyester resin in tension will not be constant, but it will depend upon the duration of loading until fracture occurs, and the ultimate strain will increase significantly with the increase of the pre-fracture loading period. This time-dependence of the ultimate strain includes the energy criterion, which means that a visco-elastic material fails with time if the work of stresses W_{m} reaches an ultimate value W_{mu}. Thus the strength criterion for the polymer matrix may be written as follows

$$W_{\mathrm{m}} = W_{\mathrm{mu}}\,. \tag{1}$$

The ultimate work of stresses at the moment of failure is an outstanding feature of the strength characteristics of any polymer matrix. In the case of linear deformation of the polymer matrix up to its complete failure the value of W_{mu} is obtained from the formula

$$W_{\mathrm{mu}} = \frac{\sigma_{\mathrm{m}}^2}{2E_{\mathrm{m}}}\,. \tag{2}$$

Let us analyse a case of simple combined compression and shear loading of the polymer matrix according to the diagram shown in Fig. 1. In this case two types of fracture are possible depending on the relation between stresses σ_1 and σ_6 (see Fig. 1).

If the ratio $|\sigma_1/\sigma_6|$ does not exceed a certain limit, the fracture will result from the principal tensile stresses and the mode of failure is schematically shown in Fig. 1a. This type of fracture will be dealt with in a more detailed way.

The polymer matrix represents a brittle material. The physical reason of brittle material fracture is the action of tensile stresses. Therefore the following practical hypothesis will be adopted for shear and compressive strength determination of the polymer matrix: the matrix fails when the specific work of the principal tensile stress reaches its ultimate value. In every particular case the specific work of the principal tensile stress is obtained from the relationship:

$$W_{\mathrm{m}}^{+}=\frac{1}{4E_{\mathrm{m}}}\left[2(1+\nu_{\mathrm{m}})\sigma_6^2+\sigma_1^2-\sigma_1\sqrt{\sigma_1^2+4\sigma_6^2}\right]. \tag{3}$$

Equalization of the value W_{m}^{+} and the ultimate value of the specific work W_{mu}^{+}, obtained from formula (2), gives us the equation of the strength envelope for the polymer matrix under combined shear and compressive loading:

$$\sigma_1^2+2(1+\nu_{\mathrm{m}})\sigma_6^2-\sigma_1\sqrt{\sigma_1^2+4\sigma_6^2}=2(\sigma_{\mathrm{m}}^{+})^2. \tag{4}$$

It should be marked that the applicability of the strength condition (4) is limited, since it becomes meaningless when the applied compressive stress values exceed the strength of the polymer matrix. The ultimate curve plotted from (4) is given in Fig. 2.

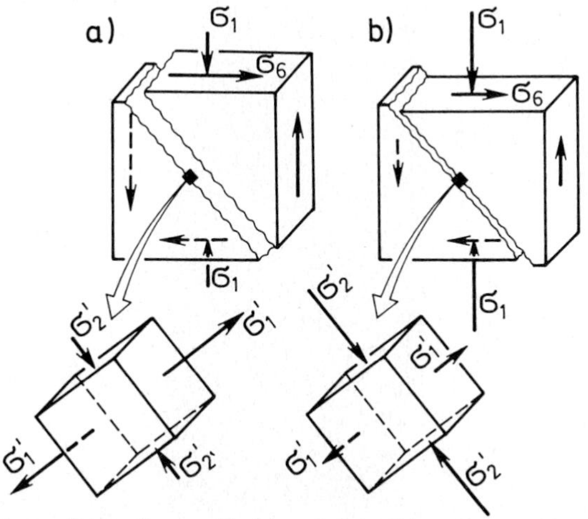

FIG. 1. Possible fracture schemes for tensile (a) and shear (b) failure of the polymer matrix under combined action of compressive and shear stresses.

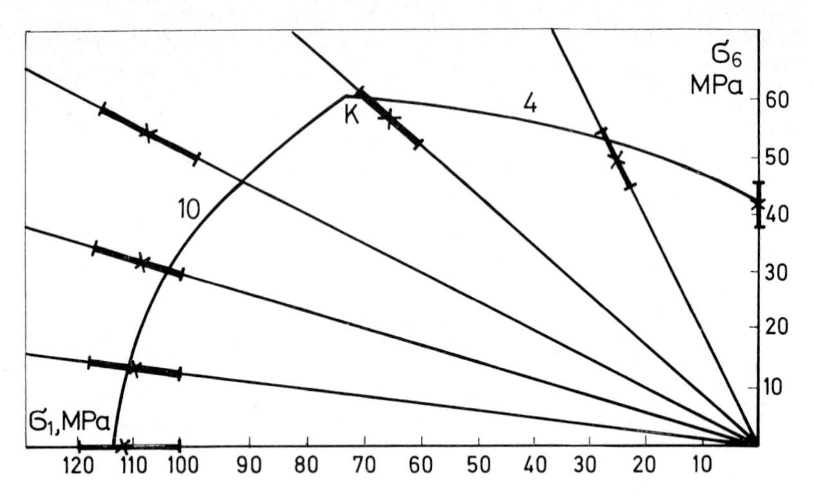

FIG. 2. Strength envelope for epoxy matrix ED-16. The numbers near the curves indicate from which formulas they have been plotted.

The uniaxial compressive strength of the polymer matrix can also be obtained from Equation (4).

Fig. 3 shows a characteristic fracture mode of the polymer matrix under compression. It is seen from Fig. 3 that this type of polymer matrix failure is typical of shear. Normal stresses σ_1 and shear stresses σ_6 act upon a sloped fracture plane oriented at an angle φ. Designating the ultimate value of compressive stress at the moment of fracture by σ_m^- and taking into consideration the expressions for σ_1 and σ_6 from (4) we get the following

FIG. 3. Mode of fracture for epoxy matrix ED-16 in compression.

relationship between the polymer matrix strength σ_m^- and the fracture angle φ:

$$2\left(\frac{\sigma_m^+}{\sigma_m^-}\right)^2 = \sin^4\varphi + \frac{1+\nu_m}{2}\sin^2 2\varphi - \sin^3\varphi\sqrt{4-3\sin^2\varphi}\,. \tag{5}$$

It follows from (5) that the given fracture angle corresponds to a definite strength value σ_m^-. The least of these values represents the actual matrix strength in compression, and the angle corresponding to this value is the actual fracture angle.

Consequently the problem is reduced to finding the maximum of function $2(\sigma_m^+/\sigma_m^-)^2$. By solving the problem and substituting the obtained angle value in (5) we get

$$\sigma_m^- = k_c\sigma_m^+\,, \tag{6}$$

where

$$k_c = \left[\frac{1}{2}\left(\sin^4\varphi + \frac{1+\nu_m}{2}\sin^2 2\varphi - \sin^3\varphi\sqrt{4-3\sin^2\varphi}\right)\right]^{-1/2} \tag{7}$$

Formula (6) provides the following relationship between the ultimate values of specific tensile and compressive properties of the polymer matrix:

$$W_{mu}^- = k_c W_{mu}^+\,. \tag{8}$$

From the above calculation it follows that fracture angle is 39° for the epoxy matrix if $\nu_m = 0.35$, which correlates with the test data (see Fig. 3). We should observe that the fracture angle φ for a polyester matrix is approximately of the same magnitude. If $\nu_m = 0.35$ and $\varphi = 39°$, we get the following relationship between the compressive and tensile strengths of the polymer matrix through using formula (6):

$$\sigma_m^- = 2.28\,\sigma_m^+ \approx 2.3\,\sigma_m^+\,. \tag{9}$$

When the stress ratio $|\sigma_1/\sigma_6|$ exceeds a certain limit the fracture proceeds as shown in Fig. 1(b). This mode of fracture takes place when the compressive stresses approach the compressive strength of the polymer matrix. Then the ultimate strength curve equation is obtained from the energy criterion, taking into account Equation (8):

$$\sigma_1^2 + 2(1+\nu_m)\sigma_6^2 + \sigma_1\sqrt{\sigma_1^2 + 4\sigma_6^2} = 2k_c^2(\sigma_m^+)^2\,. \tag{10}$$

The strength envolope based on formula (10) is plotted in Fig. 2. Therefore the complete strength envelope is seen to consist of two parts, described by formulas (4) and (10) respectively. The application area of the two formulas determines the location of point k in Fig. 2. The point coordinates are found from

$$\sigma_1 = \frac{\sigma_m^+}{\sqrt{1+\nu_m}} \sqrt{1 + k_c^2 - \sqrt{4k_c^2 + \nu_m^2(1-k_c^2)^2}}\,,$$
$$\sigma_6 = \frac{\sigma_m^+}{\sqrt{2(1-\nu_m^2)}} \sqrt{\sqrt{4k_c^2 + \nu_m^2(1-k_c^2)^2} - \nu_m(1+k_c^2)}\,. \tag{11}$$

Fig. 2 also presents the points obtained by experiment for epoxy resin EDT-10, each point representing the mean result of 6 tests.

2.2. Fibre strength

Various fibres may be used for producing reinforced plastics: glass, boron, carbon, organic material, etc. Most of them are brittle, their strength depending largely on the surface flaws. The effect of these factors is apparent from the scattering of the test data obtained from strength investigations with constant-length fibres.

The use of fibre bundles as plastics reinforcement is not infrequent. These bundles consist of hundreds of individual filaments. The testing of such bundles reveals random ruptures of separate fibres after load had been applied. The failure occurs in the fibres having a larger number of flaws. Evidently the bundle strength depends mostly on the scattering in the strength properties of some fibres. The scattering is estimated on the basis of the strength variation rate of the constituent fibres.

It should be noted that the strength variation rate has a very limited practical use since the actual behaviour conditions of a fibre bundle incorporated in a reinforced plastic differ greatly from those of an unbedded bundle. This is explained by the fact that the presence of the matrix localizes the rupture phenomenon of the individual fibres, and they tend to preserve their load-bearing capacity to some extent even after the rupture has taken place. So it may be assumed that the failure of only some monofibres does not weaken the composite strength, because there is a load transfer from the fractured fibres to the neighbouring undamaged ones due to the presence of the polymer matrix.

A question arises: how many of the fibres can be safely ruptured without reducing materially the composite strength? (The extreme case would be the rupture of all the fibres into a great number of small fragments). The answer is provided by comparing the monofibre strength distribution with the strength distribution of the compositive having a parallel layup of similar fibres. LOCKWOOD et al. (1963) give a comparison between the strength distribution of a bundle of E-glass monofibres and that of an epoxy resin composite reinforced by a bundle of the same type of E-glass fibres. The results of LOCKWOOD et al. (1963) show the composite strength to be equal with that of the fibre bundle, when the degree of fibre fracture is 50 per cent for the composite and 20 per cent for the fibre bundle.

The wide 20 to 80 per cent variation in the degree of composite fracture corresponds to a narrow range of stress variations. LOCKWOOD et al. (1963) showed that with a 15 per cent degree of fibre fracture the strengths of the composite and the fibre bundle are equal to 3290 MPa. Assuming that the ruptured fibres do not sustain any load, the mean stress in the remaining intact fibres of the composite will go up to 3870 MPa. Such a stress will create conditions under which 70 per cent of the individual fibres would fail. It is unlikely that the sample could preserve its load-bearing capacity with such a great number of its fibres having failed, unless the polymer matrix redistributed the load from the fracture points to the neighbouring fibres, the short lengths of which still posess high strength.

The fibre failure does not occur in the same cross-section plane. Therefore the mean stress in the intact fibres should not materially exceed the nominal stress value calculated from the cross-section area of the total number of fibres. This assumption appears to be right even considering stress concentrations at the ends of the ruptured fibres.

SERENSEN and ZAITSEV (1965) also showed that an avalanche fibre fracture in a glass epoxy composite starts after 10–15 per cent of fibres have fractured.

During the avalanche rupture of fibres the stresses vary in a quite narrow range. Therefore it may be assumed, that during this process the strain in a reinforced plastic remains practically the same. The ultimate strain of fibres corresponding to the starting point of the avalanche fracture will be denoted by e_{fu}. From the above it follows that the ultimate strain e_{fu} is an important strength characteristic in case of a fibre bundle. In practice this ultimate strain is equal to that of a reinforced plastic loaded in the fibre direction. This assumption provides the basis for the evaluation by means of test.

The fibre strength condition can be assumed to be

$$e_f \leqslant e_{fu} \tag{12}$$

where e_f is the actual strain of the fibre bundle.

Due to the fact that the strength of the polymer matrix and that of the bond are significantly lower than the strength of the fibres, it may be assumed in the first approximation, that the fibres are in a uniaxial state of stresses (SKUDRA and BULAVS (1978)).

2.3. *Bond strength*

Usually normal as well as shear stresses are simultaneously active on the interface in reinforced plastics. Physically substantiated criteria concerning adhesive strength have not yet been developed for such a combined state of stress. Therefore we shall make use of an approximated phenomenological approach, and the term 'bond strength' will be substituted for the physical term 'adhesive strength'.

Physical adhesion theories may be grouped into two broad classes:

(1) theories explaining adhesion as interactions of adsorption type,

(2) theories explaining the adhesion phenomenon as interaction of diffusion type.

Two different models for evaluating the bond strength between two different materials have been suggested by SKUDRA and KIRULIS (1975) and SKUDRA et al. (1977) which correspond to the above classes of division: the abrupt transfer model and the model of diffusion interlayer.

To illustrate the case of the abrupt transfer model we shall choose a smooth surface of arbitrary configuration (Fig. 4(a)) which forms the interface between two volume components A and B. The rectangular system of coordinates is chosen so that axis X_1 will be perpendicular to and axes X_2 and X_3 coincide with the tangents of the contact surface in point O.

A small volume in the region of O is selected where the contact layer may be regarded to be flat (Fig. 4(b)). We also introduce a condition that the contact layer is not merely a mathematical term, but it has a certain thickness determined by the equilibrium distance between the surface molecules of components A and B.

We further assume that the intermolecular links (Fig. 4(b)) are destroyed only in tension. The tensile action is possible only when normal tensile stresses σ_{11}, shear stresses $\pm\sigma_{12}$, $\pm\sigma_{13}$ or both combined are active. The effect of other types of stresses, compressive stress $-\sigma_{11}$ and also $\pm\sigma_{22}$, $\pm\sigma_{33}$, $\pm\sigma_{23}$ and a combination of these, cannot produce the extension of intermolecular links. Therefore under such compressive stresses the failure does not start on the contact surface, but it does so in the volume of one of the contacting materials.

To obtain the surface of the strength for the model under discussion, it is possible to employ the strength criterion proposed by GOLDENBLAT and KOPNOV (1965):

$$(\pi_{\alpha\beta}\sigma_{\alpha\beta})^a + (\pi_{\alpha\beta\gamma\delta}\sigma_{\alpha\beta}\sigma_{\gamma\delta})^b + \cdots = 1 .$$

This criterion gives us a possibility of including the anisotropy aspect in the strength characteristics of the model.

Limiting ourselves to the first two members of the series and assuming that a and b are equal to 1, like it was assumed by TSAI and WU (1971), we get

$$\pi_{\alpha\beta}\sigma_{\alpha\beta} + \pi_{\alpha\beta\gamma\delta}\sigma_{\alpha\beta}\sigma_{\gamma\delta} = 1 . \tag{a}$$

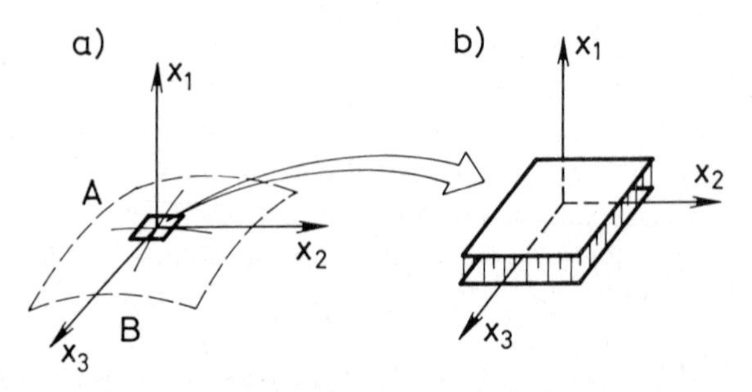

FIG. 4. A calculation scheme for bond strength determination.

The unknown coefficients in this criterion are the second range tensor elements $\pi_{\alpha\beta}$ and the fourth range tensor elements $\pi_{\alpha\beta\gamma\delta}$. These coefficients should be determined from the properties of strength symmetry for the contacting surface model.

We shall proceed to designate the strength tensor components:

$$r_{ijklmn}$$

where the indices i, j, k, l, m, n take the value of

$$i = 0, 1, \bar{1} \qquad j = 0, 2, \bar{2} \qquad k = 0, 3, \bar{3}$$
$$l = 0, 4, \bar{4} \qquad m = 0, 5, \bar{5} \qquad n = 0, 6, \bar{6}\,.$$

The given indexing takes into account the sign of stresses: $1, 2, 3, 4, 5, 6$ – positive stresses $\sigma_{11}, \sigma_{22}, \sigma_{33}, \sigma_{12}, \sigma_{13}, \sigma_{23}$; $\bar{1}, \bar{2}, \bar{3}, \bar{4}, \bar{5}, \bar{6}$ – negative stresses $-\sigma_{11}, -\sigma_{22}, -\sigma_{33}, -\sigma_{12}, -\sigma_{13}, -\sigma_{23}$.

Thus, for example, the symbol $r_{10\bar{3}400}$ stands for the strength under stresses $\sigma_{11}, -\sigma_{33}$ and σ_{12}.

With the aid of above symbols the strength symmetry properties of the contacting surface model can formally be described.

(1) If $\sigma_{11} > 0$ and $\sigma_{12} = \sigma_{13} = 0$ then, regardless of the other stresses σ_{22}, σ_{33} and σ_{23} present, the contact surface strength is equal to σ_b^u, i.e., the rupture strength of the bond

$$r_{123006} = r_{1\bar{2}0000} = r_{10300\bar{6}} = \cdots = r_{100000} = \sigma_b^u \tag{b}$$

(2) If $\sigma_{11} = 0$, $\sigma_{13} = 0$ and $\sigma_{12} \neq 0$ or $\sigma_{11} = 0$, $\sigma_{12} = 0$ and $\sigma_{13} \neq 0$ then, regardless of the other stresses present, the strength of the contact surface is τ_b^u, i.e., the shear strength of the bond:

$$r_{003050} = r_{0\bar{2}3406} = \cdots = r_{000400} = r_{000050} = \tau_b^u\,. \tag{c}$$

The bond strengths σ_b^u and τ_b^u are phenomenological values to be determined by test. They depend on the chemical structure of the two contacting materials, on the way the contact is made and on temperature. They are independent of the stress state.

(3) If $\sigma_{11} \neq 0$, $\sigma_{12} \neq 0$ and $\sigma_{13} \neq 0$, then, regardless of the other stresses present, the strength of the contact surface is a finite value which is described by the following relationship based on symmetry considerations:

$$\begin{aligned} r_{123\bar{4}\bar{5}6} &= r_{123456} = \cdots = r_{123\bar{4}56} = r_{1234\bar{5}6}\,, \\ r_{\bar{1}23\bar{4}\bar{5}6} &= r_{\bar{1}23456} = \cdots = r_{\bar{1}234\bar{5}6} = r_{\bar{1}23\bar{4}56}\,. \end{aligned} \tag{d}$$

(4) If $\sigma_{11} < 0$ and $\sigma_{12} = \sigma_{13} = 0$ then, regardless of the other stresses present,

the strength of the contact surface is infinite:

$$r_{\bar{1}23006} = r_{\bar{1}2\bar{3}00\bar{6}} = \cdots = r_{\bar{1}\bar{2}\bar{3}000} = r_{\bar{1}03006} = r_{023006} = \infty \,. \tag{e}$$

The meaning of the latter condition is that for the types of stress state under discussion, the failure does not take place on the contact surface but in the nearest volume point of one of the contacting materials.

Making use of (b)–(e) the unknown coefficients of the strength criterion $\pi_{\alpha\beta}$ and $\pi_{\alpha\beta\gamma\delta}$ are obtained. With only σ_{11} active we get the following set of equations from (a):

$$\begin{cases} \pi_{11} r_{100000} + \pi_{1111} r^2_{100000} = 1 \,, \\ -\pi_{11} r_{\bar{1}00000} + \pi_{1111} r^2_{\bar{1}00000} = 1 \,. \end{cases}$$

Solving this set with respect to π_{11} and π_{1111} we get:

$$\pi_{11} = \frac{r^2_{\bar{1}00000} - r^2_{100000}}{r_{100000} r^2_{\bar{1}00000} + r_{\bar{1}00000} r^2_{100000}} \,,$$

$$\pi_{1111} = \frac{r_{100000} - r_{\bar{1}00000}}{r_{100000} r^2_{\bar{1}00000} + r_{\bar{1}00000} r^2_{100000}} \,.$$

Considering (b) and (e) we obtain

$$\pi_{11} = \frac{1}{\sigma^{\mathrm{u}}_{\mathrm{b}}} \,, \qquad \pi_{1111} = 0 \,.$$

With only stress σ_{22} active, the set of equations obtained from (a) is

$$\begin{cases} \pi_{22} r_{020000} + \pi_{2222} r^2_{020000} = 1 \,, \\ -\pi_{22} r_{0\bar{2}0000} + \pi_{2222} r^2_{0\bar{2}0000} = 1 \,. \end{cases}$$

Taking into account (e), the solution of the set is

$$\pi_{22} = \pi_{2222} = 0 \,.$$

By analogy we establish that all the coefficients $\pi_{\alpha\beta}$ and $\pi_{\alpha\beta\gamma\delta}$, except π_{11}, π_{1212} and π_{1313}, are equal to zero, and

$$\pi_{1212} = \pi_{1313} = \frac{1}{(\tau^{\mathrm{u}}_{\mathrm{b}})^2}$$

From the obtained expressions for π_{11} and $\pi_{1212} = \pi_{1313}$ the strength criterion (a) may be written as

$$\frac{\sigma_{11}}{\sigma_b^u} + \frac{\sigma_{12}^2 + \sigma_{13}^2}{(\tau_b^u)^2} = 1 .$$

The ultimate surface obtained from this strength criterion is plotted in Fig. 5.

The following paragraph is to show, that for unidirectionally reinforced plastics

$$\sigma_{11} = \sigma_{rr} = \langle\sigma_\perp\rangle\bar{\sigma}_r ,$$

$$\sigma_{12} = \sigma_{rz} = \langle\tau_{\|\perp}\rangle\bar{\sigma}_{rz} ,$$

$$\sigma_{13} = \sigma_{r\theta} = \langle\sigma_\perp\rangle\bar{\sigma}_{r\theta} ,$$

and in this case the strength criterion is written as

$$\frac{\langle\sigma_\perp\rangle\bar{\sigma}_r}{\sigma_b^u} + \frac{\langle\sigma_\perp^2\rangle\bar{\sigma}_{r\theta}^2 + \langle\tau_{\|\perp}^2\rangle\bar{\sigma}_{rz}^2}{(\tau_b^u)^2} = 1 . \tag{13}$$

In the case of the diffusional layer the contact between the two volume components A and B is supposed to be made within a layer of finite thickness. In contrast with the abrupt transfer model, where no molecules of the components A and B are found within the contact layer, here the contact layer is likely to contain the molecules of both materials. We may imagine that the material within the limits of the diffusional contact layer is homogeneous and isotropic, its mechanical behaviour generally differing from those of the original materials. The strength surface for this type of model may be described by the criterion which states that their exists a functional relationship between the octahedric stresses at the moment of fracture:

$$\tau_{oct} = F(\sigma_{oct}) .$$

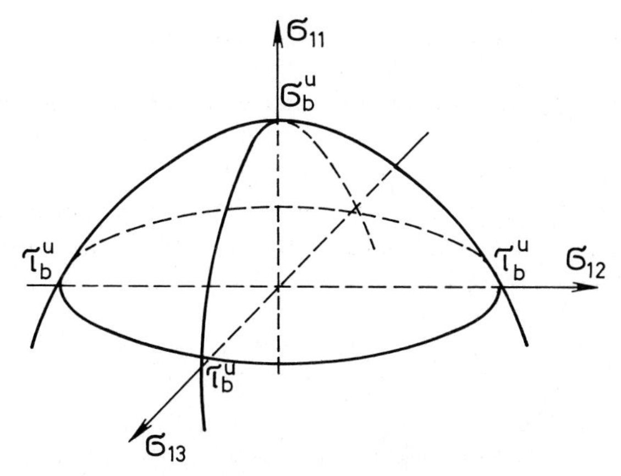

FIG. 5. Ultimate surface of bond strength.

In the first approximation we may adopt the linear relationship between σ_{oct} and τ_{oct}, and then the criterion of the bond strength is written as:

$$\tau_{oct} = a - k\sigma_{oct} . \tag{14}$$

Coefficients a and k included in (14) are obtained from the following (SKUDRA and KIRULIS 1975):

$$a = \frac{\sqrt{6}}{3}\tau_b^u , \qquad k = \frac{\sqrt{6}\tau_b^u - \sqrt{2}\sigma_b^u}{\sigma_b^u} .$$

Substituting the expressions for a and k in criterion (14) we get

$$\tau_{oct} = \frac{\sqrt{6}}{3}\tau_b^u \pm \frac{\sqrt{6}\tau_b^u - \sqrt{2}\sigma_b^u}{\sigma_b^u}\sigma_{oct} . \tag{15}$$

The formula has a plus sign in case of tensile stresses, and it is a minus quantity in case of compressive stresses. It follows from criterion (15) that compressive stresses tend to increase the bond strength, whereas tensile stresses reduce it.

3. Strength of unidirectionally reinforced plastics

3.1. Tensile strength

When the load is applied in the direction of the reinforcement, the whole load is sustained by the fibres. The strength in the direction of the reinforcement $\sigma_{\parallel}^{+}$ is calculated from the formula:

$$\sigma_{\parallel}^{+} = [(1 - v_f)E_m + v_f E_{fz}]e_{fu}^{+} . \tag{16}$$

Formula (16) is based on the so-called 'mixture law' and is supported by tests for plastics reinforced with glass, carbon, boron and other types of fibres.

Under transverse tension the filaments usually do not undergo fracture. For this reason the cause of failure may be the failure of the polymer matrix or the bond. The possible mechanical causes of failure will be analysed in greater detail. The calculation scheme is shown in Fig. 6.

The stresses in an arbitrary selected point of the polymer matrix, having the coordinates r and θ, are calculated with the aid of the following formulas (SKUDRA and BULAVS (1978)):

$$\sigma_{rr} = \langle\sigma_{\perp}\rangle\bar{\sigma}_r \tag{17}$$

$$\sigma_{\theta\theta} = \langle\sigma_{\perp}\rangle\bar{\sigma}_\theta \tag{18}$$

$$\sigma_{r\theta} = \langle\sigma_{\perp}\rangle\bar{\sigma}_{r\theta} \tag{19}$$

$$\sigma_{zz} = \langle\sigma_{\perp}\rangle\bar{\sigma}_z = \langle\sigma_{\perp}\rangle\nu_m(\bar{\sigma}_r + \bar{\sigma}_\theta) \tag{20}$$

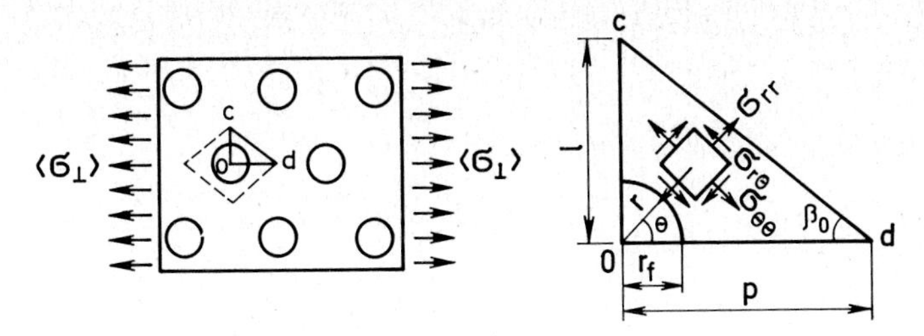

FIG. 6. A calculation scheme for a unidirectionally reinforced plastic in transverse tension.

where

$$\bar{\sigma}_r = f(l, r_f)\Big\{B_0 r^{-2} + 2C_0 - \sum_{n=2,4\ldots}^{N} [n(n-1)A_n r^{n-2} + n(n+1)B_n r^{-n+2} + (n+1)(n-2)C_n r^n + (n-1)(n+2)D_n r^{-n}]\cos n\theta\Big\} \tag{21}$$

$$\bar{\sigma}_\theta = f(l, r_f)\Big\{-B_0 r^{-2} + 2C_0 + \sum_{n=2,4\ldots}^{N} [n(n-1)A_n r^{n-2} + n(n+1)B_n r^{-n-2} + (n+1)(n+2)C_n r^n + (n-1)(n-2)D_n r^{-n}]\cos n\theta\Big\} \tag{22}$$

$$\bar{\sigma}_{r\theta} = f(l, r_f) \sum_{n=2,4\ldots}^{N} [n(n-1)A_n r^{n-2} - n(n+1)B_n r^{-n-2} + n(n+1)C_n r^n - n(n-1)D_n r^{-n}]\sin n\theta \tag{23}$$

$$f(l, r_f) = \Big\{2F_0 r_f + B_0\Big(\frac{1}{l} - \frac{1}{r_f}\Big) + 2C_0(l - r_f) + \sum_{n=2,4\ldots}^{N} \Big[nA_n(l^{n-1} - r_f^{n-1}) - nB_n\Big(\frac{1}{l^{n+1}} - \frac{1}{r_f^{n+1}}\Big) + (n+2)C_n(l^{n+1} - r_f^{n+1}) - (n-2)D_n\Big(\frac{1}{l^n} - \frac{1}{r_f^n}\Big) + nK_n r_f^{n-1} + (n+2)F_n r_f^{n+1}\Big](-1)^{n/2}\Big\}^{-1}.$$

Coefficients $B_0, C_0, A_N, B_n, C_n, D_n$ and K_n, F_n are the parameters of the corresponding stress function for both the polymer matrix and the fibres. The methods of measuring them are explained by SKUDRA and BULAVS (1978).

As seen from Fig. 6 the parameter p and l decide the way the fibres are packed and, like the fibre radius r_f, are predetermined values. The relationship between p and l is expressed

$$l = p \operatorname{tg} \beta_0 .$$

When the volume content of fibres v_f is also given, in addition to packaging, there is the following relationship between the values r_f, p and v_f:

$$r_f = \sqrt{\frac{2p^2 \operatorname{tg} \beta_0}{\pi} v_f} .$$

In calculations it is convenient to assume that $p = 1$. Then

$$l = \operatorname{tg} \beta_0 \qquad r_f = \sqrt{\frac{2 \operatorname{tg} \beta_0}{\pi} v_f} .$$

In case of a square packaging of fibres $p = l = 1$ and

$$r_f = \sqrt{\frac{2v_f}{\pi}}$$

Introducing the stress values σ_{rr}, $\sigma_{\theta\theta}$, σ_{zz} and $\sigma_{r\theta}$ obtained from (17)–(20) into the energy strength criterion and knowing that at the moment of fracture $\langle\sigma_\perp\rangle = \sigma_\perp^+$, we have

$$\frac{\bar{\sigma}_r^2 + \bar{\sigma}_\theta^2 + \bar{\sigma}_z^2}{(\sigma_m^+)^2} + \frac{\bar{\sigma}_{r\theta}^2}{(\tau_m^u)^2} - 2\nu_m \frac{\bar{\sigma}_r\bar{\sigma}_z + \bar{\sigma}_r\bar{\sigma}_\theta + \bar{\sigma}_\theta\bar{\sigma}_z}{(\sigma_m^+)^2} = \frac{1}{(\sigma_\perp^+)^2} . \tag{24}$$

It follows from this equation that

$$\sigma_\perp^+ = \frac{\sigma_m^+ \tau_m^u}{\sqrt{(\tau_m^u)^2(\bar{\sigma}_r^2 + \bar{\sigma}_\theta^2 + \bar{\sigma}_z^2) + (\sigma_m^+)^2\bar{\sigma}_{r\theta}^2 - 2\nu_m(\tau_m^u)^2(\bar{\sigma}_r\bar{\sigma}_z + \bar{\sigma}_r\bar{\sigma}_\theta + \bar{\sigma}_\theta\bar{\sigma}_z)}} . \tag{25}$$

Formula (25) is adoptable if the coordinates $r = r_{crit}$ and the angle $\theta = \theta_{crit}$ are known (see Fig. 6), where $\sigma_\perp^+$ has a minimum value.

Minimization of Equation (25) proves, that $\theta_{crit} = 0°$ and $r_{crit} = r_f$.

With $r = r_f$ and $\theta = 0°$ there is $\bar{\sigma}_{r\theta} = 0$ and Equation (25) can be written as

$$\sigma_\perp^+ = \frac{\sigma_m^+}{\sqrt{\bar{\sigma}_r^2 + \bar{\sigma}_\theta^2 + \bar{\sigma}_z^2 - 2\nu_m(\bar{\sigma}_r\bar{\sigma}_z + \bar{\sigma}_r\bar{\sigma}_\theta + \bar{\sigma}_\theta\bar{\sigma}_z)}} . \tag{26}$$

The value of the structural parameter $\bar{\sigma}_\theta$ and $\bar{\sigma}_z$ is dependent upon the anisotropy of the elastic behaviour of fibres and these parameters may be ignored in some practical cases of problem solving. Then formula (26) is reduced to

$$\sigma_\perp^+ = \frac{\sigma_m^+}{\bar{\sigma}_r\sqrt{1 - \nu_m^2}} \tag{27}$$

Dimensionless values $\bar{\sigma}_r$, $\bar{\sigma}_\theta$ and $\bar{\sigma}_z$ are very important for characterizing the structure of a reinforced plastic. It is these values that govern the stress concentration depending on the elastic properties of the matrix and the fibres, their volume content and fibres packaging. Figs. 7–9 show the dependences

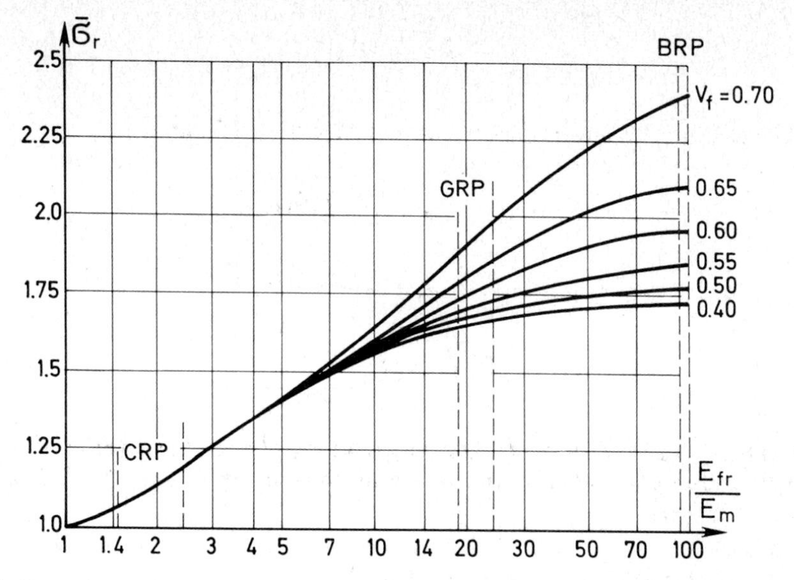

FIG. 7. Dependence of $\bar{\sigma}_r$ on the fibre volume content (v_f) and on the ratio of elasticity moduli of the fibres and the matrix.

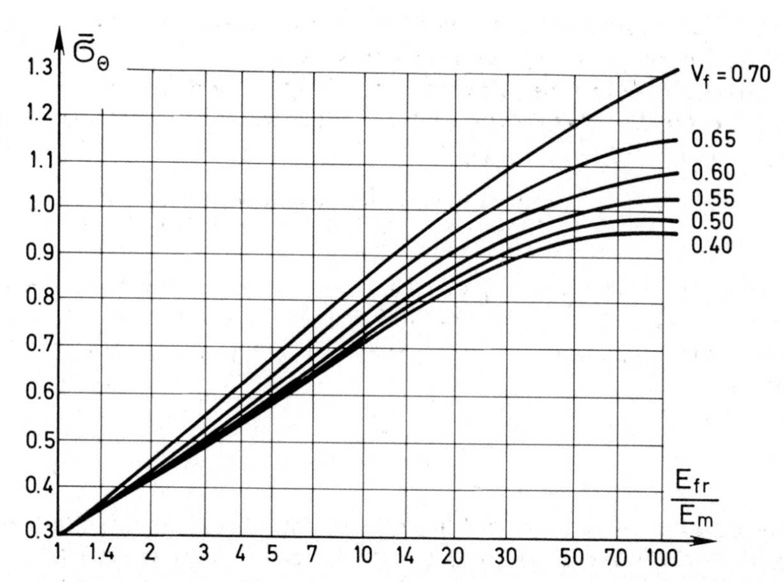

FIG. 8. Dependence of $\bar{\sigma}_\theta$ on the fibre volume content (v_f) and on the ratio of elasticity moduli of the fibres and the matrix.

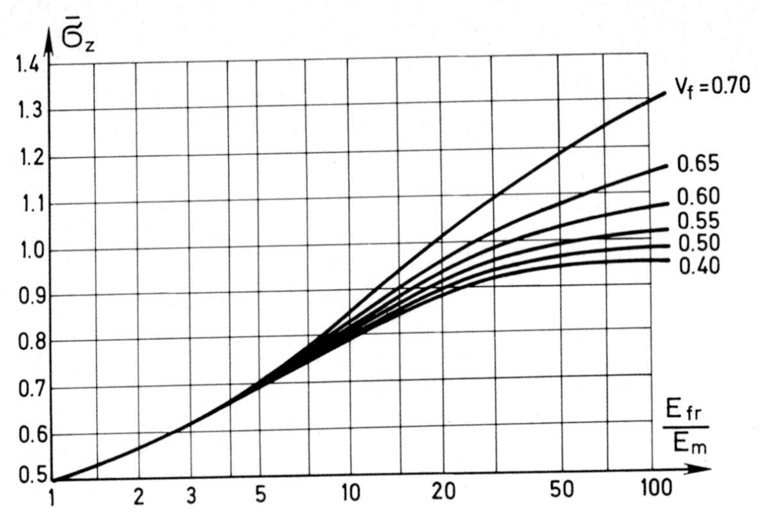

FIG. 9. Dependence of $\bar{\sigma}_z$ on the fibre volume content (v_f) and on the ratio of elasticity moduli of the fibres and the matrix.

$\bar{\sigma}_r$, $\bar{\sigma}_\theta$ and $\bar{\sigma}_z$ on the volume content of fibres v_f and on the ratio between the moduli of elasticity of the fibres and the matrix. The curves are based on formulas (21) and (22) for the square packaging of fibres when $\theta = 0°$.

Figs. 10 and 11 offer the test results and the theoretical curves derived from (26) and (27).

We shall proceed to analyse an alternative case of possible fracture of a reinforced plastic in transverse tension if the strength of the polymer matrix exceeds that of the bond between the matrix and the fibres. Criteria (13) and (15) produce results of only a slight difference. Further the more simple criterion (13) will be employed.

In transverse tension $\langle \tau_{\parallel\perp} \rangle = 0$. At the moment of rupture the mean trans-

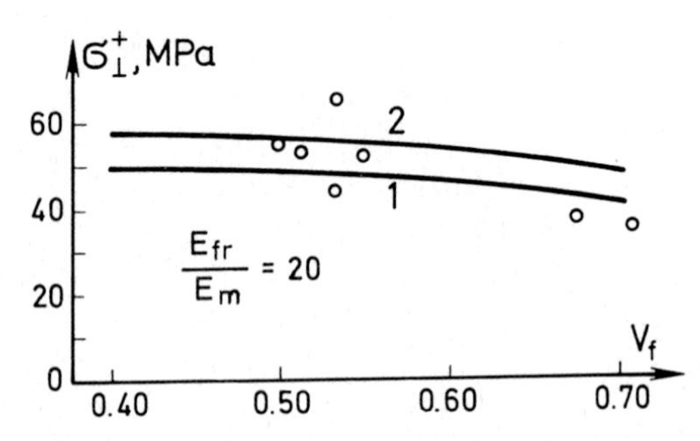

FIG. 10. Dependence of transverse tensile strength upon the volume content of glass fibres v_f. Test data have been borrowed from NOYES and JONES (1968, 1971). The theoretical Curve 1 has been plotted from (27) and Curve 2 from (26), if $\sigma_m^+ = 74.2$ MPa; $E_m = 3\,700$ MPa; $\nu_m = 0.34$; $E_f = 70\,000$ MPa; $\bar{\sigma}_r$, $\bar{\sigma}_\theta$ and $\bar{\sigma}_z$ have been obtained from curves of Figs. 7–9.

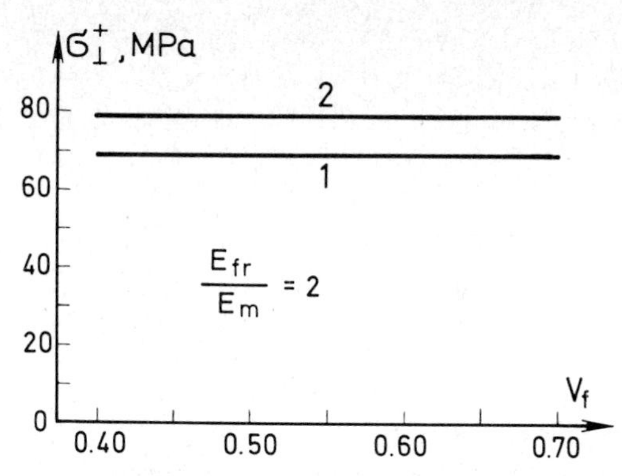

FIG. 11. Dependence of transverse tensile strength upon the volume content of carbon fibres v_f. Curve 1 has been plotted from (27) and Curve 2 from (26), where $\sigma_m^+ = 74.2$ MPa; $E_m = 3500$ MPa; $\nu_m = 0.34$; $E_{fr} = 7000$ MPa.

verse tensile stress $\langle\sigma_\perp\rangle$ is equal to the transverse tensile strength of a reinforced plastic $\sigma_\perp^+$, and criterion (13) is given by the following:

$$\frac{\bar{\sigma}_r\sigma_\perp^+}{\sigma_b^u} + \left(\frac{\bar{\sigma}_{r\theta}\sigma_\perp^+}{\tau_b^u}\right)^2 = 1\,.$$

To determine the coordinate θ_{crit} of the greatest danger point on the interface, it is necessary to minimize the strength criterion. The extreme condition for $\sigma_\perp^+(\theta)$ may be written as:

$$\frac{\partial\sigma_\perp^+(\theta)}{\partial\theta} = 0\,.$$

We see that $\theta_{crit} = 0°$. With $\theta = 0°$, $\bar{\sigma}_{r\theta} = 0$ the strength criterion takes the expression

$$\sigma_\perp^+ = \frac{\sigma_b^u}{\bar{\sigma}_r}\,. \tag{28}$$

Formula (28) enables us to calculate the strength of the reinforced plastic in transverse tension, provided its structure and the rupture strength of the bond σ_b^u between the fibres and the matrix have been given. The formula is suitable for solving a reverse problem, namely, calculating the bond strength σ_b^u if the strength $\sigma_\perp^+$ has been found by test and the structure of the material is known, i.e., $\bar{\sigma}_r$. The formula for this case is

$$\sigma_b^u = \sigma_\perp^+\bar{\sigma}_r\,. \tag{29}$$

Thus (26) can be adopted for calculating the strength of a reinforced plastic if the polymer matrix fails first and (28) will be suitable if the bond fails first. The actual strength of the composite will be the least of the two strength values.

Most convenient is the case when the polymer matrix and the bond undergo fracture simultaneously. Equalizing the right sides of the relationships (26) and (28) we get

$$\frac{\sigma_b^u}{\bar{\sigma}_r} = \frac{\sigma_m^+}{\sqrt{\bar{\sigma}_r^2 + \bar{\sigma}_\theta^2 + \bar{\sigma}_z^2 - 2\nu_m(\bar{\sigma}_r\bar{\sigma}_z + \bar{\sigma}_r\bar{\sigma}_\theta + \bar{\sigma}_\theta\bar{\sigma}_z)}}. \tag{30}$$

The type of relationship like (30) has a great practical significance. In the first place it gives us the possibility to evaluate the degree of exploitation of the polymer matrix strength with σ_b^u given, and secondly, it becomes possible to determine the bond strength σ_b^u needed for a full exploitation of the polymer matrix strength.

The constituents of a reinforced plastic are in a non-homogeneous state of stress. The failure mechanics of a material under non-homogeneous stresses is far from being well developed. This is why several assumptions were to be made in deriving formulas (26) and (28).

Criteria (1) and (13) determine the critical state of stress in a point (either of the polymer matrix or the interface). The minimum value of the criteria is found to be in the point on the fibre surface if $\theta = 0°$ or $180°$. In Fig. 12 this point is marked by letter k. Similar points are located on the surface of all the

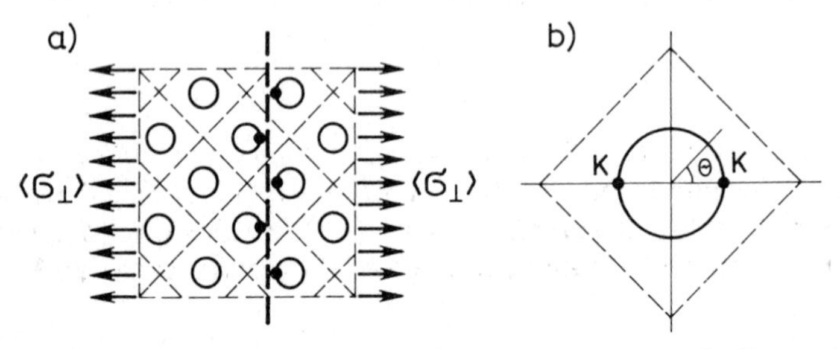

FIG. 12. Formation of the dangerous section (a) and danger points on the fibre surface (b).

fibres and the plane formed by these points, which is perpendicular to the direction of $\langle\sigma_\perp\rangle$, forms a cross section area where the failure of the material occurs. Although the material is in a non-homogeneous stress state, this particular case (i.e., that of a reinforced plastic) gives a logical and well-grounded basis to admit that in transverse tension the presence of an individual point failure evidences the failure of the whole material.

3.2. *Shear strength*

The inner structure geometry of a reinforced plastic in longitudinal shear causes the constituents to be in a non-homogeneous stress state. The stress concentration grows with an increased ratio between the shear moduli of the fibres and matrix and of the fibre volume content.

SKUDRA and BULAVS (1978) describe a situation when shear stress values in the critical areas have reached the strength value of the matrix and yet the avalanche fracture of the entire material is not initiated. This contrasts with the case of transverse tension. After the stress values have grown to be equal to the shear strength value of the matrix, a conditional flow of the polymer matrix starts in the maximum stress concentration sites, and there is a redistribution of the stress concentration sites, and there is a redistribution of the stress field. An analogical phenomenon was reported also by UEMURA and YAMAWAKI (1971). It was established by SKUDRA and BULAVS (1978) that, due to the conditional flow of the polymer matrix, the longitudinal shear strength of unidirectionally glass-fibre reinforced plastic may within the scattering range be regarded as equal to the shear strength of the polymer matrix.

The same results have been obtained by a number of authors studying the strength properties of various reinforced plastics under shear. The ratio between the longitudinal shear strength of a unidirectionally reinforced plastic and the matrix strength with varying volume contents of fibres is shown in Fig. 13. The results lead to the conclusion that the longitudinal shear strength of unidirectionally reinforced plastics are practically unaffected by the fibre volume content; it depends solely on the matrix strength. In this way the stress concentration in the matrix is kept down, as it were, and does not affect the strength properties of the whole material.

When the matrix strength is lower than the bond strength it may be said in the first approximation that the longitudinal shear strength is obtained from

$$\tau_{\parallel\perp}^{\mathrm{u}} = \tau_{\mathrm{m}}^{\mathrm{u}} \,. \tag{31}$$

For several types of reinforced plastics the failure of the composite is initiated by the failure of the fibre-matrix bond. SKUDRA and BULAVS (1978) have shown that the longitudinal shear strength of such materials depends largely upon the stress concentration and is calculated from the following formula:

$$\tau_{\parallel\perp}^{\mathrm{u}} = \frac{\tau_{\mathrm{b}}^{\mathrm{u}}}{\bar{\sigma}_{rz}} \tag{32}$$

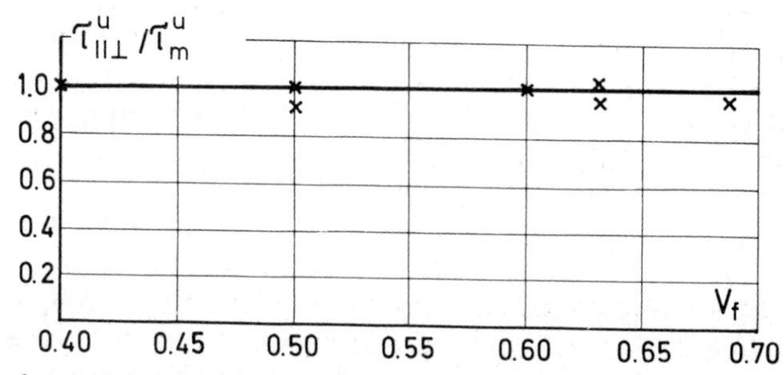

FIG. 13. Dependence of the ratio of longitudinal shear strength ($\tau_{\parallel\perp}^{\mathrm{u}}$) and matrix shear strength ($\tau_{\mathrm{m}}^{\mathrm{u}}$) upon the fibre volume content.

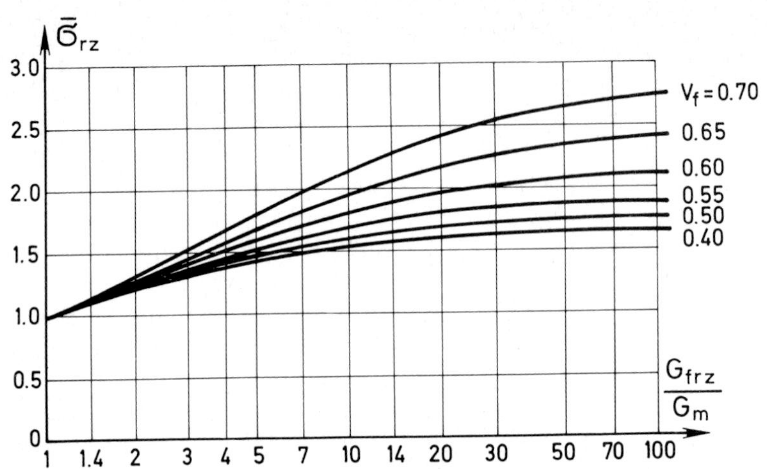

FIG. 14. Dependence of $\bar{\sigma}_{rz}$ on the fibre volume content (v_f) and on the ratio between shear moduli of the fibres and the matrix.

Fig. 14 presents the dependence of $\bar{\sigma}_{rz}$ on the fibre volume content and on the ratio between the elasticity modulus of the fibres and that of the matrix.

With the longitudinal shear strength $\tau^{u}_{\|\perp}$ established by test and the material structure $\bar{\sigma}_{rz}$ given formula (32) may be applied to determine the bond strength is shear

$$\tau^{u}_{b} = \tau^{u}_{\|\perp}\bar{\sigma}_{rz} . \tag{33}$$

3.3. Compressive strength

In a unidirectionally reinforced plastic under compression with the load applied in the fibre direction, the fibres usually fail first. The 'mixture law' can be used for strength determination.

$$\sigma^{-}_{\|} = [E_{fz}v_f + E_m(1 - v_f)]e^{-}_{fu} . \tag{34}$$

It may be assumed in the first approximation that the ultimate strain of fibres e^{-}_{fu} is equal to the ultimate strain of the reinforced plastic under compression. The dependence $\sigma^{-}_{\|}$ on the fibre volume content for a glass epoxy composite is shown in Fig. 15.

The typical mode of fracture of unidirectionally reinforced plastic under transverse compression is shown in Fig. 16.

In the first approximation we may assume that in the plane perpendicular to the fibre direction, the strengths are the same in all directions.

The stresses acting in the fracture plane are illustrated in Fig. 17. It is obvious that shear stresses $\langle\sigma'_6\rangle$ and compressive stresses $\langle\sigma'_1\rangle$ are active in the fracture plane oriented at an angle φ to the loading direction. To calculate the strength $\sigma^{-}_{\perp}$ in case of matrix failure, criterion (4) may be adopted, which for

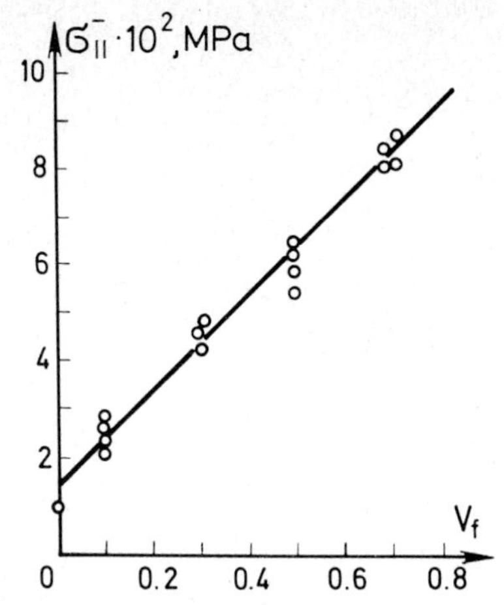

FIG. 15. Dependence of glass/epoxy strength $\sigma_{\parallel}^{-}$ upon the fibre volume content. The theoretical line has been plotted from (34) if $E_m = 3\,500$ MPa; $E_f = 70\,000$ MPa; $e_{fu}^{-} = 0.017$. Test data have been borrowed from HAYASHI and KOYAMA (1972).

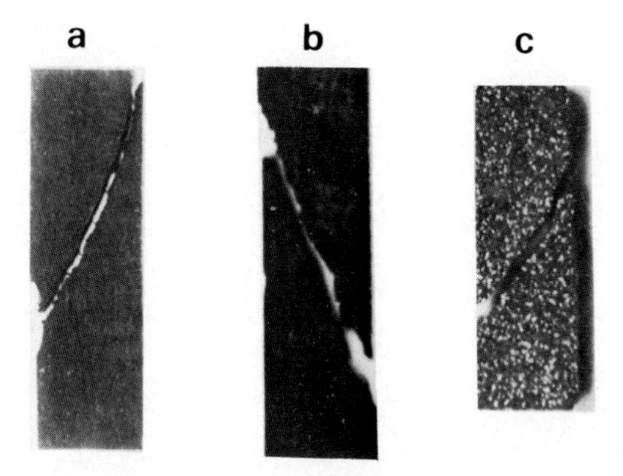

FIG. 16. Characteristic fracture modes of a unidirectionally reinforced plastic under compression applied perpendicularly to the reinforcement: (a) glass/epoxy; (b) carbon/epoxy; (c) boron-epoxy.

this particular case is written as

$$\langle\sigma_1'\rangle^2 + 2(1+\nu_{\perp\perp})\langle\sigma_6'\rangle^2 - \langle\sigma_1'\rangle\sqrt{\langle\sigma_1'\rangle^2 + 4\langle\sigma_6'\rangle^2} = 2(\sigma_{\perp}^{+})^2 \tag{35}$$

It should be noted that strength $\sigma_{\perp}^{+}$ is calculated from (26).

Inserting the expressions for $\langle\sigma_1'\rangle$ and $\langle\sigma_6'\rangle$ in the above and designating the value of the applied compressive stresses at the moment of failure by $\sigma_{\perp}^{-}$ we get

$$(\sigma_{\perp}^{-})^2[\sin^2\varphi + 2(1+\nu_{\perp\perp})\cos^2\varphi - \sin\varphi\sqrt{4-3\sin^2\varphi}]\sin^2\varphi = 2(\sigma_{\perp}^{+})^2 .$$

From the latter equation, which shows the dependence of strength $\sigma_{\perp}^{-}$ on the

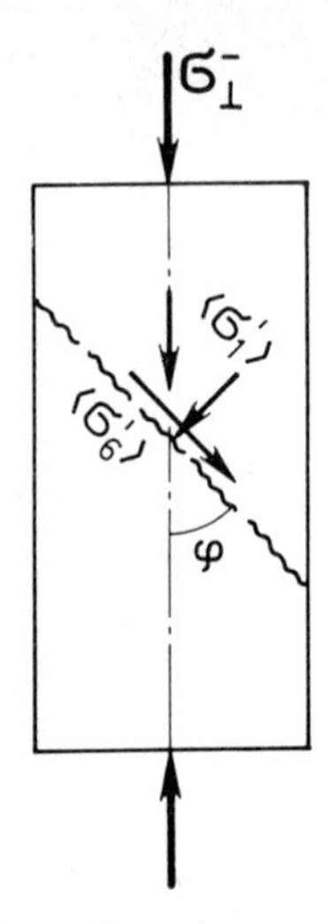

FIG. 17. A calculation scheme for strength determination.

angle of the fracture plane φ, we obtain

$$2\left(\frac{\sigma_\perp^+}{\sigma_\perp^-}\right)^2 = \sin^2\varphi[\sin^2\varphi + 2(1+\nu_{\perp\perp})\cos^2\varphi - \sin\varphi\sqrt{4-3\sin^2\varphi}]\,. \qquad \text{(a)}$$

The function $2(\sigma_\perp^+/\sigma_\perp^-)^2$ is at its maximum, but $\sigma_\perp^-$ at the minimum with the value of φ obtained from equation

$$a\sin^6\varphi + b\sin^4\varphi + c\sin^2\varphi + d = 0$$

where

$$\begin{aligned} a &= 3[3+(1+2\nu_{\perp\perp})^2]\,, \\ b &= -2[9+(1+2\nu_{\perp\perp})(5+7\nu_{\perp\perp})]\,, \\ c &= 9+(1+\nu_{\perp\perp})(11+19\nu_{\perp\perp})\,, \\ d &= -4(1+\nu_{\perp\perp})^2\,. \end{aligned} \qquad \text{(b)}$$

The magnitude of angle φ satisfying Equation (b) is the one at which the fracture plane is inclined towards the direction of loading.

The angle φ having been measured, formula (a) gives us a clue for determining transverse compression strength:

$$\sigma_\perp^- = \frac{\sqrt{2}\sigma_\perp^+}{\sqrt{\sin^4\varphi + \dfrac{1+\nu_{\perp\perp}}{2}\sin^2 2\varphi - \sin^3\varphi\sqrt{4-3\sin^2\varphi}}} \qquad (36)$$

The calculation of strength $\sigma_\perp^-$ may also be based upon Mohr's strength

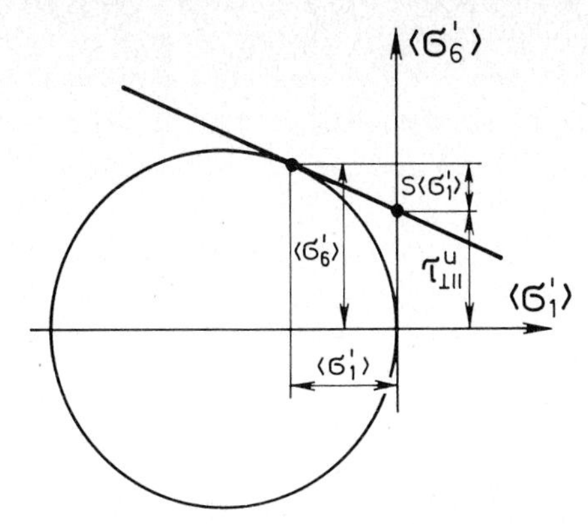

FIG. 18. The mechanical interpretation of the parameters incorporated in Mohr's strength criterion.

theory. The respective calculation scheme is shown in Fig. 18. The equation for the circumference of the Mohr's principal rings for a unidirectionally reinforced plastic, under transverse compression loading in isotropy plane, is the following:

$$\langle\sigma_6'\rangle = \tau^{u}_{\perp\perp} + s\langle\sigma_1'\rangle \tag{c}$$

where $\tau^{u}_{\perp\perp}$ is the shear strength in a transverse isotropy plane; and s is the influence rate of normal stresses acting perpendicular to the fracture plane.

Inserting the expressions for $\langle\sigma_1'\rangle$ and $\langle\sigma_6'\rangle$ into Equation (c) we get

$$\sigma_{\perp}^{-} = \frac{2\tau^{u}_{\perp\perp}}{\sin 2\varphi - s(1-\cos 2\varphi)} \tag{37}$$

There will be the maximum strength level $\sigma_{\perp}^{-}$ when the function

$$\frac{\tau^{u}_{\perp\perp}}{\sigma_{\perp}^{-}} = \frac{1}{2}[\sin 2\varphi - s(1-\cos 2\varphi)]$$

reaches its maximum, i.e., when $\operatorname{tg} 2\varphi = 1/s$.

The latter expression leads to

$$\varphi = \frac{1}{2}\operatorname{arctg}\left(\frac{1}{s}\right) \quad \text{or} \quad s = \frac{1}{\operatorname{tg} 2\varphi}.$$

Inserting the expression for s into formula (37) we have:

$$\sigma_{\perp}^{-} = \frac{2\tau^{u}_{\perp\perp}\sin 2\varphi}{1-\cos 2\varphi} = 2\operatorname{ctg}\varphi\cdot\tau^{u}_{\perp\perp} \tag{38}$$

where

$$\tau^{u}_{\perp\perp} = 2G_{\perp\perp}\langle e^{+}_{u}\rangle .$$

The shear modulus $G_{\perp\perp}$ is calculated according to the formula:

$$G_{\perp\perp} = \frac{E_{\perp}}{2(1+\nu_{\perp\perp})} .$$

Elastic characteristics $E_{\perp}$ and $\nu_{\perp\perp}$ are obtained from the formulas offered by VAN FO FI (1971) and SKUDRA and BULAVS (1978).

Inserting the expressions for $\tau^{u}_{\perp\perp}$ and $G_{\perp\perp}$ we get

$$\sigma^{-}_{\perp} = \frac{2E_{\perp}\langle e^{+}_{u}\rangle}{1+\nu_{\perp\perp}} \operatorname{ctg}\varphi = \frac{2\sigma^{+}_{\perp}}{1+\nu_{\perp\perp}} \operatorname{ctg}\varphi . \tag{39}$$

When the polymer matrix fails we calculate strength $\sigma^{+}_{\perp}$ from (26), but in case of bond failure we should resort to (28).

The inclination angle of the fracture plane φ (see Fig. 17) is determined by test. For a wide group of reinforced plastics we may assume in the first approximation that $\varphi \approx 30° = \text{const}$.

Then formula (39) looks as follows:

$$\sigma^{-}_{\perp} = 3.5 \frac{\sigma^{+}_{\perp}}{1+\nu_{\perp\perp}} . \tag{40}$$

When the polymer matrix fails, the dependence of $\sigma^{-}_{\perp}$ on the fibre volume content v_f is derived from (40) and shown in Fig. 19.

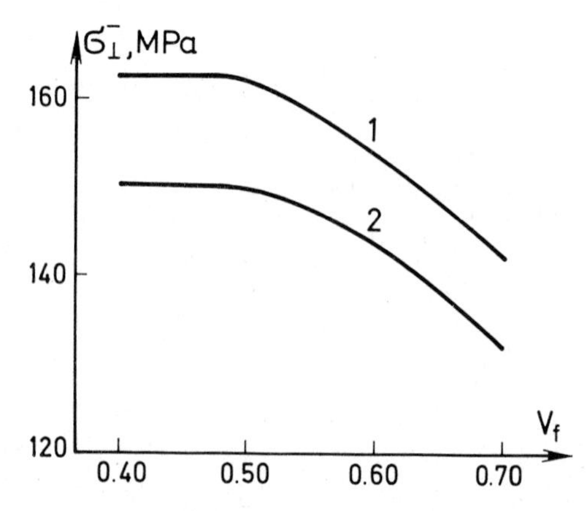

FIG. 19. Dependence of a compression strength $\sigma^{-}_{\perp}$ upon the glass fibre volume content v_f if $E_m = 3\,500$ MPa; $\nu_m = 0.35$; $\sigma^{+}_{m} = 74.2$ MPa; $E_f = 70\,000$ MPa; $\nu_f = 0.22$; $\varphi = 30°$; $\bar{\sigma}_r = 1.7$. Theoretical Curve 1 is based on (36) and Curve 2 on (40).

3.4. Strength under combined loading

Now we shall discuss a particular case of combined loading when a unidirectionally reinforced plastic is acted upon by normal stresses $\langle\sigma_\perp\rangle$, perpendicular to the direction of reinforcement, and longitudinal shear stresses $\langle\tau_{\|\perp}\rangle$. Under such a loading the strength of a reinforced plastic is usually governed by the strength of the polymer matrix or that of the bond.

First we shall deal with the case when the polymer matrix strength is lower than the bond strength. Supposing that under combined tension and shear the failure of a unidirectionally reinforced plastic occurs when the ultimate tensile stresses reach the level of the polymer matrix strength, then according to energy criterion and (26) we get:

$$\left(\frac{\langle\sigma_\perp\rangle}{\sigma_{\mathrm{m}}^{+}}\right)^2 A + \left(\frac{\langle\tau_{\|\perp}\rangle}{\tau_{\mathrm{m}}^{\mathrm{u}}}\right)^2 = 1 \qquad (41)$$

where

$$A = \bar\sigma_r^2 + \bar\sigma_\theta^2 + \bar\sigma_z^2 - 2\nu_{\mathrm{m}}(\bar\sigma_r\bar\sigma_z + \bar\sigma_r\bar\sigma_\theta + \bar\sigma_\theta\bar\sigma_z)\,.$$

Ignoring the influence of parameter $\bar\sigma_\theta$ and considering that $\bar\sigma_z = \nu_{\mathrm{m}}(\bar\sigma_\theta + \bar\sigma_z)$ we obtain

$$(1-\nu_{\mathrm{m}}^2)\left(\frac{\bar\sigma_r\langle\sigma_\perp\rangle}{\sigma_{\mathrm{m}}^{+}}\right)^2 + \left(\frac{\langle\tau_{\|\perp}\rangle}{\tau_{\mathrm{m}}^{\mathrm{u}}}\right)^2 = 1\,. \qquad (42)$$

The strength corresponding to Equation (42) is seen in Fig. 20.

If the failure of the composite is due to the bond failure, criterion (13)

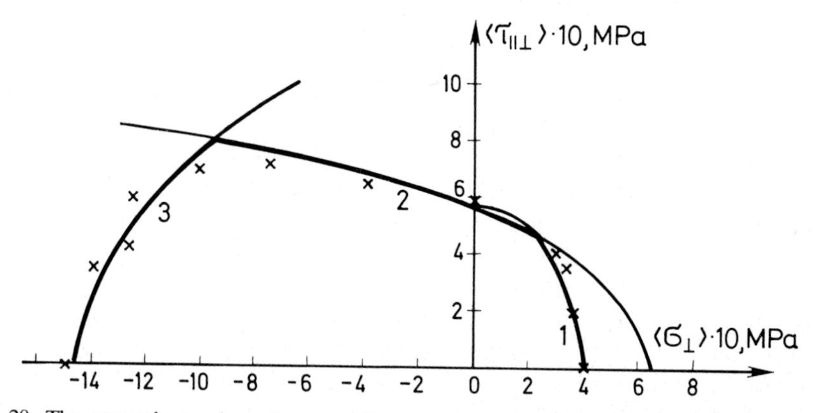

FIG. 20. The strength envelope for a unidirectionally reinforced glass/epoxy composite under combined axial and shear loading. Curve 1 is based on formula (42), Curve 2 on (45), Curve 3 on (46).

produces the following failure condition:

$$\langle\sigma_\perp\rangle(\tau_b^u)^2\bar{\sigma}_r + \langle\tau_{\|\perp}^2\rangle\sigma_b^u\bar{\sigma}_{rz}^2 = (\tau_b^u)^2\sigma_b^u . \tag{43}$$

We should note that compressive stresses $\langle\sigma_\perp\rangle$ in (43) come under a minus sign.

Dimensionless structural parameter $\bar{\sigma}_{rz}$ is obtained from the Fig. 14 or from the following:

$$\bar{\sigma}_{rz} = \frac{G_m}{G_{\|\perp}}\left[1 - \frac{r_f}{l}\left(1 - \frac{G_m}{G_{frz}}\right)\right]^{-1} \tag{44}$$

where r_f is the fibre radius; and l the fibre packaging parameter.

For a square distribution of fibres

$$\frac{r_f}{l} = 2\sqrt{\frac{v_f}{\pi}}$$

and for a hexagonal distribution

$$\frac{r_f}{l} = 2\sqrt{\frac{\sqrt{3}}{2}\cdot\frac{v_f}{\pi}}\,.$$

If the failure of the composite under combined loading is typical of longitudinal shear, the condition that the failure of the material is initiated when the specific work of the principal tensile stress within the whole repeating element reaches its ultimate level, serves as a strength criterion. Then, neglecting the effect of $\sigma_{\theta\theta}$ and σ_{zz} the strength envelope is described by the following equation (SKUDRA and BULAVS 1978):

$$\langle\sigma_\perp^2\rangle + 2(1+\nu_m)\langle\tau_{\|\perp}^2\rangle + \langle\sigma_\perp\rangle\sqrt{\langle\sigma_\perp^2\rangle + 4\langle\tau_{\|\perp}^2\rangle} = 2(\sigma_m^+)^2 \tag{45}$$

The curve plotted from (45) is in Fig. 20 drawn by the continuous line 2.

Thus the actual strength envelope for longitudinal shear and transverse tension is obtained from Equations (42) through (45). Criterion (45) satisfactorily describes the test data under a combined action of shear and compressive stresses as well. When the mean transverse compressive stresses at the fracture moment of the reinforced plastic are significantly higher than mean longitudinal shear stresses, the failure of the composite is then similar to the failure under transverse compression. To describe the test data within the interaction range of destructive stress $\langle\sigma_\perp\rangle$ and $\langle\tau_{\|\perp}\rangle$ we shall make use of the criterion which describes the specific work of the principal compressive stress. An equation of the strength envelope under a combined action of compressive and shear stresses results in:

$$\langle\sigma_\perp^2\rangle + 2(1+\nu_{\perp\perp})\langle\tau_{\|\perp}^2\rangle + \langle\sigma_\perp\rangle\sqrt{\langle\sigma_\perp^2\rangle + 4\langle\tau_{\|\perp}^2\rangle} = 2(\sigma_\perp^-)^2 \tag{46}$$

Here

$\langle\sigma_\perp\rangle$ – absolute values of compressive stresses;
$\nu_{\perp\perp}$ – Poisson's ratio for a unidirectionally reinforced plastic in an isotropy plane;
$\sigma_\perp^-$ – transverse compressive strength determined by the mechanical behaviour of the constituents according to (40).

The case when the polymer matrix fails first is shown in Fig. 20 by a continuous curve 3 based on Equation (46).

Finally we conclude that the ultimate strength curve in plane $\langle\tau_{\parallel\perp}\rangle - \langle\sigma_\perp\rangle$ consists of three regions, every one of them representing different failure mechanics. Test data shown in Fig. 20 have been borrowed from KNAPPE and SCHNEIDER (1972).

In case of bond failure the strength criterion for combined transverse compression and longitudinal shear is

$$\langle\tau^2_{\parallel\perp}\rangle\sigma^u_b\bar{\sigma}^2_{rz} - \langle\sigma_\perp\rangle(\tau^u_b)^2\bar{\sigma}_r = (\tau^u_b)^2\sigma^u_b\,.$$

3.5. General strength criteria for a plane state of stress

Multidirectionally reinforced plastics have a laminated structure and when loaded are found to be in a plane stress state. The calculation scheme for a multidirectionally reinforced laminate in a plane state of stress is shown in Fig. 21.

In the process of loading the unidirectionally reinforced plies oriented at various angles do not fail simultaneously. The failure of some plies oriented at critical angles does not always coincide with the failure of the whole material. The failure is then described as a two-stage process: at first the material loses its continuity due to matrix or bond failure in the most unfavourably oriented plies and then as the load is increased a complete failure takes place as a result of fibre failure.

To determine the moment of continuity loss and the complete failure of the

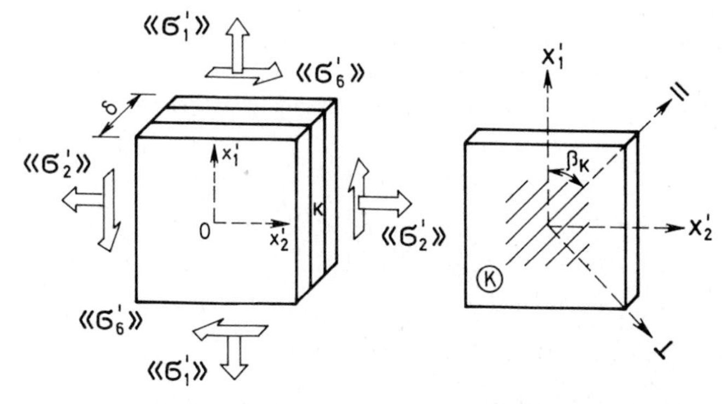

FIG. 21. A calculation scheme for a multidirectionally reinforced laminate.

material it is necessary to estimate the strength properties of individual unidirectionally reinforced plies oriented at various angles. Criteria (41)–(46) enable us to do so.

In order to simplify the calculation formulas it is assumed that the structure of the material is symmetrical to the midplane.

First we shall deal with a case when the fibres are the first to fail.

The individual lamina contain fibres oriented in various directions which are fixed by angle β.

Fig. 21(b), for instance, shows an arbitrary layer 'k'. The strength criterion of this layer is given by

$$\langle e_{\|}\rangle_k = e_{fu} \tag{47}$$

where e_{fu} is the ultimate strain of fibres under tension or compression.

Inserting the expression for $\langle e_{\|}\rangle_k$ into (47), we can write:

$$\langle\langle e_1'\rangle\rangle \cos^2\beta_k + \langle\langle e_2'\rangle\rangle \sin^2\beta_k + \langle\langle e_6'\rangle\rangle \sin\beta_k \cos\beta_k = e_{fu}\,. \tag{48}$$

Strains $\langle\langle e_1'\rangle\rangle$, $\langle\langle e_2'\rangle\rangle$ and $\langle\langle e_6'\rangle\rangle$ are calculated according to methods suggested by lamination theory provided that there is no sliding between the laminae. Taking into account the expressions of these strains (48) can be written as

$$\begin{aligned}&\langle\langle \sigma_1'\rangle\rangle(a_1'\cos^2\beta_k + a_2'\sin^2\beta_k + a_{12}'\cos\beta_k \sin\beta_k)\\ &+\langle\langle \sigma_2'\rangle\rangle(b_1'\cos^2\beta_k + b_2'\sin^2\beta_k + b_{12}'\cos\beta_k \sin\beta_k)\\ &+\langle\langle \sigma_6'\rangle\rangle(c_1'\cos^2\beta_k + c_2'\sin^2\beta_k + c_{12}'\cos\beta_k \sin\beta_k) = e_{fu}\,.\end{aligned} \tag{49}$$

Criterion (49) defines the moment of fibre failure in layer 'k'. This criterion should repeatedly be applied to all the laminae having a different fibre orientation. As a result of a multiple calculation it can be established which lamina is liable to undergo failure first, i.e., we find critical angle β_k. Knowing the value of the critical angle β_k and the structure of the laminate, criterion (49) makes it possible to determine the total strength of the laminate. This is based upon the fact that after the failure of the most loaded fibres an abrupt redistribution of stresses normally leads to an avalanche gross fracture of the composite.

So criterion (49) is a general structural strength criterion for reinforced plastics in case of an initial fibre failure. Coefficients a', b' and c' included in it are dependent upon the structure of the material and are calculated as follows:

$$a_1' = \frac{\bar A_{22}\bar A_{66} - \bar A_{26}}{\bar\Delta}; \qquad b_1' = \frac{\bar A_{16}\bar A_{26} - \bar A_{12}\bar A_{66}}{\bar\Delta}; \qquad c_1' = \frac{\bar A_{12}\bar A_{26} - \bar A_{22}\bar A_{16}}{\bar\Delta};$$

$$a_2' = \frac{\bar A_{16}\bar A_{26} - \bar A_{12}\bar A_{66}}{\bar\Delta}; \qquad b_2' = \frac{\bar A_{11}\bar A_{66} - \bar A_{16}^2}{\bar\Delta}; \qquad c_2' = \frac{\bar A_{12}\bar A_{16} - \bar A_{11}\bar A_{26}}{\bar\Delta};$$

$$a_{12}' = \frac{\bar A_{12}\bar A_{26} - \bar A_{16}\bar A_{22}}{\bar\Delta}; \qquad b_{12}' = \frac{\bar A_{12}\bar A_{16} - \bar A_{26}\bar A_{11}}{\bar\Delta}; \qquad c_{12}' = \frac{\bar A_{11}\bar A_{22} - \bar A_{12}^2}{\bar\Delta};$$

$$\bar\Delta = \bar A_{11}(\bar A_{22}\bar A_{66} - \bar A_{26}^2) - \bar A_{12}(\bar A_{12}\bar A_{66} - \bar A_{16}\bar A_{26}) + \bar A_{16}(\bar A_{12}\bar A_{26} - \bar A_{22}\bar A_{16})\,.$$

If the fracture of a reinforced plastic is caused by the matrix failure of the fibre-matrix bond failure, the general criterion is written as

$$F(\langle\sigma_\perp\rangle_k;\langle\tau_{\parallel\perp}\rangle_k)=1\,. \tag{50}$$

In the above criterion $\langle\sigma_\perp\rangle_k$ and $\langle\tau_{\parallel\perp}\rangle_k$ denote mean stresses in layer 'k' which fails first under the given loading conditions. The form of the right-hand side function in criterion (50) depends upon the bond, and upon the ratio between $\langle\sigma_\perp\rangle_k$ and $\langle\tau_{\parallel\perp}\rangle_k$.

The characteristic features of unidirectionally reinforced plastic fracture under stresses $\langle\sigma_\perp\rangle_k$ and $\langle\tau_{\parallel\perp}\rangle_k$ were discussed in Section 3.4. If there is tensile failure of the matrix under combined tension and shear (41) or (42) leads to

$$F(\langle\sigma_\perp\rangle_k;\langle\tau_{\parallel\perp}\rangle_k)=\left(\frac{\langle\sigma_\perp\rangle_k}{\sigma_{\mathrm{m}}^{+}}\right)^2 A+\left(\frac{\langle\tau_{\parallel\perp}\rangle_k}{\tau_{\mathrm{m}}^{\mathrm{u}}}\right)^2 \tag{51}$$

or

$$F(\langle\sigma_\perp\rangle_k;\langle\tau_{\parallel\perp}\rangle_k)=(1-\nu_{\mathrm{m}}^2)\left(\frac{\bar\sigma_r\langle\sigma_\perp\rangle_k}{\sigma_{\mathrm{m}}^{+}}\right)^2+\left(\frac{\langle\tau_{\parallel\perp}\rangle_k}{\tau_{\mathrm{m}}^{\mathrm{u}}}\right)^2 . \tag{52}$$

If the fracture of a unidirectionally reinforced lamina oriented at angle β_k is typical of longitudinal shear (45) yields:

$$F(\langle\sigma_\perp\rangle_k;\langle\tau_{\parallel\perp}\rangle_k)=\frac{1}{2}\left(\frac{\langle\sigma_\perp\rangle_k}{\sigma_{\mathrm{m}}^{+}}\right)^2+(1+\nu_{\mathrm{m}})\left(\frac{\langle\tau_{\parallel\perp}\rangle_k}{\sigma_{\mathrm{m}}^{+}}\right)^2+\frac{\langle\sigma_\perp\rangle_k}{2(\sigma_{\mathrm{m}}^{+})^2}\sqrt{\langle\sigma_\perp^2\rangle_k+4\langle\tau_{\parallel\perp}^2\rangle_k}\,. \tag{53}$$

In case of an initial bond failure, (43) yields:

$$F(\langle\sigma_\perp\rangle_k;\langle\tau_{\parallel\perp}\rangle_k)=\frac{\langle\sigma_\perp\rangle_k\bar\sigma_r}{\sigma_{\mathrm{b}}^{\mathrm{u}}}+\left(\frac{\langle\tau_{\parallel\perp}\rangle_k\bar\sigma_{rz}}{\tau_{\mathrm{b}}^{\mathrm{u}}}\right)^2 \tag{54}$$

The value and the sign of stresses $\langle\sigma_\perp\rangle_k$ and $\langle\tau_{\parallel\perp}\rangle_k$ for every particular case of loading is determined with the aid of (57) and (58). If stress $\langle\sigma_\perp\rangle_k$ is a compressive stress, it is a minus quantity in (53) and (54).

If under combined action of compressive and shear stresses the unidirectionally reinforced ply fails due to transverse compression, from (46) it follows:

$$F(\langle\sigma_\perp\rangle_k;\langle\tau_{\parallel\perp}\rangle_k)=\frac{\langle\sigma_\perp^2\rangle_k}{2(\sigma_\perp^-)_k^2}+(1+\nu_{\perp\perp})\frac{\langle\tau_{\parallel\perp}^2\rangle_k}{(\sigma_\perp^-)_k^2}-\frac{\langle\sigma_\perp\rangle_k}{2(\sigma_\perp^-)_k^2}\sqrt{\langle\sigma_\perp^2\rangle_k+4\langle\tau_{\parallel\perp}^2\rangle_k}\,. \tag{55}$$

It should be marked that eventual failure of the polymer matrix or bond is provided for in (55) by introducing $\sigma_\perp^-$ obtained in its turn from (40).

In a general case of plane state of stress the following stresses act in the elastic symmetry directions of the laminae (SKUDRA (1980)):

$$\langle\sigma_{\parallel}\rangle = \frac{1}{\bar{\Delta}}(\langle\langle\sigma_1'\rangle\rangle a_1 + \langle\langle\sigma_2'\rangle\rangle b_1 + \langle\langle\sigma_6'\rangle\rangle c_1), \tag{56}$$

$$\langle\sigma_{\perp}\rangle = \frac{1}{\bar{\Delta}}(\langle\langle\sigma_1'\rangle\rangle a_2 + \langle\langle\sigma_2'\rangle\rangle b_2 + \langle\langle\sigma_6'\rangle\rangle c_2), \tag{57}$$

$$\langle\tau_{\parallel\perp}\rangle = \frac{1}{\bar{\Delta}}(\langle\langle\sigma_1'\rangle\rangle a_3 + \langle\langle\sigma_2'\rangle\rangle b_3 + \langle\langle\sigma_6'\rangle\rangle c_3). \tag{58}$$

There are the following designations in the above formulas:

$$a_1 = \cos^2\beta\bar{A}_{66}(\bar{Q}_{11}\bar{A}_{22} - \bar{Q}_{12}\bar{A}_{12}) + \sin^2\beta\bar{A}_{66}(\bar{Q}_{12}\bar{A}_{22} - \bar{Q}_{22}\bar{A}_{12}) + 2\sin\beta\cos\beta\bar{A}_{66}(\bar{Q}_{16}\bar{A}_{22} - \bar{Q}_{26}\bar{A}_{12}), \tag{59}$$

$$b_1 = \cos^2\beta\bar{A}_{66}(\bar{Q}_{12}\bar{A}_{11} - \bar{Q}_{11}\bar{A}_{12}) + \sin^2\beta\bar{A}_{66}(\bar{Q}_{22}\bar{A}_{11} - \bar{Q}_{12}\bar{A}_{12}) + 2\sin\beta\cos\beta\bar{A}_{66}(\bar{Q}_{26}\bar{A}_{11} - \bar{Q}_{16}\bar{A}_{12}), \tag{60}$$

$$c_1 = \bar{A}_{11}\bar{A}_{22}(\bar{Q}_{16}\cos^2\beta + \bar{Q}_{26}\sin^2\beta + 2\bar{Q}_{66}\sin\beta\cos\beta), \tag{61}$$

$$a_2 = \sin^2\beta\bar{A}_{66}(\bar{Q}_{11}\bar{A}_{22} - \bar{Q}_{12}\bar{A}_{12}) + \cos^2\beta\bar{A}_{66}(\bar{Q}_{12}\bar{A}_{22} - \bar{Q}_{22}\bar{A}_{12}) - 2\sin\beta\cos\beta\bar{A}_{66}(\bar{Q}_{16}\bar{A}_{22} - \bar{Q}_{26}\bar{A}_{12}), \tag{62}$$

$$b_2 = \sin^2\beta\bar{A}_{66}(\bar{Q}_{12}\bar{A}_{11} - \bar{Q}_{11}\bar{A}_{12}) + \cos^2\beta\bar{A}_{66}(\bar{Q}_{22}\bar{A}_{11} - \bar{Q}_{12}\bar{A}_{12}) - 2\sin\beta\cos\beta\bar{A}_{66}(\bar{Q}_{26}\bar{A}_{11} - \bar{Q}_{16}\bar{A}_{12}), \tag{63}$$

$$c_2 = \bar{A}_{11}\bar{A}_{22}(\bar{Q}_{16}\sin^2\beta + \bar{Q}_{26}\cos^2\beta - 2\bar{Q}_{66}\sin\beta\cos\beta), \tag{64}$$

$$a_3 = \bar{A}_{66}\sin\beta\cos\beta[\bar{A}_{22}(\bar{Q}_{12} - \bar{Q}_{11}) + \bar{A}_{12}(\bar{Q}_{12} - \bar{Q}_{22})] + \bar{A}_{66}(\cos^2\beta - \sin^2\beta)(\bar{Q}_{16}\bar{A}_{22} - \bar{Q}_{26}\bar{A}_{12}), \tag{65}$$

$$b_3 = \bar{A}_{66}\sin\beta\cos\beta[\bar{A}_{11}(\bar{Q}_{22} - \bar{Q}_{12}) + \bar{A}_{12}(\bar{Q}_{11} - \bar{Q}_{12})] + \bar{A}_{66}(\cos^2\beta - \sin^2\beta)(\bar{Q}_{26}\bar{A}_{11} - \bar{Q}_{16}\bar{A}_{12}), \tag{66}$$

$$c_3 = \bar{A}_{11}\bar{A}_{22}[\sin\beta\cos\beta(\bar{Q}_{26} - \bar{Q}_{16}) + (\cos^2\beta - \sin^2\beta)\bar{Q}_{66}]. \tag{67}$$

In a general case coefficients A_{ij} and $\bar{A}_{ij}$ are obtained from the following formulae:

$$A_{ij} = \sum_{k=1}^{n}(\bar{Q}_{ij})_k(h_k - h_{k-1}), \tag{68}$$

$$\bar{A}_{ij} = \frac{A_{ij}}{\delta} = \frac{1}{\delta}\sum_{k=1}^{n}(\bar{Q}_{ij})_k(h_k - h_{k-1}), \tag{69}$$

where δ is the laminate thickness.

Elastic characteristics $\bar{Q}_{ij}$ for the laminae oriented at arbitrary angle β, are calculated as follows:

$$\bar{Q}_{11} = \frac{1}{1 - \nu_{\parallel\perp}\nu_{\perp\parallel}}\{E_\parallel \cos^4 \beta + 2[E_\parallel \nu_{\perp\parallel} + 2G_{\parallel\perp}(1 - \nu_{\perp\parallel}\nu_{\parallel\perp})] \sin^2 \beta \cos^2 \beta + E_\perp \sin^4 \beta\}, \tag{70}$$

$$\bar{Q}_{22} = \frac{1}{1 - \nu_{\parallel\perp}\nu_{\perp\parallel}}\{E_\parallel \sin^4 \beta + 2[E_\parallel \nu_{\perp\parallel} + 2G_{\parallel\perp}(1 - \nu_{\perp\parallel}\nu_{\parallel\perp})] \sin^2 \beta \cos^2 \beta + E_\perp \cos^4 \beta\}, \tag{71}$$

$$\bar{Q}_{12} = \frac{1}{1 - \nu_{\parallel\perp}\nu_{\perp\parallel}}\{[E_\parallel + E_\perp - 4G_{\parallel\perp}(1 - \nu_{\parallel\perp}\nu_{\perp\parallel})] \sin^2 \beta \cos^2 \beta + E_\parallel \nu_{\perp\parallel}(\sin^4 \beta + \cos^4 \beta)\}, \tag{72}$$

$$\bar{Q}_{66} = \frac{1}{1 - \nu_{\parallel\perp}\nu_{\perp\parallel}}\{[E_\parallel + E_\perp - 2E_\parallel \nu_{\perp\parallel} - 2G_{\parallel\perp}(1 - \nu_{\parallel\perp}\nu_{\perp\parallel})] \sin^2 \beta \cos^2 \beta + G_{\parallel\perp}(\sin^4 \beta + \cos^4 \beta)\}, \tag{73}$$

$$\bar{Q}_{16} = \frac{1}{1 - \nu_{\parallel\perp}\nu_{\perp\parallel}}\{[E_\parallel - E_\parallel \nu_{\perp\parallel} - 2G_{\parallel\perp}(1 - \nu_{\parallel\perp}\nu_{\perp\parallel})] \sin \beta \cos^3 \beta + [E_\parallel \nu_{\perp\parallel} - E_\perp + 2G_{\parallel\perp}(1 - \nu_{\parallel\perp}\nu_{\perp\parallel})] \sin^3 \beta \cos \beta\}, \tag{74}$$

$$\bar{Q}_{26} = \frac{1}{1 - \nu_{\parallel\perp}\nu_{\perp\parallel}}\{[E_\parallel - E_\parallel \nu_{\perp\parallel} - 2G_{\parallel\perp}(1 - \nu_{\parallel\perp}\nu_{\perp\parallel})] \sin^3 \beta \cos \beta + [E_\parallel \nu_{\perp\parallel} - E_\perp + 2G_{\parallel\perp}(1 - \nu_{\parallel\perp}\nu_{\perp\parallel})] \sin \beta \cos^3 \beta\}. \tag{75}$$

Elastic characteristics of a unidirectionally reinforced lamina $E_\parallel$, $E_\perp$, $\nu_{\parallel\perp}$, $\nu_{\perp\parallel}$, $G_{\parallel\perp}$, incorporated in the above relationships are determined by formulas given by VAN FO FI (1971) and SKUDRA and BULAVS (1978).

Parameter $\bar{\Delta}$ is calculated in the following way:

$$\bar{\Delta} = \bar{A}_{11}\bar{A}_{22}\bar{A}_{66} - \bar{A}_{12}^2\bar{A}_{66}. \tag{76}$$

Structural parameters $\bar{\sigma}_r$, $\bar{\sigma}_\theta$, $\bar{\sigma}_z$, $\bar{\sigma}_{rz}$, for a reinforced plastic which occur in formulas (51)–(55) are obtained from curves in Figs. 7–9 and 14.

4. Strength of cross-ply laminates

4.1. Tensile and shear strength

In the first place we shall consider uniaxial tension along axis X_1 (Fig. 22).

It has been shown already that the principal characteristics feature of failure in reinforced plastics is the fact, that the constituents of the composite do not fail simultaneously. If the fibres (which really sustain the largest part of the load) are the first to undergo failure, there is a simultaneous and complete rupture of the composite. If it is the polymer matrix which fails first, there is an appearance of microcracks before a complete rupture of the composite. The microcracks caused by the first ply failure destroy the continuity of the composite, and at this point the diagram $\langle\langle\sigma_1\rangle\rangle$–$\langle\langle e_1\rangle\rangle$ shows a characteristic

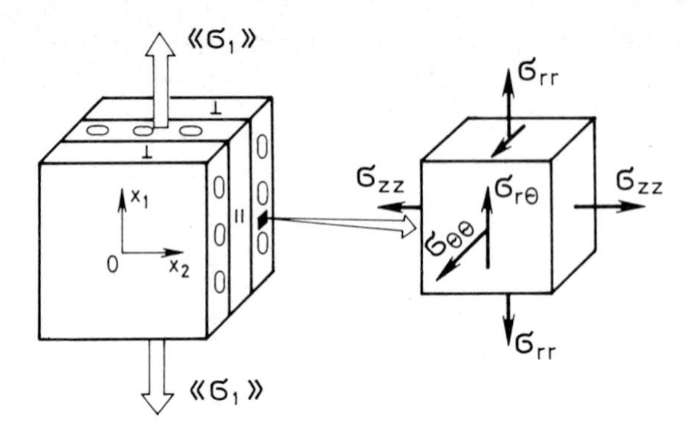

FIG. 22. A calculation scheme for an orthogonally reinforced plastic under uniaxial tension.

kink. In tension, orthogonally reinforced plastics are most sensitive to the loss of continuity if the load is applied in the direction of fibres. After the loss of continuity the whole load is practically taken by the fibres, the orientation of which coincides with the loading direction.

The calculation scheme of orthogonally reinforced plastics is shown in Fig. 22.

If tensile stress is applied along axis X_1, the loss of continuity is caused by the failure of the polymer matrix in the layer which has been reinforced perpendicularly to the loading direction (along axis X_2).

The total cross-sectional area of a composite consists of two parts, $F_\parallel$ and $F_\perp$:

$$F = F_\parallel + F_\perp .$$

Ignoring the influence of transverse effects, the relationship between the applied mean stress $\langle\langle\sigma_1\rangle\rangle$ and the stresses in orthogonally reinforced layers $\langle\sigma_\parallel\rangle$ and $\langle\sigma_\perp\rangle$ is:

$$\langle\langle\sigma_1\rangle\rangle = \frac{F_\parallel}{F}\langle\sigma_\parallel\rangle + \frac{F_\perp}{F}\langle\sigma_\perp\rangle = m_1\langle\sigma_\parallel\rangle + (1 - m_1)\langle\sigma_\perp\rangle . \tag{a}$$

Considering that $\langle\langle\sigma_1\rangle\rangle = E_1\langle\langle e_1\rangle\rangle$ and assuming that there is no sliding between the layers, i.e.,

$$\langle\langle e_1\rangle\rangle = \langle e_\parallel\rangle = \langle e_\perp\rangle$$

a formula for calculating stresses $\langle\sigma_\perp\rangle$ may be derived from (a):

$$\langle\sigma_\perp\rangle = E_\perp\langle\langle e_1\rangle\rangle .$$

Here $E_\parallel$ denotes the elastic modulus of the layer which is reinforced perpendicularly to the loading direction.

Fig. 22 shows that the polymer matrix in the layer having fibre orientation perpendicular to the loading direction is under general state of stress. Stresses σ_{rr}, $\sigma_{\theta\theta}$, σ_{zz} and $\sigma_{r\theta}$ are calculated from the following relationships providing that at the moment of continuity loss $\langle\langle e_1\rangle\rangle = \langle\langle e_u^+\rangle\rangle$:

$$\sigma_{rr} = \langle\sigma_\perp\rangle\bar{\sigma}_r = E_\perp\bar{\sigma}_r\langle\langle e_u^+\rangle\rangle\,,$$

$$\sigma_{\theta\theta} = \langle\sigma_\perp\rangle\bar{\sigma}_\theta = E_\perp\bar{\sigma}_\theta\langle\langle e_u^+\rangle\rangle\,,$$

$$\sigma_{zz} = \langle\sigma_\perp\rangle\bar{\sigma}_z = E_\perp\nu_m(\bar{\sigma}_r + \bar{\sigma}_\theta)\langle\langle e_u^+\rangle\rangle\,,$$

$$\bar{\sigma}_{r\theta} = \langle\sigma_\perp\rangle\bar{\sigma}_{r\theta} = E_\perp\bar{\sigma}_{r\theta}\langle\langle e_u^+\rangle\rangle\,.$$

In the given case the strength energy criterion of the polymer matrix is:

$$\frac{1}{(\sigma_m^+)^2}[\bar{\sigma}_r^2 + \bar{\sigma}_\theta^2 + \bar{\sigma}_z^2 - 2\nu_m(\bar{\sigma}_r\bar{\sigma}_\theta + \bar{\sigma}_r\bar{\sigma}_z + \bar{\sigma}_\theta\bar{\sigma}_z)] + \frac{\bar{\sigma}_{r\theta}^2}{(\tau_m^u)^2} = \frac{1}{E_\perp^2\langle\langle e_u^+\rangle\rangle^2}\,.$$

The latter criterion provides an opportunity of calculating the ultimate strain of an orthogonally reinforced plastic at the moment of continuity loss, when tension is applied along axis X_1:

$$\langle\langle e_u^+\rangle\rangle = \frac{\sigma_m^+\tau_m^u}{E_\perp\sqrt{(\tau_m^u)^2[\bar{\sigma}_r^2 + \bar{\sigma}_\theta^2 + \bar{\sigma}_z^2] + (\sigma_m^+)^2\bar{\sigma}_{r\theta}^2 - 2\nu_m(\tau_m^u)^2(\bar{\sigma}_r\bar{\sigma}_\theta + \bar{\sigma}_r\bar{\sigma}_z + \bar{\sigma}_\theta\bar{\sigma}_z)}}\,.$$

Dimensionless structural parameters $\bar{\sigma}_r$, $\bar{\sigma}_\theta$ and $\bar{\sigma}_{r\theta}$ are the functions of angle θ (see Fig. 6). In Section 3.1 it was specified that $\theta_{crit} = 0°$. In this case $\bar{\sigma}_{r\theta} = 0$ and the formula for $\langle\langle e_u^+\rangle\rangle$ is reduced to

$$\langle\langle e_u^+\rangle\rangle = \frac{\sigma_m^+}{E_\perp\sqrt{\bar{\sigma}_r^2 + \bar{\sigma}_\theta^2 + \bar{\sigma}_z^2 - 2\nu_m(\bar{\sigma}_r\bar{\sigma}_\theta + \bar{\sigma}_r\bar{\sigma}_z + \bar{\sigma}_\theta\bar{\sigma}_z)}}\,. \tag{77}$$

Neglecting the influence of $\bar{\sigma}_\theta$ and $\bar{\sigma}_z$ we obtain the following approximate formula for calculating strain $\langle\langle e_u^+\rangle\rangle$

$$\langle\langle e_u^+\rangle\rangle = \frac{\sigma_m^+}{E_\perp\sqrt{1 - \nu_m^2}\bar{\sigma}_r}\,. \tag{78}$$

The ultimate strain relationship which shows the dependence of the moment of continuity loss of the volume content of the fibres is given in Fig. 23.

The ultimate strain $\langle\langle e_u^+\rangle\rangle$ having been calculated the level of the mean stresses in the composite $\langle\langle\sigma_1\rangle\rangle_c$ at which the continuity loss takes place, can be obtained:

$$\langle\langle\sigma_1\rangle\rangle_c = E_1\langle\langle e_u^+\rangle\rangle \tag{79}$$

where E_1 is the elasticity modulus for an orthogonally reinforced plastic under tension applied along axis X_1.

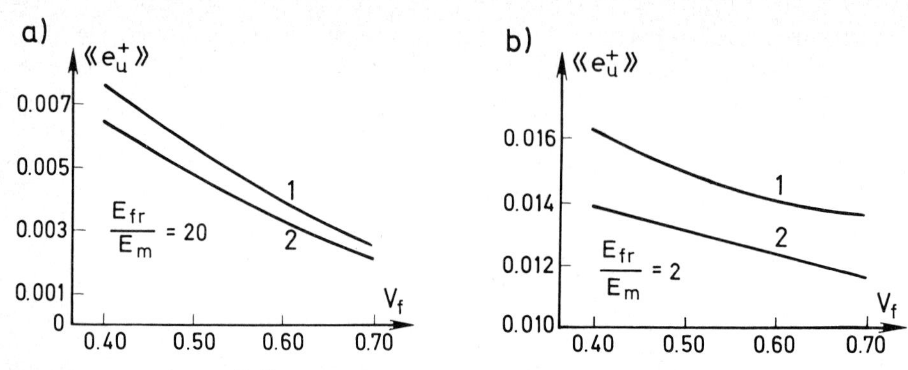

FIG. 23. Dependence of $\langle\langle e_4^+\rangle\rangle$ on the fibre volume content for a glass/epoxy (a) and a carbon/epoxy (b), Curves 1 have been plotted from formula (77) and Curves 2 from (78) with the following initial data: $\sigma_m^+ = 74.2$ MPa; $\nu_m = 0.35$; $E_m = 3\,500$ MPa; for glass fibres $E_f = 70\,000$ MPa; $\nu_f = 0.23$; for carbon fibres $E_{fr} = 70\,000$ MPa.

After continuity loss, i.e., when the layer with the fibre direction perpendicular to the applied tension has failed, the entire load is sustained by the layer reinforced in the load direction, and the strength is calculated as follows:

(a) if tension is applied in the direction of axis X_1

$$\sigma_1^+ = m_1 e_{fu}^+[(1 - v_f)E_m + v_f E_{fz}] \tag{80}$$

(b) if tension is in the direction of X_2

$$\sigma_2^+ = (1 - m_1)e_{fu}^+[(1 - v_f)E_m + v_f E_{fz}]\,.$$

The test data, with the aim of checking the relationships (79) and (80) are plotted in Fig. 24. Curve 1 illustrates the dependence $\langle\langle\sigma_1\rangle\rangle_c$ on the reinforcement embedding ratio $m_1 = F_\parallel/(F_\parallel + F_\perp)$ for a glass/epoxy composite. The test

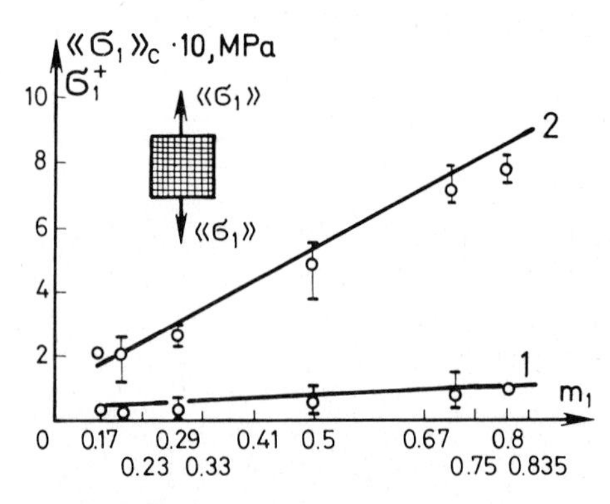

FIG. 24. Dependence of continuity limit $\langle\langle\sigma_1\rangle\rangle_c$ and the strength of an orthogonally reinforced glass/epoxy plastic unon the ratio of fibre embedding m_1. Curves 1 and 2 are based on (79) and (80) provided $v_f = 0.66$; $E_m = 3\,600$ MPa; $\nu_m = 0.36$; $\sigma_m^+ = 74$ MPa; $E_f = 70\,000$ MPa; $\nu_f = 0.22$.

data have been obtained from the paper of TSAI (1965). It should be remarked that for mean stresses $\langle\langle\sigma_1\rangle\rangle_c$ a characteristic kink is observed in the diagram $\langle\langle\sigma_1\rangle\rangle - \langle\langle e_1\rangle\rangle$.

Another possible cause of continuity loss in an orthogonally reinforced plastic is the destruction of the bond between the fibres and the polymer matrix. For making up the corresponding continuity condition, we resort again to criterion (13).

Considering that $\theta_{crit} = 0°$ and $\langle\tau_{\parallel\perp}\rangle = \bar{\sigma}s_{r\theta} = 0$, for this particular case we get

$$\langle\langle e_u^+\rangle\rangle = \frac{\sigma_b^u}{\bar{\sigma}_r E_\perp}. \tag{81}$$

So the determination of the ultimate strain $\langle\langle e_u^+\rangle\rangle$ at which the composite loses continuity, results in the two relationships (79) and (81). The actual ultimate strain is equal to the least of these. Equalization of the right sides of Equations (79) and (81) yields an equation establishing the optimum ratio between the polymer matrix strength and the bond strength.

The above relationships refer to the case when the fibre volume content v_f in both orthogonal directions is the same, i.e.,

$$v_f = \frac{V_{f\parallel}}{V_\parallel} = \frac{V_{f\perp}}{V_\perp}$$

where $V_{f\parallel}$ and $V_{f\perp}$ are volumes of fibres parallel and perpendicular to the load directions whereas $V_\parallel$ and $V_\perp$ are the volumes of the material reinforced in these directions. Under this condition, the shear strength of an orthogonally reinforced plastic in the elastic symmetry axes is practically equal to the shear strength of a unidirectionally reinforced plastic, having the same relative volume content of fibres, namely,

$$\tau_{12}^u \approx \tau_{\parallel\perp}^u = \tau_m^u .$$

4.2. Compressive strength

To determine the compressive strength of an orthogonally reinforced plastic consisting of unidirectionally reinforced elementary layers 'a' and 'b', we may use the calculation scheme shown in Fig. 25. The strength of this type of material has been studied by SKUDRA (1977).

Let us designate the value of the mean compressive stress $\langle\langle\sigma_1\rangle\rangle$ at the moment of failure by σ_1^-. The layers reinforced in the direction of the load applied ('a') and perpendicular to the load ('b') do not fail simultaneously. If the layer reinforced along axis X_1 is the first to undergo failure, the strength of an orthogonally reinforced plastic σ_1^- is determined from the formula:

$$\sigma_1^- = m_1\sigma_\parallel^- + (1 - m_1)\langle\sigma_\perp\rangle . \tag{82}$$

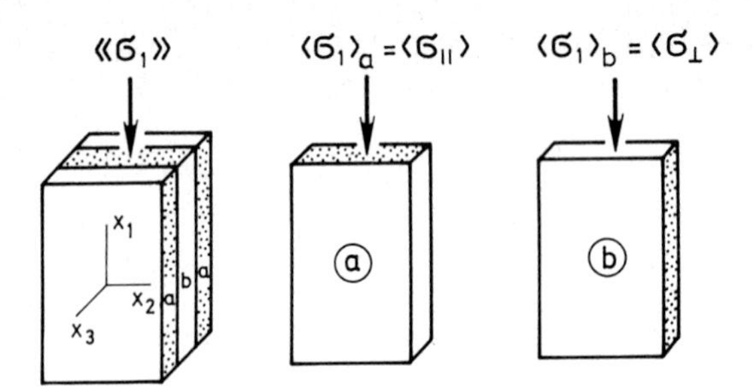

FIG. 25. A calculation scheme for an orthogonally reinforced plastic under compression.

If the layer reinforced perpendicularly to the loading direction fails first, then

$$\sigma_1^- = m_1\langle\sigma_\parallel\rangle + (1 - m_1)\sigma_\perp^- . \tag{83}$$

First we shall calculate stresses $\langle\sigma_\parallel\rangle$ and $\langle\sigma_\perp\rangle$, taking into account (34) and (40).

$$\langle\sigma_\perp\rangle = E_\perp \frac{\sigma_\parallel^-}{E_\parallel} = E_\perp e_{\mathrm{fu}}^- , \qquad \langle\sigma_\parallel\rangle = 3.5 \frac{\sigma_\perp^+}{1 + \nu_{\perp\parallel}} \cdot \frac{E_\parallel}{E_\perp} .$$

Considering the values obtained for $\langle\sigma_\perp\rangle$ and $\langle\sigma_\parallel\rangle$, formulas (82) and (83) may be written as

$$\sigma_1^- = m_1 E_\parallel e_{\mathrm{fu}}^- + (1 - m_1)E_\perp e_{\mathrm{fu}}^- = E_1 e_{\mathrm{fu}}^- , \tag{84}$$

$$\sigma_1^- = \frac{3.5\,\sigma_\perp^+}{1 + \nu_{\perp\parallel}} \left[m_1 \frac{E_\parallel}{E_\perp} + 1 - m_1 \right] . \tag{85}$$

By analogy, from (84) and (85) we can obtain the formula for the strength σ_2^-:

$$\sigma_2^- = m_1 E_\perp e_{\mathrm{fu}}^- + (1 - m_1)E_\parallel e_{\mathrm{fu}}^- = E_2 e_{\mathrm{fu}}^- , \tag{86}$$

$$\sigma_2^- = \frac{3.5\,\sigma_\perp^+}{1 + \nu_{\perp\parallel}} \left[(1 - m_1) \frac{E_\parallel}{E_\perp} + m_1 \right] . \tag{87}$$

If the fracture of a reinforced plastic is due to the fracture of the matrix, $\sigma_\perp^+$ is obtained from (26) or (27), whereas it is obtained from formula (28) if the bond fails first.

Fig. 26 illustrates the dependence of strength σ_1^- on the ratio of fibre embedding m_1 for a glss/epoxy plastic. The theoretical line is based on formula (84) and test data have been borrowed from SKUDRA and BULAVS (1978).

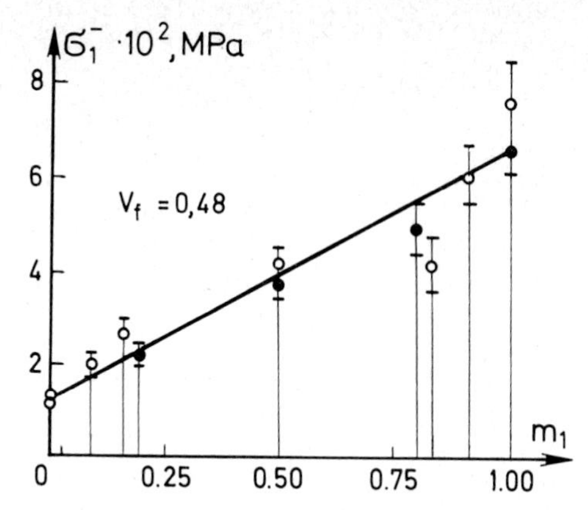

FIG. 26. Dependence of strength σ_1^- upon the fibre embedding ratio m_1 for a glass fibre orthogonally reinforced plastic (with phenolformaldehyde as a matrix). The theoretical curve has been drawn from (84) where $e_{fu}^- = 0.017$; $\sigma_m^+ = 70$ MPa; $\nu_m = 0.35$; $E_f = 70\,000$ MPa; $\nu_f = 0.22$.

4.3. *Strength under biaxial tension*

Let us consider a case of two tensile stresses $\langle\langle\sigma_1\rangle\rangle$ and $\langle\langle\sigma_2\rangle\rangle$ applied in the direction of reinforcing X_1 and X_2 as given in Fig. 27.

To determine stresses in unidirectionally reinforced layers a and b we make use of formulas (56) and (57) provided $\langle\langle\sigma_1'\rangle\rangle = \langle\langle\sigma_1\rangle\rangle$, $\langle\langle\sigma_2'\rangle\rangle = \langle\langle\sigma_2\rangle\rangle$ and $\langle\langle\sigma_6'\rangle\rangle = 0$.

For layer a angle $\beta = 0$, and relationships (59), (60), (62) and (63) which help us to calculate ratios a_1, b_1, a_2 and b_2 are written as follows:

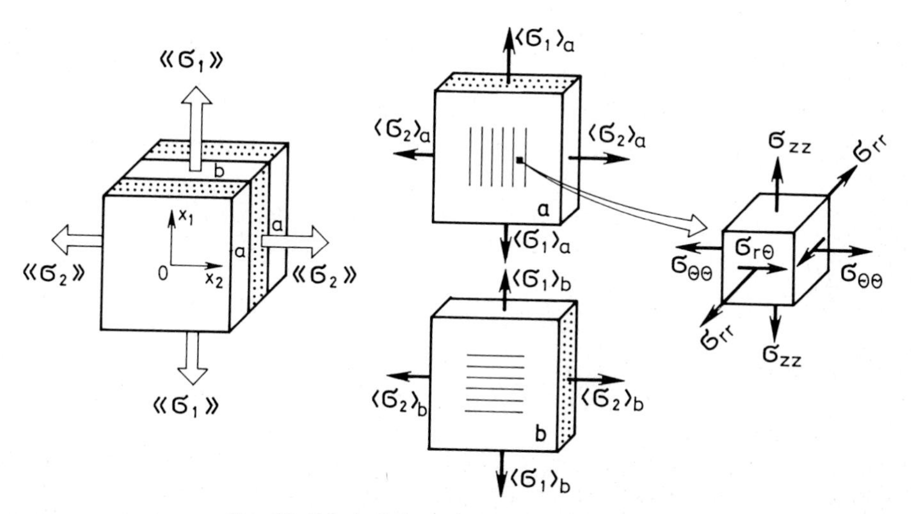

FIG. 27. A calculation scheme for biaxial tension.

$$a_1 = [(Q_{11})_a\bar{A}_{22} - (Q_{12})_a\bar{A}_{12}]\bar{A}_{66}\,,$$
$$b_1 = [(Q_{12})_a\bar{A}_{11} - (Q_{11})_a\bar{A}_{12}]\bar{A}_{66}\,,$$
$$a_2 = [(Q_{12})_a\bar{A}_{22} - (Q_{22})_a\bar{A}_{12}]\bar{A}_{66}\,,$$
$$b_2 = [(Q_{22})_a\bar{A}_{11} - (Q_{12})_a\bar{A}_{12}]\bar{A}_{66}\,.$$

By analogy, if $\beta = 90°$, we get the following expressions for layer b:

$$a_1 = [(Q_{12})_b\bar{A}_{22} - (Q_{11})_b\bar{A}_{12}]\bar{A}_{66}\,,$$
$$b_1 = [(Q_{11})_b\bar{A}_{11} - (Q_{12})_b\bar{A}_{12}]\bar{A}_{66}\,,$$
$$a_2 = [(Q_{22})_b\bar{A}_{22} - (Q_{12})_b\bar{A}_{12}]\bar{A}_{66}\,,$$
$$b_2 = [(Q_{12})_b\bar{A}_{11} - (Q_{22})_b\bar{A}_{12}]\bar{A}_{66}\,.$$

Inserting the expressions for ratios a_i and b_i into (56) and (57) we get:
(1) for layer a

$$\langle\sigma_1\rangle_a = \langle\sigma_\parallel\rangle_a = \langle\langle\sigma_1\rangle\rangle \frac{(Q_{11})_a\bar{A}_{22} - (Q_{12})_a\bar{A}_{12}}{\bar{A}_{11}\bar{A}_{22} - \bar{A}_{12}^2} + \langle\langle\sigma_2\rangle\rangle \frac{(Q_{12})_a\bar{A}_{11} - (Q_{11})_a\bar{A}_{12}}{\bar{A}_{11}\bar{A}_{22} - \bar{A}_{12}^2}\,, \tag{88}$$

$$\langle\sigma_2\rangle_a = \langle\sigma_\perp\rangle_a = \langle\langle\sigma_1\rangle\rangle \frac{(Q_{12})_a\bar{A}_{22} - (Q_{22})_a\bar{A}_{12}}{\bar{A}_{11}\bar{A}_{22} - \bar{A}_{12}^2} + \langle\langle\sigma_2\rangle\rangle \frac{(Q_{22})_a\bar{A}_{11} - (Q_{12})_a\bar{A}_{12}}{\bar{A}_{11}\bar{A}_{22} - \bar{A}_{12}^2}\,; \tag{89}$$

(2) for layer b

$$\langle\sigma_1\rangle_b = \langle\sigma_\perp\rangle_b = \langle\langle\sigma_1\rangle\rangle \frac{(Q_{22})_b\bar{A}_{22} - (Q_{12})_b\bar{A}_{22}}{\bar{A}_{11}\bar{A}_{22} - \bar{A}_{12}^2} + \langle\langle\sigma_2\rangle\rangle \frac{(Q_{12})_b\bar{A}_{11} - (Q_{22})_b\bar{A}_{12}}{\bar{A}_{11}\bar{A}_{22} - \bar{A}_{12}^2}\,, \tag{90}$$

$$\langle\sigma_2\rangle_b = \langle\sigma_\parallel\rangle_b = \langle\langle\sigma_1\rangle\rangle \frac{(Q_{12})_b\bar{A}_{22} - (Q_{11})_b\bar{A}_{12}}{\bar{A}_{11}\bar{A}_{22} - \bar{A}_{12}^2} + \langle\langle\sigma_2\rangle\rangle \frac{(Q_{11})_b\bar{A}_{11} - (Q_{12})_b\bar{A}_{12}}{\bar{A}_{11}\bar{A}_{22} - \bar{A}_{12}^2}\,. \tag{91}$$

In the given case Q_{ij} and $\bar{A}_{ij}$ are obtained in the following way:

$$(Q_{11})_a = \frac{(E_\parallel)_a}{1 - (\nu_{\parallel\perp})_a(\nu_{\perp\parallel})_a}\,, \qquad (Q_{22})_a = \frac{(E_\perp)_a}{1 - (\nu_{\parallel\perp})_a(\nu_{\perp\parallel})_a}\,,$$
$$(Q_{12})_a = \frac{(E_\parallel)_a(\nu_{\perp\parallel})_a}{1 - (\nu_{\parallel\perp})_a(\nu_{\perp\parallel})_a}\,, \qquad (Q_{11})_b = \frac{(E_\parallel)_b}{1 - (\nu_{\parallel\perp})_b(\nu_{\perp\parallel})_b}\,, \tag{92}$$
$$(Q_{22})_b = \frac{(E_\perp)_b}{1 - (\nu_{\parallel\perp})_b(\nu_{\perp\parallel})_b}\,, \qquad (Q_{12})_b = \frac{(E_\parallel)_b(\nu_{\perp\parallel})_b}{1 - (\nu_{\parallel\perp})_b(\nu_{\perp\parallel})_b}\,;$$

$$\bar{A}_{11} = m_1(Q_{11})_a + (1 - m_1)(Q_{22})_b\,,$$
$$\bar{A}_{12} = m_1(Q_{12})_a + (1 - m_1)(Q_{12})_b\,, \tag{93}$$
$$\bar{A}_{22} = m_1(Q_{22})_a + (1 - m_1)(Q_{11})_b\,.$$

Characteristics of elastic properties $E_\parallel$, $E_\perp$, $\nu_{\parallel\perp}$ and $\nu_{\perp\parallel}$ for unidirectionally reinforced layers a and b can be obtained by means of formulas offered by SKUDRA and BULAVS (1978) and VAN FO FI (1971).

First we shall analyse the conditions for continuity loss which is caused by the action of stresses $\langle\sigma_\perp\rangle_a$ and $\langle\sigma_\perp\rangle_b$. To determine the moment of continuity loss we resort to criterion (50). In the case of the polymer matrix failure $F(\langle\sigma_\perp\rangle_k; \langle\tau_{\parallel\perp}\rangle_k)$ is taken as in (51). Thus we obtain the following criterion for loss of continuity:

$$\bar{\sigma}_r^2 + \bar{\sigma}_\theta^2 + \bar{\sigma}_z^2 - 2\nu_m(\bar{\sigma}_r\bar{\sigma}_\theta + \bar{\sigma}_r\bar{\sigma}_z + \bar{\sigma}_\theta\bar{\sigma}_z) = \left(\frac{\sigma_m^+}{\langle\sigma_\perp\rangle_a}\right)^2 . \tag{94}$$

Ignoring the influence of $\bar{\sigma}_\theta$ and $\bar{\sigma}_z$ in criterion (94) we get

$$\bar{\sigma}_r\langle\sigma_\perp\rangle_a\sqrt{1-\nu_m^2} = \sigma_m^+ . \tag{95}$$

Parameters $\bar{\sigma}_r$, $\bar{\sigma}_\theta$ and $\bar{\sigma}_z$ take into account the effect of the structure of the material on stress concentration. Their dependence on the fibre volume content and on matrix-fibre modulus ratio is shown in Figs. 7–9.

Inserting $\langle\sigma_\perp\rangle_a$ into (95) as given by (89) and making some transformations we obtain the relationship between the applied mean stresses $\langle\langle\sigma_1\rangle\rangle$ and $\langle\langle\sigma_2\rangle\rangle$:

$$\langle\langle\sigma_1\rangle\rangle q_1 + \langle\langle\sigma_2\rangle\rangle q_2 = \frac{\sigma_m^+}{\bar{\sigma}_r\sqrt{1-\nu_m^2}} , \tag{96}$$

where

$$q_1 = \frac{(Q_{12})_a\bar{A}_{22} - (Q_{22})_a\bar{A}_{12}}{\bar{A}_{11}\bar{A}_{22} - \bar{A}_{12}^2} , \qquad q_2 = \frac{(Q_{22})_a\bar{A}_{11} - (Q_{12})_a\bar{A}_{12}}{\bar{A}_{11}\bar{A}_{22} - \bar{A}_{12}^2} .$$

According to (96) the material loses its continuity as a result of failure in layer a caused by transverse tension. In case the continuity of the material is lost as a result of a failing layer b, criterion (96) is substituted by

$$\langle\langle\sigma_1\rangle\rangle q_3 + \langle\langle\sigma_2\rangle\rangle q_4 = \frac{\sigma_m^+}{\bar{\sigma}_r\sqrt{1-\nu_m^2}} , \tag{97}$$

where

$$q_3 = \frac{(Q_{22})_b\bar{A}_{22} - (Q_{12})_b\bar{A}_{12}}{\bar{A}_{11}\bar{A}_{22} - \bar{A}_{12}^2} , \qquad q_4 = \frac{(Q_{12})_b\bar{A}_{11} - (Q_{22})_b\bar{A}_{12}}{\bar{A}_{11}\bar{A}_{22} - \bar{A}_{12}^2} .$$

Criteria (96) and (97) refer to the case when the strength of the polymer matrix is inferior to the bond strength.

High modulus materials, like carbon plastics, usually have a lower bond strength compared to the strength of the polymer matrices.

In order to make up a proper continuity condition, we insert $F(\langle\sigma_\perp\rangle_k; \langle\tau_{\parallel\perp}\rangle_k)$

in criterion (50) in the form it is given in (54) if $\langle\tau_{\parallel\perp}\rangle = 0$, and obtain

$$\bar{\sigma}_r \langle\sigma_\perp\rangle_a = \sigma_b^u . \tag{98}$$

Substituting $\langle\sigma_\perp\rangle_a$ in (98) as given by (89), we obtain a relationship between mean stresses $\langle\langle\sigma_1\rangle\rangle$ and $\langle\langle\sigma_2\rangle\rangle$ at the moment of continuity loss due to the bond failure between the fibres and the matrix in layer a

$$\langle\langle\sigma_1\rangle\rangle q_1 + \langle\langle\sigma_2\rangle\rangle q_2 = \frac{\sigma_b^u}{\bar{\sigma}_r} . \tag{99}$$

If the bond failure occurs in layer b, instead of (98) we write

$$\bar{\sigma}_r \langle\sigma_\perp\rangle_b = \sigma_b^u . \tag{100}$$

Inserting $\langle\sigma_\perp\rangle_b$ into (100) as given by (93), we obtain the continuity condition for layer b.

$$\langle\langle\sigma_1\rangle\rangle q_3 + \langle\langle\sigma_2\rangle\rangle q_4 = \frac{\sigma_b^u}{\bar{\sigma}_r} . \tag{101}$$

Provided the bond strength and the ultimate strain in the polymer matrix are sufficiently high, the fibres will fail first. Then the ultimate strain criterion (49) leads to the following strength criteria for the whole material if $\langle\langle\sigma_6\rangle\rangle = 0$:

(a) in the direction of X_1, i.e., if $\beta_k = 0°$

$$\langle\langle\sigma_1\rangle\rangle \bar{A}_{22} - \langle\langle\sigma_2\rangle\rangle \bar{A}_{12} = (\bar{A}_{11}\bar{A}_{22} - \bar{A}_{12}^2) e_{fu}^+ ; \tag{102}$$

(b) in the direction of X_2, i.e., if $\beta_k = 90°$

$$-\langle\langle\sigma_1\rangle\rangle \bar{A}_{12} + \langle\langle\sigma_2\rangle\rangle \bar{A}_{11} = (\bar{A}_{11}\bar{A}_{22} - \bar{A}_{12}^2) e_{fu}^+ . \tag{103}$$

Formulas (102) and (103) give a full account of layer interaction up to the moment of complete fracture. If the continuity of the material is lost, the interaction of layers is disturbed. The fibre failure may take place even after continuity loss. This happens, for instance, when vessels having a proper sealing layer are tested for interior pressure. Ignoring the transverse effects due to layer interaction, (102) and (103) lead to the following approximated strength criteria:

(a) along X_1

$$\langle\langle\sigma_1\rangle\rangle = \sigma_1^+ = m_1[(1 - v_f)E_m + v_f E_{fz}] e_{fu}^+ ; \tag{104}$$

(b) along X_2

$$\langle\langle\sigma_2\rangle\rangle = \sigma_2^+ = (1 - m_1)[(1 - v_f)E_m + v_f E_{fz}] e_{fu}^+ . \tag{105}$$

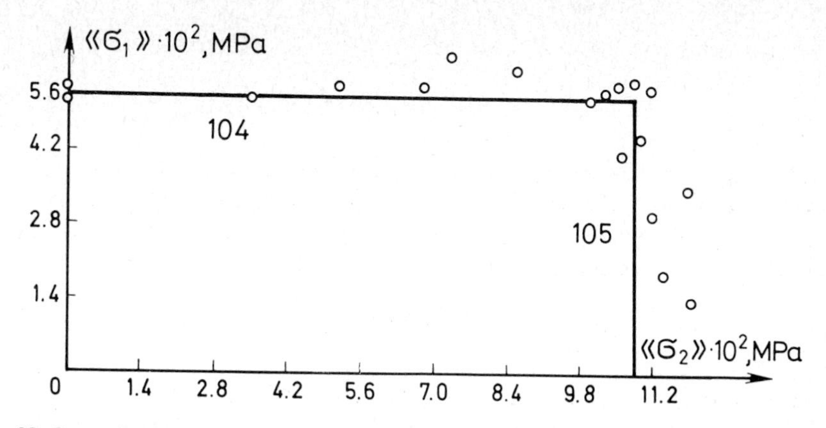

FIG. 28. Strength of orthogonally reinforced glass/epoxy plastic in biaxial tension. The numbers near the curves indicate from which formulas they have been plotted. Initial data are: $m_1 = 0.33$; $e^+_{\mathrm{fu}} = 0.03$; $v_{\mathrm{f}} = 0.60$; $E_{\mathrm{m}} = 3\,500$ MPa; $E_{\mathrm{f}} = 70\,000$ MPa.

Fig. 28 shows test data given in the paper of JONES (1969) and the theoretical lines plotted from (104) and (105).

4.4. *Strength under biaxial compression*

The loading scheme is shown in Fig. 29. The first case to be considered is when the ultimate strain of the layers in the direction of the reinforcement is smaller than the ultimate strain in the direction perpendicular to the reinforcement. Under such conditions, the fracture of the material is caused by the failure of the fibres under compression. The respective strength criteria are derived from (49), if $\langle\langle\sigma_6\rangle\rangle = 0$:

(a) in the direction of X_1, i.e., $\beta_k = 0°$

$$\langle\langle\sigma_1\rangle\rangle\bar{A}_{22} - \langle\langle\sigma_2\rangle\rangle\bar{A}_{12} = (\bar{A}_{11}\bar{A}_{22} - \bar{A}^2_{12})e^-_{\mathrm{fu}}\,; \tag{106}$$

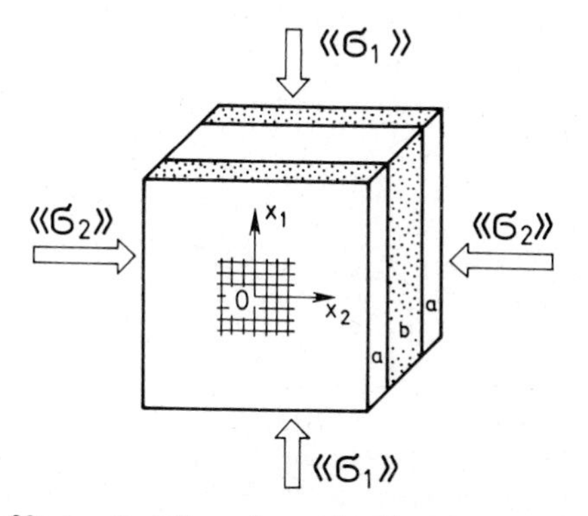

FIG. 29. A calculation scheme for biaxial compression.

(b) in the direction of X_2, i.e., $\beta_k = 90°$

$$-\langle\langle\sigma_1\rangle\rangle\bar{A}_{12} + \langle\langle\sigma_2\rangle\rangle\bar{A}_{11} = (\bar{A}_{11}\bar{A}_{22} - \bar{A}_{12}^2)e_{\bar{f}u} . \tag{107}$$

The strength envelopes based upon (55) and (56) are given in Fig. 30.

If the ultimate strain of unidirectionally reinforced layers in the direction of fibres exceeds the ultimate strain in the transverse direction, the ultimate strains of the whole composite along X_1 and X_2 will be:

$$\langle\langle e_{\bar{1}}\rangle\rangle_u = \frac{\sigma_{\bar{1}}}{E_1} . \qquad \langle\langle e_{\bar{2}}\rangle\rangle_u = \frac{\sigma_{\bar{2}}}{E_2} .$$

where E_1 and E_2 represent the elasticity moduli of the composite along axes X_1 and X_2, and strengths $\sigma_{\bar{1}}$ and $\sigma_{\bar{2}}$ are determined from (85) and (87).

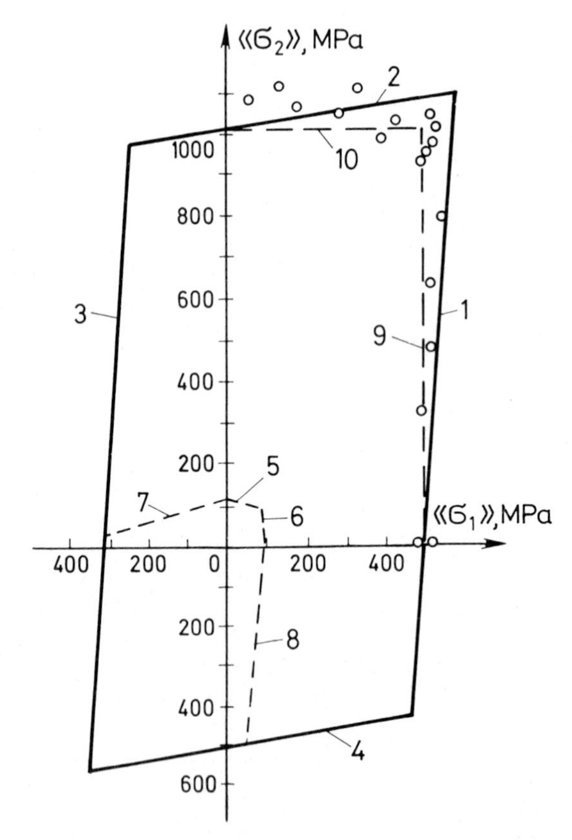

FIG. 30. Ultimate strength curves for an orthogonally reinforced plastic under biaxial load applied in the reinforcement direction. Curves 1 to 10 are based on formulas (102), (103), (106), (107), (97), (96), (115), (114), (104) and (105), respectively. Initial data are: $E_f = 70\,000$ MPa; $E_m = 3\,500$ MPa; $\nu_f = 0.23$; $\nu_m = 0.36$; $v_f = 0.50$; $m_1 = 0.333$; $e_{fu}^{+} = 0.027$; $e_{\bar{f}u} = 0.017$.

Strains in the composite in directions X_1 and X_2 are calculated from the following formulas:

$$\langle\langle e_1\rangle\rangle = \langle\langle\sigma_1\rangle\rangle \frac{\bar{A}_{22}}{\bar{A}_{11}\bar{A}_{22}-\bar{A}_{12}^2} - \langle\langle\sigma_2\rangle\rangle \frac{\bar{A}_{12}}{\bar{A}_{11}\bar{A}_{22}-\bar{A}_{12}^2}, \tag{108}$$

$$\langle\langle e_2\rangle\rangle = -\langle\langle\sigma_1\rangle\rangle \frac{\bar{A}_{12}}{\bar{A}_{11}\bar{A}_{22}-\bar{A}_{12}^2} + \langle\langle\sigma_2\rangle\rangle \frac{\bar{A}_{11}}{\bar{A}_{11}\bar{A}_{22}-\bar{A}_{12}^2}. \tag{109}$$

Insertion of the ultimate strain values into relationships (108) and (109) yields strength criteria:

(a) along X_1

$$\langle\langle\sigma_1\rangle\rangle \frac{\bar{A}_{22}}{\bar{A}_{11}\bar{A}_{22}-\bar{A}_{12}^2} - \langle\langle\sigma_2\rangle\rangle \frac{\bar{A}_{12}}{\bar{A}_{11}\bar{A}_{22}-\bar{A}_{12}^2} = \frac{3.5\,\sigma_{\mathrm{m}}^{+}\left(m_1\dfrac{E_\parallel}{E_\perp}+1-m_1\right)}{E_1(1+\nu_{\perp\parallel})\bar{\sigma}_r\sqrt{1-\nu_{\mathrm{m}}^2}}; \tag{110}$$

(b) along X_2

$$-\langle\langle\sigma_1\rangle\rangle \frac{\bar{A}_{12}}{\bar{A}_{11}\bar{A}_{22}-\bar{A}_{12}^2} + \langle\langle\sigma_2\rangle\rangle \frac{\bar{A}_{11}}{\bar{A}_{11}\bar{A}_{22}-\bar{A}_{12}^2} = \frac{3.5\,\sigma_{\mathrm{m}}^{+}\left[(1-m_1)\dfrac{E_\parallel}{E_\perp}+m_1\right]}{E_2(1+\nu_{\perp\parallel})\bar{\sigma}_r\sqrt{1-\nu_{\mathrm{m}}^2}}. \tag{111}$$

Formulas (110) and (111) describe a case when the fibre-matrix bond strength exceeds the matrix strength. If the bond strength is lower than the matrix strength, then instead of (110) and (111) we have

(a) along X_1

$$\langle\langle\sigma_1\rangle\rangle \frac{\bar{A}_{22}}{\bar{A}_{11}\bar{A}_{22}-\bar{A}_{12}^2} - \langle\langle\sigma_2\rangle\rangle \frac{\bar{A}_{12}}{\bar{A}_{11}\bar{A}_{22}-\bar{A}_{12}^2} = \frac{3.5\,\sigma_{\mathrm{b}}^{\mathrm{u}}\left(m_1\dfrac{E_\parallel}{E_\perp}+1-m_1\right)}{E_1(1+\nu_{\perp\parallel})\bar{\sigma}_r}; \tag{112}$$

(b) along X_2

$$-\langle\langle\sigma_1\rangle\rangle \frac{\bar{A}_{12}}{\bar{A}_{11}\bar{A}_{22}-\bar{A}_{12}^2} + \langle\langle\sigma_2\rangle\rangle \frac{\bar{A}_{11}}{\bar{A}_{11}\bar{A}_{22}-\bar{A}_{12}^2} = \frac{3.5\,\sigma_{\mathrm{b}}^{\mathrm{u}}\left[m_1+(1-m_1)\dfrac{E_\parallel}{E_\perp}\right]}{E_2(1+\nu_{\perp\parallel})\bar{\sigma}_r}. \tag{113}$$

4.5. *Strength under combined tension and compression*

If the material fails as a result of fibre failure under tension or compression the ultimate strength curve is described by equations (102), (103), (106) and (107) in which the signs for the applied mean stresses have been changed correspondingly.

Further we shall deal with the mechanism of continuity loss in the material.

Let us assume that stresses $\langle\langle\sigma_2\rangle\rangle$ (Fig. 29) are compressive, and $\langle\langle\sigma_1\rangle\rangle$ are tensile. Under such load the loss of continuity will result from the failure of the layer reinforced perpendicularly to the tensile stress applied. Failure in this layer may be caused by the polymer matrix or bond failure. The first case to be discussed here is when the bond strength exceeds the polymer matrix strength. In this case the moment of continuity loss is determined according to criteria (94)–(97). It follows from (97), for instance, that

$$\langle\langle\sigma_1\rangle\rangle q_3 - \langle\langle\sigma_2\rangle\rangle q_4 = \frac{\sigma_m^+}{\bar{\sigma}_r\sqrt{1-\nu_m^2}}. \tag{114}$$

If stresses $\langle\langle\sigma_1\rangle\rangle$, are compressive and $\langle\langle\sigma_2\rangle\rangle$ are tensile, then (96) yields

$$-\langle\langle\sigma_1\rangle\rangle q_1 + \langle\langle\sigma_2\rangle\rangle q_2 = \frac{\sigma_m^+}{\bar{\sigma}_r\sqrt{1-\nu_m^2}} \tag{115}$$

In the case of bound failure the corresponding continuity loss criteria are derived from (99) and (101):

$$-\langle\langle\sigma_1\rangle\rangle q_1 + \langle\langle\sigma_2\rangle\rangle q_2 = \frac{\sigma_b^u}{\bar{\sigma}_r}, \tag{116}$$

$$\langle\langle\sigma_1\rangle\rangle q_3 - \langle\langle\sigma_2\rangle\rangle q_4 = \frac{\sigma_b^u}{\bar{\sigma}_r}. \tag{117}$$

Ultimate strength and continuity loss curves for a glass/epoxy plastic under biaxial load are plotted in Fig. 30.

4.6. *Strength under combined tension and shear*

We shall start by discussing the behaviour of an orthogonally reinforced plastic load in fibre direction as shown in Fig. 31.

Mean normal stresses in layers a and b, reinforced in the directions X_1 and X_2, are calculated according to formulas (88) through (91) if $\langle\langle\sigma_2\rangle\rangle = 0$ and shear stresses according to (58) when $\beta = 0°$ and $\beta = 90°$.

$$\langle\tau_{\parallel\perp}\rangle_a = \langle\langle\sigma_6\rangle\rangle \frac{(G_{\parallel\perp})_a}{\bar{A}_{66}}, \tag{118}$$

$$\langle\tau_{\parallel\perp}\rangle_b = \langle\langle\sigma_6\rangle\rangle \frac{(G_{\parallel\perp})_b}{\bar{A}_{66}}. \tag{119}$$

When a laminate is loaded, as shown diagramatically in Fig. 31, a layer b is under less favourable conditions, namely, acted upon simultaneously by shear stresses $\langle\tau_{\parallel\perp}\rangle_b$ and tensile stresses $\langle\sigma_\perp\rangle_b$ perpendicularly to reinforcement. The failure mechanism of layer b depends upon the ratio between the applied stresses $\langle\langle\sigma_6\rangle\rangle$ and $\langle\langle\sigma_1\rangle\rangle$ and the strength ratio of the matrix and bond.

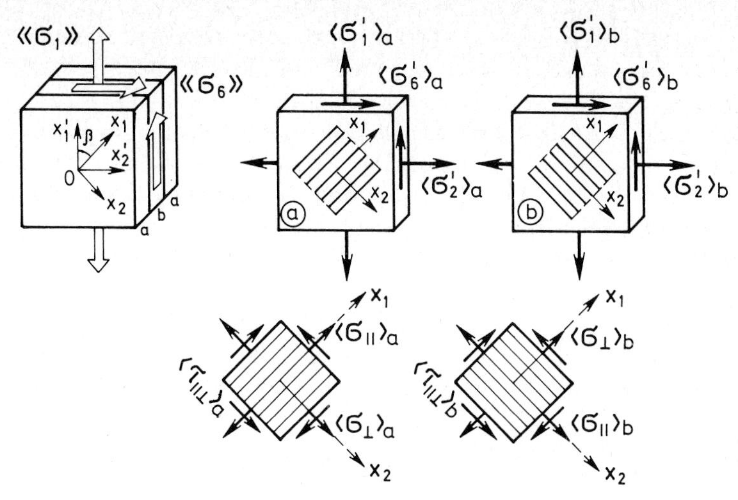

FIG. 31. A calculation scheme for an orthogonally reinforced plastic under combined tension and shear loadings.

The first case to be discussed here is when the bond strength exceeds the polymer matrix strength.

If the ratio between the applied stresses $\langle\langle\sigma_6\rangle\rangle/\langle\langle\sigma_1\rangle\rangle$ exceeds a certain limit, the fracture of lamina b is due to the failure of the polymer matrix under shear. To determine the laminate strength we apply strength criterion (50). Adopting $F(\langle\sigma_\perp\rangle_b; \langle\tau_{\|\perp}\rangle_b)$ as in relationship (53) and inserting expressions for $\langle\sigma_\perp\rangle_b$ and $\langle\tau_{\|\perp}\rangle_b$ into this criterion we get

$$\langle\langle\sigma_6\rangle\rangle = \sqrt{u_1 - u_2\langle\langle\sigma_1^2\rangle\rangle - \langle\langle\sigma_1\rangle\rangle\sqrt{u_2^2\langle\langle\sigma_1^2\rangle\rangle + \frac{2}{\nu_m}u_1u_2}}\,, \tag{120}$$

where

$$u_1 = \frac{(\sigma_m^+\bar{A}_{66})^2}{(1+\nu_m)(G_{\|\perp}^2)_b}\,, \qquad u_2 = \frac{\nu_m(q_3\bar{A}_{66})^2}{2(1+\nu_m^2)(G_{\|\perp}^2)_b}\,.$$

We should observe that parameter q_3 is obtained similar to (97) and $\bar{A}_{66}$ is calculated according to the following formula:

$$\bar{A}_{66} = m_1(G_{\|\perp})_a + (1-m_1)(G_{\|\perp})_b\,. \tag{121}$$

Note that criterion (120) expresses the polymer matrix failure under shear in layer b. After layer b has failed, shear stresses in layer a rise abruptly reaching the value

$$\langle\tau_{\|\perp}\rangle_a = \frac{\langle\langle\sigma_6\rangle\rangle}{m_1}\,. \tag{122}$$

Layer a usually fails if the stress concentration grows in such an abrupt manner thus the failure of layer b actually marks the failure of the whole laminate.

Criterion (120) is applicable if tensile stress $\langle\langle\sigma_1\rangle\rangle$ at the moment of failure increases from 0 to value $\langle\langle\sigma_1\rangle\rangle_1$. If $\langle\langle\sigma_1\rangle\rangle > \langle\langle\sigma_1\rangle\rangle_1$ the failure mechanism of the polymer matrix in layer b changes, i.e., there is rupture of the matrix. Taking into consideration (52) from criterion (50) we get the following relationship between the applied stresses $\langle\langle\sigma_6\rangle\rangle$ and $\langle\langle\sigma_1\rangle\rangle$ at the moment of failure of layer b:

$$\langle\langle\sigma_6\rangle\rangle = \sqrt{u_3 - u_3 u_4 \langle\langle\sigma_1^2\rangle\rangle}\,, \tag{123}$$

where

$$u_3 = \frac{\tau_{\mathrm{m}}^{\mathrm{u}} \bar{A}_{66}^2}{(G_{\parallel\perp}^2)_{\mathrm{b}}} \qquad u_4 = q_3^2 \frac{(1-\nu_{\mathrm{m}}^2)\bar{\sigma}_r^2}{(\sigma_{\mathrm{m}}^{+})^2}\,.$$

Equalization of (120) and (123) enables us to obtain stress value $\langle\langle\sigma_1\rangle\rangle_1$.

Criterion (123) determines the laminate strength when the value of stress $\langle\langle\sigma_1\rangle\rangle$ changes from $\langle\langle\sigma_1\rangle\rangle_1$ to $\langle\langle\sigma_1\rangle\rangle_2$.

The numerical value of stress $\langle\langle\sigma_1\rangle\rangle_2$ is calculated from the equation obtained by equalizing relationships (123) and (128).

If the failure of lamina b takes place due to the fibre-matrix bond failure, we can make use of criterion (50) also considering (44) and obtain:

$$\langle\langle\sigma_1\rangle\rangle q_3 \frac{(\bar{\sigma}_r)_{\mathrm{b}}}{\sigma_{\mathrm{b}}^{\mathrm{u}}} + \langle\langle\sigma_6^2\rangle\rangle \frac{(G_{\parallel\perp}^2)_{\mathrm{b}} \bar{\sigma}_{rz}^2}{(\tau_{\mathrm{b}}^{\mathrm{u}})^2 \bar{A}_{66}} = 1\,. \tag{124}$$

After lamina b has failed, the stress-strain state of the layers caused by the action of $\langle\langle\sigma_1\rangle\rangle$ is calculated in the following way:

$$\langle e_\parallel\rangle_{\mathrm{a}} = \langle\langle\sigma_1\rangle\rangle \frac{(S_\parallel)_{\mathrm{a}}}{m_1} + \langle\sigma_\perp\rangle_{\mathrm{a}} (S_{\parallel\perp})_{\mathrm{a}}\,,$$

$$\langle e_\perp\rangle_{\mathrm{a}} = \langle\langle\sigma_1\rangle\rangle \frac{(S_{\parallel\perp})_{\mathrm{a}}}{m_1} + \langle\sigma_\perp\rangle_{\mathrm{a}} (S_\perp)_{\mathrm{a}}\,,$$

$$\langle e_\parallel\rangle_{\mathrm{b}} = (S_\parallel)_{\mathrm{b}} \langle\sigma_\parallel\rangle_{\mathrm{b}}\,, \qquad \langle e_\parallel\rangle_{\mathrm{b}} = \langle e_\perp\rangle_{\mathrm{a}}\,,$$

$$\langle\sigma_\perp\rangle_{\mathrm{a}} m_1 + \langle\sigma_\parallel\rangle_{\mathrm{b}} (1 - m_1) = 0$$

where

$$S_\parallel = \frac{1}{E_\parallel}\,, \qquad S_\perp = \frac{1}{E_\perp}\,, \qquad S_{\parallel\perp} = -\frac{\nu_{\parallel\perp}}{E_\parallel}\,.$$

The solution of this system yields a formula for calculating normal stress acting in lamina a perpendicularly to reinforcement:

$$\langle\sigma_\perp\rangle_a = \langle\langle\sigma_1\rangle\rangle \frac{1-m_1}{m_1}(\nu_{\perp\parallel})_a \frac{(E_\parallel)_b}{E_2}. \tag{125}$$

Shear stresses which can be calculated from formula (122) are also active in lamina a.

After lamina b reinforced in the direction of X_2 has failed, the entire load is actually sustained by layer a. The strength conditions for a laminate are expressed as follows:

(a) in the case of fibre failure:

$$\langle\langle\sigma_1\rangle\rangle = m_1[v_f E_{fz} + (1-v_f)E_m]e_{fu}^+ ; \tag{126}$$

(b) in the case of the polymer matrix shear failure we get the following from criterion (50):

$$\begin{aligned}&\langle\langle\sigma_1^2\rangle\rangle(1-m_1)^2(\nu_{\perp\parallel}^2)_a(E_\parallel^2)_b + 2(1+\nu_m)E_2^2\langle\langle\sigma_6^2\rangle\rangle\\ &\quad + \langle\langle\sigma_1\rangle\rangle(1-m_1)(\nu_{\perp\parallel})_a(E_\parallel)_b\sqrt{\langle\langle\sigma_1^2\rangle\rangle(1-m_1)^2(\nu_{\perp\parallel}^2)_a(E_\parallel^2)_b + 4E_2^2\langle\langle\sigma_6^2\rangle\rangle}\\ &= 2(m_1E_2\sigma_m^+)^2 ;\end{aligned} \tag{127}$$

(c) if there is tensile failure of the polymer matrix:

$$\begin{aligned}&\langle\langle\sigma_1^2\rangle\rangle(1+\nu_m^2)[(1-m_1)(\nu_{\perp\parallel})_a(\bar\sigma_r)_a(E_\parallel)_b\tau_m^u]^2 + \langle\langle\sigma_6^2\rangle\rangle(E_2\sigma_m^+)^2\\ &= (m_1E_2\tau_m^u\sigma_m^+)^2 ;\end{aligned} \tag{128}$$

(d) in the case of bond failure:

$$\begin{aligned}&\langle\langle\sigma_1\rangle\rangle(1-m_1)m_1(\nu_{\perp\parallel})_a(E_\parallel)_b(\bar\sigma_r)_a(\tau_b^u)^2 + \langle\langle\sigma_6^2\rangle\rangle(\bar\sigma_{rz}^2)_a\sigma_b^u E_2\\ &= (\tau_b^u)^2\sigma_b^u E_2 m_1^2 .\end{aligned} \tag{129}$$

From criteria (128) and (129) it follows that both laminae have cracked and are not in condition to sustain shear stresses $\langle\langle\sigma_6\rangle\rangle$.

Criterion (128) is adoptable when $\langle\langle\sigma_1\rangle\rangle$ ranges from $\langle\langle\sigma_1\rangle\rangle_2$ to $\langle\langle\sigma_1\rangle\rangle_3$. In our specific case $\langle\langle\sigma_1\rangle\rangle_3$ is calculated according to the following formula:

$$\langle\langle\sigma_1\rangle\rangle_3 = \frac{m_1\tau_m^u}{\sqrt{u_5}}, \tag{130}$$

where

$$u_5 = (1-\nu_m^2)\left[\frac{(\bar\sigma_r)_a(1-m_1)(\nu_{\perp\parallel})_a(E_\parallel)_b\tau_m^u}{E_2\sigma_m^+}\right]^2$$

With a further increase of stress $\langle\langle\sigma_1\rangle\rangle$ from $\langle\langle\sigma_1\rangle\rangle_3$ to $\langle\langle\sigma_1\rangle\rangle_4 = \sigma_1^+$ there is fibre rupture, and the laminate strength is determined by (126).

It should be observed that the shape of the summary ultimate curve depends largely on fibre embedding rate m_1. So, for example, if

$$m_1 = \frac{1-A}{1-A\left[1-\frac{(E_\perp)_a}{(E_\parallel)_b}\right]}, \tag{131}$$

where

$$A = \frac{\sigma_m^+}{(E_\parallel \nu_{\perp\parallel}\bar{\sigma}_r)_a e_{fu}^+ \sqrt{1-\nu_m^2}}.$$

Simultaneously with the matrix failure in lamina a there is also fibre rupture, namely, $\langle\langle\sigma_1\rangle\rangle_3 = \langle\langle\sigma_1\rangle\rangle_4$.

The ultimate curves plotted from formulas (120) through (130) are shown in Fig. 32.

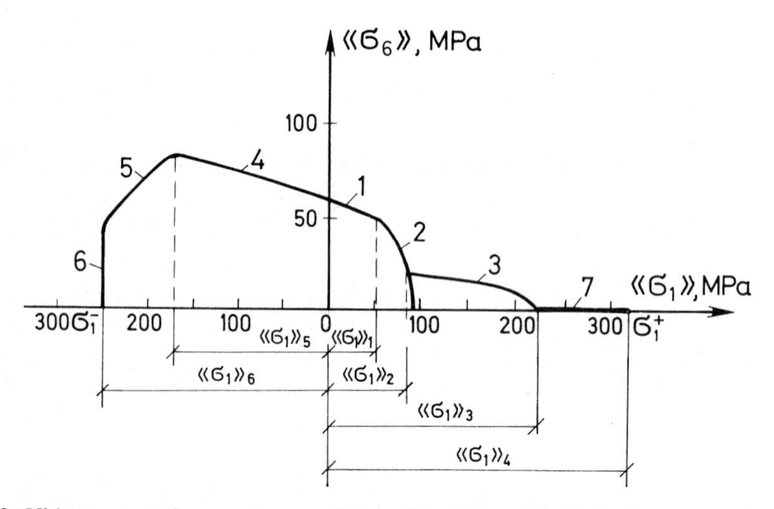

FIG. 32. Ultimate strength curves for an orthogonally reinforced (1:2) glass/epoxy under combined shear and axial loading in direction X_1 (Fig. 31). The curves have been plotted from the following formulas: Curve 1 from (120), 2 (122), 3 (128), 4 (132), 5 (133), 6 (84), and 7 from (80). Initial data are: $E_m = 3\,000$ MPa; $\nu_m = 0.36$; $E_f = 70\,000$ MPa; $\nu_f = 0.23$; $\sigma_m^+ = 70$ MPa; $\tau_m^u = 60$ MPa; $e_{fu}^+ = 0.027$; $e_{fu}^- = 0.014$; $v_f = 0.50$; $\bar{\sigma}_r = 1.7$.

4.7. Strength under combined compression and shear

Numerical values of mean stresses in laminate are obtained by analogy with combined tension and shear which was discussed in the previous chapter.

With the compressive stress changing from 0 to $-\langle\langle\sigma_1\rangle\rangle_5$ the strength of an orthogonally reinforced plastic is determined by criterion like (120):

$$\langle\langle\sigma_6\rangle\rangle = \sqrt{u_1 - u_2\langle\langle\sigma_1^2\rangle\rangle + \langle\langle\sigma_1\rangle\rangle\sqrt{u_1^2\langle\langle\sigma_1^2\rangle\rangle + \frac{2}{\nu_m}u_1u_2}}\,, \tag{132}$$

where

$$u_1 = \frac{(\sigma_m^+\bar{A}_{66})^2}{(1+\nu_m)(G_{\|\perp}^2)_a}\,, \qquad u_2 = \frac{\nu_m(q_3\bar{A}_{66})^2}{2(G_{\|\perp}^2)_b(1+\nu_m^2)}\,.$$

In our particular case this criterion determines the point of the polymer matrix failure in layer b. In practice the failure of a 'b'-type lamina is accompanied by the gross failure of the composite.

Compressive stress increasing further from $-\langle\langle\sigma_1\rangle\rangle_6$ to σ_1^-, the laminate strength is determined by criterion (133) obtained from criterion (50) taking into account relationship (55) and expressions for $\langle\sigma_\perp\rangle_b$ and $\langle\tau_{\|\perp}\rangle_b$:

$$\langle\langle\sigma_6\rangle\rangle = \sqrt{u_6 - u_7\langle\langle\sigma_1^2\rangle\rangle - \langle\langle\sigma_1\rangle\rangle\sqrt{u_7^2\langle\langle\sigma_1^2\rangle\rangle + \frac{2}{(\nu_{\perp\perp})_b}u_6u_7}}\,, \tag{133}$$

where

$$u_6 = \frac{(\sigma_\perp^-)_b^2\bar{A}_{66}^2}{[1+(\nu_{\perp\perp})_b](G_{\|\perp}^2)_b}\,, \qquad u_7 = q_3^2\frac{(\nu_{\perp\perp})_b\bar{A}_{66}^2}{2(G_{\|\perp}^2)_b[1+(\nu_{\perp\perp})_b]^2}\,.$$

To calculate stress $\langle\langle\sigma_1\rangle\rangle_5$ which limits the application range of criteria (120) and (133) it is necessary to equalize these criteria.

Criterion (133) is applicable when the compressive stress ranges from $-\langle\langle\sigma_1\rangle\rangle_5$ to $-\langle\langle\sigma_1\rangle\rangle_6 = \sigma_1^-$, where σ_1^- is obtained from formula (84).

The ultimate curve plotted from formulas (1207), (133) and (84) is seen in Fig. 32. It has been made up for a cross-ply glass plastic where the ratio of ply volumes oriented in the elastic symmetry directions X_1 and X_2 is 1:2.

Fig. 32 shows a case of axial loading applied in direction X_1.

5. Strength of angle-ply laminates

5.1. Stresses within the laminae

Angle-ply laminates are most frequently used in shells produced by winding technique. Such shells often incorporate not only helical reinforcement, oriented at angle $\pm\beta$ towards the shell axis, but also tangential reinforcement ($\beta = 90°$). The calculation scheme, shown in Fig. 33 will be used for the strength determination of a helically reinforced cylindrical shell having additional reinforcement in the tangential direction.

Let us suppose the shell to be made up of numerous, regularly alternating thin unidirectionally reinforced laminae, oriented at angles $\beta, -\beta$ and 90°

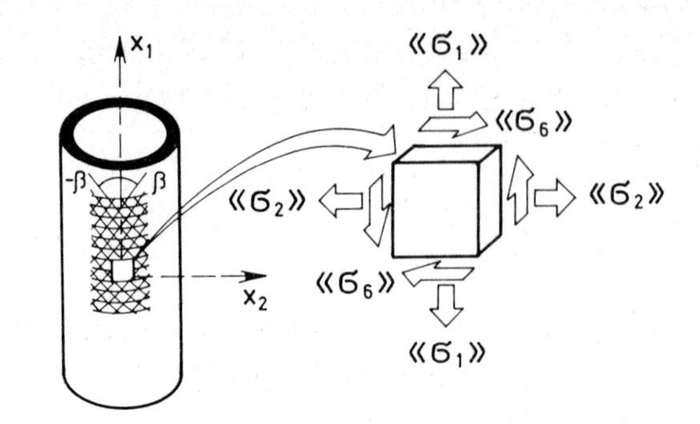

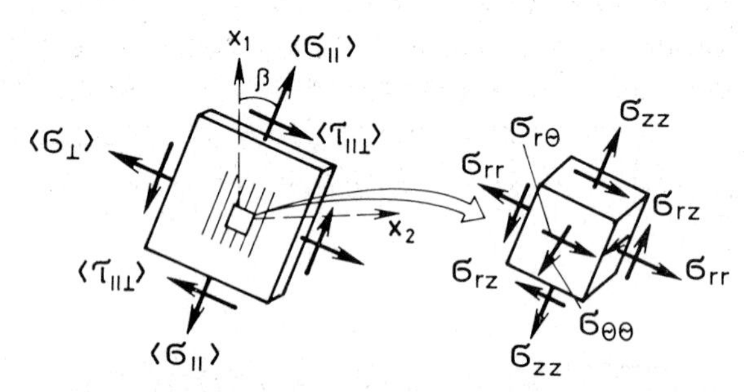

FIG. 33. A calculation scheme for a helically reinforced shell.

towards axis X_1. It may be assumed in the first approximation that the structure of the composite is symmetrical towards the middle plane. The thickness of the laminae, oriented at angles β and $-\beta$ is assumed to be equal. The total thickness of the laminae will be designated by δ_2 and the total thickness of tangentially reinforced laminae by δ_1. Between the laminate thickness δ and the thickness of individual laminae exists the following relationship: $\delta = \delta_1 + \delta_2$.

In a general case of loading, pictured in Fig. 33, stresses acting in the elastic symmetry direction are calculated according to formulas (56)–(58). In our particular case we get the following from the general formula (69) expressing $\bar{A}_{ij}$:

$$\bar{A}_{ij} = \frac{A_{ij}}{\delta_1 + \delta_2}, \tag{134}$$

where

$$A_{11} = \bar{Q}_{11}\delta_2 + Q_{22}\delta_1, \qquad A_{12} = \bar{Q}_{12}\delta_2 + Q_{12}\delta_1,$$

$$A_{22} = \bar{Q}_{22}\delta_2 + Q_{11}\delta_1, \qquad A_{66} = \bar{Q}_{66}\delta_2 + Q_{66}\delta_1.$$

5.2. *Tensile strength*

Under uniaxial tension acting along X_1 the stresses within laminae are determined by (56)–(58) if $\langle\langle\sigma_2\rangle\rangle = \langle\langle\sigma_6\rangle\rangle = 0$. The fracture of a helically reinforced plastic under tension may be initiated by the failure of the fibres, the polymer matrix or the fibre-matrix bond.

At small angles β the fracture of the composite is due to fibre failure. Then criterion (49) is suitable for the calculation of the composite strength.

At the moment of failure $e_f = e_{fu}^+$ and $\langle\langle\sigma_1\rangle\rangle = \sigma_1^+$. Then from (49) it follows (SKUDRA (1980)):

$$\sigma_1^+ = \frac{\bar{A}_{11}\bar{A}_{22} - \bar{A}_{12}^2}{\bar{A}_{22}\cos^2\beta - \bar{A}_{12}\sin^2\beta} e_{fu}^+ . \tag{135}$$

At larger angle β values the fracture of the composite will be governed by the failure of the polymer matrix or the bond. First we shall consider a case when the matrix strength is inferior to the bond strength. The composite strength calculation is made on the basis of (50). Inserting the stress expressions $\langle\sigma_\perp\rangle$ and $\langle\tau_{\parallel\perp}\rangle$ into (52) and (53) and considering that at the moment of fracture $\langle\langle\sigma_1\rangle\rangle = \sigma_1^+$, we can write (SKUDRA (1980)):

(a) for tensile failure of matrix

$$\sigma_1^+ = \frac{\bar{\Delta}\sigma_m^+\tau_m^u}{\sqrt{(1-\nu_m^2)(\bar{\sigma}_r a_2 \tau_m^u)^2 + (\sigma_m^+ a_3)^2}} ; \tag{136}$$

(b) for shear failure of the matrix

$$\sigma_1^+ = \frac{\sqrt{2}\bar{\Delta}\sigma_m^+}{\sqrt{a_2^2 + 2(1+\nu_m)a_3^2 + a_2\sqrt{a_2^2 + 4a_3^2}}} . \tag{137}$$

Fig. 34 shows the dependence of strength σ_1^+ on angle β in case of a glass/epoxy shell. Test data have been borrowed from KRAUSS and SHELLING (1962). The theoretical curves are based on (135), (136) and (137) with the following initial data: $v_f = 0.63$; $e_{fu}^+ = 0.025$; $\tau_m^u = 60$ Mpa; $\sigma_m^+ = 70$ MPa; $E_f =$ 70 000 Mpa; $E_m = 3\,500$ MPa; $\nu_m = 0.35$; $\nu_f = 0.23$; $\bar{\sigma}_r = 1.8$.

If the bond has failed (50) and (54) enables us to write:

$$\sigma_1^+ = \bar{\Delta}\left[\frac{\tau_b^u}{a_3\bar{\sigma}_{rz}} \sqrt{\left(\frac{a_2\bar{\sigma}_r\tau_b^u}{2a_3\bar{\sigma}_{rz}\sigma_b^u}\right)^2 + 1} - \frac{a_2\bar{\sigma}_r(\tau_b^u)^2}{2a_3^2\bar{\sigma}_{rz}^2\sigma_b^u}\right] . \tag{138}$$

In case of tangential tension $\langle\langle\sigma_1\rangle\rangle = \langle\langle\sigma_6\rangle\rangle = 0$, and the composite fracture is due to fibre failure.

To obtain strength σ_2^+ we resort to (49) if $\beta_k = 90°$. As $\langle\langle\sigma_2\rangle\rangle = \sigma_2^+$ at the moment of fracture (49) gives us

$$\sigma_2^+ = \frac{\bar{\Delta}}{\bar{A}_{11}\bar{A}_{66}} e_{fu}^+ = \bar{A}_{22} e_{fu}^+ . \tag{139}$$

For matrix failure (50) yields:

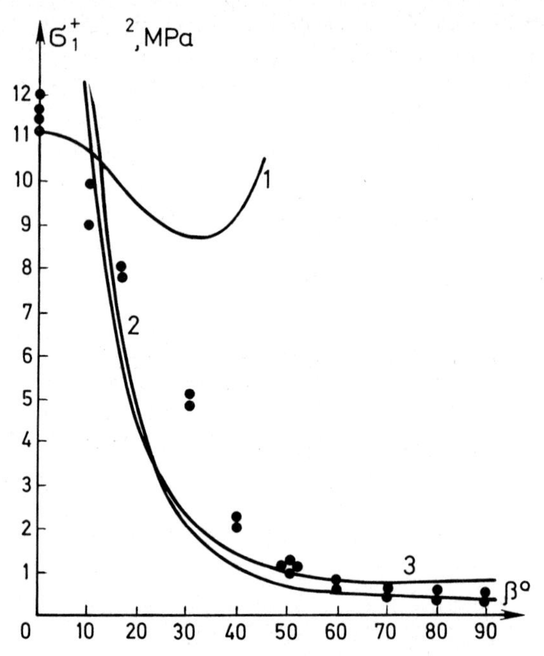

FIG. 34. Dependence σ_1^+ upon the winding angle $\pm\beta$ for a glass/epoxy shell with additional tangential reinforcement.

(a) tensile failure

$$\sigma_2^+ = \frac{\bar{\Delta}\sigma_m^+\tau_m^u}{\sqrt{(1-\nu_m^2)(\bar{\sigma}_r b_2\tau_m^u)^2 + (b_3\sigma_m^+)^2}}; \tag{140}$$

(b) shear failure

$$\sigma_2^+ = \frac{\sqrt{2}\,\bar{\Delta}\sigma_m^+}{\sqrt{b_2\sqrt{b_2^2 + 4b_3^2}}}. \tag{141}$$

If there is no tangentially reinforced lamina in the laminate, criteria (135)–(138) become simpler. In case of fibre failure (135) yields

$$\sigma_1^+ = \frac{\bar{Q}_{11}\bar{Q}_{22} - \bar{Q}_{12}^2}{\bar{Q}_{22}\cos^2\beta - \bar{Q}_{12}\sin^2\beta}\, e_{fu}^+. \tag{142}$$

In case of matrix tensile failure

$$\sigma_1^+ = \frac{\sigma_m^+\tau_m^u}{\sqrt{(1-\nu_m^2)(\bar{\sigma}_r a_4\tau_m^u)^2 + (a_5\sigma_m^+)^2}}. \tag{143}$$

In case of matrix shear failure

$$\sigma_1^+ = \frac{\sqrt{2}\sigma_m^+}{\sqrt{a_4^2 + 2(1+\nu_m)a_5^2 + a_4\sqrt{a_4^2 + 4a_5^2}}}. \tag{144}$$

If the bond fails, then

$$\sigma_1^+ = \frac{1}{2}\left(\frac{\tau_b^u}{a_5\bar{\sigma}_{rz}}\right)^2\left[\sqrt{\left(\frac{a_4\bar{\sigma}_r}{\sigma_b^u}\right)^2 + \left(\frac{2a_5\bar{\sigma}_{rz}}{\tau_b^u}\right)^2} - \frac{a_4\bar{\sigma}_r}{\sigma_b^u}\right]. \tag{145}$$

In these formulas a_4 and a_5 stand for

$$a_4 = \sin^2\beta + 2\sin\beta\cos\beta\,\frac{\bar{Q}_{12}\bar{Q}_{26} - \bar{Q}_{22}\bar{Q}_{16}}{\bar{Q}_{11}\bar{Q}_{22} - \bar{Q}_{12}^2},$$

$$a_5 = -\sin\beta\cos\beta + (\cos^2\beta - \sin^2\beta)\frac{\bar{Q}_{22}\bar{Q}_{16} - \bar{Q}_{12}\bar{Q}_{26}}{\bar{Q}_{11}\bar{Q}_{22} - \bar{Q}_{12}^2}.$$

Fig. 35 gives test data obtained by UEMURA and YAMAWAKI (1971) by testing thin-wall glass/epoxy tubes for axial tension. Fig. 35 also provides theoretical curves plotted from (142), (143) and (144).

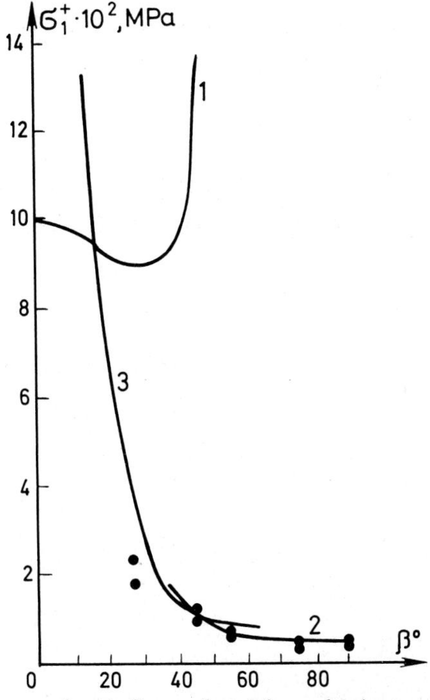

FIG. 35. Dependence σ_1^+ upon the winding angle $\pm\beta$ for a glass/epoxy shell under tension applied in direction X_1. The curves are based on formulas: Curve 1 on (142), 2 on (143), 3 on (144); initial data are: $E_f = 70\,000$ MPa, $G_f = 28\,400$ MPa; $\nu_f = 0.23$; $e_{fu}^+ = 0.027$; $E_m = 3\,000$ MPa; $\nu_m = 0.36$; $\sigma_m^+ = 70$ MPa; $\tau_m^u = 60$ MPa; $v_f = 0.50$; $\bar{\sigma}_r = 1.7$; $e_{fu}^- = 0.015$.

In the case of tension, acting along axis X_2, strength determination is carried out by the formulas, taking into account eventual fibre, matrix or bond failure:

$$\sigma_2^+ = \frac{\bar{Q}_{11}\bar{Q}_{22} - \bar{Q}_{12}^2}{\bar{Q}_{11}\sin^2\beta - \bar{Q}_{12}\cos^2\beta}\, e_{fu}^+ , \tag{146}$$

$$\sigma_2^+ = \frac{\sigma_m^+ \tau_m^u}{\sqrt{(1-\nu_m^2)(\bar{\sigma}_r a_6 \tau_m^u)^2 + (a_7 \sigma_m^+)^2}} , \tag{147}$$

$$\sigma_2^+ = \frac{\sqrt{2}\sigma_m^+}{\sqrt{a_6^2 + 2(1+\nu_m)a_7^2 + a_6\sqrt{a_6^2 + 4a_7^2}}} , \tag{148}$$

$$\sigma_2^+ = \frac{\tau_b^u}{a_7 \bar{\sigma}_{rz}}\left[\sqrt{\left(\frac{a_6 \sigma_r \tau_b^u}{2a_7 \bar{\sigma}_{rz} \sigma_b^u}\right)^2 + 1} - \frac{a_6 \bar{\sigma}_r \tau_b^u}{2a_7 \bar{\sigma}_{rz} \sigma_b^u}\right] , \tag{149}$$

where

$$a_6 = \cos^2\beta + 2\sin\beta\cos\beta\,\frac{\bar{Q}_{12}\bar{Q}_{16} - \bar{Q}_{11}\bar{Q}_{26}}{\bar{Q}_{11}\bar{Q}_{22} - \bar{Q}_{12}^2} ,$$

$$a_7 = \sin\beta\cos\beta + (\cos^2\beta - \sin^2\beta)\,\frac{\bar{Q}_{11}\bar{Q}_{26} - \bar{Q}_{12}\bar{Q}_{16}}{\bar{Q}_{11}\bar{Q}_{22} - \bar{Q}_{12}^2} .$$

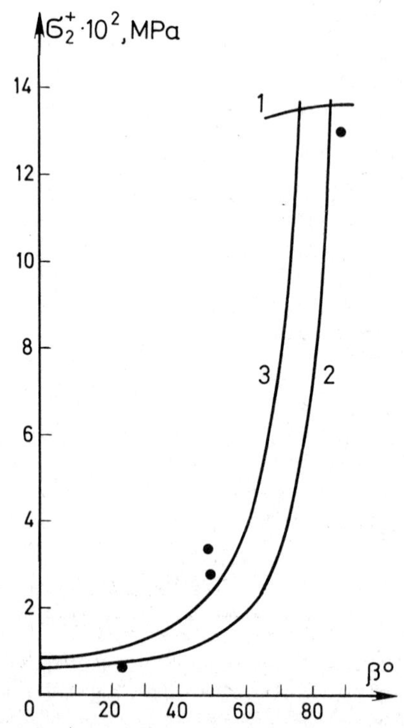

FIG. 36. Dependence of strength σ_2^+ upon the winding angle for a plastic reinforced by high-strength carbon fibres. Curve 1 has been plotted from formula (146), Curve 2 from (147) and Curve 3 from (148).

The actual strength of the composite at the given angle will be equal to the least of the three strength values. Figs. 36 and 37 include test data (UEMURA and YAMAWAKI (1971)) for a carbon/epoxy and theoretical curves based on (146), (147), (148) with the following initial data:

(a) for a composite reinforced with high-strength carbon fibres (Fig. 36): $E_{fz} = 230\,000$ MPa; $E_{fr} = 16\,100$ MPa; $G_{frz} = 20\,000$ MPa; $\nu_{fzr} = 0.30$; $e_{fu}^{+} = 0.009\,8$; $E_m = 35\,000$ MPa; $\nu_m = 0.36$; $v_f = 0.60$; $\sigma_m^{+} = 70$ MPa; $\tau_m^{u} = 60$ MPa; $\bar{\sigma}_r = 1.40$;

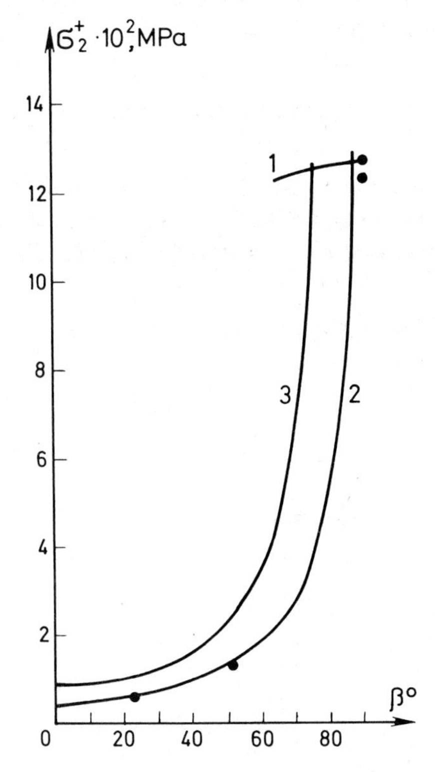

FIG. 37. Dependence of strength σ_2^{+} upon the winding angle for a plastic reinforced by high-modulus carbon fibres. Curve 1 is based on (146), Curve 2 on (147) and Curve 3 on (148).

(b) for a composite reinforced with high-modulus carbon fibres (Fig. 37): $E_{fz} = 380\,000$ MPa; $E_{fr} = 13\,300$ MPa; $G_{frz} = 20\,000$ MPa; $\nu_{fzr} = 0.40$; $e_{fu}^{+} = 0.005\,4$; $E_m = 3\,500$ MPa; $\nu_m = 0.36$; $v_f = 0.60$; $\sigma_m^{+} = 70$ MPa; $\bar{\sigma}_r = 1.30$; $\tau_m^{u} = 60$ MPa.

We shall remark that optimum relations between the matrix and the bond strengths can be established by equalizing the corresponding criteria, e.g. (143) and (145) or (147) and (149).

5.3. Compressive strength

The fibres having failed, strength criterion looks analogous to criterion (135)

$$\sigma_1^- = \frac{\bar{A}_{11}\bar{A}_{22} - \bar{A}_{12}^2}{\bar{A}_{22}\cos^2\beta - \bar{A}_{12}\sin^2\beta}\, e_{\mathrm{fu}}^- . \tag{150}$$

If the matrix fails under shear, criterion (50) yields:

$$\sigma_1^- = \frac{\sqrt{2}\bar{\Delta}\sigma_{\mathrm{m}}^+}{\sqrt{a_2^2 + 2(1+\nu_{\mathrm{m}})a_3^2 - a_2\sqrt{a_2^2 + 4a_3^2}}} . \tag{151}$$

In the case of transverse compression failure of a unidirectionally reinforced single ply from (50) and (55) it follows

$$\sigma_1^- = \frac{\sqrt{2}\bar{\Delta}\sigma_\perp^-}{\sqrt{a_2^2 + 2(1+\nu_{\perp\parallel})a_3^2 + a_2\sqrt{a_2^2 + 4a_3^2}}} . \tag{152}$$

Strength $\sigma_\perp^-$ is determined by (40). Fig. 38 shows the strength dependence on angle β for a glass/epoxy shell. Test data come from KNAUSS and SCHELLING (1969) and theoretical curves are plotted from (150), (151) and (152) with the

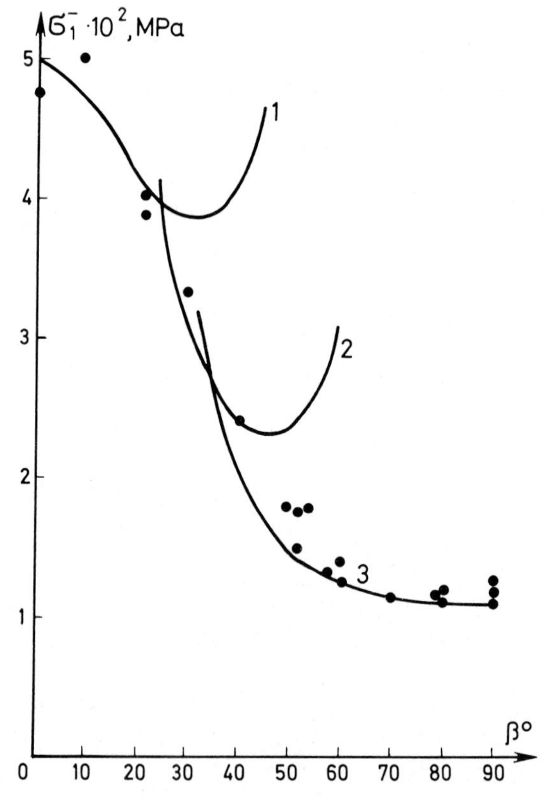

FIG. 38. Dependence of σ_1^- upon the winding angle $\pm\beta$ for a glass/epoxy shell with additional tangential reinforcement.

following initial data given:

$$v_{\rm f} = 0.63; \quad e^-_{\rm fu} = 0.011; \quad \sigma^+_{\rm m} = 72\,{\rm MPa}; \quad \tau^{\rm u}_{\rm m} = 66\,{\rm MPa}; \quad E_{\rm f} = 70\,000\,{\rm MPa};$$
$$E_{\rm m} = 3\,500\,{\rm MPa}; \quad \nu_{\rm m} = 0.35; \quad \nu_{\rm f} = 0.23; \quad \bar{\sigma}_r = 1.8\,.$$

In case of bond failure, strength is given by:

$$\sigma^-_1 = \frac{\bar{\Delta}\tau^{\rm u}_{\rm b}}{a_3\bar{\sigma}_{rz}}\left[\sqrt{\left(\frac{a_2\bar{\sigma}_r\tau^{\rm u}_{\rm b}}{2a_3\bar{\sigma}_{rz}\sigma^{\rm u}_{\rm b}}\right)^2 + 1} + \frac{a_2\bar{\sigma}_r\tau^{\rm u}_{\rm b}}{2a_3\bar{\sigma}_{rz}\sigma^{\rm u}_{\rm b}}\right]. \tag{153}$$

In a special case, when there is no tangentially reinforced lamina in the laminate lay-up, formulas (150)–(153) are written as (SKUDRA (1979b)):

$$\sigma^-_1 = \frac{\bar{Q}_{11}\bar{Q}_{22} - \bar{Q}^2_{12}}{\bar{Q}_{22}\cos^2\beta - \bar{Q}_{12}\sin^2\beta}\, e^-_{\rm fu}\,, \tag{154}$$

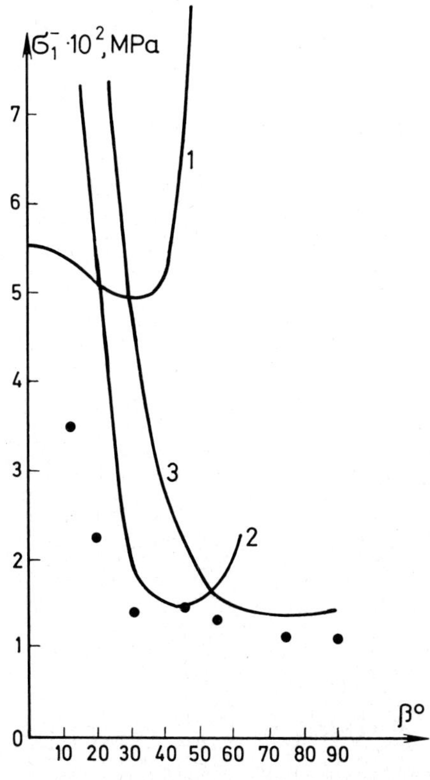

FIG. 39. Dependence of σ^-_1 upon the winding angle $\pm\beta$ for a glass/epoxy shell. Curves are based on formulas: Curve 1 on (154), 2 on (155) and 3 on (156) with the initial data as in Fig. 35.

$$\sigma_1^- = \frac{\sqrt{2}\sigma_m^+}{\sqrt{a_4^2 + 2(1+\nu_m)a_5^2 - a_4\sqrt{a_4^2 + 4a_5^2}}}, \tag{155}$$

$$\sigma_1^- = \frac{\sqrt{2}\sigma_\perp^-}{\sqrt{a_4^2 + 2(1+\nu_{\perp\perp})a_5^2 + a_4\sqrt{a_4^2 + 4a_5^2}}}, \tag{156}$$

$$\sigma_1^- = \frac{\tau_b^u}{a_5\bar{\sigma}_{rz}}\left[\sqrt{\left(\frac{a_4\bar{\sigma}_r\tau_b^u}{2a_5\bar{\sigma}_{rz}\sigma_b^u}\right)^2 + 1} + \frac{a_4\bar{\sigma}_r\tau_b^u}{2a_5\bar{\sigma}_{rz}\sigma_b^u}\right]. \tag{157}$$

Fig. 39 illustrates the dependence of σ_1^- on angle β for a glass/epoxy shell. Test data have been borrowed from UEMURA and YAMAWAKI (1971) and the theoretical curves have been plotted from (154), (155) and (156).

5.4. *Torsional strength*

The torsional strength of laminated shells has been investigated UEMURA and YAMAWAKI (1971) and SKUDRA (1979b).

An unidirectionally reinforced elementary lamina of the shell under torsion is acted upon by mean stresses $\langle\sigma_\parallel\rangle, \langle\sigma_\perp\rangle$ and $\langle\tau_{\parallel\perp}\rangle$, determinable by formulas (56)–(58) if $\langle\langle\sigma_1\rangle\rangle = \langle\langle\sigma_2\rangle\rangle = 0$. Stress $\langle\sigma_\parallel\rangle$ may cause fibre failure, whereas stresses $\langle\sigma_\perp\rangle$ and $\langle\tau_{\parallel\perp}\rangle$ may result in matrix or bond failure. Taking into account that stresses $\langle\sigma_\perp\rangle$ and $\langle\tau_{\parallel\perp}\rangle$ have little effect on the stress state of fibres, this effect will be ignored in cases of initial fibre failure. For fibre failure if $\langle\langle\sigma_1\rangle\rangle = \langle\langle\sigma_2\rangle\rangle = 0$, criterion (49) yields:

(a) when $e_{fu}^+ > e_{fu}^-$

$$\tau_{12}^u = \frac{\bar{A}_{66}}{\sin\beta\cos\beta} e_{fu}^- ; \tag{158}$$

(b) when $e_{fu}^+ < e_{fu}^-$

$$\tau_{12}^u = \frac{\bar{A}_{66}}{\sin\beta\cos\beta} e_{fu}^+ . \tag{159}$$

If the fracture of helically reinforced shells takes place owing to matrix or bond failure, the effect of stress $\langle\sigma_\parallel\rangle$ can be ignored. This is explained by the fact, that $\langle\sigma_\parallel\rangle$ causes insignificant stresses in the polymer matrix. Therefore criterion (50) will be adopted for strength determination.

Substituting the expressions for $\langle\sigma_\perp\rangle$ and $\langle\tau_{\parallel\perp}\rangle$ in relationships (52) and (53) which are adopted for expressing $F(\langle\sigma_\perp\rangle; \langle\tau_{\parallel\perp}\rangle)$ we obtain

$$\tau_{12}^u = \frac{\bar{\Delta}\sigma_m^+\tau_m^u}{\sqrt{(\tau_m^u)^2(1-\nu_m^2)(\bar{\sigma}_r c_2)^2 + (\sigma_m^+ c_3)^2}} \tag{160}$$

$$\tau_{12}^u = \frac{\sqrt{2}\bar{\Delta}\sigma_m^+}{\sqrt{c_2^2 + 2(1+\nu_m)c_3^2 + c_2\sqrt{c_2^2 + 4c_3^2}}}. \tag{161}$$

If the fracture takes place on account of bond failure, criteria (50) and taking into account (54)

$$\tau_{12}^{u} = \frac{\bar{\Delta}\tau_{b}^{u}}{c_3\bar{\sigma}_{rz}}\left[\sqrt{\left(\frac{c_2\sigma_r\tau_b^u}{2c_3\bar{\sigma}_{rz}\sigma_b^u}\right)^2 + 1} - \frac{c_2\bar{\sigma}_r\tau_b^u}{2c_3\bar{\sigma}_{rz}\sigma_b^u}\right]. \tag{162}$$

If a tangentially reinforced lamina is not incorporated in the laminate, the above formulas will be simplified.

In case of fibre failure

(a) if $e_{fu}^{+} > e_{fu}^{-}$:

$$\tau_{12}^{u} = \frac{\bar{Q}_{66}}{\sin\beta\cos\beta}\, e_{fu}^{-}\,; \tag{163}$$

(b) if $e_{fu}^{+} < e_{fu}^{-}$:

$$\tau_{12}^{u} = \frac{\bar{Q}_{66}}{\sin\beta\cos\beta}\, e_{fu}^{+}\,. \tag{164}$$

In case of polymer matrix failure

(a) in tension:

$$\tau_{12}^{u} = \frac{\sigma_m^{+}\tau_m^{u}}{\sqrt{(\tau_m^u a_8\bar{\sigma}_r)^2(1-\nu_m^2) + (\sigma_m^{+}a_9)^2}}\,; \tag{165}$$

(b) in shear:

$$\tau_{12}^{u} = \frac{\sqrt{2}\sigma_m^{+}}{\sqrt{a_8^2 + 2(1+\nu_m)a_9^2 + a_8\sqrt{a_8^2 + 4a_9^2}}}\,. \tag{166}$$

If the composite fractures as a result of bond failure, then

$$\tau_{12}^{u} = \frac{\tau_b^u}{a_9\bar{\sigma}_{rz}}\left[\sqrt{\left(\frac{a_8\bar{\sigma}_r\tau_b^u}{2a_9\bar{\sigma}_{rz}\sigma_b^u}\right)^2 + 1} - \frac{a_8\bar{\sigma}_r\tau_b^u}{2a_9\bar{\sigma}_{rz}\sigma_b^u}\right]. \tag{167}$$

In the above formulas a_8 and a_9 denote

$$a_8 = \frac{\bar{Q}_{16}}{\bar{Q}_{66}}\sin^2\beta + \frac{\bar{Q}_{26}}{\bar{Q}_{66}}\cos^2\beta - 2\sin\beta\cos\beta\,, \tag{168}$$

$$a_9 = \frac{\bar{Q}_{26} - \bar{Q}_{16}}{\bar{Q}_{66}}\sin\beta\cos\beta + (\cos^2\beta - \sin^2\beta)\,. \tag{169}$$

In determining strength τ_{12}^{u} in the case of the polymer matrix or the bond failure it is very important to consider the direction of $\langle\langle\sigma_6\rangle\rangle$ and the orientation of laminae. Direction of $\langle\langle\sigma_6\rangle\rangle$ shown in Fig. 40 will be regarded as positive.

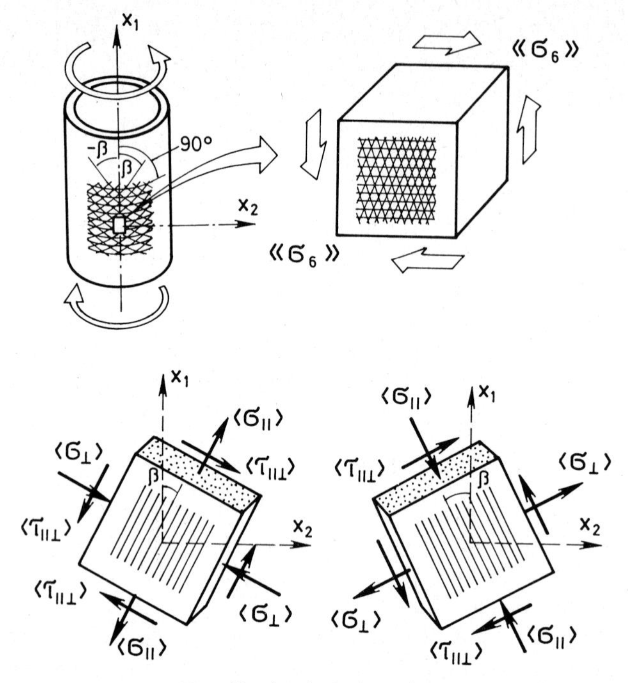

FIG. 40. A calculation scheme.

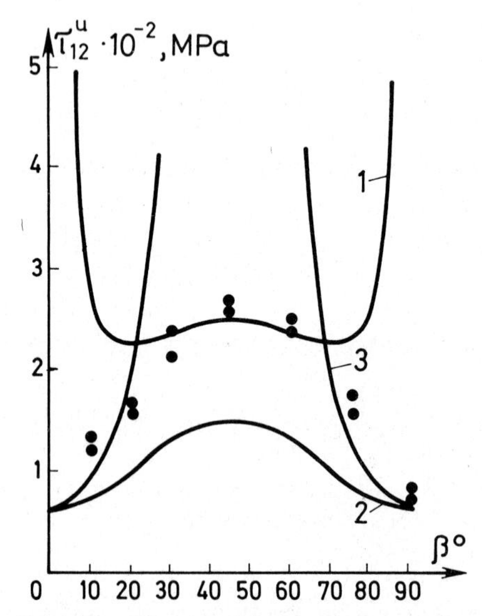

FIG. 41. Dependence of a helically reinforced glass/epoxy shell strength under torsion on winding angle $\pm\beta$. Curve 1 is based on formula (163), 2 on (165) and 3 on (166), with the initial data as in Fig. 35.

Under such loading conditions the lamina oriented at angle $+\beta$ will be acted upon by compressive stresses $\langle\sigma_\perp\rangle$ and shear stresses $\langle\tau_{\|\perp}\rangle$, whereas the lamina oriented at angle $-\beta$ will be under tensile stresses $\langle\sigma_\perp\rangle$ and shear stresses $\langle\tau_{\|\perp}\rangle$. Under combined tension and shear the strength of unidirectionally reinforced plastic will always be lower than under combined compression and shear. Therefore in calculating strength τ_{12}^{u} we should remember that the lamina oriented at $-\beta$ will be the first to undergo failure. The components of elasticity matrix $\bar{Q}_{ij}$ and ratios c_2 and c_3 – formulas (166) and (167) respectively – should be determined for angle $-\beta$ in order to observe this condition.

Fig. 41 presents the relationships obtained in tests by UEMURA and YAMAWAKI (1971) between τ_{12}^{u} and the winding angle β for a glass/epoxy shell. It also shows theoretical curves plotted from (163), (165) and (166).

5.5. *Strength under combined loading*

Strength criteria for angle-ply (helically) reinforced shells under various types of combined loading have been studied by SKUDRA (1980).

Biaxial tension

The calculation scheme has been given in Fig. 42. If the fibres oriented at angle $\pm\beta$ are the first to undergo failure, the strength criterion is written as

$$\langle\langle\sigma_1\rangle\rangle(\bar{A}_{22}\cos^2\beta - \bar{A}_{12}\sin^2\beta) + \langle\langle\sigma_2\rangle\rangle(\bar{A}_{11}\sin^2\beta - \bar{A}_{12}\cos^2\beta) = \frac{\bar{\Delta}}{\bar{A}_{66}}\, e_{fu}^{+}\,. \tag{170}$$

If tangentially oriented fibres ($\beta = 90°$) fail first, criterion (170) is simplified:

$$-\langle\langle\sigma_1\rangle\rangle\bar{A}_{12}\bar{A}_{66} + \langle\langle\sigma_2\rangle\rangle\bar{A}_{11}\bar{A}_{66} = \bar{\Delta} e_{fu}^{+}\,. \tag{171}$$

Fig. 43 contains the strength envelopes plotted from formulas (170) and (171). Theoretical curves are based on the following initial data: $v_f = 0.63$; $e_{fu}^{+} = 0.02$; $E_f = 70\,000$ MPa; $E_m = 3\,500$ MPa; $\nu_m = 0.35$; $\nu_f = 0.23$; $\bar{\sigma}_r = 1.8$; $\beta = 30°$; $\delta = 0.002$ m.

In the case under discussion, the failure of the polymer matrix or the bond is also possible. If a proper sealing layer has been incorporated in the shell structure and provided angle β does not exceed a certain limit, formulas (170) and (171) remain valid after the failure of the matrix or of the bond has taken place.

In the case of matrix failure, the strength criterion for a laminate is written as follows:

(a) tensile failure of the matrix

$$\langle\langle\sigma_1^2\rangle\rangle d_1 + 2\langle\langle\sigma_1\rangle\rangle\langle\langle\sigma_2\rangle\rangle d_2 + \langle\langle\sigma_2^2\rangle\rangle d_3 = \bar{\Delta}^2\,, \tag{172}$$

where

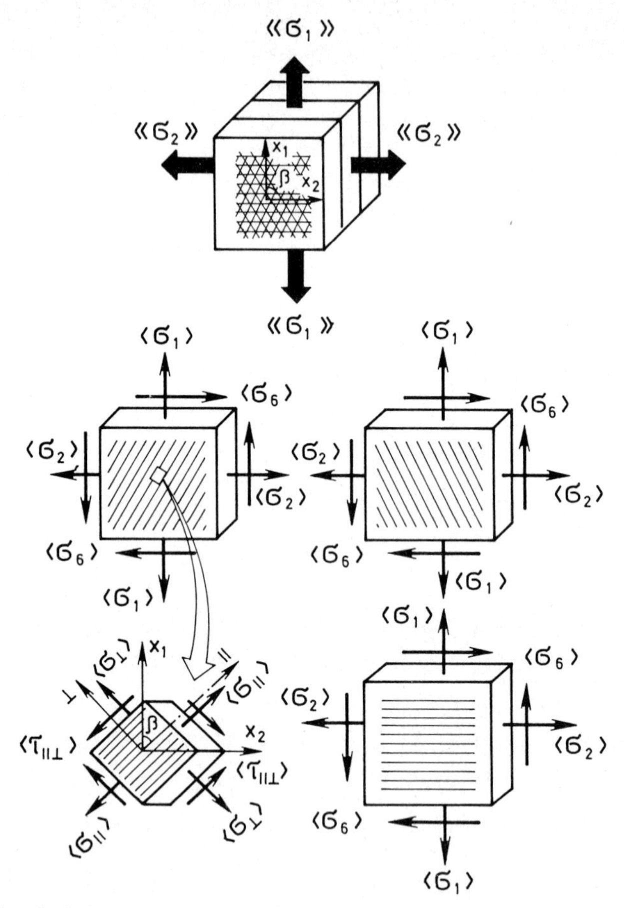

FIG. 42. A calculation scheme for a helically reinforced shell under biaxial tension.

$$d_1 = (1-\nu_{\mathrm{m}}^2)\bar{\sigma}_r^2\left(\frac{a_2}{\sigma_{\mathrm{m}}^+}\right)^2 + \left(\frac{a_3}{\tau_{\mathrm{m}}^{\mathrm{u}}}\right)^2,$$

$$d_2 = (1-\nu_{\mathrm{m}}^2)\bar{\sigma}_r^2\frac{a_2 b_2}{(\sigma_{\mathrm{m}}^+)^2} + \frac{a_3 b_3}{(\tau_{\mathrm{m}}^{\mathrm{u}})^2},$$

$$d_3 = (1-\nu_{\mathrm{m}}^2)\bar{\sigma}_r^2\left(\frac{b_2}{\sigma_{\mathrm{m}}^+}\right)^2 + \left(\frac{b_3}{\tau_{\mathrm{m}}^{\mathrm{u}}}\right)^2;$$

(b) shear failure of the matrix

$$\begin{aligned}&(\langle\langle\sigma_1\rangle\rangle a_2 + \langle\langle\sigma_2\rangle\rangle b_2)^2 + 2(1+\nu_{\mathrm{m}})(\langle\langle\sigma_1\rangle\rangle a_3 + \langle\langle\sigma_2\rangle\rangle b_3)^2\\ &\quad + (\langle\langle\sigma_1\rangle\rangle a_2 + \langle\langle\sigma_2\rangle\rangle b_2)\sqrt{(\langle\langle\sigma_1\rangle\rangle a_2 + \langle\langle\sigma_2\rangle\rangle b_2)^2 + 4(\langle\langle\sigma_1\rangle\rangle a_3 + \langle\langle\sigma_2\rangle\rangle b_3)^2}\\ &= 2(\bar{\Delta}\sigma_{\mathrm{m}}^+)^2 .\end{aligned} \tag{173}$$

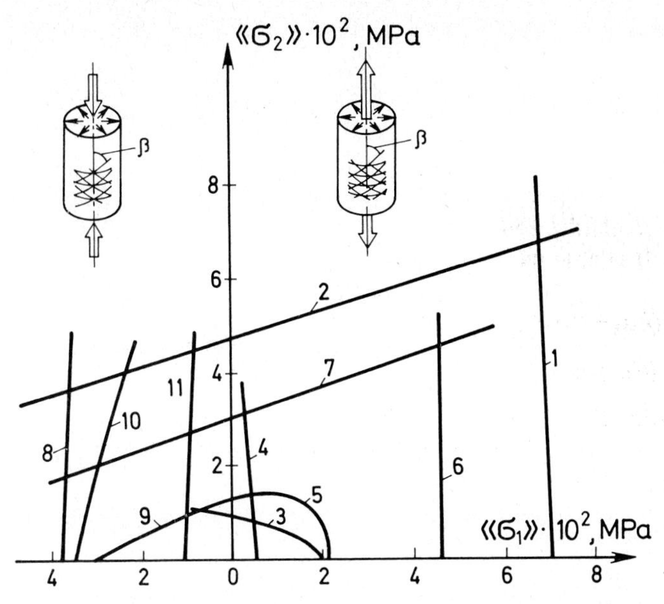

FIG. 43. Strength of a helically reinforced glass/epoxy shell under biaxial load. Ultimate curves are based on the following formulas: Curve 1 on (170); 2 on (171); 3 on (172); 4 on (172) where $\beta = 90°$; Curve 5 on (173); 6 on (170) where $E_\perp = 0$, $G_{\parallel\perp} = 0$ and $\nu_{\perp\parallel} = 0$; 7 on (169) where $E_\perp = 0$, $G_{\parallel\perp} = 0$ and $\nu_{\perp\parallel} = 0$; 8 on (175); 9 on (177); 10 on (178) and Curve 11 on (178) where $\beta = 90°$.

If the bond strength is lower than the matrix strength,

$$(\langle\langle\sigma_1\rangle\rangle a_2 + \langle\langle\sigma_2\rangle\rangle b_2)\frac{\bar{\sigma}_r}{\bar{\Delta}\sigma_\mathrm{b}^\mathrm{u}} + (\langle\langle\sigma_1\rangle\rangle a_3 + \langle\langle\sigma_2\rangle\rangle b_3)^2\left(\frac{\bar{\sigma}_{rz}}{\bar{\Delta}\tau_\mathrm{b}^\mathrm{u}}\right)^2 = 1\,. \tag{174}$$

Biaxial tension and compression

Let us suppose that tension is applied X_2 lengthwise and compression acts in the direction of X_1.

If laminate fracture occurs due to fibre failure under compression, where the fibres are oriented at angle $\pm\beta$, then the following strength criterion is used:

$$-\langle\langle\sigma_1\rangle\rangle(\bar{A}_{22}\cos^2\beta - \bar{A}_{12}\sin^2\beta) + \langle\langle\sigma_2\rangle\rangle(\bar{A}_{11}\sin^2\beta - \bar{A}_{12}\cos^2\beta) = \frac{\bar{\Delta}}{\bar{A}_{66}}e_\mathrm{fu}^-\,. \tag{175}$$

If the tangentially oriented fibres are the first to fail under tension, we obtain

$$\langle\langle\sigma_1\rangle\rangle\bar{A}_{12}\bar{A}_{66} + \langle\langle\sigma_2\rangle\rangle\bar{A}_{11}\bar{A}_{66} = \bar{\Delta}e_\mathrm{fu}^+\,. \tag{176}$$

If the laminate fracture is associated with the matrix failure under shear, the strength determination is based on the criterion (50) which in our particular case yields:

$$(-\langle\langle\sigma_1\rangle\rangle a_2+\langle\langle\sigma_2\rangle\rangle b_2)^2+2(1+\nu_m)(-\langle\langle\sigma_1\rangle\rangle a_3+\langle\langle\sigma_2\rangle\rangle b_3)^2$$
$$+(-\langle\langle\sigma_1\rangle\rangle a_2+\langle\langle\sigma_2\rangle\rangle b_2)\sqrt{(-\langle\langle\sigma_1\rangle\rangle a_2+\langle\langle\sigma_2\rangle\rangle b_2)^2+4(-\langle\langle\sigma_1\rangle\rangle a_3+\langle\langle\sigma_2\rangle\rangle b_3)^2}$$
$$=2(\bar{\Delta}\sigma_m^+)^2\,. \tag{177}$$

If unidirectionally reinforced single plies fail under transverse compression (50) and (55) enable us to write:

$$(\langle\langle\sigma_1\rangle\rangle a_2+\langle\langle\sigma_2\rangle\rangle b_2)^2+2(1+\nu_{\perp\perp})(\langle\langle\sigma_1\rangle\rangle a_3+\langle\langle\sigma_2\rangle\rangle b_3)^2$$
$$+(\langle\langle\sigma_1\rangle\rangle a_2+\langle\langle\sigma_2\rangle\rangle b_2)\sqrt{(\langle\langle\sigma_1\rangle\rangle a_2+\langle\langle\sigma_2\rangle\rangle b_2)^2+4(\langle\langle\sigma_1\rangle\rangle a_3+\langle\langle\sigma_2\rangle\rangle b_3)^2}$$
$$=2(\bar{\Delta}\sigma_\perp^-)^2\,. \tag{178}$$

The strength envelopes plotted from (175), (176) and (177) are given in Fig. 43.

The curves are based on initial data which are analogous to the case of biaxial tension considering that $e_{fu}^-=0.015$.

Axial loading and torsion

We shall start by discussing combined tension and torsion. The corresponding calculation scheme is found in Fig. 44.

In the specific case (49) enables us to obtain a strength criterion for a reinforced plastic when its fibres undergo failure:

$$\langle\langle\sigma_1\rangle\rangle\bar{A}_{66}(\bar{A}_{22}\cos^2\beta-\bar{A}_{12}\sin^2\beta)+\langle\langle\sigma_6\rangle\rangle(\bar{A}_{11}\bar{A}_{22}-\bar{A}_{12}^2)\sin\beta\cos\beta=\bar{\Delta}e_{fu}^+\,. \tag{179}$$

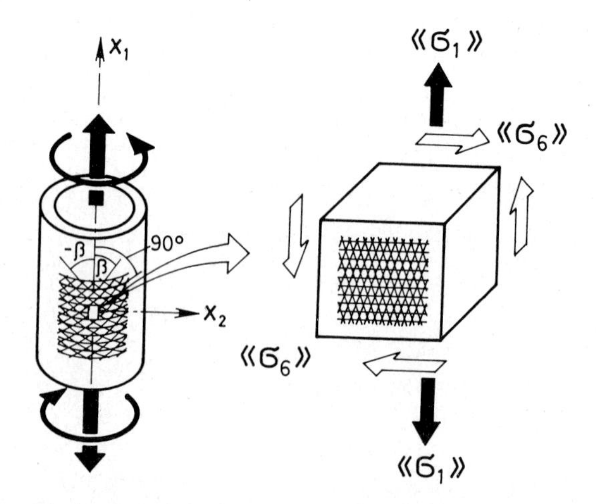

FIG. 44. A calculation scheme for combined axial load and torsion.

Further we shall analyse the polymer matrix failure. Matrix failure is also caused by a combined action of $\langle\langle\sigma_1\rangle\rangle$ and $\langle\langle\sigma_6\rangle\rangle$. Criterion (50) will be used to calculate laminate strength. First we shall refer to the type of polymer matrix failure which results from longitudinal shear. Then $F(\langle\sigma_\perp\rangle_k; \langle\tau_{\parallel\perp}\rangle_k)$ will be applicable in the form of (53).

In our case stresses $\langle\sigma_\perp\rangle$ and $\langle\tau_{\parallel\perp}\rangle$ are calculated according to relationships (57) and (58) if $\langle\langle\sigma_2\rangle\rangle = 0$.

Inserting these relationships into criterion (50) we get the following laminate strength criterion:

$$\begin{aligned}&(\langle\langle\sigma_1\rangle\rangle a_2 + \langle\langle\sigma_6\rangle\rangle c_2)^2 + 2(1+\nu_{\mathrm{m}})(\langle\langle\sigma_1\rangle\rangle a_3 + \langle\langle\sigma_6\rangle\rangle c_3)^2\\ &\quad + (\langle\langle\sigma_1\rangle\rangle a_2 + \langle\langle\sigma_6\rangle\rangle c_2)\sqrt{(\langle\langle\sigma_1\rangle\rangle a_2 + \langle\langle\sigma_6\rangle\rangle c_2)^2 + 4(\langle\langle\sigma_1\rangle\rangle a_3 + \langle\langle\sigma_6\rangle\rangle c_3)^2}\\ &= 2(\bar{\Delta}\sigma_{\mathrm{m}}^{+})^2 .\end{aligned} \tag{180}$$

When the laminate fracture is caused by tensile failure of the matrix, $F(\langle\sigma_\perp\rangle_k; \langle\tau_{\parallel\perp}\rangle_k)$ should be adopted in the form of (52). In our case criterion (50) has the following form:

$$(1-\nu_{\mathrm{m}}^2)(\bar{\sigma}_r\tau_{\mathrm{m}}^{\mathrm{u}})^2(\langle\langle\sigma_1\rangle\rangle a_2 + \langle\langle\sigma_6\rangle\rangle c_2)^2 + (\sigma_{\mathrm{m}}^{+})^2(\langle\langle\sigma_1\rangle\rangle a_3 + \langle\langle\sigma_6\rangle\rangle c_3)^2 = (\bar{\Delta}\sigma_{\mathrm{m}}^{+}\tau_{\mathrm{m}}^{\mathrm{u}})^2 . \tag{181}$$

When there is fibre-matrix bond failure, we resort to relationship (54) and get

$$\frac{\bar{\sigma}_r\tau_{\mathrm{b}}^{\mathrm{u}}}{\bar{\Delta}}(\langle\langle\sigma_1\rangle\rangle a_2 + \langle\langle\sigma_6\rangle\rangle c_2) + \frac{\bar{\sigma}_{rz}^2\sigma_{\mathrm{b}}^{\mathrm{u}}}{\bar{\Delta}^2}(\langle\langle\sigma_1\rangle\rangle a_3 + \langle\langle\sigma_6\rangle\rangle c_3)^2 = (\tau_{\mathrm{b}}^{\mathrm{u}})^2\sigma_{\mathrm{b}}^{\mathrm{u}} . \tag{182}$$

Now we proceed to analyse an angle-ply laminated shell under a combined axial compression and torsion. The calculation scheme then is similar to that in Fig. 44, with the only difference that compressive stresses will be acting instead of tensile stresses.

In the case of fibre failure, strength criterion (49) applicable for combined compression and torsion is expressed as:

$$\langle\langle\sigma_1\rangle\rangle\bar{A}_{66}(\bar{A}_{22}\cos^2\beta + \bar{A}_{12}\sin^2\beta) + \langle\langle\sigma_6\rangle\rangle(\bar{A}_{11}\bar{A}_{22} - \bar{A}_{12}^2)\sin\beta\cos\beta = \bar{\Delta}e_{\mathrm{fu}}^{-} . \tag{183}$$

Transverse compression failure may be another cause for laminate fracture, and then relationship (55) is to be used in criterion (50):

$$\begin{aligned}&(\langle\langle\sigma_1\rangle\rangle a_2 + \langle\langle\sigma_6\rangle\rangle c_2)^2 + 2(1+\nu_{\perp\perp})(\langle\langle\sigma_1\rangle\rangle a_3 + \langle\langle\sigma_6\rangle\rangle c_3)^2\\ &\quad + (\langle\langle\sigma_1\rangle\rangle a_2 + \langle\langle\sigma_6\rangle\rangle c_2)\sqrt{(\langle\langle\sigma_1\rangle\rangle a_2 + \langle\langle\sigma_6\rangle\rangle c_2)^2 + 4(\langle\langle\sigma_1\rangle\rangle a_3 + \langle\langle\sigma_6\rangle\rangle c_3)^2}\\ &= 2(\bar{\Delta}\sigma_{\perp}^{-})^2 .\end{aligned} \tag{184}$$

Strength criterion has the following expression for bond failure:

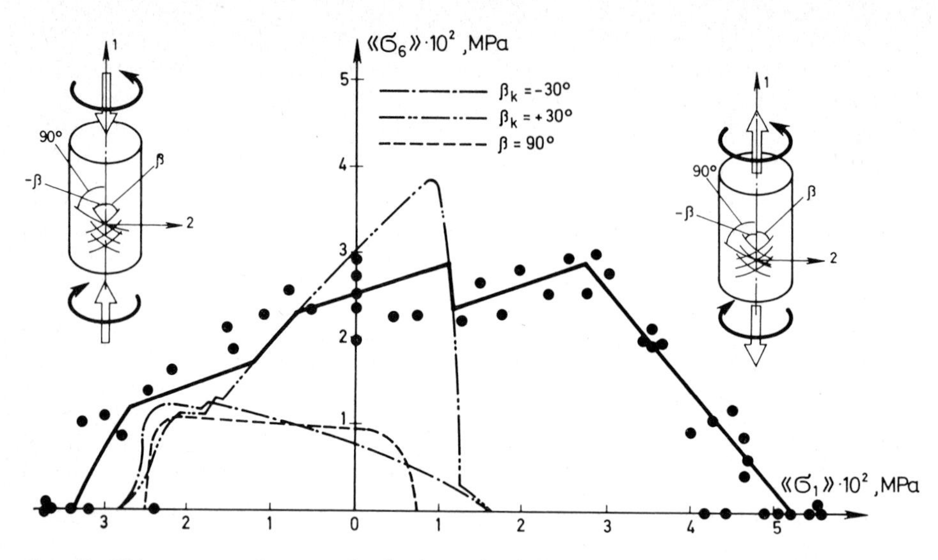

FIG. 45. Ultimate strength curves for laminae of a helically reinforced glass/epoxy shell under combined axial load and torsion.

$$-\frac{\bar{\sigma}_r(\tau_b^u)^2}{\bar{\Delta}}(\langle\langle\sigma_1\rangle\rangle a_2+\langle\langle\sigma_6\rangle\rangle c_2)+\frac{\bar{\sigma}_{rz}^2\sigma_b^u}{\bar{\Delta}^2}(\langle\langle\sigma_1\rangle\rangle a_3+\langle\langle\sigma_6\rangle\rangle c_3)^2=(\tau_b^u)^2\sigma_b^u\,. \quad (185)$$

Fig. 45 gives theoretical curves plotted from the formulas (179)–(184) taking into account the stepwise character of fracture. The initial data are: $E_f = 70\,000$ MPa; $E_m = 3500$ MPa; $v_f = 0.63$; $e_{fu}^+ = 0.020$; $e_{fu}^- = 0.011$; $\nu_m = 0.35$; $\nu_f = 0.23$; $\tau_m^u = 60$ MPa; $\sigma_m^+ = 70$ MPa; $\bar{\sigma}_r = 1.8$; $\beta = 30°$; $\delta = 0.002$ m; $\delta_1 = 0.0067$ m; $\delta_2 = 0.00133$ m. Experimental data are taken from the work of KRAUSS and SCHELLING (1969).

References

GOLDENBLAT, I.I. and V.A. KOPNOV (1968), *Strength and Plasticity Criteria of Structural Materials* (Mashinostroyenie, Moscow).

GOLDENBLAT, I.I. and V.A. KOPNOV (1965), *Polymer Mechanics* **2**, 70.

HARRIS B. (1972), *Composites* **3**, 4.

HAYASHI, T. and K. KOYAMA (1972), Theory and experiments of compressive strength of unidirectionally fiber-reinforced composite materials, in: *Oroc. Int. Conf. of Mechanical Behavior of Materials*, Kyoto, Vol. 5, 104–112.

JONES, E.R. (1969), *SPE-Journal.* **25**, 50.

KNAPPE, W. and W. SCHNEIDER (1972), *Kunststoffe*, Vol. 62, Chapter 12, 864.

KRAUSS, H. and H. SCHELLING (1969), *Kunststoffe*, Vol. 59, Chapter 12, 911.

LOCKWOOD, P.A. and G.R. MACHLAN (1963), Tensile properties of impregnated E and S 994 Rowing, in: *4th Semi-annual Polaris Glass Reinforced Plastics*, Owens Corning Fiberglass Corp., Research and Development Conference.

NOYES, I.V. and B.H. JONES (1968), Analytical design procedures for the strength and elastic properties of multilayer fibrous composites, in: *AIAA/ASME 9th Structures, Stuctural Dynamics and Materials Conference*, Palm Springs, Paper N 68-336, 16.

NOYES, I.V. and B.H. JONES (1968), Grazing and Yielding of Reinforced Plastics, in: Air Force Materials Laboratory Technical Report AFML-TR-68-51. Dayton, 23.

SERENSEN, S.V. and G.P. ZAITSEV (1965), *Polymer Mechanics*, **2**, 93.

SKUDRA, A.M. F.YA. BULAVS and K.A. ROCENS (1971), *Creep and Static Fatigue of Reinforced Plastics* (Zinatne, Riga).

SKUDRA, A.M. and F.YA. BULAVS (1978), *Structural Theory of Reinforced Plastics* (Zinatne, Riga).

SKUDRA, A.M. and B.A. KIRULIS (1975), Structural strength theory for reinforced plastics under tension, in: *Mechanics of Strained Bodies and Units* (Mashinostroyenie, Moscow), 454–459.

SKUDRA, A.M., B.A. KIRULIS and A.V. ZAHAROV (1977), Fibre-matrix bond strength in reinforced plastics, in: *Mechanics of Composite Materials* (RPI, Riga), Vol. 1, 30–37.

SKUDRA, A.A. (1977), Strength of orthogonally reinforced plastics under compression, in: *Mechanics of Composite Materials* (RPI, Riga), Vol. 1, 74ff.

SKUDRA, A.A. (1980), Strength of helically reinforced tangentially strengthened shells, in: *Mechanics of Composite Materials* (RPI, Riga), Vol. 3, 75ff.

SKUDRA, A.A. (1979), Strength of helically reinforced shells under tension, in: *Mechanics of Composite Materials* (RPI, Riga), Vol. 2, 33ff.

SKUDRA, A.A. (1979), Strength of helically reinforced shells under compression and torsion, in: *Mechanics of Composite Materials* (RPI, Riga), Vol. 2, 48ff.

TSAI, S.W. and E.M. WU (1971), *J. Composite Materials* **5**, 58–80.

TSAI, S.W. (1965), Strength Characteristics of Composite Materials, NASA CR-224.

UEMURA, M. and K. YAMAWAKI (1971), Fracture strength of helically wound composite cylinders, in: *Proceedings of the 9th Int. Symp. Space Technol. and Sci.*, Tokyo, 1971, Tokyo, 215–223.

VAN FO FI, G.A. (1971), *Theory of Reinforced Plastics* (Naukova Dumka, Kiev).

CHAPTER II

Strength (Failure) Theories and their Experimental Correlation

Robert E. Rowlands

Professor of Engineering Mechanics
and Director, Structural and Materials Testing Laboratory
University of Wisconsin
Madison, WI 53706
U.S.A.

Contents

HANDBOOK OF COMPOSITES, VOL. 3 – Failure Mechanics of Composites
Edited by G.C. SIH and A.M. SKUDRA

List of Symbols

A_i^*, A_{ij}^* Strength coefficients
A_{ij} Extensional stiffnesses
$[A^*], [A']$ Laminate properties
B_{ij} Bending-stretching stiffnesses
B_{ij}^* Strength coefficients
$[B^*], [B']$ Laminate properties
c Cohesive strength
C_{ij} Stiffnesses
$[C^*], [C']$ Laminate Properties
D_{ij} Bending stiffnesses
$[D^*], [D']$ Laminate Properties
E Elastic modulus
e Tensorial strain
$f_i, f_{ij}, F_i, F_{ij}, F_{ijk}, \bar{F}_i, \bar{F}_{ij}, \bar{F}_{ijk}$
Strength coefficients
G Shear modulus
h Laminate thickness
I_1, I_2 Stress invariants
k kth ply of a laminate as a subscript
K, K_l Elastic coefficients
K'_l Correlation coefficient
K_i Coefficients
M Bending and twisting moments
m_i Shape factors
n Load increment as a subscript
N Inplane forces
Q_{ij} Reduced stiffnesses
$\bar{Q}_{ij}$ Transformed reduced stiffnesses
S_{ij} Compliances
S Shear strength
S' Negative shear strength
S_{45} Shear strength on 45-deg. axes
$[T]$ Transformation matrix
u Ultimate as a superscript
U_i Elastic coefficients
U_D Energy of distortion
U Normal strength of 45-deg. coupon
U'_θ Compressive strength of θ-deg. coupon

u, v, w	Displacements
x, y, z	Coordinate directions
X, Y, Z, Ξ, Π	Strengths
X', Y', Z'	Compressive strengths
α	Power as a superscript
θ	Angle
κ	Curvature
γ	Shear strain
$\dot{\gamma}, \dot{\gamma}'$	Interaction factor
$\varepsilon_i, \varepsilon_{ij}$	Engineering strains
ν	Poisson's ratio
ξ, η	Coordinate directions
ρ, λ, μ	Coefficients
$\dot{\mu}$	Coefficient of friction
σ	Normal stress
σ_y	Yield stress
$\sigma_p, \sigma_q, \sigma_r$	Principal stresses
τ	Shear stress
1, 2	Principal directions of material symmetry.

1. Introduction

Extensive acceptance and utilization of composite structures requires confidence in their load carrying capacity. The ability to predict accurately the strength of a particular composite is necessary for design. Most available strength information is based on uniaxial stress states, while practical applications invariably involve at least biaxial loading. Structural composites usually consist of several anisotropic plies bonded together at different orientations. Unlike isotropic materials, the strength of composites is typically directionally dependent. Furthermore, failure of some plies of a laminated structure need not necessarily constitute rupture of the total laminate since multiple layers may provide other sustaining load paths. Natural composites such as wood and paper tend to be reasonably homogeneous throughout the thickness, rather than laminated. Other than say with plywood, the analysis is then somewhat simplified in that lamination theory is not needed. However, from a design consideration, the situation can now be more critical since there need not be alternate load-carrying paths once failure initiates.

Recognizing the numerous combinations of constituent materials, plus the various possible stacking constructions of a laminate, economics and expedience necessitates the ability to predict strength accurately under combined loading conditions and with a minimum of testing. Advantageous as allowable composite strength predictions are, present theoretical approaches require experimental verification.

Motivation for and results of biaxial testing of orthotropic and composite materials are discussed herein relative to existing theories. The Chapter reviews briefly the more commonly employed composite failure criteria and compares predictions with experimental observation. Sufficient theory is included to make the Chapter self-contained. Since strength theories are conceived primarily to predict onset, not mode, of failure, the macroscopic viewpoint will predominate. However, some mode-of-failure discussion is included. Various utilized and proposed biaxial specimens and testing techniques are reviewed, and their advantages and disadvantages noted. Relative agreement between predicted and experimentally observed results is demonstrated for those theories whose stress-strain information is generated numerically to failure. The present emphasis is on the correlation between prediction and observation. Possible considerations for future research are included.

While careful efforts have been taken to be complete, those theories and testing concepts which have been or are expected to be most important are emphasized.

2. Synopsis of failure theories

Particularly for preliminary design purposes, it is important that a reliable strength prediction be available for various combinations of normal plus shear loading. To capitalize on the wide range of possible constituent materials and laminate constructions, the analysis should employ minimal lamina input data. To obtain this data should, preferably, require only relatively simple testing such as can be conducted in any normally-equipped laboratory. If relevant, the design analysis ought to account for residual stresses (such as due to fabrication or curing), as well as interlaminar, nonlinear and inelastic effects. The theory should be no more complicated than it is useful. In the conceptual design stage, it is inconceivable to physically test all possible composite materials and laminate constructions. However, having decided upon one or two candidate laminates, it would then be advantageous to have a reliable strength theory based upon a limited number of laminate tests.

Several of the anisotropic strength or failure theories are extensions of isotropic yield criteria. Much of the early work (before 1950) on the strength of anisotropic materials was conducted independently in the areas of wood technology and the mechanical response of single crystals and metals. Representative of this activity were the interaction equations developed to describe the strength of wood as a function of grain orientation (HANKINSON (1921), NORRIS (1939), RABINOVICH (1946), ASHKENAZI (1966), and KOLLMANN and COTE (1968)). JENKINS (1920) extended the isotropic maximum stress theory to predict failure of wood.

The early work on metals appears to have been done in parallel to the above, yet independently. The VON MISES' (1928) energy of distortion or octahedral shear stress yield theory has significantly influenced the basis of many of the composite strength criteria. HILL (1950) generalized Von Mises' formulation to include anisotropy. MARIN (1957) subsequently suggested a formulation similar to Hill's by presupposing coincident directions of principal stress and material. In 1965, Hill's theory was adapted by Azzi and Tsai as a strength criterion for composites. HOFFMAN (1967) altered Hill's criterion to provide for unequal tensile and compressive strengths by introducing linear stress terms in the functional form. Norris' extension of the distortion-energy theory was modified by FISCHER (1960, 1967) to predict the strength of glass reinforced composites.

A variation of the maximum-stress theory for unidirectional composites was proposed by STOWELL and LIU (1961) in which the failure stress of the fibers was taken as the limiting strength of the lamina in the fiber direction, while the limiting transverse and shear stresses are those of the matrix. As pointed out by KELLY and DAVIS (1965), utilization of bulk matrix properties for transverse and shear stresses ignores any fiber-matrix interaction.

The St. Venant's maximum strain theory of isotropic mechanics was extended to include anisotropy (PETIT (1967), REED (1970), PETIT and WADDOUPS (1969)).

GOL'DENBLAT and KOPNOV (1965), and ASHKENAZI (1959, 1965, 1967, 1970) appear to have initiated the attempts to develop a strength theory which is invariant with respect to the coordinate system. Subsequent efforts to this aspect have been made by MALMEISTER (1966), TSAI and WU (1971), and TENNYSON et al. (1978, 1980).

A strength criterion predicated on the strain energy to failure has been proposed recently by SANDHU (1974b). Whereas most of the strength predictions assume a linear stress-strain relationship in the associated lamination analysis, those by SANDHU, and PETIT–WADDOUPS (1969) provide for nonlinear lamina material response.

The composite strength theories discussed above are based on lamina material properties and a lamina failure criterion. PUPPO and EVENSEN (1972) proposed a laminate failure criterion using interaction factors and the concept of principal axes of strength. WU and SCHEUBLEIN (1974) also proposed a direct laminate strength theory.

Although considerable effort has been devoted to the formulation of composite strength theories and to their correlation with test data, no single analysis has emerged which is fully adequate. The analyses of Norris, Tsai–Hill, Hoffman, Fischer and Cowin are relatively easy to use but are valid rigorously only under special orthotropy. This hampers their direct application to complete laminates. The Tsai–Wu and Tennyson criteria are invariant with respect to the axes of reference, but use of these analyses is inconvenienced by the need to evaluate interaction coefficients under biaxial states of stress.

3. Relevant analytical background

3.1. Orthotropic elasticity

While several of the strength criteria do not in themselves actually address whether the material is elastic or inelastic, composite lamination theory does involve constitutive response. The classical lamination theory employed to predict strength of stacked plies normally assumes a linear-elastic, orthotropic constitutive behavior. Many composites are not linearly elastic, but to provide for a more general stress-strain relationship can run the risk of creating an analysis more complicated than useful.

The generalized Hooke's law can be written as (ASHTON et al. (1969), FUNG (1977), JONES (1975), TSAI et al. (1968), VINSON and CHOU (1975), CHRISTENSEN (1979))

$$\sigma_{ij} = C_{ijkl} e_{kl} , \tag{1}$$

where σ_{ij} is the stress tensor, e_{kl} is the strain tensor, and C_{ijkl} is a 4th order stiffness or material tensor. The tensorial e and engineering ε strains are related to the three cartesian displacement components u, v and w by

$$\begin{aligned} &e_x = \varepsilon_x = \frac{\partial u}{\partial x}, \qquad e_y = \varepsilon_y = \frac{\partial v}{\partial y}, \qquad e_z = \varepsilon_z = \frac{\partial w}{\partial z}, \\ &2e_{xy} = \varepsilon_{xy} = \gamma_{xy} = \frac{\partial u}{\partial y} + \frac{\partial v}{\partial x}, \\ &2e_{yz} = \varepsilon_{yz} = \gamma_{yz} = \frac{\partial v}{\partial z} + \frac{\partial w}{\partial y}, \\ &2e_{zx} = \varepsilon_{zx} = \gamma_{zx} = \frac{\partial w}{\partial x} + \frac{\partial u}{\partial z}. \end{aligned} \tag{2}$$

Hooke's law can be rewritten as

$$e_{ij} = S_{ijkl} \sigma_{kl} \tag{3}$$

where S_{ijkl} is a 4th order compliance tensor. It is often convenient to write Hooke's law in contracted notion as

$$\sigma_i = C_{ij} \varepsilon_j \quad \text{and} \quad \varepsilon_i = S_{ij} \sigma_j \tag{4}$$

where the stress components σ_i and the strain components ε_j are related through either the stiffness matrix C_{ij} or the compliance matrix S_{ij}. The relationships between the contracted and tensor notations are noted in Table 1. Symmetry reduces the number of independent stiffness or compliance components to 21 for even the most general anisotropic material in three-dimensions. If the response is orthotropic such that 1-, 2- and 3-directions are the three axes of material symmetry, then there are only nine independent material coefficients.

The plies of most structural components are subjected to plane stress. The constitutive relationship for an orthotropic lamina in a state of plane stress ($\sigma_3 = \tau_{13} = \tau_{23} = 0$) may be written as

$$\{\sigma_{ij}\} = \begin{Bmatrix} \sigma_1 \\ \sigma_2 \\ \tau_{12} \end{Bmatrix}_k = \begin{bmatrix} Q_{11} & Q_{12} & 0 \\ Q_{12} & Q_{22} & 0 \\ 0 & 0 & Q_{66} \end{bmatrix}_k \begin{Bmatrix} \varepsilon_1 \\ \varepsilon_2 \\ \gamma_{12} \end{Bmatrix}_k \tag{5}$$

TABLE 1. Comparison between tensor, engineering and contracted notation for stresses and strains.

Stresses		Strains		
Tensor	Contracted	Tensor	Engineering	Contracted
σ_{11}	σ_1	e_{11}	ε_{11}	ε_1
σ_{22}	σ_2	e_{22}	ε_{22}	ε_2
σ_{33}	σ_3	e_{33}	ε_{33}	ε_3
$\sigma_{23} = \tau_{23}$	σ_4	$2e_{23}$	$\gamma_{23} = \varepsilon_{23}$	ε_4
$\sigma_{31} = \tau_{31}$	σ_5	$2e_{31}$	$\gamma_{31} = \varepsilon_{31}$	ε_5
$\sigma_{12} = \tau_{12}$	σ_6	$2e_{12}$	$\gamma_{12} = \varepsilon_{12}$	ε_6

where the components of the reduced stiffness matrix Q_{ij} are

$$\begin{aligned}
Q_{11} &= E_{11}/(1-\nu_{12}\nu_{21})\,,\\
Q_{22} &= E_{22}/(1-\nu_{12}\nu_{21})\,,\\
Q_{12} &= \nu_{21}E_{11}/(1-\nu_{12}\nu_{21})\,,\\
Q_{66} &= G_{12}\,,\\
Q_{16} &= Q_{26} = 0\,,\\
\nu_{12}E_{22} &= \nu_{21}E_{11}\,.
\end{aligned} \tag{6}$$

E_{11}, E_{22}, ν_{12} and G_{12} are the four independent elastic constants of the lamina with respect to its axes (1–2) of material symmetry. Transforming Equation (5) into the orthotropic laminate x–y axes system results in

$$\{\sigma_{xy}\} = \begin{Bmatrix} \sigma_x \\ \sigma_y \\ \tau_{xy} \end{Bmatrix}_k = \begin{bmatrix} \bar{Q}_{11} & \bar{Q}_{12} & \bar{Q}_{16} \\ \bar{Q}_{12} & \bar{Q}_{22} & \bar{Q}_{26} \\ \bar{Q}_{16} & \bar{Q}_{26} & \bar{Q}_{66} \end{bmatrix}_k \begin{Bmatrix} \varepsilon_x \\ \varepsilon_y \\ \gamma_{xy} \end{Bmatrix}_k \tag{7}$$

where $\bar{Q}_{ij}$ are the transformed reduced stiffnesses given by CHRISTENSEN (1979)

$$\begin{aligned}
\bar{Q}_{11} &= U_1 + U_2\cos(2\theta) + U_3\cos(4\theta)\,,\\
\bar{Q}_{22} &= U_1 - U_2\cos(2\theta) + U_3\cos(4\theta)\,,\\
\bar{Q}_{12} &= U_4 - U_3\cos(4\theta)\,,\\
\bar{Q}_{66} &= U_5 - U_3\cos(4\theta)\,,\\
\bar{Q}_{16} &= -\tfrac{1}{2}U_2\sin(2\theta) - U_3\sin(4\theta)\,,\\
\bar{Q}_{26} &= -\tfrac{1}{2}U_2\sin(2\theta) + U_3\sin(4\theta)
\end{aligned} \tag{8}$$

and the elastic coefficients U are

$$U_1 = \tfrac{1}{8}(3Q_{11} + 3Q_{22} + 2Q_{12} + 4Q_{66}),$$

$$U_2 = \tfrac{1}{2}(Q_{11} - Q_{22}),$$

$$U_3 = \tfrac{1}{8}(Q_{11} + Q_{22} - 2Q_{12} - 4Q_{66}), \tag{9}$$

$$U_4 = \tfrac{1}{8}(Q_{11} + Q_{22} + 6Q_{12} - 4Q_{66}),$$

$$U_5 = \tfrac{1}{8}(Q_{11} + Q_{22} - 2Q_{12} + 4Q_{66}).$$

Small k represents the kth layer of the laminate and θ is the angle measured counter clockwise from the laminate positive x-axis to the positive 1-axis of the lamina. Equation (5) can be inverted to yield

$$\{\varepsilon_{ij}\} = \begin{Bmatrix} \varepsilon_1 \\ \varepsilon_2 \\ \gamma_{12} \end{Bmatrix}_k = \begin{bmatrix} S_{11} & S_{12} & 0 \\ S_{12} & S_{22} & 0 \\ 0 & 0 & S_{66} \end{bmatrix}_k \begin{Bmatrix} \sigma_1 \\ \sigma_2 \\ \tau_{12} \end{Bmatrix}_k \tag{10}$$

where the lamina compliance matrix S_{ij} is given by

$$[S] = [Q]^{-1} \tag{11}$$

and

$$S_{11} = 1/E_{11}; \qquad S_{22} = 1/E_{22};$$
$$S_{12} = -\nu_{12}/E_{11} = -\nu_{21}/E_{22}; \qquad S_{66} = 1/G_{12}. \tag{12}$$

Also, under plane stress,

$$\varepsilon_3 = S_{13}\sigma_1 + S_{23}\sigma_2, \tag{13}$$

and

$$\gamma_{23} = \gamma_{31} = 0. \tag{14}$$

Similarly, Equation (10) may be transformed to the principal material axes x–y of the laminate to become

$$\{\varepsilon_{ij}\} = \begin{Bmatrix} \varepsilon_x \\ \varepsilon_y \\ \gamma_{xy} \end{Bmatrix}_k = \begin{bmatrix} \bar{S}_{11} & \bar{S}_{12} & \bar{S}_{16} \\ \bar{S}_{12} & \bar{S}_{22} & \bar{S}_{26} \\ \bar{S}_{16} & \bar{S}_{26} & \bar{S}_{66} \end{bmatrix}_k \begin{Bmatrix} \sigma_x \\ \sigma_y \\ \tau_{xy} \end{Bmatrix}_k \tag{15}$$

where, like the transformed stiffnesses, the transformed compliances $\bar{S}_{ij}$ depend only on E_{11}, G_{12}, E_{22}, ν_{12} and θ. Using the transformation relationships, the stresses and strains relative to the lamina 1–2 axes are related to those in terms of the laminate $x-y$ axes, and vice-versa, by

$$\{\sigma_{12}\}=[T]\{\sigma_{xy}\}, \qquad \{e_{12}\}=[T]\{e_{xy}\} \tag{16}$$

and

$$\{\sigma_{xy}\}=[T]^{-1}\{\sigma_{12}\}, \qquad \{e_{xy}\}=[T]^{-1}\{e_{12}\} \tag{17}$$

where

$$[T]=\begin{bmatrix} \cos^2\theta & \sin^2\theta & 2\sin\theta\cos\theta \\ \sin^2\theta & \cos^2\theta & -2\sin\theta\cos\theta \\ -\sin\theta\cos\theta & \sin\theta\cos\theta & (\cos^2\theta-\sin^2\theta) \end{bmatrix}. \tag{18}$$

The reduced stiffnesses of Equation (5) can also be expressed in terms of the compliances of Equation (10) as follows:

$$\begin{aligned} Q_{11}&=\frac{S_{22}}{S_{11}S_{22}-S_{12}^2}, & Q_{22}&=\frac{S_{11}}{S_{11}S_{22}-S_{12}^2}, \\ Q_{12}&=-\frac{S_{12}}{S_{11}S_{22}-S_{12}^2}, & Q_{66}&=\frac{1}{S_{66}}. \end{aligned} \tag{19}$$

For isotropic materials,

$$\begin{aligned} &E_{11}=E_{22}=E, && \nu_{12}=\nu_{21}=\nu, && G_{12}=G, \\ &S_{11}=S_{22}=\frac{1}{E}, && S_{12}=-\frac{\nu}{E}, && S_{66}=\frac{2(1+\nu)}{E}=\frac{1}{G}, \\ &Q_{11}=Q_{22}=\frac{E}{1-\nu^2}, && Q_{12}=\frac{\nu E}{1-\nu^2}, && Q_{66}=\frac{E}{2(1+\nu)}=G \end{aligned} \tag{20}$$

3.2. Lamination theory

From classical plate theory, the laminate strains ε_{xy} at any distance z from the middle surface are given by

$$\{\varepsilon\}=\{\varepsilon^0\}+z\{\kappa\} \tag{21}$$

where ε^0 represents the midplane (membrane) strain and κ_x, κ_y and κ_{xy} are the components of curvature. Substitution of Equation (21) into Equation (7), and integrating through the thickness h of the plate, produces the following expression:

$$\left[\frac{N}{M}\right]=\left[\begin{array}{c|c} A & B \\ \hline B & D \end{array}\right]\left[\frac{\varepsilon^0}{\kappa}\right], \tag{22}$$

where

$$(A_{ij};\, B_{ij};\, D_{ij})=\int_{-h/2}^{h/2} \bar{Q}_{ij}^{k}(1, z, z^2)\, dz \tag{23}$$

are the extensional, coupling and bending laminate stiffnesses, respectively. N and M are the resultant force and moment vectors acting per unit of length of the laminate, i.e.,

$$(N_x, N_y, N_{xy};\, M_x, M_y, M_{xy})=\int_{-h/2}^{h/2} (\sigma_x^k, \sigma_y^k, \tau_{xy}^k;\, z\sigma_x^k, z\sigma_y^k, z\tau_{xy}^k)\, dz\,. \tag{24}$$

In general, the matrices A_{ij}, B_{ij} and D_{ij} are all complete and symmetric, as defined in Equation 23. If the laminate layup is symmetrical in geometry and material properties about $z=0$, then $B_{ij}=0$ and the membrane-flexing coupling vanishes. Stiffness components A_{11}, A_{22}, A_{12}, A_{66}, D_{11}, D_{22}, D_{12} and D_{66} are positive definite. $A_{16}=A_{26}=D_{16}=D_{26}=0$ for laminates made up entirely of plies at 0- or 90-degrees to the laminate axes. For angle-ply laminates ($\pm\theta$) fabricated from a large number of alternating lamina, D_{16}, D_{26}, A_{16} and A_{26} can be quite small compared to the other stiffness components.

Equations (22) can be inverted to give (Ashton et al. (1969))

$$\left\{\frac{\varepsilon^0}{\kappa}\right\}=\left[\begin{array}{c|c} A' & B' \\ \hline C' & D' \end{array}\right]\left\{\frac{N}{M}\right\}, \tag{25}$$

and

$$\left\{\frac{\varepsilon^0}{M}\right\}=\left[\begin{array}{c|c} A^* & B^* \\ \hline C^* & D^* \end{array}\right]\left\{\frac{N}{\kappa}\right\}, \tag{26}$$

where

$$\begin{aligned} &[A^*]=[A^{-1}]\,, \qquad [B^*]=-[A^{-1}][B]\,, \\ &[C^*]=[B][A^{-1}]\,, \qquad [D^*]=[D]-[B][A^{-1}][B]\,, \end{aligned} \tag{27}$$

$$\begin{aligned} &[A']=[A^*]-[B^*][D^{*-1}][C^*]\,, \qquad [B']=[B^*][D^{*-1}]\,, \\ &[C']=-[D^{*-1}][C^*]\,, \qquad [D']=[D^{*-1}]\,. \end{aligned} \tag{28}$$

Equations (25) through (28) demonstrate that for non-symmetrical laminates such that $B \neq 0$, in-plane loading $\{N\}$ produces bending as well as extension, and moments $\{M\}$ produce in-plane deformations as well as flexure.

For any laminate of known lamina elastic properties (E_{11}, E_{22}, ν_{12}, G_{12}) and subjected to forces N and moments M, the strains in any ply (relative to the laminate axes) are evaluated from Equations (6), (21), (23), (25), (27), and (28). These strains can be transposed to the principal lamina axes by the second of Equations (16), while the stresses relative to the lamina axes are then obtainable from Equation (5). Having evaluated the stress and strain history in any ply of the laminate, the imminency of lamina failure can be evaluated relative to any criterion based on stress, strain or energy. While the discussion thus far has assumed constant elastic properties, these could be replaced by tangent moduli or slopes of piecewise linear stress-strain curves.

For natural orthotropic material such as wood, paper or bone, and for unidirectional laminates, the material properties are fairly uniform through the thickness. Equations (23) can then be integrated directly and there will be no membrane-flexural coupling, i.e., $B = B' = C' = B^* = C^* = 0$. A strength criterion can of course be applied directly to the entire thickness of a material whose properties are homogeneous throughout the thickness. Equations (23) and (24) involve summations over the various laminae for laminates containing individual plies whose properties in a direction vary from ply to ply. Such variations can occur either because the various plies are not aligned with each other, or because individual plies are of different materials.

Since the classical lamination theory does not provide for interlaminar effects, the analysis predicts identical response for all stacking sequence variations of the same laminate construction.

4. Isotropic failure theories

4.1. General comments

The objective of the Chapter is to discuss strength theories and associated testing of composite materials. Nevertheless, it is informative to briefly review various strength predictions used with isotropic media for at least two reasons: (1) to place the general topic into appropriate perspective, and (2) because many of the criteria used for composites are extensions of concepts employed with isotropic materials. The formulation of a strength (failure) or design theory for conventional homogeneous and isotropic materials is considerably less complex than the composite counterpart since constitutive and strength properties are now independent of orientation and position. This simplifies the analysis as no lamination theory is necessary. Moreover, the directional independence permits one to work in terms of the convenient principal stresses and strains.[1] Many isotropic and homogeneous structural materials are crystal-

[1] Since principal directions of stress, strain or material symmetry need not coincide in orthotropic media, the concept of principal stresses and strains finds little use with composites.

line in nature and their strength criteria tend to predict one or other of fracture or the initiation of yielding. The comments on the following theories are taken from articles by TIMOSHENKO (1953, 1956), NADAI (1950, 1963), LIEBOWITZ (1968), MCCLINTOCK and ARGON (1966), JAEGER (1969), BORESI et al. (1978), POPOV (1968), FUNG (1965), and the excellent survey report by SANDHU (1972).

4.2. Galileo's Theory, 1638

By subjecting rock specimens to tension, Galileo observed that strength depends on cross-sectional area but is independent of length. He concluded that failure would occur when the "absolute resistance to fracture," i.e., critical stress, was attained. This appears to be the first suggestion of the maximum normal stress theory for predicting fracture of brittle materials.

4.3. Coulomb's Theory, 1773

In that rock continued to be the principal material of construction, Coulomb noticed uniaxially compressed rocks developed cracks which were inclined to the axis of loading. The cracks were not along the 45° direction of the plane of maximum shear. This lead Coulomb to suggest failure occurred when the shear stress τ on a plane (which would become the failure plane) becomes equal to the sum of the cohesive strength c of the material and the friction force $-\dot{\mu}\sigma$, where $\dot{\mu}$ is the coefficient of friction and σ is the normal stress on the plane, i.e.,

$$|\tau| = c - \dot{\mu}\sigma . \tag{29}$$

Failure was predicted to occur on the plane inclined at $45° - \frac{1}{2}\arctan\dot{\mu}$. While Equation (29) is not normally obeyed by crystalline materials, it does yield reasonably good results with soils of low cohesion.

4.4. Maximum Normal Stresses Theory – Rankine, Lamé, Clapeyron, 1858

This theory postulates that failure occurs when the principal stresses[2] satisfy the expression

$$(\sigma_p^2 - \sigma_0^2)(\sigma_q^2 - \sigma_0^2)(\sigma_r^2 - \sigma_0^2) \geqslant 0 , \tag{30}$$

where σ_0 is a characteristic value for the material and is obtained from a uniaxial tensile test. Failure is predicted if $\sigma_p = \sigma_0$, $\sigma_q = \sigma_0$ or $\sigma_r = \sigma_0$, separately. However, many results contradict Equation (30). The maximum normal stress theory continues to be used in the form $\sigma_p \geqslant \sigma_0$ to predict fracture of isotropic, brittle materials. The theory is analytically simple to use but does not account for the magnitudes of the other principal stresses. It is also not directly

[2] It is convenient to denote principal stresses by $\sigma_p \geqslant \sigma_q \geqslant \sigma_r$ so as not to confuse them with stresses σ_1, σ_2, σ_3 in the directions of material symmetry of orthotropic materials.

useful for anisotropic materials since principal directions of stress and strength need not then occur. One advantage of the concept is that no linear or elastic assumptions are involved.

4.5. Maximum Principal Strain Theory – Saint Venant, 1837

The theory predicts that failure (by yielding or fracture) occurs under any state of stress when the maximum strain reaches some limiting critical value for the material and as determined from a simple test. Since the criterion is usually used to predict allowable stress or load application, the concept has normally assumed linear-elastic response. As discussed subsequently, this criterion has been extended to orthotropic and composite materials.

4.6. Maximum Shear-Stress Theory – Tresca, 1870

Yielding is predicted by this theory to initiate when

$$[(\sigma_p-\sigma_r)^2-\sigma_y^2][(\sigma_q-\sigma_p)^2-\sigma_y^2][(\sigma_r-\sigma_q)^2-\sigma_y^2]=0\,. \tag{31}$$

This criterion represents three sets of parallel planes, each set being normal to a coordinate plane. These six planes define a hexagonal prism in σ_p, σ_q and σ_r stress space with its axis making equal angles with these coordinate axes. The theory satisfactorily predicts the onset and orientation (at 45° with respect to σ_p and σ_r) of yielding in structural materials such as mild steel. The quantity σ_y is the yield stress as obtained under uniaxial loading.

4.7. Total Strain Energy Theory – Beltrami, 1885

Failure by yielding is predicted to occur at a point when the total strain energy reaches some critical value for the particular material and as obtained from some simple test such as under pure tension, i.e.,

$$(1+\nu)\{(\sigma_p-\sigma_q)^2+(\sigma_q-\sigma_r)^2+(\sigma_r-\sigma_p)^2\}+(1-2\nu)(\sigma_p+\sigma_q+\sigma_r)^2=3\sigma_y^2 \tag{32}$$

where ν is Poisson's ratio and σ_y is the yield stress under uniaxial testing. This theory has not proved satisfactory because considerable strain energy can be stored under hydrostatic pressure without actually contributing to failure. This total energy concept has been improved upon by removing the volume change so as to involve only the energy of distortion, Section 4.9.

4.8. Mohr's Theory, 1900

Mohr hypothesized that of all planes having the same normal stress, failure will most likely occur on that also having the greatest shear stress. This means that of the three Mohr's circles at a point, only the largest circle connecting σ_p and σ_r must be considered and failure is independent of σ_q. By generating and superimposing a sufficient number of such principal circles, each related to a failure state of the material, envelopes of these circles can be drawn with which to predict the failure state for any arbitrary stress condition. The shear τ and normal stress σ on the plane of failure can be related as

$$|\tau| = F(\sigma) \tag{33}$$

where the function of σ, $F(\sigma)$, is determined experimentally and is symmetric about the σ-axis. $F(\sigma)$ cannot be negative without contradicting the fact that large hydrostatic pressures do not of themselves produce yielding. Experiments by Von Karman indicated that for large negative values of σ, Mohr's failure envelope tends to become parallel to the σ-axis. This theory is not used as frequently with metals as are Tresca's maximum shear stress theory or Von Mises' distortional energy prediction (Section 4.9). However, it does find favor in soil mechanics.

4.9. Distortional Energy Theory – Huber, Hencky, Von Mises, Nadai, Novozhilov, ~1920

Nothing that the hydrostatic state of stress changes only the volume and not the shape of a material, it is postulated that yielding occurs due to the energy of distortion. By dropping the hydrostatic contribution $1/3(\sigma_p + \sigma_q + \sigma_r)$ of Equation (32), this criterion can be written as

$$(\sigma_p - \sigma_q)^2 + (\sigma_q - \sigma_r)^2 + (\sigma_r - \sigma_p)^2 = 2\sigma_y^2 , \tag{34}$$

where σ_y is again the yield stress under uniaxial tension. Under plane-stress, $\sigma_r = 0$ and Equation (34) becomes

$$\left(\frac{\sigma_p}{\sigma_y}\right)^2 - \frac{\sigma_p \sigma_q}{\sigma_y^2} + \left(\frac{\sigma_q}{\sigma_y}\right)^2 = 1 . \tag{35}$$

As to be observed subsequently, several of the anisotropic failure criteria are of this same general form.

5. Anisotropic failure criteria

5.1. General comments

With macroscopically homogeneous but orthotropic materials, development of a strength theory has often involved extending one of the isotropic analyses to account for anisotropy. It may be reasonable with homogeneous materials to base structural strength on the initial combination of loads which causes the postulated failure (strength) envelope or criterion to be reached. Although strength of isotropic materials is often predicted on initial failure of any of the possible modes, i.e., normal or shear, limiting shear of a metal can produce yielding without necessarily causing catastrophic fracture. For laminates, a criterion is typically applied on a ply-by-ply basis and the load-carrying capability of the entire composite is then predicted using the lamination theory of Section 3. A laminate is sometimes assumed to fail analytically when the strength criterion of any one of its laminae is reached. While load redistribution usually occurs within a laminate upon actual failure of an individual ply, failure of a single ply need not necessarily cascade into total fracture of the structure. Composite materials can also sometimes retain longitudinal capacity even after transverse or shear failure has occurred.

While analytical flow and fracture criteria of isotropic materials may bear a reasonable correlation to actual mechanical behavior, this is less so with non-metallic composites. Yielding normally does not occur in fiber-reinforced plastics in the same sense as in metals. Nevertheless, many of the orthotropic strength theories are anisotropic extensions of isotropic yield criteria.

Several of the more commonly used composite strength theories are now discussed. Theoretical-experimental correlations, along with some of the testing approaches, will be presented subsequently. Although some of the failure theories were originally formulated in three-dimensions, most laminates are subjected to plane-stress and so only that form of the criteria are included here.

In addition to the specific references cited below, various anisotropic failure theories are reviewed by FRANKLIN (1968), SENDECKYJ (1972), SANDHU (1972), WU (1974a, 1974b), and VICARIO and TOLAND (1975).

5.2. Maximum Strain Theory

This criterion is an extension of St. Venant's maximum principal strain theory (Section 4.5) to anisotropic media. For an orthotropic lamina, the strain components must be referred to the principal material axes; therefore, three strain components can exist in the criterion. As originally formulated, linear elastic response is assumed to failure such that the criterion becomes a strength prediction in terms of applied loads or stresses. A ply of a laminate is considered to have failed when either its longitudinal, transverse or shear strain reaches a limiting value as determined from simple one-dimensional stress experiments. The minimum common envelope of the superposition of the interaction failure diagrams of all the individual plies related to the principal material axes of a laminate becomes the failure diagram for the laminate.

From Equation (10) and with the strains equal to the experimentally limiting values ε_1^u, ε_2^u, the maximum strain criterion becomes:

$$\begin{Bmatrix} \varepsilon_1^u \\ \varepsilon_2^u \end{Bmatrix} = \begin{bmatrix} S_{11} & S_{12} \\ S_{12} & S_{22} \end{bmatrix} \begin{Bmatrix} \sigma_1 \\ \sigma_2 \end{Bmatrix} \tag{36}$$

for the case of $\tau_{12} = 0$. Upon rearranging,

$$\sigma_2 = \frac{\varepsilon_1^u}{S_{12}} - \frac{S_{11}}{S_{12}}\sigma_1 \quad \text{and} \quad \sigma_2 = \frac{\varepsilon_2^u}{S_{22}} - \frac{S_{12}}{S_{22}}\sigma_1 \,. \tag{37}$$

Equations (37) represent two straight lines in the σ_1–σ_2 coordinate space which define failure of an orthotropic lamina. Utilization of limiting strains in tension and compression results in this failure envelope consisting of two straight lines in each quadrant. Superposition of the laminate failure envelope for varying values of τ_{xy} produces the laminate failure surface.

The linear maximum-strain strength criterion has been programmed for the digital computer (REED (1970)). The analysis is referred to as SQ-5. This program handles unsymmetrical laminates ($B_{ij} \neq 0$) subjected to inplane and bending loads. It employs limiting values of lamina longitudinal, transverse, and shear strains in tension and compression, and average lamina elastic properties (E_1, E_2, ν_{12}, G_{12}) as basic input data. Both mechanical and thermally induced loads are provided for. As output, the SQ-5 program provides complete lamina and laminate stresses and strains; laminate elastic properties (E_{xx}, E_{yy}, ν_{xy}, G_{xy}, A, B, D) and thermal coefficients of expansions; curvatures; and the coordinates of the corners of the laminate interaction diagram. Failed plies are identified and the modes of failure indicated.

5.3. Petit–Waddoups Theory

Many of the orthotropic strength criteria assume a linear elastic analysis. Those analyses which are extensions of isotropic yield theories inherently assume that mathematical yielding[3] of a ply, as predicted by some criterion, is synonymous with the ultimate rupture of that lamina. This results in either a linear stress-strain curve for the laminate, or a piecewise-linear laminate flow curve if constituent laminae fail prior to total laminate failure. Many practical composite systems actually exhibit extensive nonlinear mechanical response in shear and transverse to the reinforcement, resulting in nonlinear laminate mechanical behavior. Moreover, individual plies may have considerable load carrying capability beyond their linear range, and inaccurate strength predictions can result from a linear lamina analysis.

PETIT and WADDOUPS (1969) extended the traditional anisotropic maximum-strain strength prediction concept (Section 5.2) to provide for nonlinear mechanical response of individual plies. They represented the actual lamina stress-strain behavior by piecewise linear curves. With this method, average stresses are applied incrementally to the laminate. On application of the $(n+1)$th average laminate stress resultant increment $[\Delta N]_{n+1} = (\Delta N_x, \Delta N_y, \Delta N_{xy})_{n+1}$, the average laminate strain increment $[\Delta\varepsilon]_{n+1} = (\Delta\varepsilon_x, \Delta\varepsilon_y, \Delta\gamma_{xy})_{n+1}$ is computed using Equation (22) and assuming a symmetric laminate ($B = B' = 0$),

$$[\Delta\varepsilon]_{n+1} = [A]_n^{-1}[\Delta N]_{n+1} \qquad (38)$$

where $[A]_n^{-1}$ is the laminate compliance matrix at the end of nth load increment.[4] These laminate strain increments $[\Delta\varepsilon]_{n+1}$ are added to the previous

[3] As discussed previously, materials such as fiber-reinforced plastics, wood, etc., do not actually yield in the sense that metals do.

[4] x, y, z are principal material axes of the laminate, while 1, 2, and 3 are principal material axes of a general lamina.

laminate strains to obtain the current laminate strains, i.e.,

$$[\varepsilon]_{n+1} = [\varepsilon]_n + [\Delta\varepsilon]_{n+1} . \tag{39}$$

Individual lamina strains ε_1, ε_2 and γ_{12} are computed from the laminate strains using the strain transformations of Equation (16). These lamina strains are used in turn to determine the elastic constants from the basic lamina stress-strain curves, thereby evaluating $[A]^{-1}_{n+1}$ for use in Equation (38) at the next $(n+2)$th load increment.

The above procedure is continued until one of the lamina fails. The latter occurs when its strains attain the limiting values as determined from simple uniaxial tests. When the lamina strain in the transverse direction (transverse to fibers or grain) reaches the ultimate strain, the lamina is assumed to remain capable of carrying load only parallel to the fibers or in shear. Upon shear degradation, a lamina remains capable of carrying load parallel or transverse to the fibers. A lamina is assumed to have failed completely and to be incapable of carrying any load once longitudinal degradation occurs.

Upon failure of a particular lamina, the loads on the laminate are not relaxed. Degradation is assumed to be confined to the failed ply and its share of the load is transferred to the other laminae. This is achieved by making the tangent modulus of the failed lamina highly negative under incremented loading. When unloading of the lamina is complete, its tangent modulus is set to zero and laminate loading continues until either the stiffness $[A]$ becomes singular or achieves a negative sign along its principal diagonal.

The Petit–Waddoups analysis employs a constant compliance during each load increment and assumes the different failure modes are independent of each other. The tangent moduli employed in the compliance calculations depend only on one strain component, e.g., the tangent modulus in the fiber direction is assumed to be uninfluenced by transverse or shear strains. The analysis is confined to plane anisotropic laminates having midplane symmetry ($B_{ij} = 0$) and subjected to uniaxial or proportional biaxial loading. As is typical, classical lamination theory is utilized which implies interlaminar effects are omitted. Identical laminate stress-strain response and strength are therefore again predicted for all stacking variations of a particular composite lay-up. In actuality, stacking sequence variations of a specific lay-up can exhibit significant changes in strength.

As input information, the Petit–Waddoups method requires lamina uniaxial tensile and compressive stress-strain information parallel and transverse to the fibers, shear stress-strain data, plus the variation in the major Poisson's ratio with ε_1. This is considerable more input data than required for the maximum strain theory of Section 5.2. The complete analysis has been programmed for the digital computer (PETIT (1967)). The numerical program output provides both lamina and laminate stress-strain data up to laminate failure. Failed lamina are identified when and how they failed. Current laminate stiffness and compliance components, moduli and Poisson's ratios are printed out for each load level. The analysis is easy, versatile and expedient to employ, although care must be exercised when entering the basic lamina information.

5.4. Maximum Stress Theory

JENKINS (1920) extended the concept of the maximum normal or principal stress theory to predict the strength of planar orthotropic materials such as wood. With this theory, the stresses acting in a material are resolved into the directions of material symmetry (σ_1, σ_2, τ_{12}). It is then postulated that failure will occur when one (or all) of these stresses attains a respective maximum value X, Y, S. The latter strengths are obtained experimentally under uniaxial loading. This criterion can be stated that failure will not occur as long as the following prevail,

$$-X' < \sigma_1 < X, \qquad -Y' < \sigma_2 < Y, \qquad -S' < \tau_{12} < S \tag{40}$$

where X, X', Y, Y' and S, S' are the uniaxial tensile and compressive normal and shear strengths, respectively, with respect to the in-plane directions of material symmetry. $S = S'$ for the principal material axes so that shear strength is independent of the sign of τ_{12}. There is absolutely no interaction among modes of failure with this theory. Rupture is actually predicted by the least of the three expressions of Equation (40).

If one considers a unidirectionally reinforced laminate subjected to uniaxial tension σ at some angle θ to the fibers, then the maximum allowable loading according to this prediction is the smallest of the following:

$$\sigma = \frac{X}{\cos^2\theta}, \qquad \sigma = \frac{Y}{\sin^2\theta} \quad \text{or} \quad \sigma = \frac{S}{\sin\theta\cos\theta}. \tag{41}$$

For comparison, if strength were predicted according to the maximum strain theory of Section 5.2, then the expressions corresponding to Equations (41) become (JONES (1975))

$$\sigma = \frac{X}{\cos^2\theta - \nu_{12}\sin^2\theta}, \qquad \sigma = \frac{Y}{\sin^2\theta - \nu_{21}\cos^2\theta} \quad \text{or} \quad \sigma = \frac{S}{\sin\theta\cos\theta}. \tag{42}$$

The only difference between these maximum stress and maximum strain predictions is the inclusion of the Poisson's ratio terms in Equations (42).

Like the maximum strain criterion, the popularity of this maximum stress theory tends to be based more on simplicity of concept and ease of use than on validity or rationale.

5.5. Hill Theory

Under large deformation, the crystal structure of even initially isotropic metals becomes aligned which renders the behavior to be anisotropic. This lead HILL

(1950) to propose an orthotropic yield criterion. Under plane-stress, his theory predicts yielding would initiate when the magnitudes of the stresses reach the following condition

$$\left(\frac{\sigma_1}{X}\right)^2 + \left(\frac{\sigma_2}{Y}\right)^2 - \left(\frac{1}{X^2} + \frac{1}{Y^2} - \frac{1}{Z^2}\right)\sigma_1\sigma_2 + \left(\frac{\tau_{12}}{S}\right)^2 = 1\,, \tag{43}$$

where X, Y and Z are the yield strengths under uniaxial loading in each of the three directions of material symmetry, and S is the in-plane shearing yield strength with respect to the 1–2 axes of material symmetry. Hill assumed that the yield stresses are the same in tension and compression, i.e., there is no Bauschinger effect. While it is not common to use Equation (43) with composites, this concept does form the basis of several composite strength criteria. For isotropy, $X=Y=Z=\sigma_y$ and Equation (43) reduces to

$$\left(\frac{\sigma_1}{\sigma_y}\right)^2 - \frac{\sigma_1\sigma_2}{\sigma_y^2} + \left(\frac{\sigma_2}{\sigma_y}\right)^2 + \left(\frac{\tau_{12}}{S}\right)^2 = 1 \tag{44}$$

which is the same as the distortional energy (Von Mises) yield theory of Equation (35). Unlike the maximum stress or strain criteria, Equation (43) contains interaction among the stresses and therefore involves combined modes of failure.

5.6. *Marin Theory*

MARIN (1957) proposed an extension of the distortional energy criterion to account for differences in tensile and compressive strengths in orthotropic materials. For plane-stress, the theory predicts failure will occur if the following is satisfied:

$$K_1\sigma_1 + K_2\sigma_2 + \sigma_1^2 + K_3\sigma_1\sigma_2 + \sigma_2^2 \geq K_4 \tag{45}$$

where

$$\begin{aligned} &K_1 = X' - X\,, \qquad K_2 = \frac{XX'}{Y} - Y\,,\\ &K_3 = 2 - \frac{XX' - S[X' - X - X'(X/Y) + Y]}{S^2}\,, \qquad K_4 = XX'\,. \end{aligned} \tag{46}$$

The theory assumes that the principal directions of stress and orthotropy coincide so there is no way to account for shear stresses which may exist along the material axes. No compressive strength Y' in the σ_2-direction is introduced.

5.7. *Norris Theory*

With wood again being the orthotropic material of interest, NORRIS (1950) postulated failure would occur under plane-stress if any one of the following equations is satisfied:

$$\left(\frac{\sigma_1}{X}\right)^2 + \left(\frac{\sigma_2}{Y}\right)^2 - \frac{\sigma_1\sigma_2}{XY} + \left(\frac{\tau_{12}}{S}\right)^2 \geqslant 1\,,$$
$$\left(\frac{\sigma_1}{X}\right)^2 \geqslant 1 \quad \text{or} \quad \left(\frac{\sigma_2}{Y}\right)^2 \geqslant 1\,. \tag{47}$$

X and Y are either tensile or compressive strengths consistent with the sign of σ_1 and σ_2. As with several of the orthotropic strength criteria, the first of Equations (47) is similar in form to that of Hill's orthotropic yield theory of Equations (43). The last two expressions of Equations (47) are the same as the maximum stress theory (Section 5.4). However, these last two equations actually represent a degenerate situation of original combined-stress polynomials similar to the first of Equations (47) and for the complete three-dimensional state of stress. Although the form of Equations (47) was undoubtedly influenced by Hill's yield criterion, Norris' theory is employed to predict ultimate strength under combined loading by using uniaxial strength values for X, Y and S.

Norris' theory is one of the few strength criteria used for solid wood, but continues to be employed quite extensively with that material (GOODMAN and BODIG (1971), and KOBETZ and KRUEGER (1976)). Its application to man-made fiber-reinforced laminated composites is less prevalent.

5.8. Tsai–Hill Theory

As do most strength theories used with laminated composites, the Tsai–Hill analysis considers the lamina to be the basic unit of laminate technology.[5] Laminate moduli, and stiffness and compliance matrices are evaluated from the four independent lamina elastic properties using the standard transformation and integration procedures (Section 3). The laminate strains due to any applied load are calculated knowing these laminate elastic properties. These laminate strains are transformed to provide ε_{ij} for each lamina, from which σ_{ij} are computed. The imminence of failure of each ply is then evaluated by the Tsai–Hill failure criterion

$$\left(\frac{\sigma_1}{X}\right)^2 - \frac{\sigma_1\sigma_2}{X^2} + \left(\frac{\sigma_2}{Y}\right)^2 + \left(\frac{\tau_{12}}{S}\right)^2 = 1 \tag{48}$$

where X, Y and S are again the uniaxial strengths parallel and perpendicular to fibers, and in shear. Equation (48) was obtained from Equation (43) by assuming $Y = Z$ for fiber-reinforced composites (AZZI and TSAI (1965)). The similarity in form between the Tsai–Hill and Norris theories is noteworthy.

[5] Like many of the analyses, the Tsai–Hill theory is not invariant with respect to coordinate rotations and is consequently restricted to specially orthotropic materials. This contributes to the criterion not being well suited for direct application to complete laminates.

In practice, loading of a laminate is incremented until failure in one or more of the individual plies is indicated according to the above criterion. As employed by this writer, each ply failure is then investigated to determine whether degradation is resin or fiber induced. This decision is made as follows. If a ply fails (its state of stress reaches the Tsai–Hill failure surface) but its value of σ_1 remains less than the unaxial strength of the lamina in the fiber direction, then fracture is determined to be due to resin failure. On the other hand, if σ_1 of a failed ply exceeds the corresponding strength of the basic lamina, then fiber failure is assumed. Upon resin deterioration, the lamina is assumed to unload transversely and in shear, and to remain capable of supporting longitudinal stresses only. This is accomplished by equating E_{22} and G_{12} to close to zero.[6] If the lamina degradation is fiber in nature, total ply rupture is assumed and E_{11}, E_{22} and G_{12} are all made extremely small (~100 psi). As with most strength theories, degradation is assumed to be confined to the failed ply only. Upon modifying the elastic properties of the failed ply to reflect its damage, the effective laminate compliance and stiffness matrices are recalculated and the sustained laminate loads are carried by the remaining plies. The next load increment is applied to the laminate and the process is repeated until the laminate is considered to have failed. With this formulation, a laminate is normally considered to have failed once fiber failures have occurred in at least two plies.

The analysis version used by this writer distinguishes between tensile and compressive ply strengths parallel and transverse to the fibers. However, average tensile and compressive elastic lamina properties (E_1, E_2, ν_{12}, G_{12}) are utilized. The programmed analysis is capable of providing the complete failure surface (referred to the laminate material axes) for any increments of applied shear stress.[7] For loading along any vector in stress space, the total laminate stress-strain information is indicated to failure. Degraded plies are so indicated, but failures modes are only indicated to the extent that they are fiber or matrix in nature. Whether the matrix failures are associated with limiting transverse normal or shear conditions is not explicitly defined.

It should be emphasized that whereas the maximum stress and strain theories involve several separate equations and no interaction of the stresses or failure modes, the Tsai–Hill quadratic criterion does provide for interaction. However, the interaction is fixed. Subsequent theories will be reviewed which have independent interaction among all stress components. TSAI and WU (1971) subsequently demonstrated shortcomings of the Tsai–Hill type expressions due to Equation (48) not being invariant.

5.9. *Gol'denblat and Kopnov Theory*

GOL'DENBLAT and KOPNOV (1965) seem to be the first to recognize that a

[6] Say to 100 psi; to equate the moduli to zero would result in the compliance matrix becoming singular.

[7] Computer program prepared by B.W. Cole, Amoco Chemical Corporation, Naperville, IL 60540, USA.

strength theory should be independent of the coordinate system. Under plane-stress, their criterion is as follows:

$$f_1\sigma_1 + f_2\sigma_2 + \sqrt{f_{11}\sigma_1^2 + 2f_{12}\sigma_1\sigma_2 + f_{22}\sigma_2^2 + f_{66}\tau_{12}^2} = 1\,, \tag{49}$$

where

$$\begin{aligned} f_1 &= \frac{1}{2}\left(\frac{1}{X} - \frac{1}{X'}\right), \qquad f_2 = \frac{1}{2}\left(\frac{1}{Y} - \frac{1}{Y'}\right), \\ f_{11} &= \frac{1}{4}\left(\frac{1}{X} + \frac{1}{X'}\right)^2, \qquad f_{22} = \frac{1}{4}\left(\frac{1}{Y} + \frac{1}{Y'}\right)^2 \\ f_{12} &= \frac{1}{8}\left\{\left(\frac{1}{X} + \frac{1}{X'}\right)^2 + \left(\frac{1}{Y} + \frac{1}{Y'}\right)^2 - \left(\frac{1}{S_{45}} + \frac{1}{S'_{45}}\right)^2\right\} \\ f_{66} &= \left(\frac{1}{S}\right)^2 \end{aligned} \tag{50}$$

and S_{45} and S'_{45} are the shear strengths at 45° to the axes of material symmetry. Equation (49) accounts for independent interaction through f_{12} as well as providing for differences in the tensile and compressive shear strengths S_{45} and S'_{45}, respectively. An inconvenience of the theory is that either biaxial normal tests or 45° shear tests must be conducted to evaluate the S_{45} and S'_{45} terms of coefficient f_{12}.

Equation (49) was found to correlate quite well with biaxial experimental data in the first and fourth quadrant as obtained from loaded fiberglass tubes (PROTASOV and KOPNOV (1965)).

5.10. Ashkenazi Theory

Again inspired by the anisotropic behavior of wood and rolled metal. ASHKENAZI (1959, 1965, 1967, 1970, 1976) also formulated an orthotropic strength criterion based on an extension of Von Mises' plasticity theory. Under plane stress, Ashkenazi's theory predicts failure will occur at a material point when the stresses satisfy or exceed the following expression:

$$\left(\frac{\sigma_1}{X}\right)^2 + \left(\frac{\sigma_2}{Y}\right)^2 + \left(\frac{\tau_{12}}{S}\right)^2 + 2f_{12}^*\sigma_1\sigma_2 = 1\,. \tag{51}$$

He suggested evaluating the interaction coefficient f_{12}^* by the use of a 45-degree specimen such that

$$\sigma_1 = \sigma_2 = \tau_{12} = \tfrac{1}{2}U\,, \tag{52}$$

where U is the longitudinal strength of the 45-degree off-axis coupon. The value of f_{12}^* is obtained from Equations (51) and (52),

$$f_{12}^{*} = \frac{1}{2}\left[\frac{4}{U^2} - \frac{1}{X^2} - \frac{1}{Y^2} - \frac{1}{S^2}\right]. \tag{53}$$

With this value of f_{12}^{*}, Equation (51) can be rewritten as

$$\left(\frac{\sigma_1}{X}\right)^2 + \left(\frac{\sigma_2}{Y}\right)^2 + \left(\frac{\tau_{12}}{S}\right) + \left[\frac{4}{U^2} - \frac{1}{X^2} - \frac{1}{Y^2} - \frac{1}{S^2}\right]\sigma_1\sigma_2 = 1. \tag{54}$$

The similarity between the form of Equation (54) and those of Hill, Norris and Tsai–Hill is obvious. Ashkenazi found that for non-highly anisotropic materials such as plywood, the second degree polynomial of Equation (54) is adequate. However, for very highly directional materials, the following fourth-order polynomial is suggested:

$$\begin{aligned}
&\left\{\frac{\sigma_1^2}{X} + \frac{\sigma_2^2}{Y} + \sigma_1\sigma_2\left(\frac{4}{U} - \frac{1}{X} - \frac{1}{Y} - \frac{1}{S}\right) + \frac{\tau_{12}^2}{S}\right\}^2 \\
&+ 2\frac{\sigma_1\sigma_2 - \tau_{12}^2}{S}\left[\sigma_1\sigma_2\left(\frac{1}{X} + \frac{1}{Y}\right) + \frac{\sigma_1^2}{X} + \frac{\sigma_2^2}{Y}\right] \\
&-(\sigma_1\sigma_2 - \tau_{12}^2)[\sigma_1\sigma_2(\lambda + \mu) + \lambda\sigma_1^2 + \mu\sigma_2^2] \\
&+ \rho(\sigma_1\sigma_2 - \tau_{12}^2)(\sigma_1 + \sigma_2) - (\sigma_1^2 + \sigma_2^2 + \sigma_1\sigma_2 + \tau_{12}^2) = 0.
\end{aligned} \tag{55}$$

Coefficients λ, μ and ρ must be determined experimentally under three different biaxial states of stress. Equation (55) is valid for $(\sigma_1 + \sigma_2) \geqslant 0$. A similar expression applies when $(\sigma_1 + \sigma_2) \leqslant 0$ but the tensile strengths must now be replaced by compressive strengths and the coefficients λ, μ, ρ will acquire new values. This criterion involves the experimental determination of thirteen quantities, six of which must be evaluated under biaxial states of stress. Such extensive testing distracts from the usefulness of Equation (55).

ASHKENAZI and PEKKER (1970) subsequently proposed the slightly simpler expression

$$\begin{aligned}
&\left\{\frac{\sigma_1^2}{X} + \frac{\sigma_2^2}{Y} + \sigma_1\sigma_2\left(\frac{4}{U} - \frac{1}{X} - \frac{1}{Y} - \frac{1}{S}\right) + \frac{\tau_{12}^2}{S}\right\}^2 \\
&= \sigma_1^2 + \sigma_2^2 + \sigma_1\sigma_2 + \tau_{12}^2 = I_1^2 - I_2
\end{aligned} \tag{56}$$

where I_1, I_2 are stress invariants. Equation (56) comes from Equation (55) by setting

$$\lambda = \frac{2}{XS}, \qquad \mu = \frac{2}{YS} \quad \text{and} \quad \rho = 0. \tag{57}$$

ASHKENAZI and PEKKER (1970) correlated Equation (56) with data from internally pressurized glass-composite tubes subjected to longitudinal tension, compression, and torsion.

5.11. Malmeister Theory

MALMEISTER (1966) postulated a strength theory of the form

$$F_{ij}\sigma_{ij} + F_{ijkl}\sigma_{ij}\sigma_{kl} + \cdots = 1 . \tag{58}$$

Under plane-stress, he proposed the expression

$$A_1^*\sigma_1 + A_2^*\sigma_2 + A_{11}^*\sigma_1^2 + 2A_{12}^*\sigma_1\sigma_2 + A_{22}^*\sigma_2^2 + A_{66}^*\tau_{12}^2 = 1 , \tag{59}$$

where

$$\begin{aligned} A_1^* &= \frac{1}{X} - \frac{1}{X'}, \quad A_{11}^* = \frac{1}{XX'}, \quad A_{66}^* = \frac{1}{S^2}, \\ A_2^* &= \frac{1}{Y} - \frac{1}{Y'}, \quad A_{22}^* = \frac{1}{YY'} . \end{aligned} \tag{60}$$

By assuming testing is conducted in the principal stress directions ($\tau_{12} = \tau_{pq} = 0$) and that the ultimate stresses reduce to pure shear upon rotating the axes through 45°, i.e., $\sigma_p = -\sigma_q = S'_{45}$, Equation (59) becomes

$$A_1^*\sigma_p + A_2^*\sigma_q + A_{11}^*\sigma_p^2 + 2A_{12}^*\sigma_p\sigma_q + A_{22}^*\sigma_q^2 = 1 , \tag{61}$$

so that

$$2A_{12}^* = \frac{A_1^* - A_2^*}{S'_{45}} + A_{11}^* + A_{22}^* - \frac{1}{(S'_{45})^2} . \tag{62}$$

Quantities X, Y and S remain the normal and shear strengths and the prime denotes compression. Equations (59) and (61) account directly for different strengths in tension and compression.

Although the criteria by Gol'denblat and Kopnov, Ashkenazi and Malmeister do not find widespread application as such in the western world, variations of Equation (58) have been employed extensively in the Soviet Union (UPITIS et al. (1974), UPITIS and RIKARD (1976), PLUME and MAKSIMOV (1978), SOKOLOV et al. (1978), RIKARDS et al. (1979), SOKOLOV (1979), and MAKSIMOV et al. (1979)).

5.12. Hankinson Formula

R.L. HANKINSON (1921) arrived at the following expression for describing the strength of wood

$$U'_\theta = \frac{X'Y'}{X' \sin^2\theta + Y' \cos^2\theta} , \tag{63}$$

where X' and Y' are the compressive strengths parallel and transverse to the grain, and U'_θ is the compressive strength in a direction inclined at an angle θ to the grain. From test data on spruce, fir and pine, Hankinson observed that this empirical expression applied to the compressive elastic limit as well as to the compressive ultimate strength. ROWSE (1923), and GOODMAN and BODIG (1971) have since demonstrated the usefulness of this theory to wood. The Hankinson formula bears some similarities to the maximum stress and strain expressions of Equations (41) and (42). Recognizing the difficulties in obtaining reliable values of the shear strength, it is significant that Equation (63) does not involve S.

Although the Hankinson formula of Equation (63) is not a general anisotropic failure criterion for completely arbitrary loading, it does play an important role in a recent theory by COWIN (1979), Section 5.14. Applications of Equation (63) to man-made fiber-reinforced composites are unknown to the author.

5.13. Tsai–Wu Theory

In an effort to more adequately predict experimental results, TSAI and WU (1971) proposed a lamina failure criterion having additional stress terms not appearing in theories such as the Hill analysis. Influenced by the work of Soviet scientists, they assumed a failure surface in stress space of the form

$$f(\sigma) = F_i\sigma_i + F_{ij}\sigma_i\sigma_j = 1\,, \quad i, j = 1, 2, \ldots, 6\,. \tag{64}$$

The F_i and F_{ij} are second and fourth order lamina strength tensors. The linear stress terms account for possible differences in tensile and compressive strengths. The quadratic stress terms are similar to those in the Tsai–Hill formulation, and describe an ellipsoid in stress space. The F_{ij} $(i \neq j)$ terms are new. Off-diagonal terms of the strength tensor provide independent interactions among the stress components. Under plane-stress conditions, this failure criterion becomes

$$F_1\sigma_1 + F_2\sigma_2 + F_6\sigma_6 + F_{11}\sigma_1^2 + F_{22}\sigma_2^2 + 2F_{12}\sigma_1\sigma_2 + F_{66}\sigma_6^2 = 1\,, \tag{65}$$

where

$$\begin{aligned} &F_1 = \frac{1}{X} - \frac{1}{X'}\,, \qquad F_2 = \frac{1}{Y} - \frac{1}{Y'}\,, \qquad F_6 = \frac{1}{S} - \frac{1}{S'}\,, \\ &F_{11} = \frac{1}{XX'}\,, \qquad F_{22} = \frac{1}{YY'}\,, \qquad F_{66} = \frac{1}{SS'}\,. \end{aligned} \tag{66}$$

X, Y, S and X', Y', S' remain the lamina longitudinal, transverse and shear strengths in tension and compression, respectively.[8] These strength values are

[8] In this theory, as with those of maximum stress, Gol'denblat, Malmeister, and Hankinson, etc., compressive strengths X', Y' and S' are taken as positive numbers.

not sufficient to determine coefficients such as F_{12}. For its determination, biaxial tests are required. The latter have to be selected carefully to obtain accurate values for such interaction terms (WU (1972)). Stability conditions require that $F_{ii}F_{jj}-F_{ij}^2 \geqslant 0$ (repeated indices do not imply summations). WU (1972) has shown that for graphite epoxy, F_{12} may be considered zero if it falls within $\pm 0.6 \times 10^{-4}$. Shear strength in principal material directions is independent of sign, i.e., $S' = S$ and $F_6 = 0$ for these orientations.

The Tsai–Wu tensor strength theory is more general than say the Tsai–Hill analysis. Specific advantages include:

(1) invariant under rotation of coordinates,

(2) transforms according to established tensorial laws,

(3) symmetrical strength properties akin to those of stiffnesses and compliances, and

(4) provides independent interactions among stress components.

In strength theories such as the Tsai–Hill analysis, the stress interactions are fixed or implied (not independent). In the maximum stress or maximum strain criteria, simultaneous equations are required and interactions are not included. Like the modified Tsai–Hill criterion, the Tsai–Wu analysis employs tensile and compressive lamina strengths. While it is a stress criterion, linear-elastic lamina response is typically assumed in the accompanying lamination theory. The criterion predicts the imminency of failure but nothing about the failure mode (tensile or compressive longitudinal, transverse or shear). In the formulation employed by this writer, fiber or matrix failure and the ply of the occurrence are predicted analogously as with the previously described Tsai–Hill analysis. The most inconvenient aspect of this theory is the determination of F_{12}.

COWIN (1979) subsequently formulated a similar theory although he again restricted the interaction by expressing F_{12} in terms of uniaxial normal and shear strengths. TENNYSON et al. (1978, 1980) extended the Tsai–Wu concept to include cubic terms.

The Tsai–Wu theory has received quite extensive use, e.g., HERAKOVICH and O'BRIEN (1979) have recently employed the three-dimensional form of Equation (65) to analyze damage zones in a composite.

5.14. Cowin Theory

COWIN (1979) again proposed the criterion of Equation (65),

$$F_1\sigma_1 + F_2\sigma_2 + F_{11}\sigma_1^2 + 2F_{12}\sigma_1\sigma_2 + F_{22}\sigma_2^2 + F_{66}\sigma_{12}^2 = 1 \tag{67}$$

where F_1, F_2, F_{11}, F_{22} and F_{66} are given by Equations (66). However, by using a Hankinson-type expression (Equation (63)), he proposed the interaction coefficient to be given by

$$F_{12} = \sqrt{F_{11}}\sqrt{F_{22}} - \frac{1}{2S^2}\,. \tag{68}$$

This analysis has been demonstrated for only specially orthotropic materials and the interaction between stresses is again restrictive. However, that no biaxial testing is required does make the criterion attractive from a user's viewpoint. Cowin applied his analysis to bone.

5.15. Tennyson Theory

TENNYSON et al. (1978, 1980) extended Tsai–Wu analysis of Section 5.13 to include cubic terms, i.e.,

$$F_1\sigma_1 + F_2\sigma_2 + F_{11}\sigma_1^2 + 2F_{12}\sigma_1\sigma_2 + F_{22}\sigma_2^2 + F_{66}\tau_{12}^2 + 3F_{112}\sigma_1^2\sigma_2 + 3F_{221}\sigma_2^2\sigma_1 + 3F_{166}\sigma_1\tau_{12}^2 + 3F_{266}\sigma_2\tau_{12}^2 = 1\,. \tag{69}$$

Coefficients F_1, F_2, F_{11}, F_{22} and F_{66} continue to be given by Equations (66). Tennyson evaluates the interaction coefficients F_{12}, F_{112}, F_{221}, F_{166} and F_{266} from biaxial test data and constraint equations.

5.16. Hoffman Theory

HOFFMAN (1967) proposed the following Tsai-type condition to predict rupture of brittle orthotropic materials and which accounts for different strengths in tension and compression:

$$F_1\sigma_1+F_2\sigma_2+F_{11}\sigma_1^2+2B_{12}^*\sigma_1\sigma_2+F_{22}\sigma_2^2+F_{66}\tau_{12}^2=1\,. \tag{70}$$

Coefficients F_1, F_2, F_{11}, F_{22} and F_{66} are given by Equations (66), while

$$B_{12}^* = -\frac{1}{2XX'}\,. \tag{71}$$

The expressions of Equations (65), (67) and (70) are identical except for the interaction terms. The Tsai–Wu analysis remains invariant under rotation of coordinates while those by Cowin and Hoffman assume specially orthotropic materials. Although determination of F_{12} of Equation (65) necessitates biaxial data, the fixed interaction coefficients of the Cowin and Hoffman criteria require only uniaxial test information. It is not uncommon for persons to employ Equation (70) but with $B_{12}^*=0$ (NARAYANASWAMI and ADELMAN (1977)).

5.17. Fischer Theory

Another orthotropic strength criterion of the Norris- or Hill-type is the following proposed by FISCHER (1960, 1967):

$$\left(\frac{\sigma_1}{X}\right)^2 + \left(\frac{\sigma_2}{Y}\right)^2 + \left(\frac{\tau_{12}}{S}\right)^2 - K\frac{\sigma_1\sigma_2}{XY} = 1\,, \tag{72}$$

where

$$K = \frac{E_{11}(1+\nu_{21}) + E_{22}(1+\nu_{12})}{2\sqrt{E_{11}E_{22}(1+\nu_{12})(1+\nu_{21})}} \tag{73}$$

Equation (72) assumes specially orthotropic materials. For $K = 1$, Equation (72) reduces to the Norris theory, Equation (47). If the elastic constants are not known, K may be determined from strength data (CHAMIS (1969)), Section 5.18.

5.18. Chamis Theory

CHAMIS (1969) proposed a composite failure criterion which, in addition to constituent content, is intended to account for fabrication variables such as filament and void content and distribution, fiber-matrix bonding, residual stresses, etc. Whereas Fischer introduced one correlation coefficient, Chamis introduced two such coefficients. The lamina failure theory is expressed as

$$f(\sigma, K_{l12}) = 1 - \left[\left(\frac{\sigma_{1\alpha}}{X_\alpha}\right)^2 + \left(\frac{\sigma_{2\beta}}{Y_\beta}\right)^2 + \left(\frac{\tau_{12}}{S}\right)^2 - K'_{l12\alpha\beta}K_{l12}\frac{\sigma_{1\alpha}\sigma_{2\beta}}{|X_\alpha||Y_\beta|}\right] \tag{74}$$

where $f > 0$ indicates no failure, $f = 0$ onset of failure, and $f < 0$ signifies the failure condition has been exceeded. The coefficient K_{l12} is defined as

$$K_{l12} = \frac{(1 + 4\nu_{12} - \nu_{13})E_{22} + (1 - \nu_{23})E_{11}}{[E_{11}E_{22}(2 + \nu_{12} + \nu_{13})(2 + \nu_{21} + \nu_{23})]^{1/2}} \tag{75}$$

and $K'_{l12\alpha\beta}$ is a theory–experiment correlation coefficient. Subscripts α and β denote tension or compression and l signifies a layer property. For an isotropic material, Equation (74) becomes the Von Mises criterion. The $K'_{l12\alpha\beta}$ coefficient accounts for different tensile and compressive strengths, in addition to providing for variable interaction between stresses. It can have different values in the different quadrants. Equation (74) is applicable to any generally orthotropic material exhibiting elastic symmetry. This criterion can be combined with lamination theory and applied to multidirectional laminates.

5.19. Sandu Theory

A composite strength analysis proposed by SANDHU (1974a, b) differs from the more conventional formulations in the following respects:

(1) the nonlinear lamina stress-strain responses to failure are represented analytically by cubic-spline functions, the tangent moduli of which are employed to evaluate lamina and laminate stiffnesses and compliances during load increments;

(2) in keeping with the nonlinear provision, ply degradation is based on an energy to failure criterion, and

(3) equivalent strain increments are defined.

On application of the $(n+1)$th laminate load increment, initial laminate and lamina strain increments are computed (Equation (38)) using a compliance matrix $[A]_n^{-1}$ based upon tangent moduli occurring at the end of the previous nth load increment. The strain increments are added to those at the end of the previous nth increment to obtain current strains. PETIT and WADDOUPS' (1969) earlier use of the lamina strains ε_1 and ε_2 developed under biaxial loading to determine the tangent moduli from stress-strain data generated under simple uniaxial loading is not totally rigorous. Sandhu attempts to compensate for the biaxial effect now occurring by defining equivalent strains. Using these corrected strains, average elastic properties of the plies are determined and a new $[A]^{-1}$ is computed with which to evaluate new strain increments. This procedure is repeated until the difference between values of the strain increments in two consecutive cycles is small to the order of approximation desired. Increments in lamina stresses are obtained from these improved strain increments and tangent moduli [Equation (5) written in terms of increments]. The final stress and strain increments are added to those at the end of the previous nth increment to obtain current values and the strain energy. Repetition of this procedure under incremental loading generates the response of the laminate until the accumulated strain energy of a lamina reaches its limiting value.

Laminae ultimately reach a state of degradation determined by the assumed criterion, which, under plane-stress conditions, is

$$f(\sigma, \varepsilon) = K_i \left[\int^{\hat{\varepsilon}_i} \sigma_i \, \mathrm{d}\varepsilon_i \right]^{m_i} = 1 \quad (i = 1, 2, 6) \tag{76}$$

where $\hat{\varepsilon}_1$, $\hat{\varepsilon}_2$ and $\hat{\varepsilon}_6 = \gamma_{12}$ are the current strain components, and repeated indices imply summation. The failure criterion is based on total strain energy, including the effects of hydrostatic loading. The latter is included to account for the heterogeneous deviatoric response produced in composites by hydrostatic loading. Strain energies parallel and transverse to the fibers, and in shear are assumed to be independent. In the absence of experimental data, it is assumed typically that $m_1 = m_2 = m_6 = 1$. By testing under simple load conditions, the K's become

$$K_i = \left[\int^{\varepsilon_i^{\mathrm{u}}} \sigma_i \, \mathrm{d}\varepsilon_i \right]^{-1} \quad (i = 1, 2, 6) \tag{77}$$

where $\varepsilon_{1,2}^{\mathrm{u}}$ and $\varepsilon_6^{\mathrm{u}} = \gamma_{12}^{\mathrm{u}}$ are the ultimate normal (tensile or compressive) and shear strains. When a lamina reaches the defined state of failure, laminate loads are maintained but the lamina is assumed to unload. Under either transverse or shear impairment (characterized by matrix failure), the affected lamina is assumed to transfer both its transverse and shear loads to the remaining laminae. It continues to carry longitudinal load. In the case of fiber failure, total unloading of the failed ply is assumed. Transfer of loads under incremental laminate loading is achieved by making the affected moduli initially negative. On complete unloading of the damaged ply, the moduli are equated to zero. This process of load transfer under incremental loading is assumed to continue until general failure of the laminate occurs.

Sandhu's development is similar in many respects to that by PETIT and WADDOUPS (1969). However, the latter bases lamina failure on maximum strains rather than on energy, and approximates the nonlinear stress-strain data by piecewise linear curves. Like the Petit analysis, Sandhu's method requires the complete nonlinear lamina tensile and compressive stress-strain data under longitudinal, transverse and shear loading as input. Both analyses assume total laminate failure to occur when the A_{ij} matrix becomes singular or when a negative sign appears along the main diagonal of A_{ij}. Complete laminate stresses and strains to failure are provided. Failure modes (resin or fiber) and the degradated plies are identified. A slight inconvenience of the Sandhu analysis is the experimental evaluation of shape factors m_1, m_2 and m_6. To date, these three factors have been assumed equal to one due to lack of adequate biaxial experimental information.

5.20. Griffith–Baldwin Theory

Numerous anisotropic failure analyses are patterned after Von Mises' isotropic yield criterion but which are not explicitly restricted to elastic response. GRIFFITH and BALDWIN (1962), assuming failure to be independent of hydrostatic stress, extended the distortional-energy concept to include general orthotropy. For plane stress, the elastic distortional energy U_D is

$$U_D = \frac{\sigma_1^2}{3}\left(S_{11} - \frac{S_{12}+S_{13}}{2}\right) + \frac{\sigma_2^2}{3}\left(S_{22} - \frac{S_{12}+S_{23}}{2}\right) + \frac{\sigma_1\sigma_2}{3}\left(2S_{12} - \frac{S_{11}+S_{22}+S_{13}+S_{23}}{2}\right) + S_{66}\tau_{12}^2 . \tag{78}$$

Fracture is assumed to occur when U_D of Equation (78) reaches some critical value as obtained from a simple uniaxial test. Although the theory is based on identical tensile and compressive strengths and Hookean behavior to rupture, Equation (78) correlated quite well with some experimental data.

5.21. Puppo–Evensen Theory

Probably partly to account for interlaminar effects, PUPPO and EVENSEN (1972) postulated an anisotropic strength theory applicable directly to the laminate as a whole, not just the individual laminae. Such an approach eliminates the need to compute stress distributions in the individual plies. They pointed out that the analytical interaction function describing the strength of a multidirectional laminate must be a tensor statement. Hill-type criteria do not fulfill this requirement and hence are generally applicable only to individual plies of a multidirectional laminate. Puppo and Evensen appear to have been motivated by observed behavior of orthotropic composites which lies between the following two extreme cases: that of isotropic materials on one hand and non-interacting fabric-like materials exhibiting minimal shear strength on the other hand. Predictions for ideal non-interacting fabric materials are identical to those obtained from 'netting analysis' (shear effects ignored). The authors assert that no single continuous closed surface of the Hill-type is capable of transition between these two extreme cases, but rather the criterion must be stated in terms of several equations representing intersecting surfaces in stress space.

A failure criterion of tensor form applicable to homogeneous or laminated anisotropic materials was formulated by introducing two new concepts: interaction factors and principal strength axes. For the prevalent plane-stress case, the interaction factor $\dot{\gamma}$ is defined to be

$$\dot{\gamma} = \frac{3S^2}{XY} \tag{79}$$

where X, Y and S are the basic normal and shear strengths determined relative to a general x, y coordinate system. The principal strength axes of a material are those for which the interaction parameter $\dot{\gamma}$ is a relative minimum with respect to rotation. For an ideal noninteracting fabric material, $\dot{\gamma}$ approaches zero, the failure curve is a rectangle and the principal strength axes are parallel and perpendicular to the fibers. For an isotropic material obeying the Von Mises criterion, $\dot{\gamma} = 1$.

Under plane stress, Puppo and Evensen proposed failure occurs when the stress vector reaches the following interacting surfaces:

for $\dot{\gamma}' < 1$

$$\begin{aligned} \left(\frac{\sigma_\xi}{\Xi}\right)^2 - \dot{\gamma}'\frac{\sigma_\xi\sigma_\eta}{(\Pi)^2} + \dot{\gamma}'\left(\frac{\sigma_\eta}{\Pi}\right)^2 + \left(\frac{\tau_{\xi\eta}}{S_{\xi\eta}}\right)^2 = 1\,, \\ \dot{\gamma}'\left(\frac{\sigma_\xi}{\Xi}\right)^2 - \dot{\gamma}'\frac{\sigma_\xi\sigma_\eta}{(\Xi)^2} + \left(\frac{\sigma_\eta}{\Pi}\right)^2 + \left(\frac{\tau_{\xi\eta}}{S_{\xi\eta}}\right)^2 = 1\,. \end{aligned} \tag{80}$$

Similar type expressions exist for $\dot{\gamma}' > 1$. The specific interaction factor $\dot{\gamma}'$ is that with respect to the principal strength axes ξ, η; i.e.,

$$\dot{\gamma}' = \frac{3S_{\xi\eta}^2}{(\Xi)(\Pi)} \tag{81}$$

where Ξ, Π and $S_{\xi\eta}$ are the laminate normal and shear strengths relative to the principal strength axes ξ and η. For an isotropic material, Equation (80) degenerates to Von Mises' distortional energy expression. Under hydrostatic stress, the Puppo–Evensen failure surface is never reached. This implies no material failure occurs even though distortions could exist.

This criterion can be applied to laminated or homogeneous anisotropic materials. The principal strength axes do not necessarily coincide with the axes of mechanical orthotropy, nor must they even be orthogonal to each other. When the strength properties are defined with respect to a set of axes of orthotropy, it is not necessary to compute the principal strengths, nor must one locate a set of principal strength axes. The magnitude of $\dot{\gamma}$ must be considered, however, when choosing the particular form of the strength criterion.

5.22. Wu–Scheublein Theory

Like PUPPO and EVENSEN, WU and SCHEUBLEIN (1974) sought a direct laminate strength criterion. They proposed representing the laminate failure surface by the following tensor polynomial[9]

$$f(\sigma) = \bar{F}_i\sigma_i + \bar{F}_{ij}\sigma_i\sigma_j + \bar{F}_{ijk}\sigma_i\sigma_j\sigma_k = 1 . \tag{82}$$

For an orthotropic laminated composite under plane-stress, considerations of path independency, symmetry, redundancy and the assumption of identical plus and minus shear strengths with respect to the axes of material symmetry reduces the number of coefficients of Equation (82) to ten. These are

$$\bar{F}_1, \bar{F}_2; \bar{F}_{11}, \bar{F}_{22}, \bar{F}_{66}, \bar{F}_{12}; \bar{F}_{122}, \bar{F}_{112}, \bar{F}_{266} \text{ and } \bar{F}_{166} .$$

Of the ten laminate coefficients, experimental values of $\bar{F}_1$, $\bar{F}_2$, $\bar{F}_{11}$, $\bar{F}_{22}$ and $\bar{F}_{66}$ can be evaluated directly from uniaxial laminate tension, compression and shear tests analogously to the lamina situation of Section 5.13. The remaining strength terms $\bar{F}_{12}$, $\bar{F}_{112}$, $\bar{F}_{122}$, $\bar{F}_{266}$, $\bar{F}_{166}$ of Equation (82) characterize independent interactions among various stress components. To ensure that inherent material scatter of the laminate properties does not mask evaluations of these remaining strength coefficients, they must be evaluated under prescribed optimum stress ratios. Unfortunately, these stress ratios depend upon the sought coefficients, necessitating iterative procedures for their determination. The coefficients $\bar{F}_{12}$, $\bar{F}_{112}$ and $\bar{F}_{122}$ represent strength interaction in the σ_x-σ_y laminate stress space. They can be determined only through iteration as a group and under optimum ratios of stress biaxiality. Similarly, $\bar{F}_{166}$ and $\bar{F}_{266}$ represent strength interactions in σ_x-τ_{xy} and σ_y-τ_{xy} space, respectively, and their determinations must be made iteratively and under particular ratios of stress biaxiality. As a starting point, and to minimize the number of iterations required (hence the extent of experimental testing needed), Wu suggests that initial estimates of $\bar{F}_{12}$, $\bar{F}_{122}$, $\bar{F}_{112}$, $\bar{F}_{166}$ and $\bar{F}_{266}$ be obtained by curve fitting Equation (82) to the laminate failure surface generated from the tensor lamina criterion of Equation (65) and classical lamination theory (Section 3). Basic lamina strengths and elastic properties would be employed in such an analysis. These initial estimates of the laminate coupling strength coefficients can then be used to approximate the optimum stress ratios under which these same

[9] This cubic concept is not totally unlike, and preceded, that by Tennyson (Section 5.15).

tensor coefficients should be evaluated. While the process can be repreated to obtain adequately accurate values for these interaction terms, this concept of obtaining initial values should, hopefully, minimize the number of iterations required.

This proposed hybrid laminate strength criterion is relatively complete mathematically, but utilizes rather extensive experimental testing. At present, the optimum ratio for conducting the biaxial experiments at best provides guidance for the iterations and cannot eliminate iteration. Like the Tsai–Wu lamina criterion, accurate evaluation of the interaction strength coefficients presupposes the ability to physically load composite tubes under virtually any condition of biaxiality. Few organizations can make this claim.

5.23. Closure

Just as anisotropic materials need more elastic constants to describe their constitutive behavior than do isotropic materials, the former media require more strength quantities than do the latter. Designing with orthotropic plies necessitates knowing longitudinal and transverse strengths, plus shear strength S. Normal tensile X, Y and compressive $X'Y'$ strengths are often sufficiently different that both must be included. Moreover, usually $S \neq S'$ except with respect to axes of material symmetry.

Neither the maximum stress nor maximum strain theories involve interactions or mixed-mode failure: most anisotropic criteria do, Table 2. The lamina strength theories of Table 2 are of the form $F_i\sigma_i + (F_{ij}\sigma_i\sigma_j)^\alpha = 1$ with $\alpha = 1$ or $\frac{1}{2}$. $F_i = 0$ in several cases. The analyses are written here with respect to the axes of material symmetry such that $S = S'$ and $F_6 = 0$. Although many of the theories of Table 2 have a dependent interaction term F_{12}, that of the Tsai–Wu criterion is independent and must be evaluated under biaxial loading. This need to determine F_{12} from biaxial data can be inconvenient, especially if the necessary testing facilities are not available. The analyses by Cowin, Hoffman and Malmeister are similar to that by Tsai–Wu except with them F_{12} is evaluated explicitly in terms of uniaxial strengths. F_{12} of the Cowin and Malmeister predictions involve shear strengths, but that of the Hoffman theory does not. Recognizing the difficulties in determining reliable values of the interaction coefficient of the Tsai–Wu theory could motivate F_{12} being evaluated instead according to the analysis by either of Cowin, Malmeister or Hoffman. This is particularly so if such a computed value of F_{12} satisfies the inequality $F_{12}^2 < F_{11}F_{22}$. That these latter analyses are not necessarily invariant with respect to coordinate rotation is not denied. If $F_{12}^2 < F_{11}F_{22}$ predicts a relatively small F_{12}, it is not uncommon to ignore the interaction term altogether. The criterion equations by Cowin, Hoffman, Malmeister and Tsai–Wu are then the same. In the absence of other data, TSAI and HAHN (1980) have suggested taking $F_{12} = -0.5\sqrt{F_{11}F_{22}}$. While recognizing the difficulties in evaluating F_{12}, one must realize that relatively small changes in F_{12} can significantly affect the predicted strength (e.g., TSAI and HAHN 1980).

TABLE 2. Lamina strength theories having stress interaction.

Theory	Criterion equation	F_1	F_2	F_{11}	F_{12}	F_{22}	F_{66}
Ashkenazi	$F_{ij}\sigma_i\sigma_j = 1$	–	–	$\frac{1}{X^2}$	$\frac{1}{2}\left[\frac{4}{U^2} - \frac{1}{X^2} - \frac{1}{Y^2} - \frac{1}{S^2}\right]$	$\frac{1}{Y^2}$	$\frac{1}{S^2}$
Chamis	$F_{ij}\sigma_i\sigma_j = 1$	–	–	$\frac{1}{X^2}$	$-\frac{K'_l K_l}{2XY}$	$\frac{1}{Y^2}$	$\frac{1}{S^2}$
Cowin	$F_i\sigma_i + F_{ij}\sigma_i\sigma_j = 1$	$\frac{1}{X} - \frac{1}{X'}$	$\frac{1}{Y} - \frac{1}{Y'}$	$\frac{1}{XX'}$	$\sqrt{F_{11}F_{22}} - \frac{1}{2S^2}$	$\frac{1}{YY'}$	$\frac{1}{S^2}$
Fischer	$F_{ij}\sigma_i\sigma_j = 1$	–	–	$\frac{1}{X^2}$	$-\frac{K}{2XY}$	$\frac{1}{Y^2}$	$\frac{1}{S^2}$
Gol'denblat–Kopnov	$F_i\sigma_i + \sqrt{F_{ij}\sigma_i\sigma_j} = 1$	$\frac{1}{2}\left(\frac{1}{X} - \frac{1}{X'}\right)$	$\frac{1}{2}\left(\frac{1}{Y} - \frac{1}{Y'}\right)$	$\frac{1}{4}\left(\frac{1}{X} + \frac{1}{X'}\right)^2$	$\frac{1}{8}\left[\left(\frac{1}{X} + \frac{1}{X'}\right)^2 + \left(\frac{1}{Y} + \frac{1}{Y'}\right)^2 - \left(\frac{1}{S_{45}} + \frac{1}{S'_{45}}\right)^2\right]$	$\frac{1}{4}\left(\frac{1}{Y} + \frac{1}{Y'}\right)^2$	$\frac{1}{S^2}$
Hoffman	$F_i\sigma_i + F_{ij}\sigma_i\sigma_j = 1$	$\frac{1}{X} - \frac{1}{X'}$	$\frac{1}{Y} - \frac{1}{Y'}$	$\frac{1}{XX'}$	$-\frac{1}{2XX'}$	$\frac{1}{YY'}$	$\frac{1}{S^2}$
Malmeister	$F_i\sigma_i + F_{ij}\sigma_i\sigma_j = 1$	$\frac{1}{X} - \frac{1}{X'}$	$\frac{1}{Y} - \frac{1}{Y'}$	$\frac{1}{XX'}$	$\frac{1}{2}\left[\frac{1}{S'_{45}}(F_1 - F_2) + F_{11} + F_{22} - \left(\frac{1}{S'_{45}}\right)^2\right]$	$\frac{1}{YY'}$	$\frac{1}{S^2}$
Marin	$F_i\sigma_i + F_{ij}\sigma_i\sigma_j = 1$	$\frac{1}{X} - \frac{1}{X'}$	$\frac{XX' - YY}{XX'Y}$	$\frac{1}{XX'}$	$\frac{1}{XX'} - \frac{XX' - S[X' - X - X'(X/Y) + Y]}{2S^2XX'}$	$\frac{1}{XX'}$	–
Norris	$F_{ij}\sigma_i\sigma_j = 1;\ \sigma_1^2 = X^2\ \&\ \sigma_2^2 = Y^2$	–	–	$\frac{1}{X^2}$	$-\frac{1}{2XY}$	$\frac{1}{Y^2}$	$\frac{1}{S^2}$
Tsai–Hill	$F_{ij}\sigma_i\sigma_j = 1$	–	–	$\frac{1}{X^2}$	$-\frac{1}{2X^2}$	$\frac{1}{Y^2}$	$\frac{1}{S^2}$
Tsai–Wu	$F_i\sigma_i + F_{ij}\sigma_i\sigma_j = 1$	$\frac{1}{X} - \frac{1}{X'}$	$\frac{1}{Y} - \frac{1}{Y'}$	$\frac{1}{XX'}$	$\leqslant \pm\sqrt{F_{11}F_{22}}$ and determined under biaxial stress	$\frac{1}{YY'}$	$\frac{1}{S^2}$

$$K_l = \frac{(1 + 4\nu_{12} - \nu_{13})E_{22} + (1 - \nu_{23})E_{11}}{[E_{11}E_{22}(2 + \nu_{12} + \nu_{13})(2 + \nu_{21} + \nu_{23})]^{1/2}}$$

K'_l experimentally determined correlation coefficient

$$K = \frac{E_{11}(1 + \nu_{21}) + E_{22}(1 + \nu_{12})}{2\sqrt{E_{11}E_{22}(1 + \nu_{12})(1 + \nu_{21})}}$$

6. Testing techniques

While analytical composite strength predictions are highly advantageous, lack of total confidence in any particular approach necessitates testing of actual materials under combined loading. Criteria such as those proposed by Tsai, Tennyson, Wu and Scheublein also require biaxial failure information for evaluation of their interaction strength coefficients. Of the various specimens employed to evaluate the strengths of composites under combined loading, the thin-walled tube subjected to internal and external pressure, axial load and torsion is superior. Laminated tubular specimens whose properties virtually duplicate those of flat plates can be fabricated (COLE and PIPES (1974), ROWLANDS (1975)). That the stress field is determinate is advantageous. However, extreme care must be exercised to load the specimen without introducing extraneous stresses. Hydraulic loading of composite tubes is suitable, but attention to seals and tabs is necessary. Provisions must be made if large deformations occur. Suitable mechanical loading of tubes is difficult.

Examples of biaxial tube testing facilities are illustrated by the photographs of Figs. 1, 2 and 3. The equipment of Fig. 1 was fabricated by Southwest

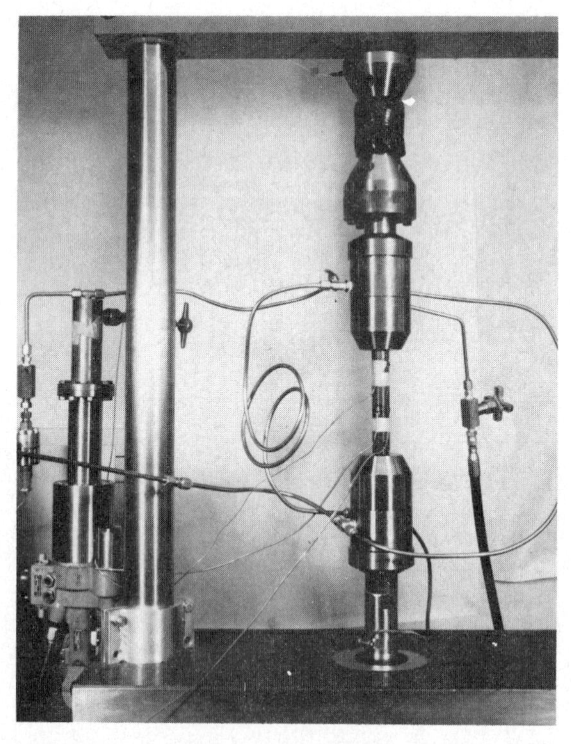

FIG. 1. Biaxial testing system: Southwest Research Inistitute, San Antonio, Texas, U.S.A. (NAGY and LINDHOLM (1973)).

FIG. 2. Composite biaxial testing system: IIT Research Institute, Chicago, Illinois, U.S.A. (COLE and PIPES (1974)).

FIG. 3. Apparatus for subjecting paper or wood-veneer tubes to torsion, or torsion plus longitudinal tension: Forest Products Laboratory, Madison, Wisconsin, U.S.A. (SETTERHOLM, BENSON and KUENZI (1968)).

Research Institute of San Antonio, Texas (NAGY and LINDHOLM (1973)), that of Fig. 2 is located at the IIT Research Institute, Chicago (COLE and PIPES (1974)), and that of Fig. 3 is at the Forest Products Laboratory, Madison, Wisconsin (SETTERHOLM, BENSON and KUENZI (1968)). The electro-hydraulic machine of Fig. 1 is capable of testing tubes under torsion, internal pressure and tension/torsion. The machine has axial load capacity of 44 500 newtons (10 000 pounds) and a torque capacity of 770 newton-meters (6800 inch-pounds). The loads are servo-controlled with feedback provided by either load, strain or displacement. The system employs hydraulic, self-compensating grips. The facility of Fig. 2 can apply axial, hoop or shear stress, simultaneously or independently and under a wide range of biaxial stress ratios. The composite tube to be tested is located inside of the cut-away steel cylinder, and together they slide into the thick-walled pressure vessel (white in the photograph) to the left foreground. The specimen tube is loaded hydraulically through conforming tabs. The tabs are hydraulically deformed radially and compatibly with the composite tube. As with the previous system, the result is a virtual elimination of end constraints and a tube of extended 'effective' length. The composite tube and tab pressures are servo-controlled. The hydraulic supply, control and monitoring components are visible behind the biaxial fixture. Through a finite-element stress analysis, DANIEL et al. (1980) have recently modified the original design and loading of the end tabs of the specimens for the system of Fig. 2, and an improved apparatus is now being prepared. Hydraulic loading of composite tubes has also been discussed by STOTLER et al. (1971), GRIMES et al. (1972), SANDHU et al. (1973), HOTTER et al. (1974), PROTASOV and KOPNOV (1965), and ASHKENAZI and PEKKER (1970), among others.

With the apparatus of Fig. 3, paper and wood-veneer tubes of 5 cm (2 inches) diameter were tested in torsion with their principal axes of material symmetry parallel, perpendicular and at 45° to the longitudinal axis of the tube (SETTERHOLM, BENSON and KUENZI (1968)). The tubes were supported internally to prevent buckling. Shear strains were recorded by an optical-mechanical gage.

The off-axis coupon is also an effective and relatively inexpensive combined loading specimen that has been employed for both unidirectional and multidirectional composites. However, the available biaxial ratios are limited. Load introduction must not constrain the shear deformation, otherwise an unknown state of stress results. In practice, uniaxial loading of long, narrow off-axis composite coupons produces virtually pure uniaxial stress in the middle of the test section. The off-axis ring specimen enjoys the advantage that it is totally unconstrained and has an infinite length-to-width ratio (ROWLANDS (1978b)). Off-axis composite specimens suffer in that not all fibers are continuous, but rather some intersect the edges of the specimen. Interlaminar stresses introduced by these free fibers at the boundaries can be detrimental strengthwise for laminates. An advantageous feature of tabbed tubular specimens is the effective absence of fiber ends. Other things being equal, adverse interlaminar stresses can produce lower strengths in an off-axis laminate than in a tubular specimen of identical construction and under the same stress biaxiality.

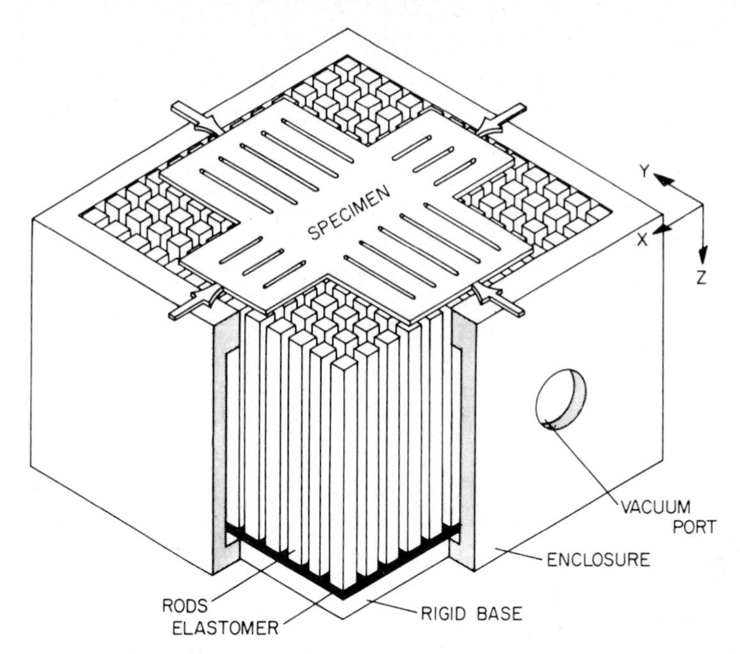

FIG. 4. Fixture for biaxial loading of thin paper or wood specimens, including compression-compression: Forest Products Laboratory, Madison, Wisconsin, U.S.A. (GUNDERSON and ROWLANDS (1983)).

The device of Fig. 4 is employed to subject thin paper and wood specimens to uniaxial and biaxial in-plane loads, including compression-compression (GUNDERSON and ROWLANDS (1983)). Both on- and off-axis specimens are utilized, the latter providing shear as well as normal stresses. Under compression loading, an ingenious lateral restraining system developed by GUNDERSON (1983) utilizes a vacuum to maintain the specimen flat against an array of vertical, flexible supporting rods visible in the Figure. Results confirm that the rods contribute negligible in-plane restraint.

Cross-beam specimens have been used for biaxial testing of composites. This geometry can suffer from detrimental stress concentrations and an indeterminate stress field. The bulge-plate biaxial specimen is limited to information in the tension-tension quadrant. Advantages and disadvantages of the tube, off-axis coupon, cross-beam and bulge plate specimens are discussed in detail by COLE and PIPES (1974).

7. Experimental correlation

Figs. 5 through 18 contains results of composite failure criteria and experimental observations. Details of the theoretical analyses are contained in Sections 3 and 5. The present examples are intended only to be representative and

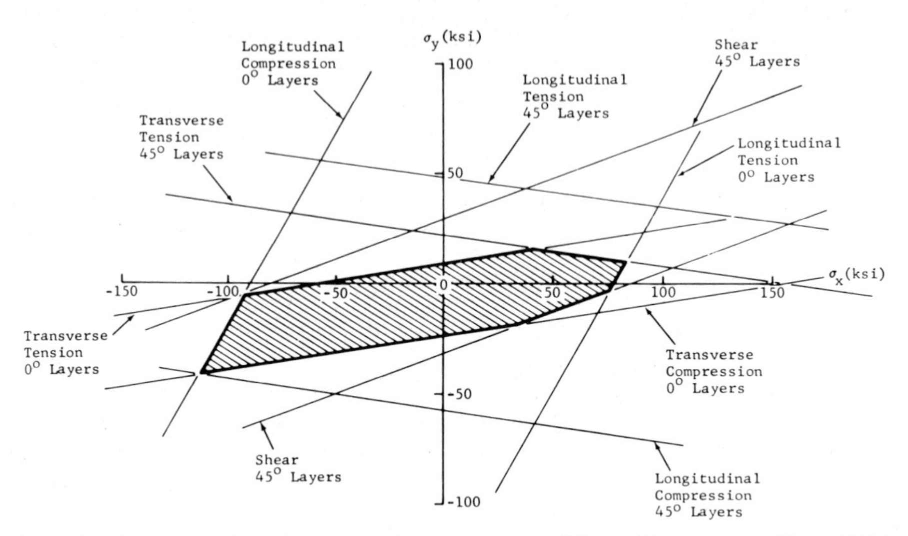

FIG. 5. Laminae and laminate interaction diagrams for [0/±45/0] graphite composite (REED (1970)).

numerous additional data are available in the literature. Some of the theories are recent developments and their acceptance and utilization have not as yet always progressed beyond the respective developer's affiliation. Fiscal and proprietary considerations can also cause an organization to use habitually one particular composite analysis, rather than investigate the suitability of others. Confidence in strength predictions requires comparison with reliable experimental results obtained from numerous composite systems and under widely varying conditions of biaxial loading. A relative paucity of such experimental results remains.

Most laminated composite strength predictions are formulated on the basis of individual ply response. A failure envelope based on a lamina strength criterion and material properties exists for each ply. The minimum interaction diagram of the respective plies of the complete composite, and referred to the laminate axes, becomes the failure surface of the particular laminate. Loading of the various laminae is determined according to lamination theory, providing for load redistribution as individual plies fail. Fig. 5 demonstrates the various failure mode cutoffs of the individual plies for a [0/±45/0] graphite composite ($\tau = 0$). The shaded region in the center is the minimum commonality and would be the interaction diagram for the entire laminate. Once the stress vector reaches the boundary of the shaded region, failure occurs in one of the indicated plies. The interaction diagram of Fig. 5 was obtained by REED (1970) using the linear maximum strain criterion, hence the straight boundaries. Fig. 5 simply contains the failure modes and stresses for the individual plies. It does not provide for any transfer of load from failed plies to the remaining laminate. With homogeneous anisotropic materials, a strength criterion is applied to the material as a whole and lamination theory is not required.

Fig. 5 shows interaction diagrams for zero inplane shearing stress. The full three-dimensional failure surface is that generated by all possible combinations of σ_x, σ_y and τ_{xy} which produce failure. Such a surface is viewed conveniently in terms of interaction diagrams at varying constant values of τ_{xy}. Fig. 6 has such strength interaction surfaces predicted for a $[0/\pm 60]_s$ boron-epoxy (Narmco 5505 resin) laminate at $\tau_{xy} = 0$, 69 MN/m^2 (10 ksi) and 138 MN/m^2 (20 ksi). Like Fig. 5, the maximum strain criterion was employed to construct Fig. 6(a). Interaction diagrams for this laminate, but using the Tsai–Hill and Tsai–Wu criteria are plotted in Figs. 6(b) and 6(c). As the shear stress increases, the Tsai–Hill and Tsai–Wu analyses predict decreasing envelopes in terms of σ_x and σ_y. Except for the sharp corners predicted in the first and third quadrants by the Tsai–Hill theory, these latter two predictions are quite similar. However, the failure surface predicted by the maximum strain analysis differs significantly. Predicted laminate strengths for loading in the first and third quadrants and for large $\pm\sigma_x$ are appreciably greater for the maximum strain analysis, while those for large $\pm\sigma_y$ are smaller for this latter theory as compared with the corresponding values by Tsai–Hill and Tsai–Wu analyses.

Correlation by COLE and PIPES (1974) of theoretical interaction predictions

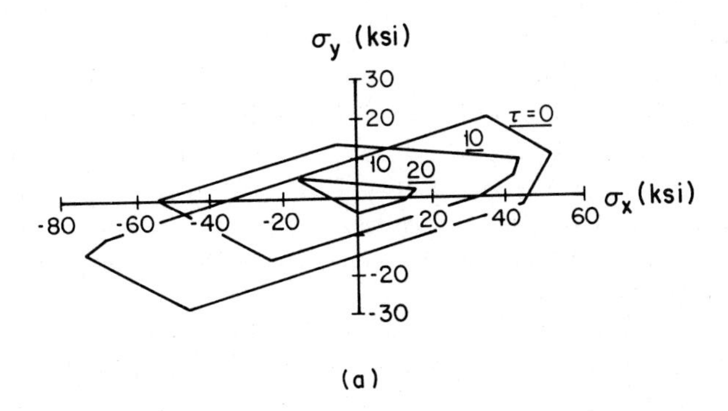

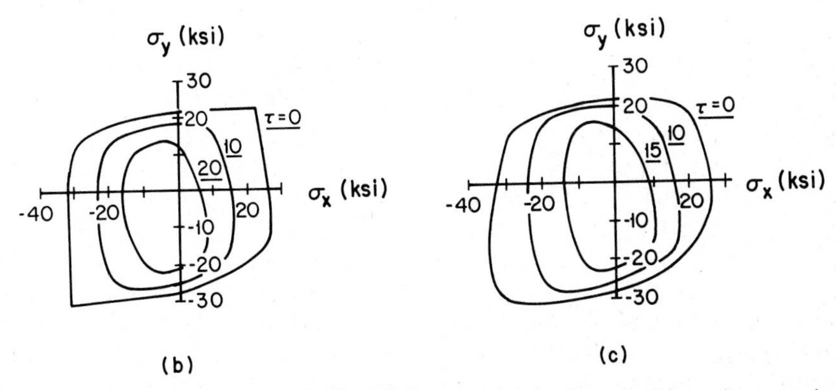

FIG. 6. Design allowable surfaces for $[0/\pm 60]_s$ boron-epoxy laminate by (a) maximum-strain, (b) Tsai–Hill, and (c) Tsai–Wu analyses (COLE and PIPES (1974)).

with observed biaxial data for several boron-epoxy laminates are contained in Figs. 7 through 10. The data are for $[0/\pm45]_s$, $[0/\pm60]_s$, $[0/\pm45/90]_s$ and $[0/90]_s$ constructions, respectively. The responses are plotted for only the first quadrant and for zero shearing stress. The laminate strength envelopes based on each of the Tsai–Hill, Tsai–Wu and maximum-strain lamina criteria are included. The experimental biaxial results (closed circles) were obtained by COLE and PIPES (1974) using the equipment of Fig. 2. As in Fig. 6. the Tsai–Hill and Tsai–Wu predictions agree closely with each other, the latter tending to be slightly more conservative. The maximum-strain analyses vary considerably from the others. While there is appreciable scatter in the experimental data, none of the employed analyses sufficiently adequately predicts the observed behavior in these cases. Coincidently, the best correlation between theory and experiment occurs in Fig. 7 in a region where all three numerical predictions agree closely with each other.

In the classical lamination theory by which the strains are evaluated at any applied load, the maximum strain, Tsai–Hill and Tsai–Wu analyses are all predicted on the basis of constant material properties to failure. Linear-elastic response is again assumed in computing the lamina stresses from these strains. Composites often actually respond nonlinearly and predictions based on linear material assumptions can be questionable. COLE and PIPES (1974) considered this dilemma by using design allowable lamina input data in the analyses of Fig. 6 which were less than the actual strengths. The predictions of Fig. 6 should consequently be interpreted as design allowable interaction plots rather than ultimate data. The ultimate strength predictions by Tsai–Hill and Tsai–Wu of Figs. 7 through 10 are based on ultimate lamina properties for the boron composite. Those indicated for the maximum strain analysis were based on boron ply properties corresponding to essentially 67% of the observed ultimate values. The laminate interaction curves obtained by this manner were then multiplied by 1.5 to produce the plots shown in the figures. Again, this is a concept for applying a linear theory to describe a nonlinear response. The idea has been further investigated by ROWLANDS (1978a). In the analytical predictions of Figs. 6 through 10, theoretical laminate failure is considered to be

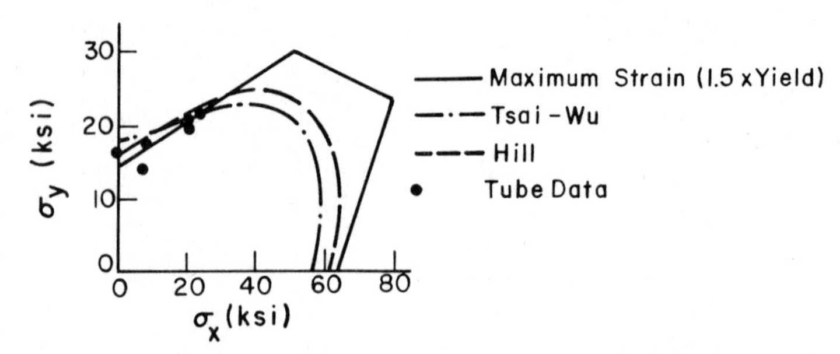

FIG. 7. Predicted and observed ultimate strengths of $[0/\pm45]_s$ boron-epoxy composite (COLE and PIPES (1974)).

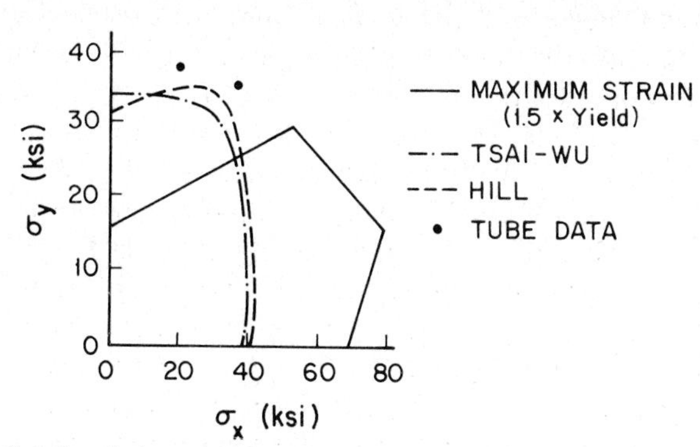

FIG. 8. Predicted and observed ultimate strengths of $[0/\pm 60]_s$ boron-epoxy composite (COLE and PIPES (1974)).

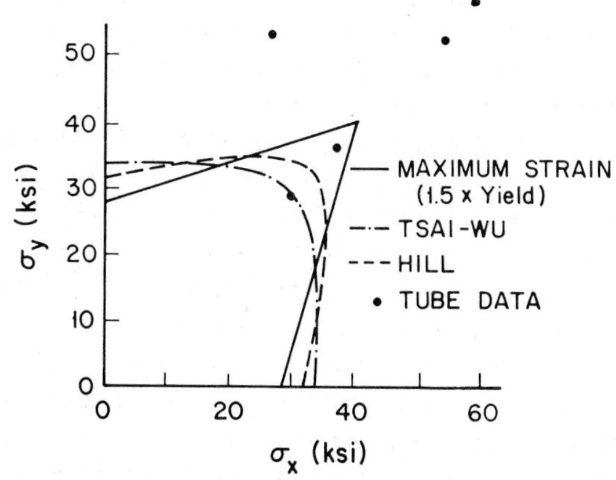

FIG. 9. Predicted and observed ultimate strengths of $[0/\pm 45/90]_s$ boron-epoxy composite (COLE and PIPES (1974)).

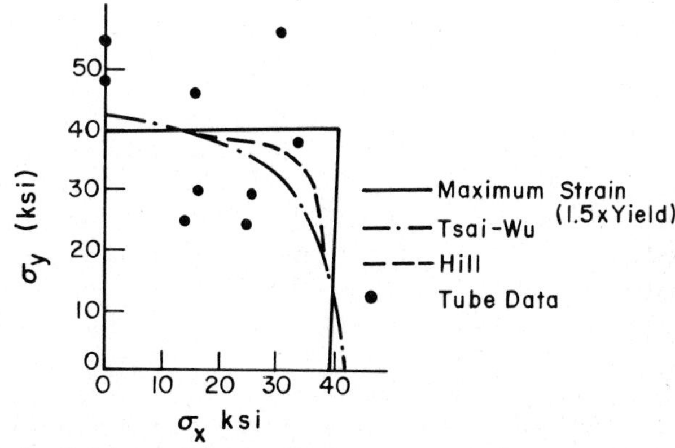

FIG. 10. Predicted and observed ultimate strengths of $[0/90]_s$ boron-epoxy composite (COLE and PIPES (1974)).

coincident with initial failure of any ply in the composite. The value of the Tsai–Wu cross-coupling term F_{12} was sufficiently small compared to the other coefficients of the criterion of Equation (65) to be assumed equal to zero in these analyses.

Some representative observed and predicted strengths of $[0/\pm45]_s$ boron-epoxy laminates subjected to varying ratios of biaxiality ($\tau_{xy} \neq 0$) are compared in Table 3 (COLE and PIPES (1974)). The same lamina input information was used in the theoretical analyses of this Table as was employed for those of Fig. 6. The nominally applied stress ratio is indicated in the top row, followed by the observed stress ratios at failure. Predictions by each of the maximum-strain, Tsai–Hill and Tsai–Wu analyses are indicated. Experimental data obtained from both tubular and off-axis laminate coupon specimens are included. The results are averages of several tests. The observed strengths of the off-axis coupons are consistantly lower than those of the tubular specimens, and incidently are much closer to the predicted laminate design values. This is regarded as coincidental. That the observed strengths are from one to two times larger than the predicted allowables is compatible with utilization of input data equal to 67% of measured lamina strengths.

The correlation of observed and predicted strengths for envelope and sack paper have been obtained by DERUVO, CARLSSON and FELLERS (1980), Fig. 11. These data were obtained by subjecting tubular specimens to longitudinal tension and internal pressure. The machine direction (MD) of the paper was either parallel or transverse to the axis of the tube. The open and closed symbols of Fig. 11 represent tests using different size tubes.

Fig. 12 illustrates the strength envelope ($\tau_{xy} = 0$) for a $[0_2/\pm45/90]_s$ boron-epoxy laminate as predicted by the SANDHU (1974b) analysis. The dashed line of Fig. 12 indicates the strength based on initial (resin) failure, while the solid line denotes the envelope assuming final (fiber) failure. The accompanying experimental data (solid circles) of Fig. 12 originated with STOTLER, VAN PUTTEN and DICKERSON (1971). Prediction based on initial failure tends to provide a lower

TABLE 3. Comparison of predicted allowable and observed strengths of $[0/\pm45]_s$ boron-epoxy composites under nonzero shear stresses (COLE and PIPES (1974)).

	Stress ratios	
Nominal ($\sigma_x : \sigma_y : \tau_{xy}$)	13.9:1:3.73	3:1:1.73
Tubes (ksi)	57:4.4:15.5	51:17:30
Off-axis coupons	41:3.0:11	22:7.5:13
Max-strain[a]	40:2.8:10.6	24:8:14
Tsai–Hill[a]	30:2.2:8	20:6.7:11.5
Tsai–Wu[a]	30:2.2:8.2	22:7.0:12

[a] Predictions based on input data equal to 67% of measured lamina strengths.

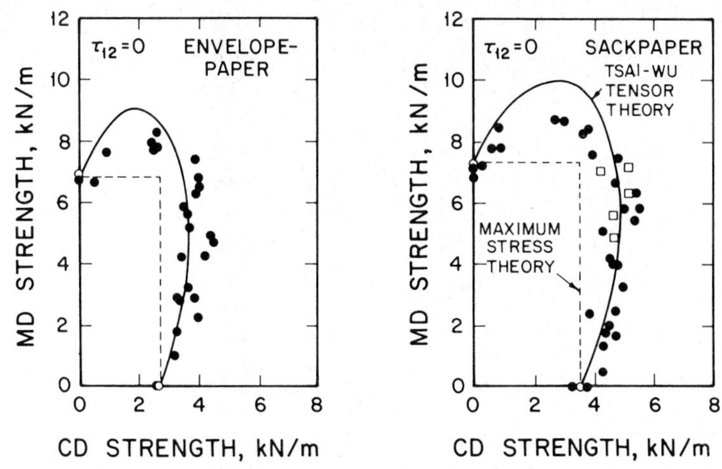

FIG. 11. Biaxial strength (force per unit width) of envelope and sack paper (DERUVO, CARLSSON and FELLERS (1980)).

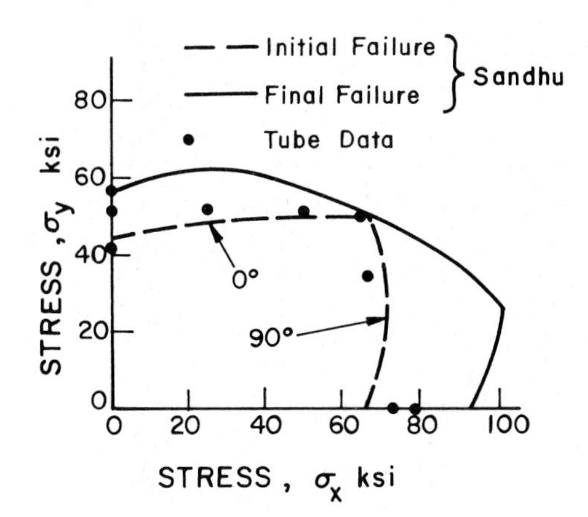

FIG. 12. Strength envelope for $[0_2/\pm45/90]_s$ boron-epoxy laminate (SANDHU (1974b)).

bound and in this respect perhaps represents a conservative, but safe, analysis for design.

Unlike the maximum-strain, Tsai–Hill and Tsai–Wu analyses, those by Sandhu and Petit–Waddoups provide for nonlinear lamina stress-strain responses, Section 5. Fig. 13 illustrates the ability of these analyses to predict the nonlinear stress-strain results of a [±45] boron-epoxy laminate subjected to uniaxial stress σ_x. Also contained in Fig. 13 is the stress-strain profile as predicted by HAHN and TSAI (1973). This latter theory provides for nonlinearity of the shear stress-strain response. However, it neither distinguishes between tensile and compressive behaviors of the lamina, nor does it treat failure.

Analyses such as the Tsai–Hill or Tsai–Wu assume constant lamina elastic properties to failure. However, once a ply fails and unloads, the stiffness of the

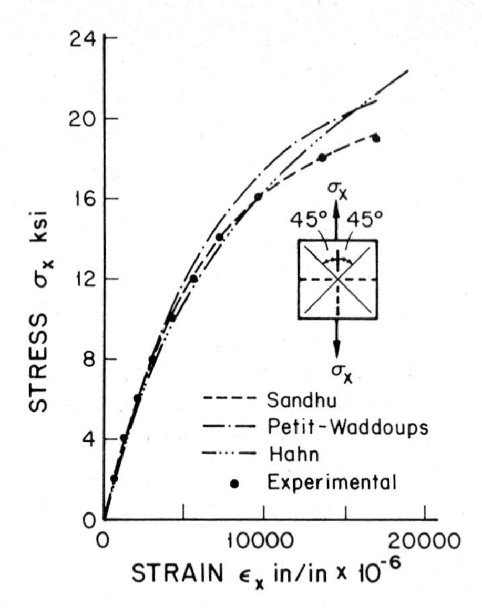

FIG. 13. Nonlinear response of $[\pm 45]_s$ laminate (SANDHU (1974b)).

remaining laminate is reduced. The effect is a piecewise-linear laminate stress-strain curve. Slope changes in the laminate constitutive relation therefore occur when a ply fails and unloads. This is illustrated in Fig. 14 which shows the stress-strain response to failure as predicted by the Sandhu (open circles), Petit–Waddoups (open triangles) and Tsai–Wu (open squares) analyses. The composite is a 30-degree off-axis $[0_2/\pm 45]_s$ graphite laminate. The Sandhu and

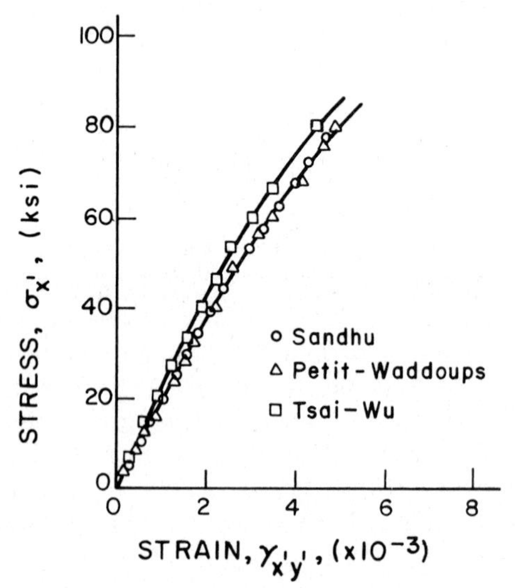

FIG. 14. Analytically predicted responses for 30-degree off-axis $[0_2/\pm 45]_s$ graphite composite (ROWLANDS (1975)).

Petit–Waddoups theories predict virtually identical response, the Tsai–Wu theory being not too different. While the prediction by the latter theory is actually piecewise linear, it has been smoothed by the solid line. For this case, the Tsai–Wu analysis predicts individual ply failures occur at 324 MN/m^2 (47 ksi), 412 MN/m^2 (60 ksi) and 550 MN/m^2 (80 ksi). Actual flat and ring 30-degree off-axis specimens of this laminate failed at 275 MN/m^2 (40 ksi) and 345 MN/m^2 (50 ksi), respectively. The x'-y' axes of Fig. 14 are parallel and perpendicular to the coupon loading axes.

Fig. 15 illustrates the degree to which CHAMIS' analysis (1969) can be made to agree with biaxial experimental data (open circles). Results are for a unidirectional graphite composite. The sensitivity of the analysis to relatively small variations of the correlation coefficient $K'_{l12\alpha\beta}$ is noteworthy.

The interaction curves for three mutually orthogonal planes passing through the failure surface of a glass-fiber composite $[90/\pm 30/90]_s$ laminate are illustrated in Fig. 16 (HOTTER, SCHELLING and KRAUSS (1974)). Fig. 16a is the response in the σ_x-σ_y plane ($\tau_{xy} = 0$), Fig. 16b that in the σ_y-τ_{xy} plane ($\sigma_x = 0$) while that in the σ_x-τ_{xy} plane ($\sigma_y = 0$) is displayed in Fig. 16c. The experimental biaxial information (solid circles) was obtained from tubular specimens, while the theoretical predictions are based on the Puppo and Evensen failure criterion. Whereas analyses such as the maximum-strain, Tsai–Hill, Tsai–Wu and Sandhu use a lamina criterion, the Puppo and Evensen analysis treats the laminate as a whole, without computing the stresses in the individual plies. The theoretical predictions based upon tensile input ultimate laminate strength are denoted by solid lines in Figs. 16b and c; those based on compressive strength are represented by a broken line. It appears to make little consequence whether the tensile or compressive strength is employed in the theory, and indeed the analysis predicts the observed results quite well.

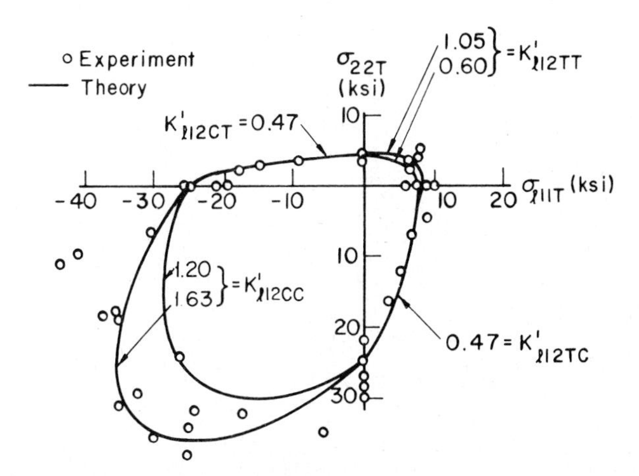

FIG. 15. Correlation of experimental and theoretical strength envelope for unidirectional graphite composite (CHAMIS (1969)).

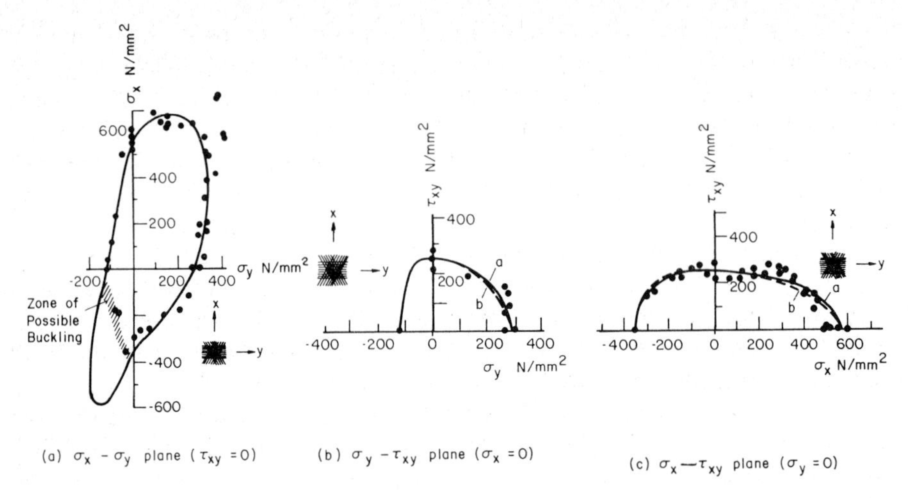

FIG. 16. Interaction diagrams for othogonal planes passing through failure surface of $[90/\pm30/90]_s$ glass-epoxy composite: Lines represent Puppo–Evensen prediction (HOTTER, SCHELLING and KRAUSS (1974)).

Another comparative example of theoretically predicted and experimentally observed (circles) biaxial failure results is that of the laminate criterion by Wu and Scheublein (Equation (82)) applied to a $[0/90/0/90]_s$ graphite-epoxy laminate, Fig. 17. For this construction the x- and y-axes are interchangeable. Data of the solid circles were obtained experimentally, whereas the open circles were located by symmetry from the solid circles. The solid line of Fig. 17 was obtained by fitting the seven-degrees-of-freedom polynomial of Equation

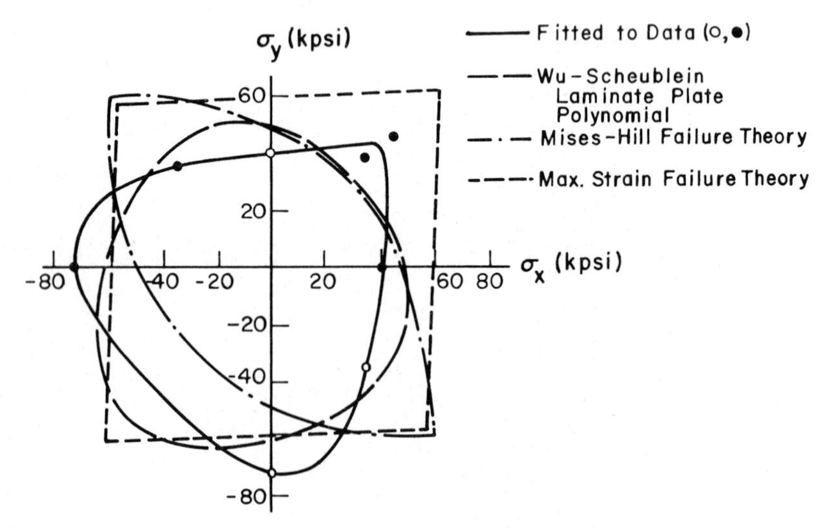

FIG. 17. Observed and predicted strengths for $[0/90/0/90]_s$ graphite-epoxy laminate (WU and SCHEUBLEIN (1974)).

(82) to the eight experimental data points. The dashed line was obtained by curve fitting the laminate polynomial of Equation (82) to the interaction profile generated by the lamina Tsai–Wu criterion of Equation (65) which employs lamina input values and classical lamination theory. The other two theoretical predictions of Fig. 17 are the traditional Tsai–Hill and maximum-strain profiles generated from the same lamina data as was the dashed line labelled Wu–Scheublein.

Correlation of quadratic and cubic criteria by TENNYSON, NANYARO and WHARRAM (1980) with data from graphite-epoxy tubes is illustrated in Fig. 18. In this case, both initial and final failure are much better predicted by the cubic analysis. Although this example suggests stacking sequence has little influence on the strength, such is normally not true.

RAHMAN (1981) recently found that each of the Norris, Tsai–Hill, Tsai–Wu, Hoffman, Fischer and Cowin theories produced similar results for double-bolted mechanical fasteners in sitka spruce. The initiation of observed damage was predicted reasonably well and ultimate fastener strength was predicted accurately by these theories in this case provided fracture was of the opening-mode form.

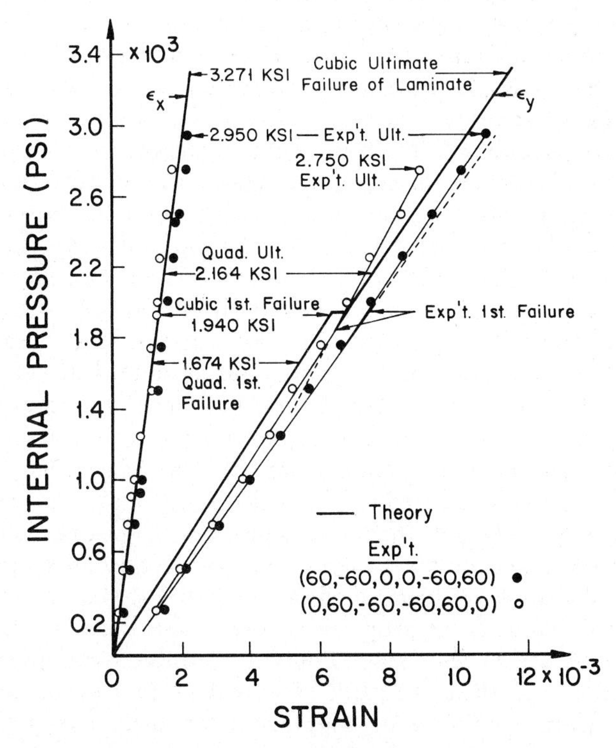

FIG. 18. Comparison of cubic and quadratic failure models with observed strengths of graphite-epoxy tube data (TENNYSON, NANYARO and WHARRAM (1980)).

8. Summary, discussion and conclusion

The more prevalent failure theories are discussed and their predictions compared with experimental observation. Both natural (wood, paper) and man-made composites are considered. Relevant isotropic analyses are included since their extensions form the essence of many of the composite strength criteria. The necessary aspects of orthotropic elasticity, lamination theory and mechanical testing are present for completeness.

Strength (failure) criteria tend to be phenomenological and empirical in nature, not mechanistic. Therefore, no unique mathematical representation of strength exists for a given composite. The analyses predict primarily the onset, but not mode, of failure. As with isotropic criteria, composite strength theories are predicted on the basis of limiting stress, strain or energy. Most analyses assume a lamina criterion, employing lamina input information and lamination theory to evaluate the stresses and strains in the various plies. Linear-elastic lamina behavior is typically assumed, and criteria such as the maximum strain and Griffith–Baldwin are also based on Hookean response. Significant considerations of any strength or design theory are the ability to predict accurately actual behavior and ease of application. The choice of a particular criterion may depend on the extent of experimentation required, one's preference in basic approach, whether the material is homogeneous or laminated, and whether or not the response is inelastic and/or nonlinear.

None of the available anisotropic strength criteria represents observed results sufficiently accurately to be employed confidently by itself in design. Several of the theories suffer from the inconvenience of requiring biaxial information as basic input data. Of the analyses cited here, the Puppo and Evensen prediction agrees as well as any with experimental observation. This theory is a direct laminate analysis which makes no constitutive assumption and does not involve lamination theory. However, introduction of the interaction factor and strength axes can render this approach less convenient to use than some of the other analyses. The Tsai–Wu lamina theory enjoys a good theoretical basis and its predictions often agree quite well with experiment. Nevertheless, the need to evaluate the independent interaction coefficient F_{12} from biaxial data can be a serious inconvenience. The analyses by Cowin, Hoffman and Malmeister are similar to that by Tsai–Wu except that in each of the former cases F_{12} is evaluated explicitly in terms of simple uniaxial strengths. No biaxial testing is needed. If the inequality $F_{12}^2 \leqslant F_{11}F_{22}$ predicts a relatively small F_{12}, it is not uncommon to ignore the interaction term altogether. The expressions by Tsai–Wu, Cowin, Hoffman and Malmeister are then the same.

In addition to constitutive properties, most of the strength analyses require as input information at least tensile and compressive longitudinal and transverse strengths, and shear strength. These data can all be obtained from uniaxial specimens, including possibly tubes for shear strength. For biaxial testing, a thin-walled tube subjected to internal and external pressure, longitudinal load and torsion is the most satisfactory specimen. Care is required to eliminate end conditions, and the tube diameter must be sufficient to achieve reasonably flat laminate conditions and to minimize normal stresses. The off-axis specimen is an expedient biaxial specimen for limited information. Other than ensuring that the shear strain is able to develop, this specimen is easier than the tube to prepare and to load. However, interlaminar stresses at the longitudinal edges of laminated off-axis specimens can be detrimental strength-wise.

The following are among the greatest deficiencies of current composite strength analyses (theoretical or experimental): lack of a simple but representative lamination theory which realistically accounts for interlaminar, and nonlinear effects; and insufficient reliable experimental results. The latter is in part associated with experimental complexities. Additional topics which warrant further effort include the following:

(1) When a particular ply of a laminate fails, evaluate how and to what extent it unloads and exactly how this influences the remaining laminate, i.e., how its loads are transferred to the active plies.

(2) Determine what constitutes total laminate failure, i.e., which and how many plies must fail (and in what modes) before the laminate should be considered to have failed.

(3) Determine what the effects of residual stresses are relative to strength criteria.

(4) Extend strength theories to include inelastic and nonlinear lamina response.

(5) Develop a simple but mathematically rigorous failure criterion whose interaction coefficients can be evaluated easily and with confidence.

Most biaxial composite failure investigations have been confined to static (or quasi-static) loadings and without notches or intended cracks. Analytically, laminae are normally assumed to be a homogeneous continuum. Certainly, consideration of biaxially loaded composites under cyclic and impact conditions, extreme temperatures, or in conjunction with fracture mechanics concepts merit additional pursuit. Application of biaxial strength studies to hybrid composites should be undertaken and few examples involving metal or ceramic constituents appear to be available in the literature.

Acknowledgments

The author gratefully acknowledges the cooperation received from the following persons: C.C. Chamis, NASA Lewis Research Center, Cleveland, Ohio, U.S.A.; I.M. Daniel, Illinois Institute of Technology, Chicago, Illinois, U.S.A.; C. Fellers, Swedish Forest Products Research Laboratory, Stockholm, Sweden; D.E. Gunderson, Forest Products Laboratory, Madison, Wisconsin, U.S.A.; U.S. Lindholm, Southwest Research Institute, San Antonio, Texas, U.S.A.; D.L. Reed, General Dynamics Corporation, Forth Worth, Texas, U.S.A.; R.S. Sandhu, Wright–Patterson Air Force Base, Ohio, U.S.A.: H. Schelling, Institute for Structural and Construction Research, Stuggart, Germany; V.C. Setterholm, Forest Products Laboratory, Madison, Wisconsin, U.S.A.; R.C. Tennyson, Institute of Aerospace Studies, University of Toronto, Toronto, Ontario, Canada; and E.M. Wu, Lawrence Livermore Laboratory, Livermore, California, U.S.A. Many of these persons kindly provided photographs. This chapter was proficiently typed by Mrs. M.M. Lynch.

References

ASHKENAZI, E.K. (1959), On the problem of strength anisotropy of construction materials, *Soviet Physics – Technical Physics* **4**(3), 333–338 (English translation).

ASHKENAZI, E.K. (1965), Problems of the anisotropy of strength, *Mekhanika Polimerov* **1**(2), 79–92 (Polymeric Mechanics **1**(2), 60–70).

ASHKENAZI, E.K. (1966), Strength of anisotropic wooden and synthetic materials, *Isdaniia Lesnaya Promishlennost*, Moscow.

ASHKENAZI, E.K. (1967), Geometry of strength theory, *Mekhanika Polimerov* **3**(4), 703–707 (Polymer Mechanics **3**(4), 466–468).

ASHKENAZI, E.K. and F.P. PEKKER (1970), Experimental testing of the applicability of a fourth degree polynomial describing surfaces of critical planer stress distributions in glass-reinforced plastics, *Mekhanika Polimerov* **6**(2), 284–294 (Polymeric Mechanics, **6**(2), 251–258).

ASHKENAZI, E.K., O.S. MYL'NIKOVA and R.S. RAIKHEL'GAUZ (1976), Further study on the strength geometry of anisotropic materials, *Mekhanika Polimerov* **12**(2), 269–278 (Polymer Mechanics **12**(2), 232–240).

ASHTON, J.E., J.C. HALPIN and P.H. PETIT (1969), *Primer on Composite Materials: Analysis* (Technomic Publishing, Stamford, CT).

AZZI, V.D. and S.W. TSAI (1965), anisotropic strength of composites, *Experimental Mechanics* **5**(9), 283–288.

BORESI, A.P., O.M. SIDEBOTTOM, F.B. SEELY and J.O. SMITH (1978), *Advanced Strength of Materials* (Wiley, New York).

CHAMIS, C.C. (1969), Failure criteria for filamentary composites, Composite Materials: Testing and Design, ASTM STP 460, 336–351.

CHRISTENSEN, R.M. (1979), *Mechanics of Composite Materials* (Wiley Interscience, New York).

COLE, B.W. and R.B. PIPES (1974), Filament Composite Laminates subjected to Biaxial Stress Fields, U.S. Air Force Technical Report AFFDL TR-73-115, Dayton, OH, U.S.A.

COWIN, S.C. (1979), On the strength anisotropy of bone and wood, *Trans. ASME J. Appl. Mech.* **46**(4), 832–838.

DANIEL, I.M., T. LIBER, R. VANDERBY and G.M. KOLLER (1980), Analysis of Tubular Specimen for Biaxial Testing of Composite Laminates, *Proc. 3rd Int'l. Conf. on Composite Materials, ICCM-3*, Paris, France.

DERUVO, A., L. CARLSSON and C. FELLERS (1980), The biaxial strength of paper, *TAPPI* **63**(5), 133–136.

FISCHER, L. (1960), How to predict structural behavior of reinforced plastic laminates, *Modern Plastics* **37**(2), 121–128, 208, 209.

FISCHER, L. (1967), Optimization of orthotropic laminates, *Engineering for Industry* **89**(3), Series B, 399–402.

FRANKLIN, H.G. (1968), Classical theories of failure of anisotropic materials, *Fiber Sci. Tech.* **1**(2), 137–150.

FUNG, Y.C. (1965), *Foundations of Solid Mechanics* (Prentice-Hall, Englewood Cliffs, NJ).

FUNG, Y.C. (1977), *A First Course in Continuum Mechanics* (Prentice-Hall, Englewood Cliffs, NJ).

GOL'DENBLAT, I. and V.A. KOPNOV (1965), Strength of glass-reinforced plastics in the complex stress state, *Mekhanika Polimerov* **1**(2), 70–78 (Polymer Mechanics **1**(2), 54–59).

GOODMAN, J.R. and J. BODIG (1971), Orthotropic strength of wood in compression, *Wood Sci.* **4**(2), 83–94.

GRIFFITH, J.E. and W.M. BALDWIN (1962), Failure theories for generally orthotropic materials, *Developments Theoret. and Appl. Mech.*, Vol. I, 410–420.

GRIMES, G.C., P.H. FRANCIS, G.E. COMMERFORD and G.K. WOLFE (1972), An Experimental Investigation of the Stress Level at which Significant Damage Occurs in Graphite Plastic Composites, U.S. Air Force Technical Report AFML-TR-72-40, Dayton, OH, U.S.A.

GUNDERSON, T.E. (1983), Edgewise compression testing of paperboard – A new concept of lateral support, *APPITA* **37**(2), 137–141.

GUNDERSON, D.E. and R.E. ROWLANDS (1983), Determining paperboard strength-tension, compression and shear, *Proc. Internat. Paper Physics Conf.* (TAPPI, Atlanta, GA) 253–263.

HAHN, H.T. and STEPHEN W. TSAI (1973), Nonlinear elastic behavior of unidirectional composite laminae, *J. Composite Materials* **7**, 102–108; *also* HAHN, H.T., Nonlinear Behavior of Laminated Composites, *J. Composite Materials* **7**, 257–270.

HANKINSON, R.L. (1921), Investigation of Crushing Strength of Spruce at Varying Angles of Grain, U.S. Air Service Information Circular, Vol. III, No. 259 (Material Section Report, No. 130, McCook Field, Dayton, OH, U.S.A.).

HERAKOVICH, C.T. and D.A. O'BRIAN (1979), Failure Analysis of an Idealized Composite Damage Zone, presented at the MFPG Symposium on Advanced Composites: Design and Applications (NBS, Gaithersburg, MD, U.S.A.).

HILL, R. (1950), *The Mathematical Theory of Plasticity* (Oxford University Press, London).

HOFFMAN, O. (1967), The brittle strength of orthotropic materials, *J. Composite Materials* **1**, 200–206.

HOTTER, U., H. SCHELLING and H. KRAUSS (1974), An experimental study to determine failure envelope of composite tubular specimens under combined loads and comparison with classical criteria, *Proc. of the NATO AGARD Meeting on Advanced Composites*, No. 163, Munich, Germany.

JAEGER, J.C. (1969), *Elasticity, Fracture and Flow* (Methuen, London).

JENKINS, C.F. (1920), Materials of Construction used in Aircraft and Aircraft Engines, Report to the Great Britain Aeronautical Research Committee.

JONES, R.M. (1975), *Mechanics of Composite Materials* (McGraw-Hill, New York).

KELLY, A. and DAVIS, G.J. (1965), *Metal Rev.* **10**(1).

KOBETZ, R.W. and G.P. KRUEGER (1976), Ultimate strength design of reinforced timber-biaxial stress failure criteria, *Wood Sci.* **8**(4), 252–261.

KOLLMANN, F. and W.A. COTE (1968), *Principles of Wood Science and Technology*, Vol. 1, Solid Wood (Springer-Verlag, New York).

LIEBOWITZ, H. (1968), *Fracture* (Academic Press, New York).

MAKSIMOV, R.D., E.A. SOKOLOV and E.Z. PLUME (1979), Strength of an organic-glass fiber-reinforced textolite in plane stress, *Mekhanika Kompozitnykh Materialov* **15**(6), 1021–1026 (Mechanics of Composite Materials **15**(6), 702–707).

MALMEISTER, A.K. (1966), Geometry of theories of strength, *Mekhanika Polimerov* **2**(4), 519–534 (Polymer Mechanics **2**(4), 324–331).

MARIN, J. (1957), Theories of strength for combined stresses and non-isotropic materials, *J. Aeronautical Sci.* **24**(4), 265–268.

MCCLINTOCK, F.A. and A.S. ARGON (1966), *Mechanical Behavior of Materials* (Addison-Wesley, New York).

NADAI, A. (1950, 1963), *Theory of Flow and Fracture of Solids*, Vols. I and II (McGraw-Hill, New York).

NAGY, A. and U.S. LINDHOLM (1973), Hydraulic Grip System for Composite Tube Specimens, U.S. Air Force Technical Report, AFML-TR-73-239, Dayton, OH.

NARAYANASWAMI, R. and H.M. ADELMAN (1977), Evaluation of the tensor polynomial and Hoffman strength theories for composite materials, *J. Composite Materials* **11**, 366–377.

NORRIS, C.B. (1939), The elastic theory of wood failure, *Trans. of the ASME* **61**, 259–261.

NORRIS, C.B. (1950), Strength of Orthotropic Materials Subjected to Combined Stress, U.S. Forest Products Laboratory Report #1816.

PETIT, P.H. (1967), Ultimate Strength of Laminated Composites, General Dynamics Fort Worth Report FZM-4977 (U.S. Air Force Contract No. AF33615-5257).

PETIT, P.H. and M.E. WADDOUPS (1969), A method of predicting the nonlinear behavior of laminated composites, *J. Composite Materials* **3**, 2–19.

PLUME, E.Z. and R.D. MAKSIMOV (1978), Determining the strength-surface tensor components of anisotropic materials, *Mekhanika Polimerov* **14**(1), 51–54 (Polymer Mechanics **14**(1), 42–45).

POPOV, E.P. (1968), *Introduction to Mechanics of Solids* (Prentice Hall, Englewood Cliffs, NJ).

PROTASOV, V.D. and V.A. KOPNOV (1965), Study of the strength of glass-reinforced plastics in the plane, *Mekhanika Polimerov* **1**(5), 39–44 (Polymer Mechanics **1**(5), 26–28).

PUPPO, A.H. and H.A. EVENSEN (1972), Strength of anisotropic materials under combined stresses, *AIAA J.* **10**(4), 468–474.

RABINOVICH, A.L. (1946), On elastic constants and strength of anisotropic materials, *Trudy Tsentralnyiaero-gidrodimanicheskii Institut* **582**.

RAHMAN, M.U. (1981), Stress and Strength Analysis of Double-Bolted Mechanical Fasteners in Orthotropic Materials, Ph.D. thesis, Dept. of Engineering Mechanics, University of Wisconsin, Madison, WI.

REED, D.L. (1970), Point Stress Laminated Analysis, General Dynamics Report No. FZM-5494 (Fort Worth, TX, U.S.A.) to USAF Materials Laboratory (Contract No. F33615-69-C-1494).

RIKARDS, R.B., G.A. TETERS and Z.T. UPITIS (1979), Models of the failure of composites having various reinforcement structures, *Mekhanika Kompozitnykh Materialov*, **15**(2), 222–227 (Mechanics of Composite Materials **15**(2), 162–167).

ROWLANDS, R.E. (1975), Analytical-Experimental Correlation of the Biaxial State of Stress in Composite Laminates, U.S. Air Force Technical Report AFFDL-TR-75-11, Dayton, OH.

ROWLANDS, R.E. (1978a), Analytical-experimental correlation of polyaxial states of stress in Thornel-epoxy laminates, *Experimental Mechanics* **18**(7), 253–260.

ROWLANDS, R.E. (1978b), Effect of stacking sequence on strength of T-300/5208 composites, *Fibre Sci. Technology* **11**, 329–334.

ROWSE, R.C. (1923), The Strength of Douglas-fir in Compression at Various Angles to the Grain, B.S. Thesis in Civil Engineering (Washington University, St. Louis, MO).

SANDHU, R.S. (1972), A Survey of Failure Theories of Isotropic and Anisotropic Materials, U.S. Air Force Technical Report No. AFFDL-TR-72-71, Wright Patterson AFB, OH.

SANDHU, R.S., J.B. MONFORT, F.E. HUSSONG et al. (1973), Laminate Tubular Specimens Subjected to Biaxial Stress States (Glass/Epoxy), Technical Report U.S. Air Force AFFDL-TR-73-7, Vol. 1.

SANDHU, R.S. (1974a), Nonlinear Response of Unidirectional and Angle-ply Laminates, AIAA Paper No. 74-380, Presented at the 15th AIAA-ASME Structural Dynamics and Materials Conference (Las Vegas, NV).

SANDHU, R.S. (1974b), Ultimate Strength Analysis of Symmetric Laminates, U.S. Air Force Report AFFDL-TR-73-137, Dayton, OH.

SENDECKYJ, G.P. (1972), A Brief Survey of Empirical Multiaxial Strength Criteria for Composites, ASTM STP 497, Amer. Soc. for Testing Materials, pp. 41–51.

SETTERHOLM, V.C., R. BENSON and E.W. KUENZI (1968), Methods for measuring edgewise shear properties of paper, *TAPPI* **51**(5), 196–202.

SOKOLOV, E.A., A.F. KREGER and R.D. MAKSIMOV (1978), Comparative analysis of strength anisotropy of glass- and organic-fiber plastic laminates, *Mekhanika Polimerov* **14**(5), 841–847 (Polymer Mechanics **14**(5), 676–681).

SOKOLOV, E.A. (1979), Experimental evaluation of the strength anisotropy of a unidirectionally reinforced organic fiber-reinforced plastic, *Mekhanika Kompozitnykh Materialov* **15**(5), 799–803 (Mechanics of Composite Materials **15**(5), 520–523).

STOTLER, C.L., D.J. VAN PUTTEN and E.D. DICKERSON (1971), Boron/Epoxy Wing Skins, F-100D Aircraft, Structural Design and Analysis, U.S. Air Force AFML-TR-71-29, Contract F33615-69-C-1445, North American Rockwell, Los Angeles, U.S.A.

STOWELL, E.Z. and T.S. LIU (1961), On the mechanical behavior of fiber-reinforced crystalline materials, *J. Mech. Phys. Solids* **9**, 242–260.

TENNYSON, R.C., D. MACDONALD and A.P. NANYARO (1978), Evaluation of the tensor polynomial failure criterion for composite materials, *J. Composite Materials* **12**, 63–75.

TENNYSON, R.C., A.P. NANYARO and G.E. WHARRAM (1980), Application of the cubic polynomial strength criterion to the failure analysis of composite materials, *J. Composite Materials Supplement* **14**, 28–41.

TIMOSHENKO, S.P. (1953), *History of Strength of Materials* (McGraw-Hill, New York).

TIMOSHENKO, S.P. (1956), *Strength of Materials*, Part II (Van Nostrand, Princeton, NJ).

TSAI, S.W., J.C. HALPIN and N.J. PAGANO (1968), Composite Materials Workshop (Technomic, Stamford, CT), 217–253.

TSAI, S.W. and E.M. WU (1971), A General theory of strength for anisotropic materials, *J. Composite Materials* **5**, 58–80.

TSAI, S.W. and H.T. HAHN (1980), *Introduction to Composite Materials* (Technomic, Westport, CT).

UPITIS, Z.T., Y.A. BRAUNS and R.B. RIKARDS (1974), Determination of strength-surface tensor components by the method of least squares, *Mekhanika Polimerov* **10**(3), 552–554 (Polymer Mechanics **10**(3), 472–474).

UPITIS, Z.T. and R.B. RIKARD (1976), Dependence of the strength of a composite material on the reinforcement structure in a plane-stressed state, *Mekhanika Polimerov* **12**(6), 1018–1024 (Polymer Mechanics **12**(6), 889–895).

VICARIO, A.A. and R.H. TOLAND (1975), Failure criteria and failure analysis of composite structural components, in: C.C. CHAMIS, ed., *Composite Materials*, Vol. 7, Structural Design and Analysis, Part I (Academic Press, New York and London), 52–97.

VINSON, J.R. and T.W. CHOU (1975), *Composite Materials and their Use in Structures* (Wiley, New York).

VON MISES, R. (1928), Mechanik der plastischen Formanderung von Kristallen, *Z. Angewandte Mathematik und Mechanik* **8**, 161–185.

WU, E.M. (1972), Optimal experimental measurements of anisotropic failure tensors, *J. Composite Materials* **6**, 472–489.

WU, E.M. and J.K. SCHEUBLEIN (1974), Laminated Strength – A Direct Characterization Procedure, Composite Materials: Testing and Design (3rd Conf.) ASTM STP #546 (Amer. Soc. for Testing Materials, Philadelphia, U.S.A.), 188–206.

WU, E.M. (1974a), Strength and Fracture of Composites, in: L.J. BROUTMAN, ed., *Composite Materials*, Vol. 5 – Fatigue and Fracture (Academic Press, New York and London), 191–248.

WU, E.M. (1974b), Phenomenological Anisotropic Failure Criteria, in: G.P. SENDECKYJ, ed., *Composite Materials*, Vol. 2 – Mechanics of Composite Materials (Academic Press, New York and London), 353–431.

CHAPTER III

Dynamics of Composites with Cracks

G.C. SIH

Institute of Fracture and Solid Mechanics
Lehigh University
Bethlehem, PA 18015
U.S.A.

Contents

HANDBOOK OF COMPOSITES, VOL. 3 – Failure Mechanics of Composites
Edited by G.C. SIH and A.M. SKUDRA

1. Introduction

Advanced composite materials are now being widely used for aerospace applications. These materials are usually made of high-modulus graphite fibers embedded in an epoxy resin matrix so as to achieve a high strength-to-weight ratio. Their efficient and reliable use in design, however, are sensitive to the way with which they are fabricated. Unlike the behavior of metal alloys, composites may be anisotropic as well as nonhomogeneous. The microstructure arrangements and hence the macromechanical properties composites can vary over a wide range if procedures are not developed to assure near perfect control of the fabrication process. Quality control must be enforced; otherwise, no confidence can be placed in characterizing the mechanical behavior of composites.

The ability to model the behavior of multi-phase materials is extremely important if composites are to achieve practical application in high performance structures. It is insufficient simply to analyze material properties through anisotropic and/or nonhomogeneous theories in continuum mechanics. The modes of deformations would be seriously limited if the damage processes of fiber breaking, matrix cracking and/or debonding at the fiber-to-matrix interfaces are not accounted for. For the more complicated structural units such as laminated composites, delamination between the layers is another failure mechanism that needs to be considered. It is not always possible to accurately model the load transmission characteristics across the interface of two dissimilar media. For this reason alone, the correlation between actual composite response and predictions from analytical models is to say the least still very much lacking. It is problematic to distinguish the gross response of a composite with inherent fabricated inperfections in the form of voids or cracks from those occurred during loading. Unless the damage process is properly analyzed, test specimen data serve little or no useful purpose to composite structure design. This is mainly because each composite specimen itself tends to behave as a structure.

Dynamic effects introduce additional complexities to composite material analyses. The reflection and refraction of stress waves in an anisotropic and/or nonhomogeneous system would prohibit any solutions in simple form. Effective models, however, can be developed in situations where certain parameters tend to outweigh others. The problem can then be reduced to manageable properties. The discipline of fracture mechanics can be effective when the local failure mode leading to global instability is clearly identifiable. Often, matrix cracking or fiber breaking may be the dominant failure mode that leads to the loss of structure integrity. Material damage prior to global instability may or may not have to be considered depending on the loading rate and other variables. SIH (1979) has adopted this approach and provided numerous examples on thru-laminar cracking, delamination of layered system, etc., where the effects of mixed mode fracture[1] were significant. Caution, however, must be exercised in applying the fracture toughness K_{1c} parameter developed for metals to composites. The concept being restricted to homogeneous materials is valid only when K_{1c} is relatively insensitive to changes in loading direction, specimen geometry and microstructure orientation. There would be no advantage to use fracture mechanics if the so defined toughness parameter behaved as strength that changes with the relative position of load with respect to fibers. Hence the K_{1c} approach serves little or no useful purpose to the design of composite systems. The necessary information is the allowable load and net section size of the structure component.

Successful development of predictive models depends not only on effective stress analyses[2] but also on a suitable failure criterion. The opinions on how they should be combined remain diversified and often appear to be a matter of personal appreciation. The objective of this chapter is to emphasize the need for effective dynamic stress solutions local to defects which are the likely sites of failure initiation. Illustrations will be provided to show how the strain energy density criterion of SIH (1973) can be coupled with stress solutions for predicting crack behavior in nonhomogeneous media. This criterion[3] is valid for all materials modeled within the framework of continuum mechanics. One of its salient features is the strain energy density function $\mathrm{d}W/\mathrm{d}V$ can in general be expressed in the form S/r, where S is the strain energy density factor and r the distance from the site of failure initiation. In linear elasticity for sufficiently small r, S varies only with respect to the angular variable. This $1/r$ singular character[4] prevails φ regardless of the configuration of defects and nonhomogeneous nature of the composite. In addition, $\mathrm{d}W/\mathrm{d}V$ or S provides a unique physical interpretation of material damage assessed in terms of excessive distortion and/or dilatation at the macroscopic scale level. For more details, refer to SIH and CHEN (1981a).

2. Impact on composites with defects

Composites are known to be vulnerable to impact. Designers are in need of a better understanding of the response of advance composites to various types of

[1] In a fiber-reinforced composite, the growth direction of the defect is not known and depends on the relative orientation of load with the composite microstructure.

[2] The question of whether a composite system should be modeled as a homogeneous and anisotropic medium or nonhomogeneous and isotropic medium depends on the microstructure composition and loading rates and has been discussed by SIH et al. (1975).

[3] It has been used recently by SIH and MADENCI (1983) to study the influence of loading rates on subcritical crack growth in plastically deformed methods.

[4] Any failure criteria based on stress parameters are problematic because the order of stress singularities can depend on the defect configuration and material nonhomogeneity. This introduces inconsistencies for describing different failure modes in a composite.

impact as described in the ASTM Special Technical Publication 568 (1973). This Section provides a number of fundamental dynamic stress solutions for composites weakened by initial defects or cracks.

2.1. Elastodynamic equations

In what follows, the elastodynamic equations will be used to obtain effective solutions for several composite systems subjected to plane extensional, axisymmetrical and torsional loadings.

2.1.1. Plane extensional loading

Suppose that a composite system is modeled in two dimensions and subjected to loading in the xy-plane. Such a condition can be referred to as plane strain[5] in elasticity such that the composite is constrained from deformation in the z-direction. The x and y displacement components u_x and u_y in each of the constituents of the composite can be expressed in terms of two scalar potentials $\varphi(x, y, t)$ and $\psi(x, y, t)$:

$$u_x = \frac{\partial \varphi}{\partial x} + \frac{\partial \psi}{\partial y}, \qquad u_y = \frac{\partial \varphi}{\partial y} - \frac{\partial \psi}{\partial x}, \tag{1}$$

where the z-component displacement u_z is zero. Through the stress and displacement relations, it can be shown that the in-plane stress σ_x, σ_y and τ_{xy} take the forms

$$\begin{aligned}
\sigma_x &= 2G\left[\frac{\nu}{1-2\nu}\nabla^2\varphi + \frac{\partial^2\varphi}{\partial x^2} + \frac{\partial^2\psi}{\partial x \partial y}\right], \\
\sigma_y &= 2G\left[\frac{\nu}{1-2\nu}\nabla^2\varphi + \frac{\partial^2\phi}{\partial y^2} - \frac{\partial^2\psi}{\partial x \partial y}\right], \\
\tau_{xy} &= G\left(2\frac{\partial^2\varphi}{\partial x \partial y} + \frac{\partial^2\psi}{\partial y^2} - \frac{\partial^2\psi}{\partial x^2}\right),
\end{aligned} \tag{2}$$

with ν and G being, respectively, the Poisson's ratio and shear modulus. The Laplacian operator ∇^2 as in Equations (2) stands for $\partial^2/\partial x^2 + \partial^2/\partial y$. For plane strain,

$$\begin{aligned}
\sigma_z &= \nu(\sigma_x + \sigma_y) = \frac{2\nu G}{1-2\nu}\nabla^2\varphi, \\
\tau_{xz} &= \tau_{yz} = 0.
\end{aligned} \tag{3}$$

The satisfaction of the equations of motion yields a pair of wave equations

$$c_1^2\nabla^2\varphi = \frac{\partial^2\varphi}{\partial t^2}, \qquad c_2^2\nabla^2\psi = \frac{\partial^2\psi}{\partial t^2}, \tag{4}$$

[5] Plane stress solution is not of interest in crack analysis, because the state of affairs near a crack is always plane strain, SIH (1971).

in which c_1 and c_2 are the dilatational and shear wave speeds given by

$$c_1 = \left[2\left(\frac{1-\nu}{1-2\nu}\right)\frac{G}{\rho}\right]^{1/2}, \qquad c_2 = \left(\frac{G}{\rho}\right)^{1/2}. \tag{5}$$

Here, ρ is the mass density. The additional subscripts m, f and c will be used for identifying those quantities associated with the matrix, fiber and composite, respectively.

2.1.2. Axisymmetric deformation

If the loading, crack and composite configuration possess axisymmetry, the deformation can be expressed more conveniently in terms of the cylindrical polar coordinates (r, φ, z). A pair of potentials $\varphi(r, z, t)$ and $\psi(r, z, t)$ can be defined as before for the radial and axial displacements

$$u_r = \frac{\partial \varphi}{\partial r} - \frac{\partial \psi}{\partial z}, \qquad u_z = \frac{\partial \varphi}{\partial z} + \frac{\partial \psi}{\partial r} - \frac{\psi}{r}, \tag{6}$$

where $u_\theta = 0$. The longitudinal direction of the composite is taken along the z-axis. The corresponding stress state can be found as

$$\begin{aligned}
\sigma_r &= 2G\left[\frac{\partial}{\partial r}\left(\frac{\partial \varphi}{\partial r} - \frac{\partial \psi}{\partial z}\right) + \frac{\nu}{1-2\nu}\nabla^2\varphi\right],\\
\sigma_\theta &= 2G\left[\frac{1}{r}\left(\frac{\partial \varphi}{\partial r} - \frac{\partial \psi}{\partial z}\right) + \frac{\nu}{1-2\nu}\nabla^2\varphi\right],\\
\sigma_z &= 2G\left[\frac{\partial}{\partial z}\left(\frac{\partial \varphi}{\partial z} + \frac{\partial \psi}{\partial r} + \frac{\psi}{r}\right) + \frac{\nu}{1-2\nu}\nabla^2\varphi\right],\\
\tau_{rz} &= G\left[\frac{\partial}{\partial z}\left(2\frac{\partial \varphi}{\partial r} - \frac{\partial \psi}{\partial z}\right) + \frac{\partial}{\partial r}\left(\frac{\partial \psi}{\partial r} + \frac{\psi}{r}\right)\right].\\
\tau_{r\theta} &= \tau_{\theta z} = 0
\end{aligned} \tag{7}$$

The operator ∇^2 in the (r, θ, z) system stands for $\partial^2/\partial r^2 + (1/r)\partial/\partial r + \partial^2/\partial z^2$. Similarly, the governing wave equations became

$$\begin{aligned}
c_1^2\left(\frac{\partial^2 \varphi}{\partial r^2} + \frac{1}{r}\frac{\partial \varphi}{\partial r} + \frac{\partial^2 \varphi}{\partial z^2}\right) &= \frac{\partial^2 \varphi}{\partial t^2},\\
c_2^2\left(\frac{\partial^2 \psi}{\partial r^2} + \frac{1}{r}\frac{\partial \psi}{\partial r} - \frac{\psi}{r^2} + \frac{\partial^2 \psi}{\partial z^2}\right) &= \frac{\partial^2 \psi}{\partial t^2}.
\end{aligned} \tag{8}$$

Refer to Equations (4) for c_1 and c_2.

2.1.3. Torsional shear

Let the cylindrical composite body be twisted dynamically about the z-axis such that the only displacement component is in the θ-direction:

$$\begin{aligned} u_\theta &= w(r, z, t), \\ u_r &= u_z = 0. \end{aligned} \tag{9}$$

The two non-zero stress components are

$$\tau_{r\theta} = G\left(\frac{\partial w}{\partial r} - \frac{w}{r}\right), \qquad \tau_{\theta z} = G\frac{\partial w}{\partial z}. \tag{10}$$

This leads to the wave equation

$$c_2^2\left(\frac{\partial^2 w}{\partial r^2} + \frac{1}{r}\frac{\partial w}{\partial r} - \frac{w}{r^2} + \frac{\partial^2 w}{\partial z^2}\right) = \frac{\partial^2 w}{\partial t^2}, \tag{11}$$

where c_2 is given by the second of Equations (4).

2.2. Plane crack parallel to fibers

Consider a unidirectional fiber-reinforced composite as illustrated in Fig. 1. A defect in the shape of a plane crack with width $2a$ is situated in the matrix with properties G_m, ν_m and ρ_m while the fiber has properties G_f, ν_f and ρ_f. The combination of fiber and matrix has the properties[6] G_c, ν_c and ρ_c. Both in-plane normal and shear impact will be treated such that the superimposed solution gives loading in an arbitrary direction.

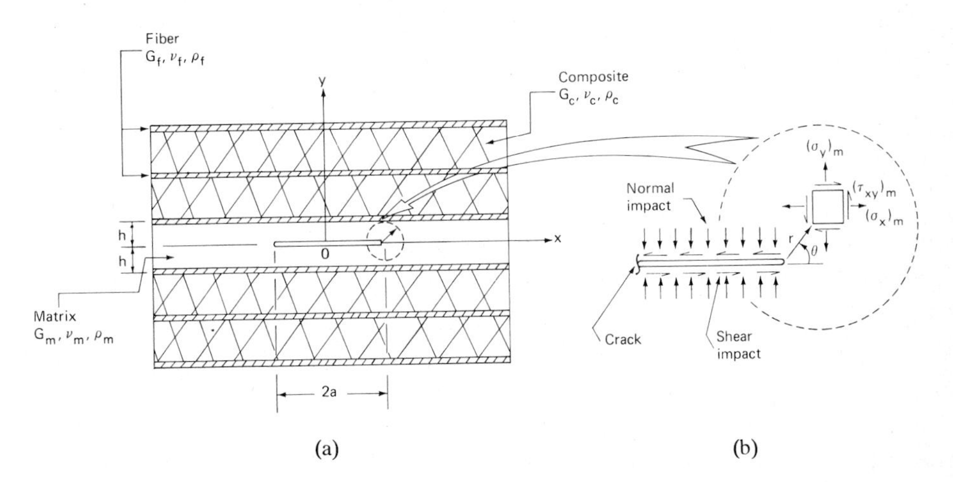

FIG. 1. Impact of unidirectional composite with cracked matrix. (a) Crack parallel to fibers. (b) Details of local region.

[6] The elastic constant (G_c, ν_c) of the composite can be found from those of the constituents (G_m, ν_m) and (G_t, ν_f). The specific relations have been given by SIH and CHEN (1981a).

2.2.1. Normal impact

The defect in the matrix suddenly experiences a normal impact of magnitude σ_0. This can be specified mathematically by the heaviside unit step function $H(t)$. Referring to the coordinate system in Fig. 1(a), the following boundary conditions prevail for $t > 0$

$$\left.\begin{aligned}(\sigma_y)_m &= -\sigma_0 H(t)\\ (\tau_{xy})_m &= 0\end{aligned}\right\} \quad 0 \leqslant |x| < a\ ; \quad y = 0\,. \tag{12}$$

Symmetry across the x-axis further imposes the conditions

$$(u_y)_m = (\tau_{xy})_m = 0\,, \quad |x| \geqslant a\ ; \quad y = 0\,. \tag{13}$$

The cracked matrix layer of thickness $2h$ is assumed to be bonded perfectly to the neighboring material that behaves as a composite with the average properties G_c, ν_c and ρ_c. The conditions for displacement and stress continuity at $y = \pm h$ for all values of x and time t are

$$\left.\begin{aligned}(u_x)_m = (u_x)_c\ ; \quad (u_y)_m = (u_y)_c\\ (\sigma_y)_m = (\sigma_y)_c\ ; \quad (\tau_{xy})_m = (\tau_{xy})_c\end{aligned}\right\} \quad \text{for all } x \text{ and } y = \pm h \tag{14}$$

SIH and CHEN (1981a) have solved this problem by application of an integral transform method developed by SIH et al. (1972). The dynamic crack tip stress field was found to possess the inverse square root singularity in terms of the radial distance r measured from the crack tip while the angular distribution is expressed by θ. The results are

$$\begin{aligned}(\sigma_x)_m &= \frac{k_1(t)}{\sqrt{2r}} \cos\tfrac{1}{2}\theta(1 - \sin\tfrac{1}{2}\theta \sin\tfrac{3}{2}\theta) + \cdots\,,\\ (\sigma_y)_m &= \frac{k_1(t)}{\sqrt{2r}} \cos\tfrac{1}{2}\theta\,(1 + \sin\tfrac{1}{2}\theta \sin\tfrac{3}{2}\theta) + \cdots\,,\\ (\tau_{xy})_m &= \frac{k_1(t)}{\sqrt{2r}} \cos\tfrac{1}{2}\theta \sin\tfrac{1}{2}\theta \cos\tfrac{3}{2}\theta + \cdots\end{aligned} \tag{15}$$

in which the stress intensity factor $k_1(t)$ varies as a function of time, i.e.

$$k_1(t) = f_1[(c_2)_m t/a]\sigma_0\sqrt{a}\,. \tag{16}$$

The higher order terms of r in Equations (15) are not significant because their magnitudes tend to become negligibly small as r approaches the crack tip, Fig. 1(b). The stress component $(\sigma_z)_m$ is equal to $\nu_m[(\sigma_x)_m + (\sigma_y)_m]$.

The numerical values of $f_1[(c_2)_m t/a]$ in Equation (16) are displayed graphically in Figs. 2 and 3 for different ratios of G_c/G_m and a/h. Nonhomogeneities

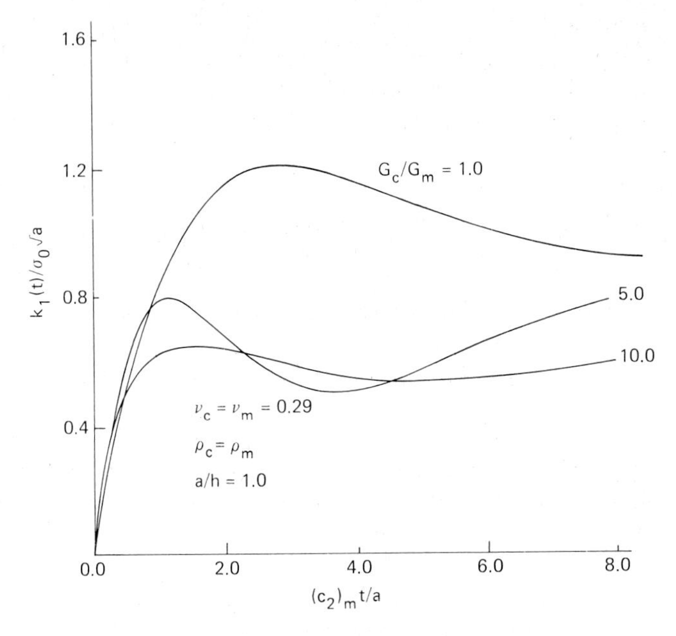

FIG. 2. Normal impact stress intensity factor versus time for different shear moduli ratio.

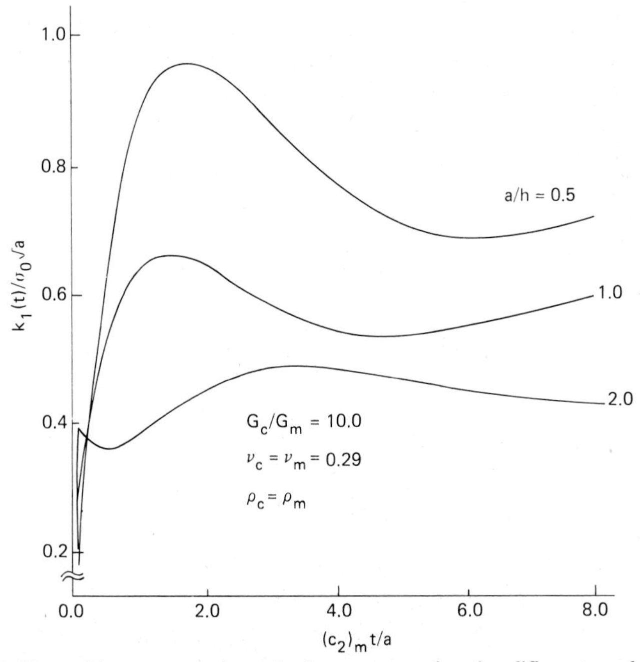

FIG. 3. Normal impact stress intensity factor versus time for different crack length to fiber spacing ratio.

arising from the Poisson's ratio and mass density are small. Hence, it is justified to assume $\nu_c \simeq \nu_m = 0.29$ and $\rho_c \simeq \rho_m$. For $a/h = 1.0$, Fig. 2 gives a plot of the normalized stress intensity factor $k_1(t)/\sigma_0\sqrt{a}$ against the dimensionless time parameter $(c_2)_m t/a$. All the curves tend to rise to a maximum and then decrease in magnitude with the homogeneous case $G_c = G_m$ giving the highest peak. The influence of the crack length to the matrix layer thickness is exhibited in Fig. 3 for $T_c/G_m = 10$. The dynamic stress intensity factor $k_1(t)$ for a longer crack with $a/h = 2.0$ is lower than for a shorter crack with $a/h = 0.5$. There is the tendency for the crack to stabilize as it becomes longer. The kink in the curve for $a/h = 2.0$ and small t represents the arrival of the reflected waves from the interface. It does not show up in the curve for $a/h = 0.5$ as in this case the interface is further away from the crack and the reflected waves are relatively weaker in strength.

2.2.2. Shear impact

The corresponding problem of shear impact with magnitude τ_0 has also been treated by SIH and CHEN (1981a). Specified for $t > 0$ are

$$\left.\begin{aligned}(\tau_{xy})_m &= -\tau_0 H(t)\\ (\sigma_y)_m &= 0\end{aligned}\right\} \quad 0 \leqslant |x| < a\ ;\ y = 0\,, \tag{17}$$

together with the conditions of skew-symmetry:

$$(u_x)_m = (\sigma_y)_m = 0\,, \quad |x| \geqslant a\ ;\ y = 0\,. \tag{18}$$

The assumptions of perfect bonding remain the same as those given in Equations (14).

With reference to the local polar coordinates r and θ in Fig. 1(b), the singular terms of the local stresses can be expressed in terms of the stress intensity factor $k_2(t)$:

$$\begin{aligned}(\sigma_x)_m &= -\frac{k_2(t)}{\sqrt{2r}} \sin\tfrac{1}{2}\theta(2 + \cos\tfrac{1}{2}\theta\cos\tfrac{3}{2}\theta) + \cdots\,,\\ (\sigma_y)_m &= \frac{k_2(t)}{\sqrt{2r}} \sin\tfrac{1}{2}\theta\cos\tfrac{1}{2}\theta\cos\tfrac{3}{2}\theta + \cdots\,,\\ (\tau_{xy})_m &= \frac{k_2(t)}{\sqrt{2r}} \cos\tfrac{1}{2}\theta(1 - \sin\tfrac{1}{2}\theta\sin\tfrac{3}{2}\theta) + \cdots\end{aligned} \tag{19}$$

in which $k_2(t)$ is given by

$$k_2(t) = f_2[(c_2)_m t/a]\tau_0\sqrt{a}\,. \tag{20}$$

The same ratios of a/h and G_c/G_m are used to compute the numerical values of $f_2[(c_2)_m t/a]$. Comparing the results in Fig. 4 for $k_2(t)$ with those in Fig. 3 for

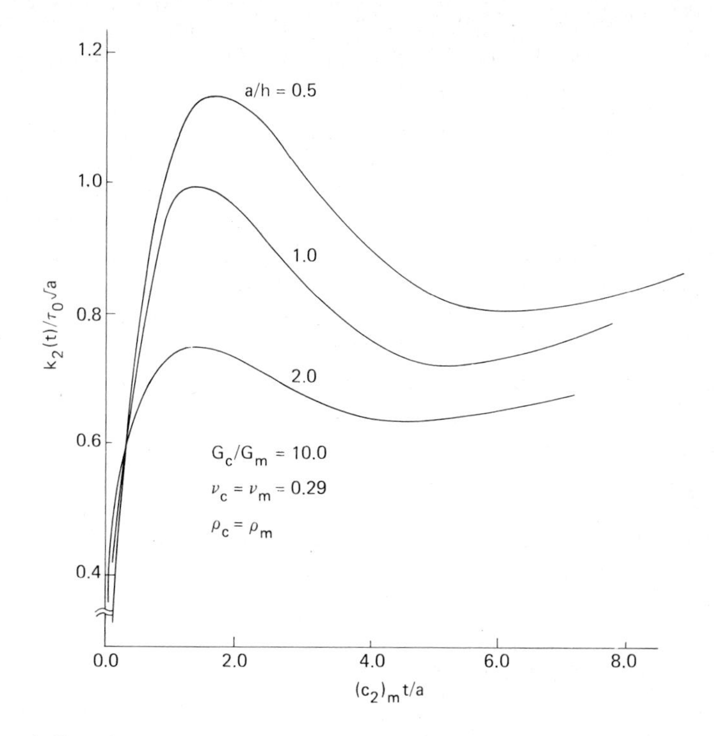

FIG. 4. Shear impact stress intensity factor versus time for different crack length to fiber spacing ratio.

$k_1(t)$, the curves possess the same general trend except that $k_2(t) > k_1(t)$ in magnitude. Shear impact is seen to be more detrimental to matrix cracking.

In general, both normal and shear impact may occur simultaneously. Failure initiation will depend on a combination of $k_1(t)$ and $k_2(t)$ reaching some critical condition, say

$$F[k_1(t), k_2(t)] = F_c \,. \tag{21}$$

The specific form of Equation (21) will be discussed subsequently in conjunction with the strain energy density criterion.

2.3. *Plane crack normal to fibers*

Another possible orientation of the defect is for a crack positioned normal to the fibers, Fig. 5. The remaining conditions are the same as those specified in Section 2.2 for a crack parallel to the fibers. The fibers are now parallel to the y-axis and hence the continuity of displacements and stresses in Equations (14) is referred to $x = \pm h$ and all values of y.

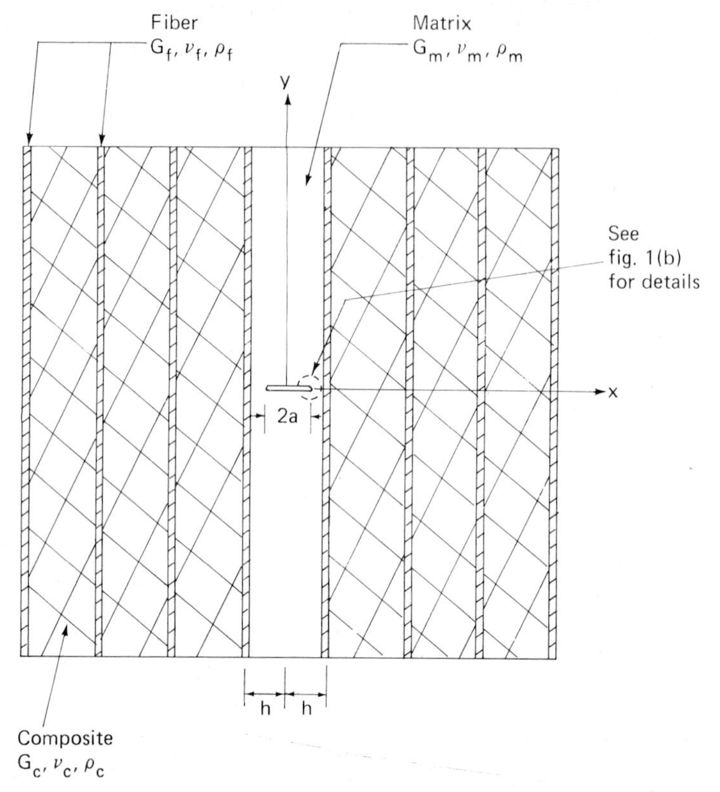

FIG. 5. Crack normal to fibers under impact.

2.3.1. Normal tractions

Since the crack is still in matrix material, the functional relationship of the dynamic crack tip stress field is the same as that given to Equations (15). The intensity of this stress field is changed as the crack is isolated by 90° since this will affect the dynamic load transmission characteristics, i.e.,

$$k_1(t) = g_1[(c_2)_m t/a]\sigma_0\sqrt{a}\,. \tag{22}$$

The crack boundary conditions are the same as those specified by Equations (12).

By letting $\rho_c \simeq \rho_m$, $\nu_c \simeq \nu_m = 0.29$ and $G_c/G_m = 10$, numerical values of $g_1[(c_2)_m t/a]$ are given in Fig. 6 for a/h = 0.2, 0.4, . . . , 0.8. For the same values of a/h and G_c/G_m, the curves in Fig. 6 are greater in magnitude than those in Fig. 3. Unlike static loading, the crack tip region can be stressed more severely for a crack normal rather than parallel to the fibers. This is because the tips of a normally-oriented crack being closer to the fibers of a given composite experience more intensive reflected waves.

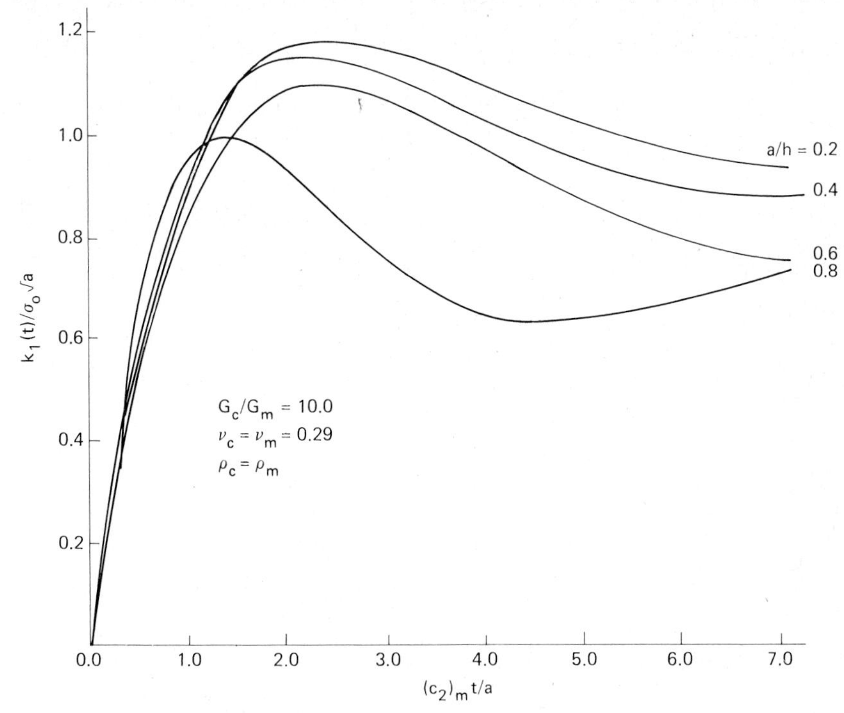

FIG. 6. Normal impact stress intensity versus time for crack normal to fibers.

2.3.2. *Shear tractions*

Applying the conditions in Equations (17) and (18) to the crack in Fig. 6, SIH and CHEN (1981a) have also shown that $k_2(t)$ in Equation (19) becomes

$$k_2(t) = g_2[(c_2)_m t/a]\tau_0\sqrt{a}\ . \tag{23}$$

Fig. 7 again indicates that the local dynamic stresses oscillate with time with a maximum peak occurring during the early state of wave reflection from the crack tip and interface. The interpretation of the results in Equation (23) is similar to those in Equation (22) and will not be repeated.

The stress intensity factor results of Sections 2.2 and 2.3 may be superimposed to yield the general solution for a crack tilted at an arbitrary angle with reference to the fibers.

2.4. *Circular defect between broken fibers*

The wave equations given by Equations (8) and (11) apply to elastodynamic problems possessing axisymmetry or rotational symmetry. This pertains to geometry of fibers with circular cross sections subjected to uniform stretching or torsion. A likely site of failure initiation is illustrated in Fig. 8(a) where the

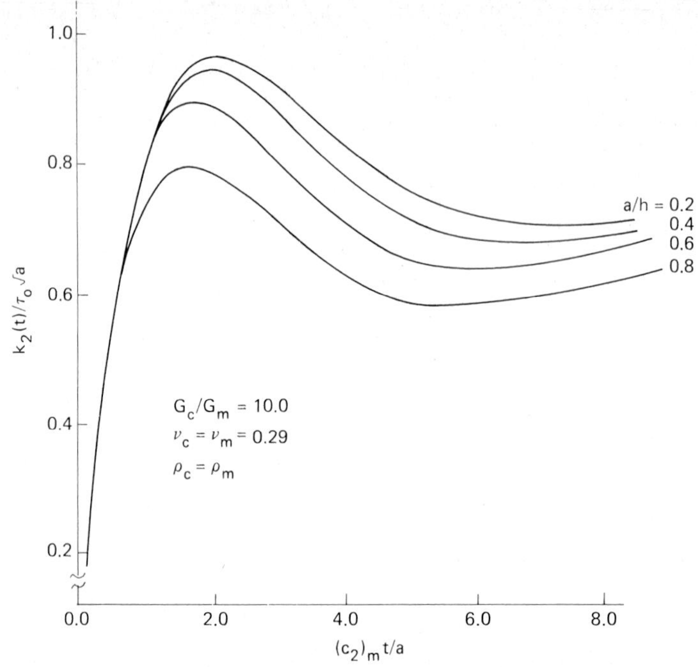

FIG. 7. Shear impact stress intensity factor versus time for crack normal to fibers.

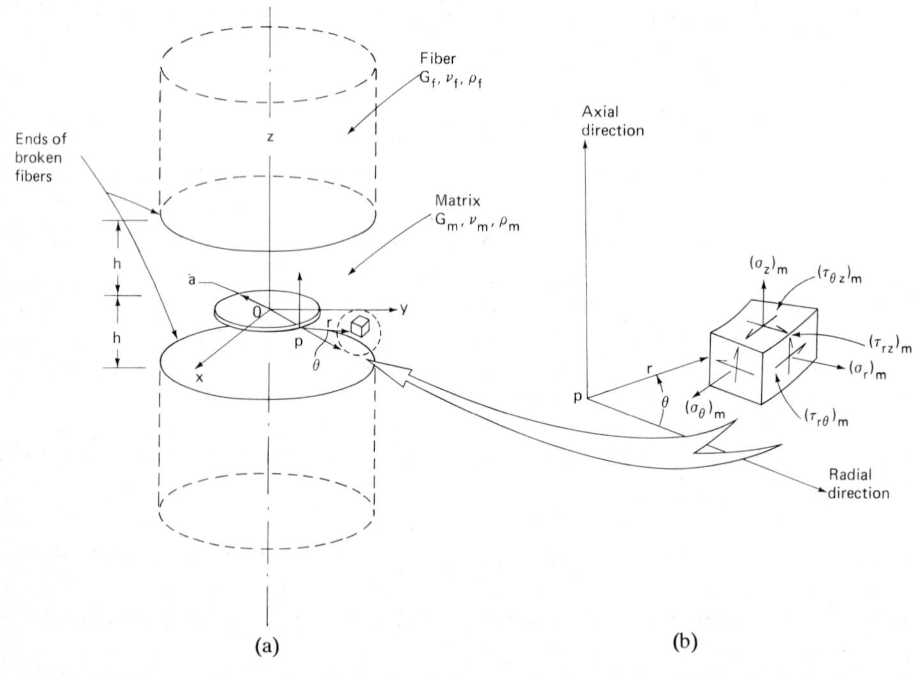

FIG. 8. Sudden stretching and torsion of crack between broken fibers. (a) Circular defect between broken fibers. (b) Stress components in polar cylindrical coordinates.

ends of two broken circular fibers are spaced at a distance $2h$ apart. A defect in the shape of a circle with radius a is in between the gap as shown with reference to the rectangular Cartesian coordinates (x, y, z) in Fig. 8(a).

2.4.1. *Sudden stretching*

The circular defect is suddenly stretched in the z-direction such that the displacement field as described in Equations (6) is independent of θ with $u_\theta = 0$. The required boundary conditions on the plane $z = 0$ are

$$\left.\begin{aligned}(\sigma_z)_\text{m} &= -\sigma_0 H(t)\\ (\tau_{rz})_\text{m} &= 0\end{aligned}\right\} \quad 0 \leqslant r < a;\ z = 0\,, \tag{24}$$

while symmetry requires that

$$(u_z)_\text{m} = (\tau_{rz})_\text{m} = 0\,, \quad r \geqslant a;\ z = 0\,. \tag{25}$$

The displacement and stresses at the interfaces where the matrix meets with the ends of the fibers are assumed to be continuous, i.e.,

$$\left.\begin{aligned}(u_r)_\text{m} &= (u_r)_\text{f}\,; \quad (u_z)_\text{m} = (u_z)_\text{f}\\ (\sigma_z)_\text{m} &= (\sigma_z)_\text{f}\,; \quad (\tau_{rz})_\text{m} = (\tau_{rz})_\text{f}\end{aligned}\right\} \quad \text{for all } r \text{ and } z = \pm h\,. \tag{26}$$

By means of the integral transform method, SIH and CHEN (1981a) have found the local stresses on the element in Fig. 8(b) as

$$\begin{aligned}
(\sigma_r)_\text{m} &= \frac{k_1(t)}{\sqrt{2r}} \cos\tfrac{1}{2}\theta(1 - \sin\tfrac{1}{2}\theta \sin\tfrac{3}{2}\theta) + \cdots\,,\\
(\sigma_\theta)_\text{m} &= \frac{k_1(t)}{\sqrt{2r}} 2\nu_m \cos\tfrac{1}{2}\theta + \cdots,\\
(\sigma_z)_\text{m} &= \frac{k_1(t)}{\sqrt{2r}} \cos\tfrac{1}{2}\theta(1 + \sin\tfrac{1}{2}\theta \sin\tfrac{3}{2}\theta) + \cdots,\\
(\tau_{rz})_\text{m} &= \frac{k_1(t)}{\sqrt{2r}} \cos\tfrac{1}{2}\theta \sin\tfrac{1}{2}\theta \cos\tfrac{3}{2}\theta + \cdots\,,\\
(\tau_{\theta z})_\text{m} &= (\tau_{r\theta})_\text{m} = 0\,.
\end{aligned} \tag{27}$$

Note that the r and θ dependence of the dynamic stresses for a circular defect are the same as those in Equations (15) for a plane defect. The condition of plane strain again prevails near the crack border since the relation $(\sigma_\theta)_\text{m} = \nu_\text{m}[(\sigma_r)_\text{m} + (\sigma_z)_\text{m}]$ holds for Equations (27). The stress intensity factor $k_1(t)$ is time dependent

$$k_1(t) = \tfrac{1}{2}\pi p_1[(c_2)_\text{m} t/a]\sigma_0\sqrt{a}\,. \tag{28}$$

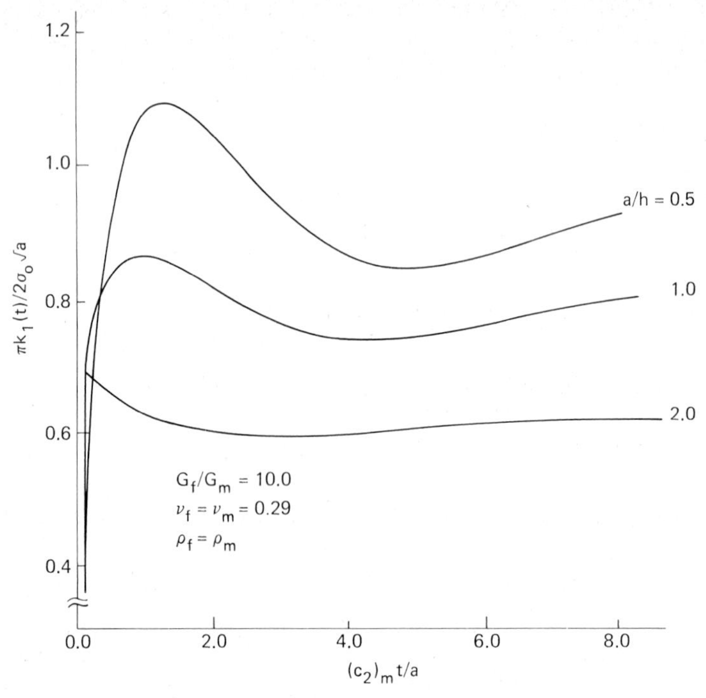

FIG. 9. Normal impact stress intensity versus time for defect in between fibers.

Assuming that $\rho_m = \rho_f$, $\nu_m \simeq \nu_f - 0.29$ and $G_f/G_m = 10.0$, Fig. 9 gives a plot of $p_1[(c_2)_m t/a]$ versus time for $a/h = 0.5$, 1.0 and 2.0. The value of $k_1(t)$ is seen to rise very quickly at first reaching a peak and then decreases to the static solution $k_1 = 2\sigma_0\sqrt{a}/\pi$. As the defect size becomes larger than the gap, the peak values of $k_1(t)$ occur at smaller time t. This is because the reflected waves from the ends of the fibers arrive sooner.

2.4.2. *Sudden twisting*

Let the composite system in Fig. 8(a) be twisted suddenly such that both the fibers and matrix experience only angular displacement u_θ, Equation (9). The wave Equation (11) is to be solved according to the conditions

$$\begin{aligned} (\tau_{\theta z})_m &= -\left(\frac{r_1}{a}\right)\tau_1\sqrt{a}\,, \quad 0 \leq r < a;\ z = 0\,, \\ (u_\theta)_m &= 0\,, \quad r \geq a;\ z = 0\,, \end{aligned} \tag{29}$$

where r_1 represents the radial distance in the xy-plane. The shear stress variation in Equation (29) corresponds to the sudden application of a torque. Continuity of shear stress and angular displacement are given by

$$\left.\begin{aligned} (u_\theta)_m &= (u_\theta)_f \\ (\tau_{\theta z})_m &= (\tau_{\theta z})_f \end{aligned}\right\} \quad \text{for all } r \text{ and } z = \pm h\,. \tag{30}$$

Referring to the stress element near the defect in Fig. 8(b), only the two shear stress components $(\tau_{r\theta})_m$ and $(\tau_{\theta z})_m$ are nonzero:

$$(\tau_{r\theta})_m = -\frac{k_3(t)}{\sqrt{2r}}\sin\tfrac{1}{2}\theta + \cdots, \qquad (\tau_{\theta z})_m = \frac{k_3(t)}{\sqrt{2r}}\cos\tfrac{1}{2}\theta + \cdots \tag{31}$$

with $k_3(t)$ being given by

$$k_3(t) = \tfrac{3}{4}\pi p_3[(c_2)_m t/a]\tau_1\sqrt{a}\,. \tag{32}$$

In the limit as t becomes infinitely large, $p_3[(c_2)_m t/a]$ tends to unity and Equation (32) reduces to the static solution of $k_3(t) = 3\pi\tau_1\sqrt{a}/4$. Fig. 10 gives the numerical values of the normalized stress intensity factor $k_3(t)$ for $\rho_m = \rho_f$ and $G_f/G_m = 10.0$. For a suddenly applied twist, $k_3(t)$ increases to a maximum within a short period of time and then the curve levels off. This character is seen for all three values of $a/h = 0.5$, 1.0 and 2.0.

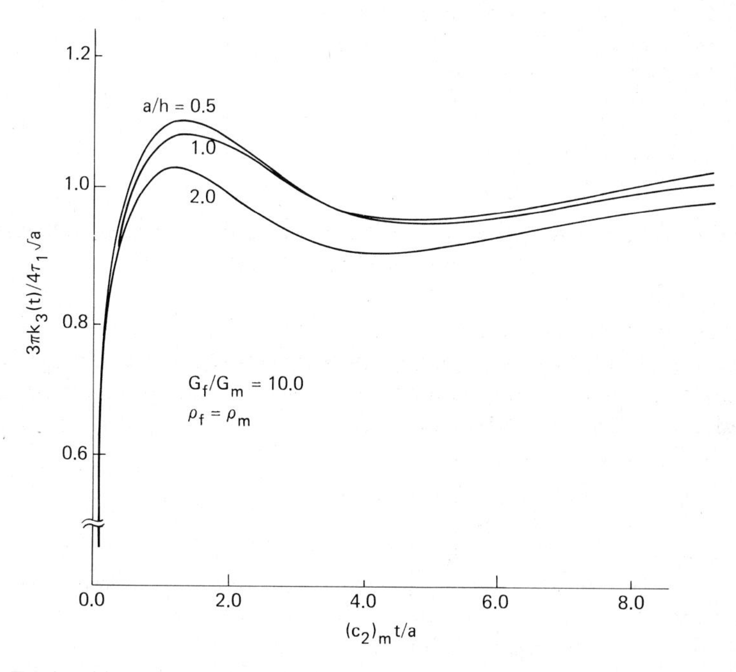

FIG. 10. Torsional impact stress intensity factor versus time for defect in between broken fibers.

2.5. *Partially cracked fiber*

Fig. 11 shows a single strand of fiber with properties G_f, ν_f and ρ_f embedded in a matrix with properties G_m, ν_m and ρ_m such that $\rho_m \simeq \rho_f$ and $\nu_m \simeq \nu_f = 0.29$. The fiber of radius b contains an imperfection in the form of a circular crack of radius $a \leq b$. The dynamic stress intensity within the fiber due to a normal and torsional impact will be considered.

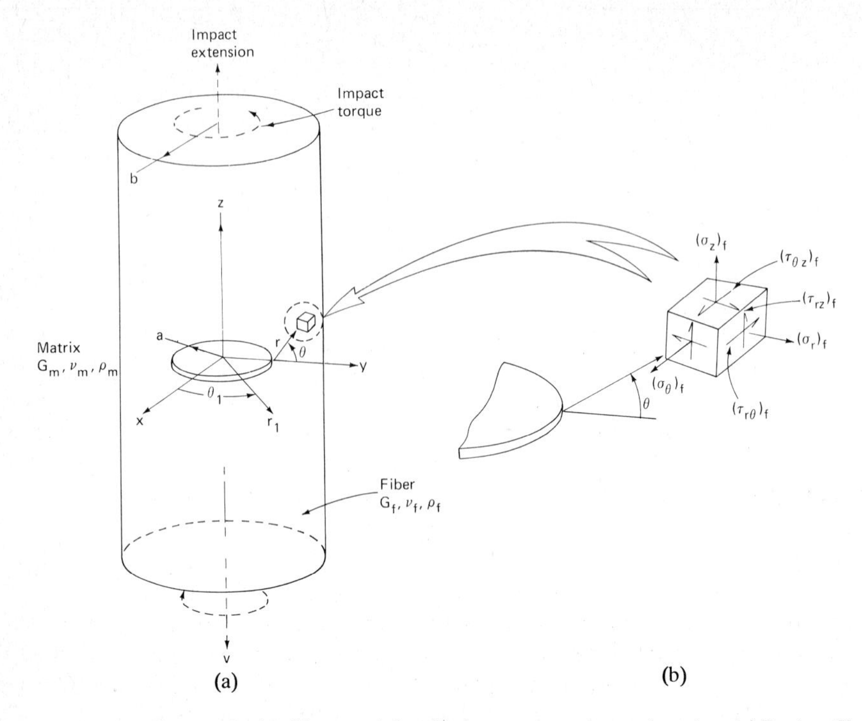

FIG. 11. Broken fibers embedded in material under impact extension and torsion. (a) Broken fiber in matrix. (b) Crack border stresses.

2.5.1. Fiber under normal impact

The defect in the fiber suddenly experiences a normal impact of magnitude σ_0. With reference to the coordinate system (x, y, z) in Fig. 11(a), the following conditions are to be observed:

$$\left.\begin{array}{l}(\sigma_z)_m = -\sigma_0 H(t) \\ (\tau_{rz})_m = 0\end{array}\right\} \quad 0 \leqslant r < a \; ; \; z = 0 \tag{33}$$

and

$$(u_z)_m = (\tau_{rz})_m = 0 \, , \quad r \geqslant a; z = 0 \, . \tag{34}$$

The fiber is bonded perfectly to the matrix along the surface $r = b$. This implies that the displacements and stresses are continuous:

$$\left.\begin{array}{l}(u_r)_m = (u_r)_f \; ; \quad (u_z)_m = (u_z)_f \\ (\sigma_r)_m = (\sigma_r)_f \; ; \quad (\tau_{rz})_m = (\tau_{rz})_f\end{array}\right\} \quad \text{for all } z \text{ and } r = b \, . \tag{35}$$

Near the crack border, the local stresses on the element in Fig. 11(b) are singular in terms of r and they are given by

$$(\sigma_r)_f = \frac{k_1(t)}{\sqrt{2r}} \cos\tfrac{1}{2}\theta(1 - \sin\tfrac{1}{2}\theta \sin\tfrac{3}{2}\theta) + \cdots,$$

$$(\sigma_\theta)_f = \frac{k_1(t)}{\sqrt{2r}} 2\nu_f \cos\tfrac{1}{2}\theta + \cdots,$$

$$(\sigma_z)_f = \frac{k_1(t)}{\sqrt{2r}} \cos\tfrac{1}{2}\theta(1 + \sin\tfrac{1}{2}\theta \sin\tfrac{3}{2}\theta) + \cdots, \tag{36}$$

$$(\tau_{rz})_f = \frac{k_1(t)}{\sqrt{2r}} \cos\tfrac{1}{2}\theta \sin\tfrac{1}{2}\theta \cos\tfrac{3}{2}\theta + \cdots,$$

$$(\tau_{\theta z})_f = (\tau_{r\theta})_f = 0\,.$$

In Equations (36), the coefficient $k_1(t)$ is common to all the stresses:

$$k_1(t) = \tfrac{1}{2}\pi q_1[(c_2)_f t/a]\sigma_0\sqrt{a}\,. \tag{37}$$

As expected, the dynamic stress intensity factor increases with the ratio a/b. Fig. 12 displays the variations of $q_1[(c_2)_f t/a]$ with $(c_2)_f t/a$ for $G_f/G_m = 10.0$ and $a/b = 0.2, 0.4, \ldots, 0.8$. The value of $k_1(t)$ increases more rapidly as a/b approaches unity, i.e., when the defect is close to breaking the fiber.

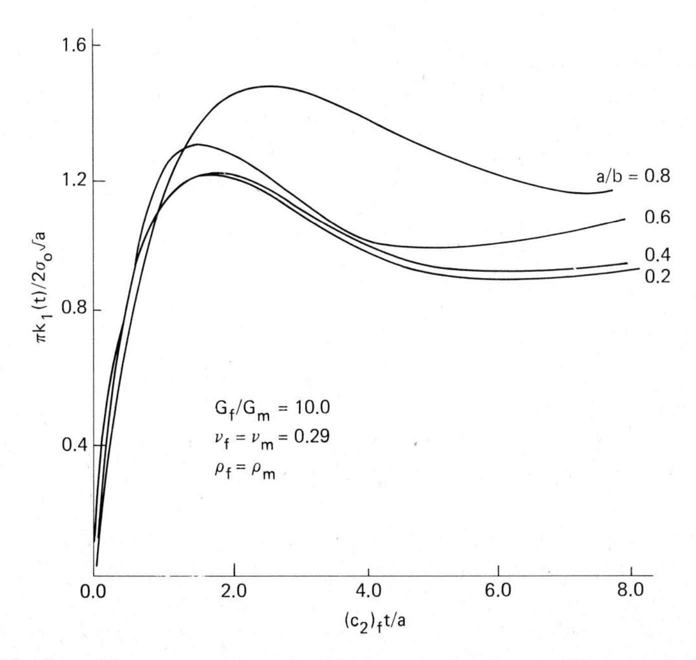

FIG. 12. Normal impact stress intensity factor versus time for partially cracked fiber.

2.5.2. Fiber under torsional impact

A suddenly applied torque corresponds to a shear that varies linearly with the radial distance r_1 defined in Fig. 11(a). On the plane $z = 0$, the required conditions are

$$\begin{aligned}(\tau_{\theta z})_f &= -\left(\frac{r_1}{a}\right)\tau_1 H(t), \quad 0 \leqslant r < a;\ z = 0, \\ (u_\theta)_f &= 0, \quad r \geqslant a;\ z = 0,\end{aligned} \tag{38}$$

while continuity gives

$$\left.\begin{aligned}(u_\theta)_m &= (u_\theta)_f \\ (\tau_{\theta z})_m &= (\tau_{\theta z})_f\end{aligned}\right\} \quad \text{for all } z \text{ and } r = b. \tag{39}$$

Equations (38) and (39) together with Equations (9) to (11) lead to the asymptotic stress solution:

$$\begin{aligned}(\tau_{r\theta})_f &= -\frac{k_3(t)}{\sqrt{2r}} \sin \tfrac{1}{2}\theta + \cdots, \\ (\tau_{\theta z})_f &= \frac{k_3(t)}{\sqrt{2r}} \cos \tfrac{1}{2}\theta + \cdots.\end{aligned} \tag{40}$$

in which $k_3(t)$ takes the form

$$k_3(t) = \tfrac{3}{4}\pi q_3[(c_2)_f t/a]\tau_1\sqrt{a}\,. \tag{41}$$

Comparing the results of Equation (41) given in Fig. 13 with those of $k_1(t)$ in

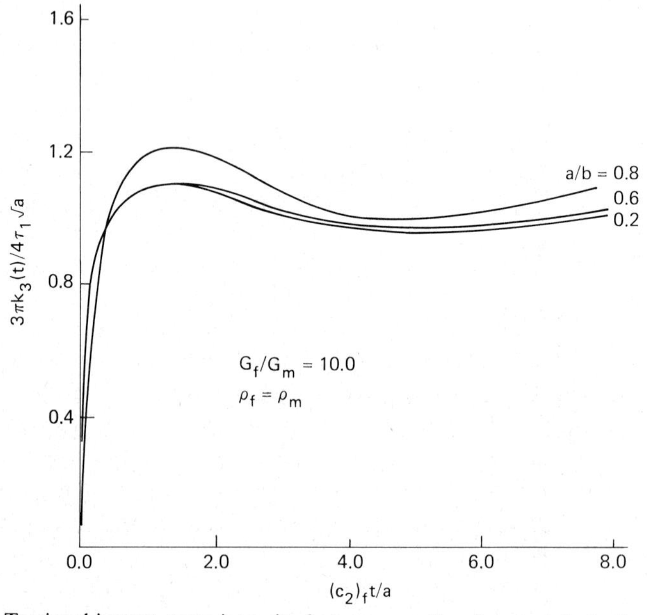

FIG. 13. Torsional impact stress intensity factor versus time for partially broken fiber.

Fig. 12, they behave in a similar fashion and no additional discussion is required.

3. Moving cracks in composites

Once the crack driving force exceeds the resistance of the material, the crack is set in motion. Its path of propagation is irregular depending on the material nonhomogeneity loading and component geometry. Because of the inability to model the local nonhomogeneity in a composite, it is extremely difficult to determine the available energy that is being converted to create free crack surfaces as a function of time. In this discussion, the analysis is simplified to assume that a crack will move only in one phase of the composite material, i.e., the matrix as shown in Fig. 14(a). The assumption of a constant velocity crack will also be invoked so as to further simplify the mathematical treatment. In reality, crack acceleration is such a quick event that for all practical purposes, it can be neglected.

3.1. Basic equations

The motion of the crack will be referred to a set of stationary coordinates (X, Y, Z). A crack of fixed length $2a$ is assumed to travel at a constant velocity c. A set of moving coordinates (x, y, z) is attached to the crack such that they are related to (X, Y, Z) by the Galilean transformation

$$x = X - ct, \qquad y = Y, \qquad z = Z. \tag{42}$$

Because of the steady state nature of the problem, the time variable t can be eliminated in the formulation.

3.1.1. Extensional motion of crack

For loads that are directed only in the XY or xy-plane, the displacement field is two-dimensional in that the Z or z-component u_Z can be set to zero for the case of plane strain:

$$u_X = \frac{\partial \varphi}{\partial X} + \frac{\partial \psi}{\partial Y}, \qquad u_Y = \frac{\partial \varphi}{\partial Y} - \frac{\partial \psi}{\partial X}, \qquad u_Z = 0. \tag{43}$$

The corresponding stress components are the same as those in Equations (2) except that the space variables X and Y will be used instead of x and y. A pair of wave equations in terms of the potentials $\varphi(X, Y, t)$ and $\psi(X, Y, t)$ will also be obtained. Making use of the transformation in Equation (42), the wave equations may be expressed in terms of the moving coordinates x and y:

$$\lambda_1^2 \frac{\partial^2 \varphi}{\partial x^2} + \frac{\partial^2 \varphi}{\partial y^2} = 0, \qquad \lambda_2^2 \frac{\partial^2 \psi}{\partial x^2} + \frac{\partial^2 \psi}{\partial y^2} = 0, \tag{44}$$

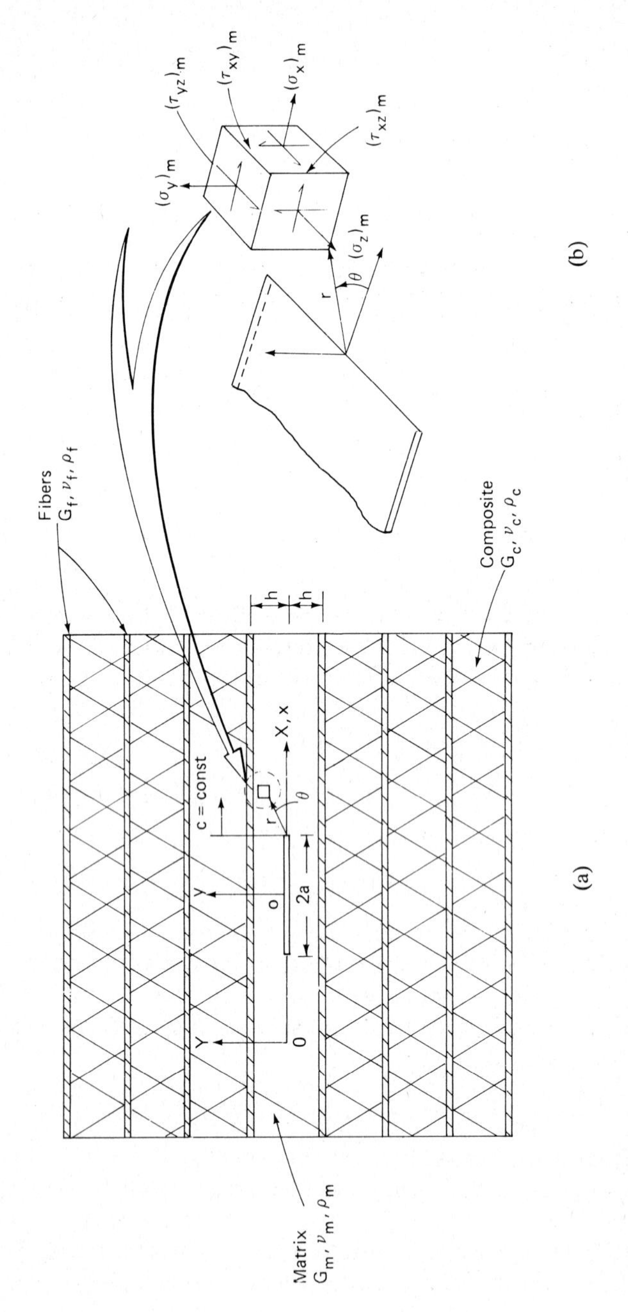

FIG. 14. Matrix cracking under extensional and anti-shear loading. (a) Constant velocity cracking of matrix. (b) Stress components in moving coordinates.

in which λ_1 and λ_2 are defined as

$$\lambda_1 = [1-(c/c_1)^2]^{1/2}, \qquad \lambda_2 = [1-(c/c_2)^2]^{1/2}. \tag{45}$$

Once the initial and boundary conditions are specified, Equations (44) can be solved.

3.1.2. Anti-plane shear motion of crack

Under anti-plane shear, the composite in Fig. 14(a) is subjected to a load such that all material elements experience displacement in the Z or z-direction only, i.e.,

$$u_X = u_Y = 0, \qquad u_Z = w(X, Y, t). \tag{46}$$

The non-vanishing stress components are

$$\tau_{XZ} = G\frac{\partial w}{\partial X}, \qquad \tau_{YZ} = G\frac{\partial w}{\partial Y}, \tag{47}$$

which may be substituted into the equations of motion to yield a wave equation of the form given by Equation (11). Application of Equation (42) transforms the independent variables from (X, Y, t) into (x, y):

$$\lambda_2^2\frac{\partial^2 w}{\partial x^2} + \frac{\partial^2 w}{\partial y^2} = 0, \tag{48}$$

where λ_2 is given by the second of Equations (45).

3.2. Crack extending under tensile load

Suppose that a unidirectional composite in Fig. 14(a) contains a running crack of length $2a$ in one of the matrix layers with properties G_m, ν_m and ρ_m. The surrounding material with properties G_c, ν_c and ρ_c is loaded with a tensile stress such that the following conditions prevail

$$\left.\begin{aligned}(\sigma_y)_m &= -\sigma_0\\ (\tau_{xy})_m &= 0\end{aligned}\right\} \quad 0 \leqslant |x| < a;\ y = 0 \tag{49}$$

and

$$(u_y)_m = (\tau_{xy})_m = 0, \quad |x| \geqslant a;\ y = 0. \tag{50}$$

The matrix layer is attached perfectly to the surrounding material at $y = \pm h$. This implies that

$$\left.\begin{array}{ll}(u_x)_m = (u_x)_c\,; & (u_y)_m = (u_y)_c \\ (\sigma_y)_m = (\sigma_y)_c\,; & (\tau_{xy})_m = (\tau_{xy})_c\end{array}\right\} \quad \text{for all } x \text{ and } y = \pm h\,. \tag{51}$$

SIH and CHEN (1982) have solved this problem with the aid of the integral transform method and found the local stress field in terms of the polar coordinates

$$r = [(x-a)^2 + y^2]^{1/2}\,, \qquad \theta = \tan^{-1}\left[\frac{y}{x-a}\right] \tag{52}$$

as defined in Fig. 14(b). For plane strain, the local stresses take the forms

$$\begin{aligned}
(\sigma_x)_m &= \frac{k_1(c)}{\sqrt{2r}H[(\lambda_1)_m, (\lambda_2)_m]}\{[1+(\lambda_2)_m^2][2(\lambda_1)_m^2 + 1 - (\lambda_2)_m^2]f[(\lambda_1)_m] \\
&\qquad - 4(\lambda_1)_m(\lambda_2)_m f[(\lambda_2)_m]\} + \cdots\,, \\
(\sigma_y)_m &= \frac{k_1(c)}{\sqrt{2r}H[(\lambda_1)_m, (\lambda_2)_m]}\{4(\lambda_1)_m(\lambda_2)_m f[(\lambda_2)_m] - [1+(\lambda_2)_m^2]f[(\lambda_1)_m]\} + \cdots\,, \\
(\sigma_z)_m &= \frac{k_1(c)}{\sqrt{2r}H[(\lambda_1)_m, (\lambda_2)_m]}\nu_m\{[1+(\lambda_2)_m^2][2(\lambda_1)_m^2 - (\lambda_2)_m^2]f[(\lambda_1)_m]\} + \cdots\,, \\
(\tau_{xy})_m &= \frac{k_1(c)}{\sqrt{2r}H[(\lambda_1)_m, (\lambda_2)_m]}[2(\lambda_1)_m[1+(\lambda_2)_m^2]\{g[(\lambda_1)_m] - g[(\lambda_2)_m]\}] + \cdots\,, \\
(\tau_{xz})_m &= (\tau_{yz})_m = 0\,.
\end{aligned} \tag{53}$$

In Equations (53), the function $H[(\lambda_1)_m, (\lambda_2)_m]$ stands for

$$H[(\lambda_1)_m, (\lambda_2)_m] = 4(\lambda_1)_m(\lambda_2)_m - [1+(\lambda_2)_m^2]^2\,. \tag{54}$$

The angular distribution of the dynamic stresses are described by the two functions

$$\begin{aligned}
f^2(\lambda) + g^2(\lambda) &= \sec\theta(1+\lambda^2\tan^2\theta)^{-1/2}\,, \\
f^2(\lambda) = g^2(\lambda) &= \sec\theta(1+\lambda^2\tan^2\theta)^{-1}\,.
\end{aligned} \tag{55}$$

The dynamic stress singularity is the same as the static case, i.e., $1/\sqrt{r}$ while the angular distribution of the stresses is distorted by the crack speed v through the parameters λ_1 and λ_2 for the matrix.

Numerical values of the dynamic stress intensity factor

$$k_1(c) = R_1[c/(c_2)_m]\sigma_0\sqrt{a} \tag{56}$$

have been reported by SIH and CHEN (1982) through the function $R_1[c/(c_2)_m]$ which is plotted in Fig. 15 for $G_c/G_m = 10.0$, $\nu_m = \nu_c = 0.25$ and $\rho_m = \rho_c$. The

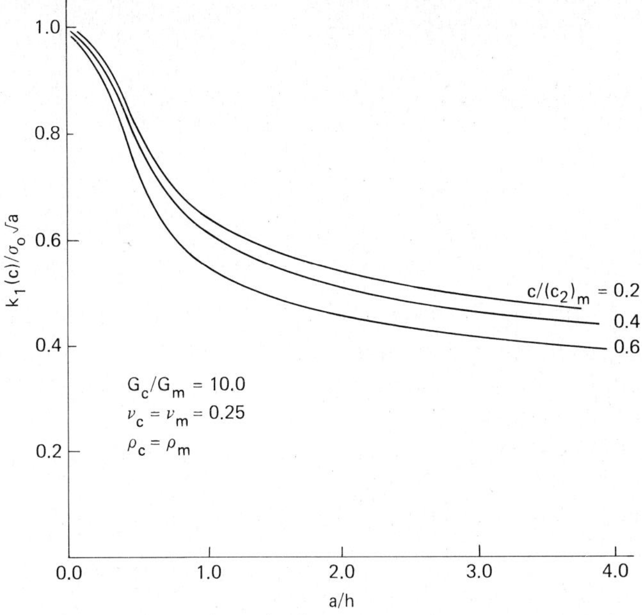

FIG. 15. Normalized in-plane stress intensity factor versus crack length to fiber spacing ratio for different crack velocities.

dynamic stress intensity factor k_1 is seen to decrease as the ratio a/h increases. The maximum value of $k_1 = \sigma_0\sqrt{a}$ corresponds to the homogeneous case when the matrix layer height $2h$ is so large that the properties of the composite G_c, ν_c and ρ_c exert no influence on the crack tip stresses. The effect material nonhomogeneity thus tends to reduce the local stress intensity depending on a/h and the crack speed to shear wave velocity ratio $c/(c_2)_m$. A greater reduction in k_1 occurs at higher crack speeds.

3.3. *Running crack under anti-plane shear*

Let the crack in Fig. 14(a) be driven by an anti-plane shear load such that

$$\begin{aligned} (\tau_{yz})_m &= -\tau_1\,, \quad 0 \leqslant |x| < a;\, y = 0\,, \\ w_m &= 0\,, \quad |x| \geqslant a;\, y = 0\,. \end{aligned} \tag{57}$$

Continuity of displacements and stresses at $y = \pm h$ are specified as

$$\left.\begin{aligned} w_m &= w_c \\ (\tau_{yz})_m &= (\tau_{yz})_c \end{aligned}\right\} \quad \text{for all } x \text{ and } y = \pm h\,. \tag{58}$$

Defining a set of polar coordinates r and θ as indicated in Fig. 14(b) with

$$r = [(x-a)^2 + (\lambda_2)_m^2 y^2]^{1/2}, \qquad \theta = \tan^{-1}\left[\frac{(\lambda_2)_m y}{x-a}\right], \tag{59}$$

the asymptotic dynamic shear stresses are found to be

$$\begin{aligned} (\tau_{xz})_m &= -\frac{k_3}{\sqrt{2r}} \sin \tfrac{1}{2}\theta + \cdots, \\ (\tau_{yz})_m &= \frac{k_3}{\sqrt{2r}} \cos \tfrac{1}{2}\theta + \cdots. \end{aligned} \tag{60}$$

Although the functional relationships of $(\tau_{xz})_m$ and $(\tau_{yz})_m$ in terms of θ are the same as those for the static case, the definition of θ in Equations (59) shows that the angular variations of the anti-plane shear stresses depend on $(\lambda_2)_m$ or $c/(c_2)_m$. The stress intensity factor

$$k_3(c) = R_3[c/(c_2)_m]\tau_1\sqrt{a} \tag{61}$$

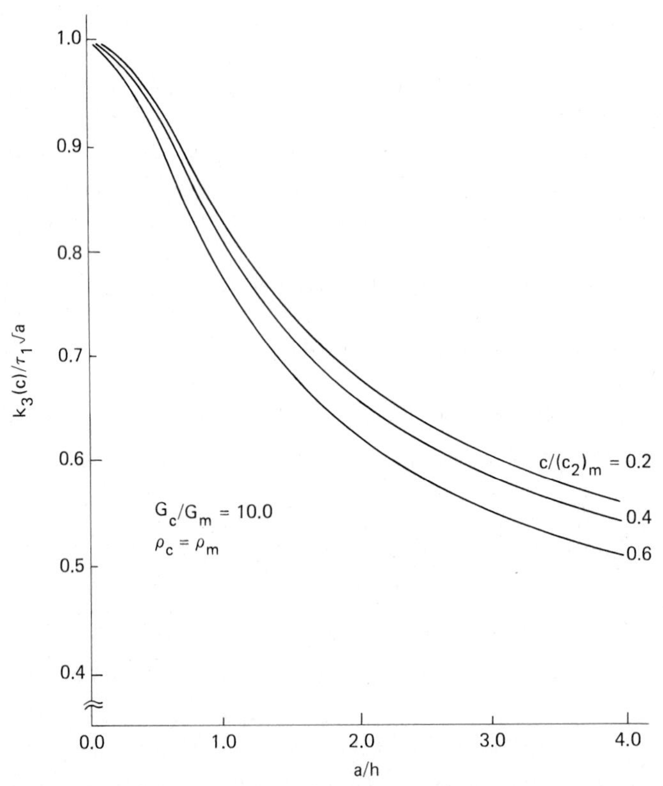

FIG. 16. Anti-plane shear stress intensity factor versus crack length to fiber spacing ratio for different crack velocities.

also depends on $c/(c_2)_m$ and the results are shown numerically in Fig. 16 for $G_c/G_m = 10.0$ and $\rho_m = \rho_c$. The variations of $k_3/\tau_1\sqrt{a}$ with a/h for different values of $c/(c_2)_m$ are the same as those shown in Fig. 15 for k_1.

3.4. *Effect of orthotropy on crack motion*

For unidirectional composites with high volume fractions, SIH et al. (1975) have shown that homogeneous anisotropy is a good assumption. The principal directions of orthotropy 1, 2 and 3 shall coincide with X, Y and Z such that the crack moves in the weaker direction of the composite, i.e., the 1- or X-direction as illustrated in Fig. 17. ARCISZ and SIH (1982) have presented solutions for six different unidirectional composite systems. This work will be discussed briefly.

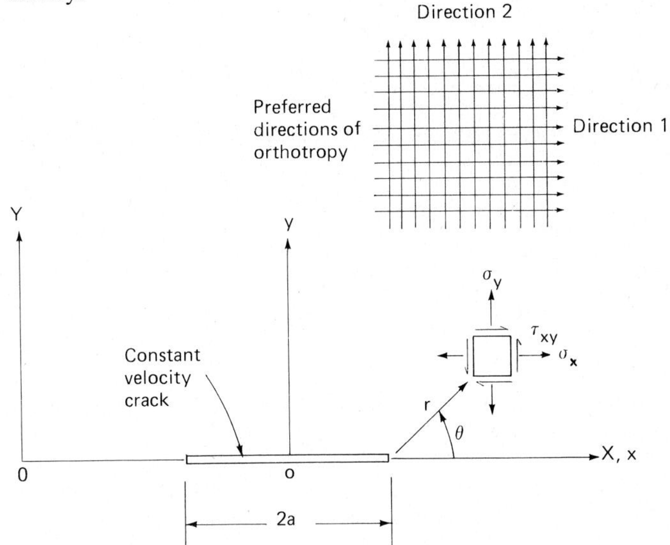

FIG. 17. Constant velocity crack in orthotropic medium.

3.4.1. *Orthotropic theory of elasticity*

Consider a state of plane strain referred to a set of stationary coordinates (X, Y) as follows:

$$u_X = u(X, Y, t), \qquad u_Y = v(X, Y, t), \qquad u_Z = 0. \tag{62}$$

It follows that for the orthotropic body in Fig. 17, the stresses can be expressed in terms of the strains by the relations

$$\begin{aligned}\sigma_X &= c_{11}e_{XX} + c_{12}e_{YY},\\ \sigma_Y &= c_{12}e_{XX} + c_{22}e_{YY},\\ \sigma_Z &= \nu_{31}\sigma_X + \nu_{32}\sigma_Y,\\ \tau_{XY} &= 2G_{12}e_{XY},\\ \tau_{XZ} &= \tau_{YZ} = 0.\end{aligned} \tag{63}$$

The elastic constants c_{11}, c_{12}, etc., are defined as

$$c_{11} = \frac{E_1}{\Delta}(1 - \nu_{23}\nu_{32}),$$

$$c_{12} = \frac{E_1}{\Delta}(\nu_{21} + \nu_{23}\nu_{31}) = \frac{E_2}{\Delta}(\nu_{12} + \nu_{13}\nu_{32}), \tag{64}$$

$$c_{22} = \frac{E_2}{\Delta}(1 - \nu_{13}\nu_{32}),$$

in which Δ is the determinant

$$\Delta = \begin{vmatrix} 1, & -\nu_{12}, & -\nu_{13} \\ -\nu_{21}, & 1, & -\nu_{23} \\ -\nu_{31}, & -\nu_{32}, & 1 \end{vmatrix}. \tag{65}$$

The equations of motion expressed in terms of $u(X, Y, t)$ and $v(X, Y, t)$ are given by

$$c_{11}\frac{\partial^2 u}{\partial X^2} + G_{12}\frac{\partial^2 u}{\partial Y^2} + (c_{12} + G_{12})\frac{\partial^2 v}{\partial X \partial Y} = \rho\frac{\partial^2 u}{\partial t^2},$$

$$G_{12}\frac{\partial^2 v}{\partial X^2} + c_{22}\frac{\partial^2 v}{\partial Y^2} + (c_{12} + G_{12})\frac{\partial^2 u}{\partial X \partial Y} = \rho\frac{\partial^2 v}{\partial t^2}. \tag{66}$$

Referring Equations (66) to a system of moving coordinates (x, y), it can be easily shown that

$$(c_{11} - \rho c^2)\frac{\partial^2 u}{\partial x^2} + G_{12}\frac{\partial^2 u}{\partial y^2} + (c_{12} + G_{12})\frac{\partial^2 v}{\partial x \partial y} = 0,$$

$$(c_{12} + G_{12})\frac{\partial^2 u}{\partial x \partial y} + (G_{12} - \rho c^2)\frac{\partial^2 v}{\partial x^2} + c_{22}\frac{\partial^2 v}{\partial y^2} = 0, \tag{67}$$

with c being the crack velocity and ρ the mass density.

3.4.2. Plane extension

Equations (67) were solved by ARCISZ and SIH (1982) for a moving crack subjected to applied stresses symmetrical with reference to the x-axis. The boundary conditions are

$$\left.\begin{aligned} \sigma_y &= -\sigma_0 \\ \tau_{xy} &= 0 \end{aligned}\right\} \quad 0 \leq |x| < a;\ y = 0 \tag{68}$$

while symmetry requires that

$$\left.\begin{aligned} v &= 0 \\ \tau_{xy} &= 0 \end{aligned}\right\} \quad |x| \geqslant a;\ y = 0\,. \tag{69}$$

By means of Fourier transform, the solution depends on two real roots η_1 and η_2 satisfying the equation

$$\begin{aligned} \eta^4 - \frac{1}{c_{22}G_{12}}\{[c_{22}(c_{11} - \rho c^2) + G_{12}(G_{12} - \rho c^2) \\ - (c_{12} + G_{12})^2]\eta^2 + (c_{11} - \rho c^2)(G_{12} - \rho c^2)\} = 0 \end{aligned} \tag{70}$$

such that

$$\sqrt{c_{11}c_{22}} - G_{12} \geqslant c_{12} + G_{12}\,. \tag{71}$$

The crack velocity must satisfy the inequality

$$\rho c^2 < \left[G_{12}, \frac{c_{11}c_{22} + G_{12}^2 - (c_{12} + G_{12})^2}{c_{22} + G_{12}}\right]_{\min}. \tag{72}$$

Without going into details, the dynamic stresses near the moving crack tip take the forms

$$\begin{aligned} \sigma_x &= \frac{k_1}{2\sqrt{r}}\frac{1}{M_4}[(c_{12}\eta_1 M_2 - c_{11})\sqrt{1 + \mu_1^{-1}\cos\theta} \\ &\quad + (c_{12}\eta_2 M_3 - c_{11})M_1\sqrt{1 + \mu_2^{-1}\cos\theta}] + \cdots, \\ \sigma_y &= \frac{k_1}{2\sqrt{r}}\frac{1}{M_4}[(c_{22}\eta_1 M_2 - c_{12})\sqrt{1 + \mu_1^{-1}\cos\theta} \\ &\quad + (c_{22}\eta_2 M_3 - c_{12})M_1\sqrt{1 + \mu_2^{-1}\cos\theta}] + \cdots, \\ \tau_{xy} &= -\frac{k_1}{2\sqrt{r}}\frac{G_{12}}{M_4}[(\eta_1 + M_2)\sqrt{1 - \mu_1^{-1}\cos\theta} \\ &\quad + (\eta_2 + M_3)M_1\sqrt{1 - \mu_2^{-1}\cos\theta}] + \cdots \end{aligned} \tag{73}$$

in which

$$\mu_j = \sqrt{\cos^2\theta + \eta_j^2\sin^2\theta}\,, \quad j = 1, 2\,. \tag{74}$$

Refer to Fig. 17 for the definition of r and θ. The quantities M_j ($j = 1, 2, \ldots, 4$) stand for

$$\begin{aligned} M_1 &= -\frac{\eta_2}{\eta_1}\frac{c_{11} - \rho c^2 + \eta_1^2 c_{22}}{c_{11} - \rho c^2 + \eta_2^2 c_{22}}, & M_2 &= \frac{c_{11} - \rho c^2 - \eta_1^2 G_{12}}{\eta_1(c_{12} + G_{12})}, \\ M_3 &= \frac{c_{11} - \rho c^2 - \eta_2^2 G_{12}}{\eta_2(c_{12} + G_{12})}, & M_4 &= c_{22}M_2\eta_1 - c_{12} + M_1(c_{22}\eta_2 M_3 - c_{12})\,. \end{aligned} \tag{75}$$

The stress component σ_z can be obtained readily from σ_x and σ_y in accordance with the third of Equations (63). The stress intensity factor k_1 in Equations (73) is defined to be the same as that of a stationary crack in a uniform stress field:

$$k_1 = \sigma_0 \sqrt{a}\,. \tag{76}$$

For an isotropic material, Equations (75) simplify to

$$M_1 = -\frac{2\eta_1\eta_2}{1+\eta_2^2}, \quad M_2 = \eta\,, \quad M_3 = \frac{1}{\eta_2}, \quad M_4 = G\left[\frac{4\eta_1\eta_2}{1+\eta_2^2} - (1+\eta_2^2)\right] \tag{77}$$

where η_1 and η_2 equal to λ_1 and λ_2 as defined in Equations (45).

Once the elastic constants are known, the numerical values of the stresses in Equations (73) can be easily computed. For the E-glass epoxy unidirectional composite with 56.5% fiber volume fraction, the constants are

$$\begin{aligned} &E_1 = 6.13\times 10^3\ \text{ksi}\,, \qquad \nu_{12} = \nu_{13} = 0.27\,,\\ &E_2 = 1.42\times 10^3\ \text{ksi}\,, \qquad \nu_{23} = 0.34\,,\\ &G_{12} = 0.80\times 10^3\ \text{ksi}\,. \end{aligned} \tag{78}$$

The normalized values of the parameters c_{ij} $(i, j = 1, 2)$ are

$$c_{11}/G_{12} = 12.190\,, \qquad c_{12}/G_{12} = 1.155\,, \qquad c_{22}/G_{12} = 3.139\,, \tag{79}$$

where $c_{11} > c_{22}$ and $G_{12} = G_{23} < G_{23}$. The crack is aligned to move parallel with the fibers. The influence of orthotropy will be discussed subsequently in connection with the strain energy density criterion. No new information can be gained from the stress intensity factor expression k_1 in Equation (76) since it is identical to the static expression.

4. Dynamic behavior of composite laminates

A laminate is a stack of thin plates bonded together to act as an integral structural element, Fig. 18. The overall mechanical behavior of the laminate is determined from the properties of the constituent layers depending on how the loads are transmitted through the individual layers. In this discussion, the interfaces are bonded perfectly where the displacements and stresses are continuous and no delamination occurs. A defect in the shape of a through crack is assumed to exist in the laminate that can be stretched or bent suddenly. The dynamic laminate theories of SIH and CHEN (1981b, 1981c) will be employed.

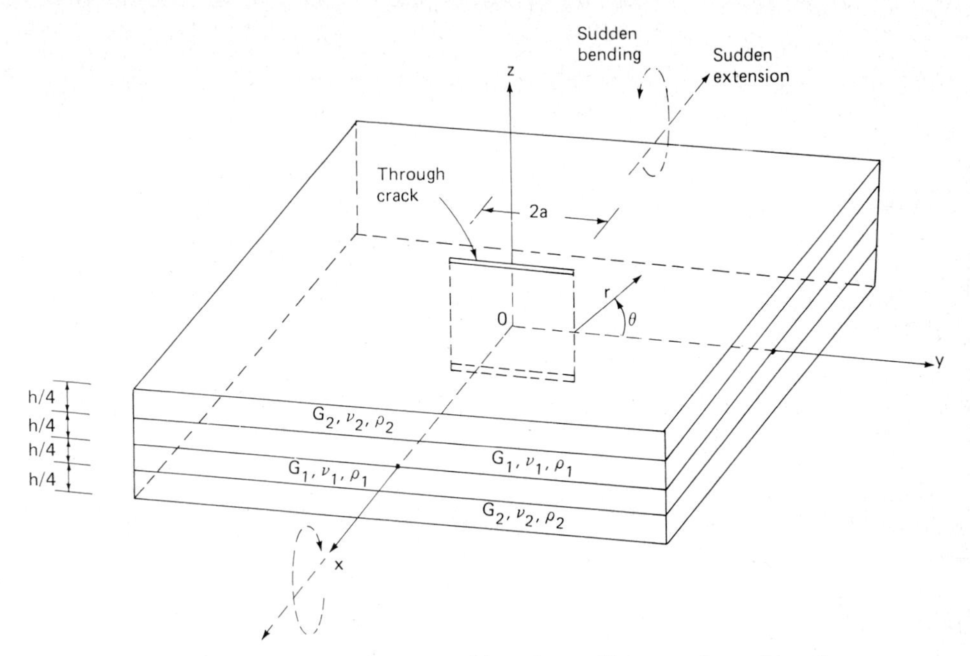

FIG. 18. Through crack in laminate subjected to sudden extension and bending.

4.1. *Extensional motion of laminate plate*

When a laminate is stretched suddenly, stress waves will be generated and propagated throughout the system. The coupling between the extensional and thickness reflective waves is essential in a laminate composite structure and cannot be neglected. This is particularly significant in the neighborhood of a crack-like defect where the stress field is three-dimensional in character. Without loss in generality, the four-layer laminate system in Fig. 17 will be considered.

4.1.1. *Theory of Sih and Chen: plate extension*

SIH and CHEN (1981c) have assumed the following time-dependent displacement field

$$u_x = u(x, y, t), \qquad u_y = v(x, y, t), \qquad u_z = \frac{2z}{h} w(x, y, t) \tag{80}$$

and developed a dynamic extensional laminate plate theory. The coordinates (x, y, z) are placed such that z is directed normal to the laminate surface with h being the total thickness. It is customary to formulate plate theories in terms of the stress resultants (Fig. 19),

$$(N_x, N_y, N_z, N_{xy}) = \int_{-h/2}^{h/2} (\sigma_x, \sigma_y, \sigma_z, \tau_{xy})\, dz \tag{81}$$

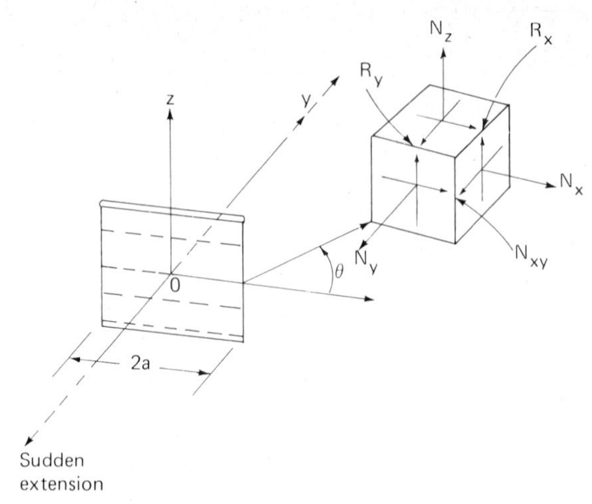

FIG. 19. Stress resultants and transverse shears near crack in four-layered laminate extended suddenly.

and the transverse shears

$$(R_x, R_y) = \int_{-h/2}^{h/2} (\tau_{xz}, \tau_{yz}) z \, dz \,. \tag{82}$$

Substituting Equations (80) into the stress-displacement relations and applying Equations (81) and (82) to the laminate structure in Fig. 18, it is found that[7]

$$\begin{aligned}
N_x &= \tfrac{1}{2}h\left[(\alpha+2\beta)\frac{\partial u}{\partial x} + \alpha\frac{\partial v}{\partial y}\right] + \alpha\kappa w\,,\\
N_y &= \tfrac{1}{2}h\left[(\alpha+2\beta)\frac{\partial v}{\partial x} + \alpha\frac{\partial u}{\partial y}\right] + \alpha\kappa w\,,\\
N_z &= (\alpha+2\beta)\kappa^2 w + \tfrac{1}{2}\alpha\kappa h\left(\frac{\partial u}{\partial x} + \frac{\partial v}{\partial y}\right),\\
N_{xy} &= \tfrac{1}{2}\beta h\left(\frac{\partial u}{\partial y} + \frac{\partial v}{\partial x}\right)
\end{aligned} \tag{83}$$

and

$$R_x = \tfrac{1}{48}\gamma h^2 \frac{\partial w}{\partial x}\,, \qquad R_y = \tfrac{1}{48}\gamma h^2 \frac{\partial w}{\partial y}\,. \tag{84}$$

The constant $\kappa = \pi/\sqrt{12}$ in Equations (83) and (84) accounts for coupling between the extensional and thickness stress waves. With ρ_1 and ρ_2 as the mass density of the inner and outer layers of the laminate, the equations governing the three unknowns u, v and w are

[7] The theory can be easily extended to a laminate made of anisotropic layers.

$$\beta\nabla^2 u + (\alpha+\beta)\frac{\partial}{\partial x}\left(\frac{\partial u}{\partial x}+\frac{\partial v}{\partial y}\right) + \frac{2\alpha\kappa}{h}\frac{\partial w}{\partial x} = (\rho_1+\rho_2)\frac{\partial^2 u}{\partial t^2},$$

$$\beta\nabla^2 v + (\alpha+\beta)\frac{\partial}{\partial y}\left(\frac{\partial u}{\partial x}+\frac{\partial v}{\partial y}\right) + \frac{2\alpha\kappa}{h}\frac{\partial w}{\partial y} = (\rho_1+\rho_2)\frac{\partial^2 v}{\partial t^2}, \tag{85}$$

$$\gamma\nabla^2 w - \frac{48}{h^2}(\alpha+2\beta)\kappa^2 w - \frac{24\alpha\kappa}{h}\left(\frac{\partial u}{\partial x}+\frac{\partial v}{\partial y}\right) = (\rho_1+7\rho_2)\frac{\partial^2 w}{\partial t^2}$$

in which $\nabla^2 = \partial^2/\partial x^2 + \partial^2/\partial y^2$ and

$$\alpha = 2\left(\frac{G_1\nu_1}{1-2\nu_1}+\frac{G_2\nu_2}{1-2\nu_2}\right), \qquad \beta = G_1 + G_2, \qquad \gamma = G_1 + 7G_2. \tag{86}$$

The system of Equations (85) have been solved by SIH and CHEN (1981c) for a through crack.

4.1.2. *Sudden stretching*

Let the through crack in Fig. 18 experience a sudden stretching of the amount N_0 which is maintained constant for all time therefore. Specified are then the conditions

$$\left.\begin{aligned} N_y &= -N_0 H(t) \\ N_{xy} &= 0 \end{aligned}\right\} \quad 0 \leqslant |x| < a;\ y = 0 \tag{87}$$

and

$$v = R_y = 0, \quad |x| \geqslant a;\ y = 0. \tag{88}$$

In plate theory, the local stress field is expressed in terms of the resultant stresses:

$$\begin{aligned} N_x &= \frac{k_1(t)}{\sqrt{2r}}\cos\tfrac{1}{2}\theta(1-\sin\tfrac{1}{2}\theta\sin\tfrac{3}{2}\theta)+\cdots, \\ N_y &= \frac{k_1(t)}{\sqrt{2r}}\cos\tfrac{1}{2}\theta\,(1+\sin\tfrac{1}{2}\theta\sin\tfrac{3}{2}\theta)+\cdots, \\ N_{xy} &= \frac{k_1(t)}{\sqrt{2r}}\sin\tfrac{1}{2}\theta\cos\tfrac{1}{2}\theta\cos\tfrac{3}{2}\theta+\cdots, \\ R_x &= R_y = O(1). \end{aligned} \tag{89}$$

The transverse shears R_x and R_y are nonsingular. In Equations (89), $k_1(t)$ is the resultant stress intensity factor

$$k_1(t) = n_1[(c_2)_1 t/a]N_0\sqrt{a} \tag{90}$$

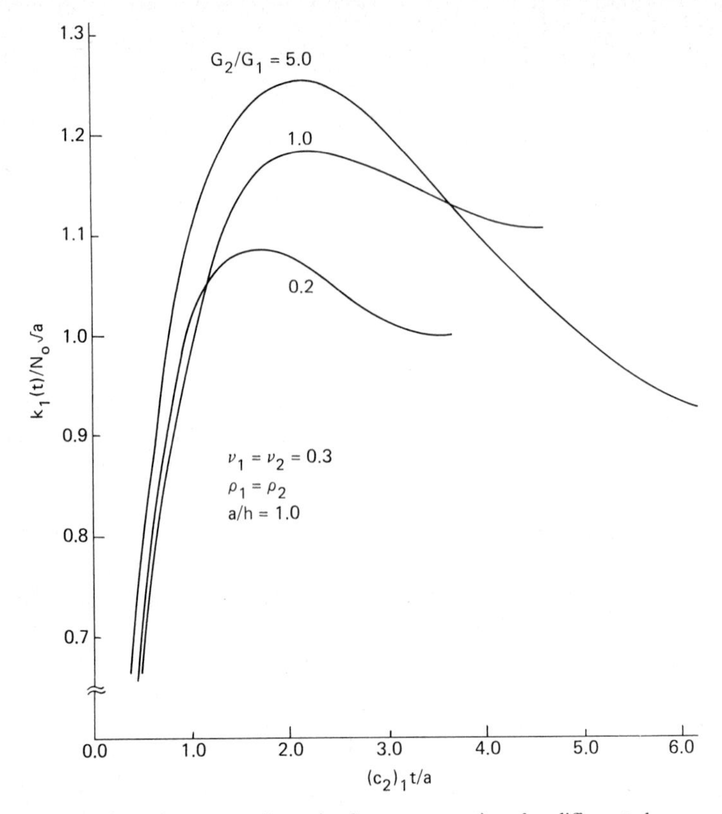

FIG. 20. Normalized resultant stress intensity factor versus time for different shear moduli ratio.

which varies as a function of time. Values of $n_1[(c_2), t/a]$ for $a/h = 1.0$, $\nu_1 = \nu_2 = 0.3$ and $\rho_1 = \rho_2$ are obtained and plotted graphically in Fig. 20. Three different relative stiffnesses of the inner and outer layers reflected by the ratio G_2/G_1 are considered. The case $G_2/G_1 = 1.0$ corresponds to the homogeneous case where the laminate is made of a single material. Note that $k_1(t)$ rises to a peak and then decays to the static solution. The maximum value of $k_1(t)$ is greater than the homogeneous solution for $G_2 > G_1$ and smaller for $G_2 < G_1$. The dynamic stress intensity near a through crack can thus be reduced by having the outer layers to be softer than the inner layers.

4.2. *Bending motion of laminate plate*

The corresponding problem of a laminate plate subjected to dynamic bending loads may be treated in the same way. SIH and CHEN (1981b) have employed variational calculus and derived a system of differential equations with the accompanying boundary conditions for treating laminates with defects under bending.

4.2.1. Theory of Sih and Chen: plate bending

In contrast with Equations (80), in-plane displacements will be odd in z as given by

$$u_x = zu(x, y, t), \qquad u_y = zv(x, y, t), \qquad u_z = w(x, y, t). \tag{91}$$

Defined in plate bending are the moments M_x, M_y and M_{xy} per unit length of the laminate plate edge:

$$(M_x, M_y, M_{xy}) = \int_{-h/2}^{h/2} (\sigma_x, \sigma_y, \tau_{xy}) z \, dz. \tag{92}$$

The shearing forces are

$$(Q_x, Q_y) = \int_{-h/2}^{h/2} (\tau_{xz}, \tau_{yz}) \, dz. \tag{93}$$

Refer to Fig. 21 for notations. The quantities in Equations (92) and (93) are related to u, v and w in Equations (91) as

$$\begin{aligned} M_x &= D_0\left(\frac{\partial u}{\partial x} + \nu_0 \frac{\partial v}{\partial y}\right), \\ M_y &= D_0\left(\frac{\partial v}{\partial y} + \nu_0 \frac{\partial u}{\partial x}\right), \\ M_{xy} &= \tfrac{1}{2}(1 - \nu_0) D_0 \left(\frac{\partial v}{\partial x} + \frac{\partial u}{\partial y}\right) \end{aligned} \tag{94}$$

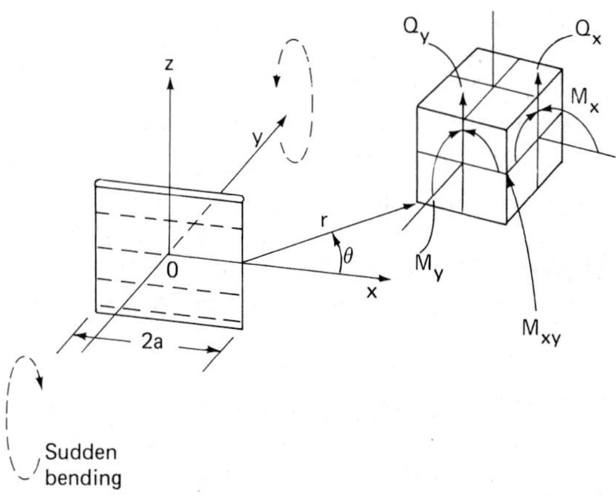

FIG. 21. Bending moments and shear stress forces near crack in four-layered laminate bent suddenly.

and

$$Q_x = \tfrac{1}{12}\pi^2 hG_0\left(u + \frac{\partial w}{\partial x}\right),$$

$$Q_y = \tfrac{1}{12}\pi^2 hG_0\left(v + \frac{\partial w}{\partial y}\right). \tag{95}$$

The parameters D_0, ν_0 and G_0 are given by

$$D_0 = D_1 + D_2\,, \qquad \nu_0 = \frac{D_1\nu_1 + D_2\nu_2}{D_0}\,, \qquad G_0 = \tfrac{1}{2}(G_1 + G_2) \tag{96}$$

in which

$$D_1 = \tfrac{1}{48}\frac{G_1h^3}{(1-\nu_1)}\,, \qquad D_2 = \tfrac{7}{48}\frac{G_2h^3}{(1-\nu_2)}\,. \tag{97}$$

The three unknowns, u, v and w, can be solved from the system of equations derived by SIH and CHEN (1981b):

$$\tfrac{1}{2}(1-\nu_0)D_0\nabla^2u + \tfrac{1}{2}(1+\nu_0)D_0\frac{\partial}{\partial x}\left(\frac{\partial u}{\partial x} + \frac{\partial v}{\partial y}\right) - \tfrac{1}{12}\pi^2 hG_0\left(u + \frac{\partial w}{\partial x}\right)$$

$$= \tfrac{1}{96}h^3(\rho_1 + \rho_2)\frac{\partial^2 u}{\partial t^2}\,,$$

$$\tfrac{1}{2}(1-\nu_0)D_0\nabla^2v + \tfrac{1}{2}(1+\nu_0)D_0\frac{\partial}{\partial y}\left(\frac{\partial u}{\partial x} + \frac{\partial v}{\partial y}\right) - \tfrac{1}{2}\pi^2 hG_0\left(v + \frac{\partial w}{\partial y}\right) \tag{98}$$

$$= \tfrac{1}{96}h^3(\rho_1 + 7\rho_2)\frac{\partial^2 v}{\partial t^2}\,,$$

$$\tfrac{1}{12}\pi^2 hG_0\left(\nabla^2 w + \frac{\partial u}{\partial x} + \frac{\partial v}{\partial y}\right) + q = \tfrac{1}{2}h(\rho_1 + \rho_2)\frac{\partial^2 w}{\partial t^2}$$

with q being the lateral load applied to the laminate in the z-direction.

4.2.2. *Sudden bending*

The laminate plate with a through crack of width $2a$ is now suddenly bent such that

$$\left.\begin{aligned} M_y &= -M_0H(t) \\ M_{xy} &= 0 \end{aligned}\right\} \quad 0 \leqslant |x| < a,\ y = 0 \tag{99}$$

and

$$v = Q_y = 0\,, \quad |x| \geqslant a,\ y = 0\,. \tag{100}$$

In the limit as $r \to 0$, the moments near the crack tip are found to be singular and they are of the forms

$$M_x = \frac{K_1(t)}{\sqrt{2r}} \cos \tfrac{1}{2}\theta \,(1 - \sin \tfrac{1}{2}\theta \sin \tfrac{3}{2}\theta) + \cdots ,$$

$$M_y = \frac{K_1(t)}{\sqrt{2r}} \cos \tfrac{1}{2}\theta (1 + \sin \tfrac{1}{2}\theta - \sin \tfrac{3}{2}\theta) + \cdots , \tag{101}$$

$$M_{xy} = \frac{K_1(t)}{\sqrt{2r}} \cos \tfrac{1}{2}\theta \sin \tfrac{1}{2}\theta \cos \tfrac{3}{2}\theta + \cdots ,$$

$$Q_x = Q_y = \mathrm{O}(1) .$$

The moment intensity factor

$$K_1(t) = m_1[(c_2), t/a]M_0\sqrt{a} \tag{102}$$

is normalized and plotted as a function of the dimensionless time parameter (c_2), t/a as shown in Fig. 22 for $a/h = 1.0$, $\nu_1 = \nu_2 = 0.3$ and $\rho_1 = \rho_2$. Comparison

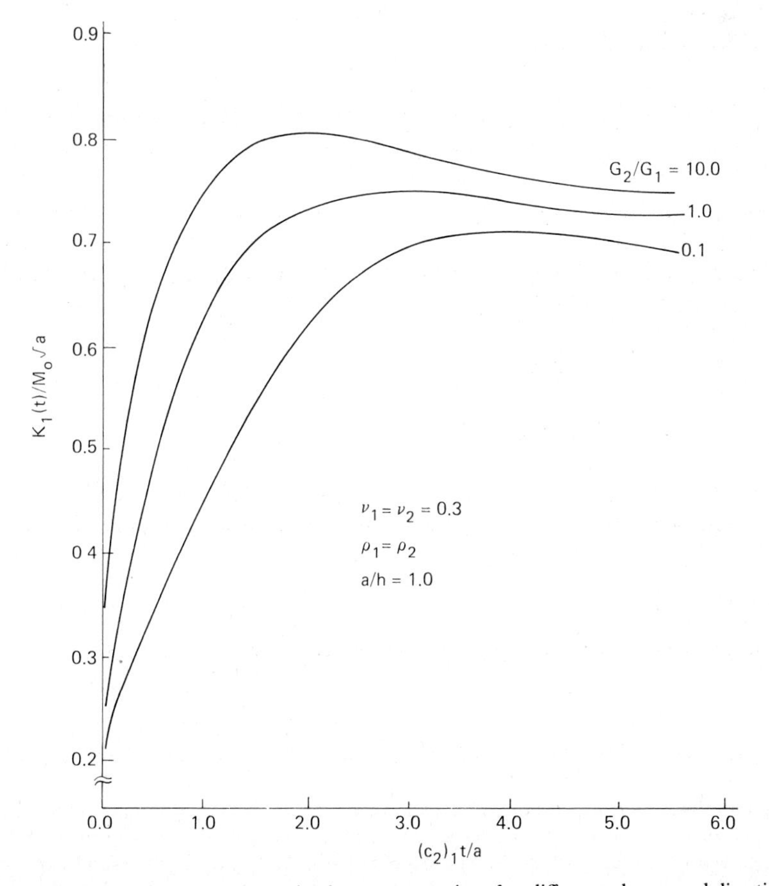

FIG. 22. Normalized moment intensity factor versus time for different shear moduli ratio.

of the results for $G_2/G_1 = 10.0$, 1.0 and 0.1 reveal that the weighted average of the moment intensity factor can be reduced in magnitude if the shear moduli of the outer layer are smaller than those of the inner layers.

5. Strain energy density criterion

The selection of an appropriate criterion to describe composite failure behavior has been problematic. Different viewpoints still prevail even in the underlying basic philosophy. Much of the concerns arise because of the inherent non-homogeneous and anisotropic character of composite materials. Their behavior is sensitive to changes in loading rates, component geometry and material construction. It is therefore essential not to concoct agreement between theory and experiment for special cases but rather to emphasize generality and consistency. A sound criterion should not be restricted by failure modes nor by material types. For these reasons, the strain energy density criterion has been chosen in the present discussion.

5.1. Energy density concept

A basic assumption of the strain energy density approach is that real materials can be divided and subdivided into volume elements whose properties are measurable from specimen tests. These elements can fail at different threshold levels depending on the rate at which the energy in each volume element is released. The time sequence and geometric location of the broken elements determine the failure modes. Within the framework of continuum mechanics, this energy density, $\mathrm{d}W/\mathrm{d}V$, can be computed from the stresses σ_{ij} and strains e_{ij} for any material:

$$\frac{\mathrm{d}W}{\mathrm{d}V} = \int_0^{e_{ij}} \sigma_{ij}\, \mathrm{d}e_{ij}\,. \tag{103}$$

The threshold values of $\mathrm{d}W/\mathrm{d}V$ associated with yielding and/or fracture can be found experimentally from standard tests.[8] In a uniaxial tensile test, $\mathrm{d}W/\mathrm{d}V$ is simply the area under the true stress-strain curve:

$$\frac{\mathrm{d}W}{\mathrm{d}V} = \int_0^{e} \sigma\, \mathrm{d}e\,. \tag{104}$$

Since not all of the energy is always available for creating macrofracture

[8] Since the energy stored in a unit volume of material depends on the loading rate, specimen size and geometry and material type, appropriate adjustments should be made on the threshold or critical values of $\mathrm{d}W/\mathrm{d}V$ or $(\mathrm{d}W/\mathrm{d}V)_c$ from experimental data taken under different conditions.

surface, the amount that is dissipated by heat or plastic deformation, say $(\mathrm{d}W/\mathrm{d}V)_p$, must be subtracted from the total critical value $(\mathrm{d}W/\mathrm{d}V)_c$, i.e.,

$$\left(\frac{\mathrm{d}W}{\mathrm{d}V}\right)_c^* = \left(\frac{\mathrm{d}W}{\mathrm{d}V}\right)_c - \left(\frac{\mathrm{d}W}{\mathrm{d}V}\right)_p . \tag{105}$$

SIH and MADENCI (1983) have shown that the resistance of material to fracture accompanied by yielding varies during subcritical crack growth in metals. The same applies to composites should yielding or other forms of energy dissipation mechanisms occur at the microscopic level.

5.1.1. Energy density factor

Without loss in generality, an energy density factor S may be defined:

$$\frac{\mathrm{d}W}{\mathrm{d}V} = \frac{S}{r}, \tag{106}$$

where r is a linear distance referenced from a point of possible failure initiation site. Fig. 23 shows that such a point could be a crack tip in the matrix, fiber or interface representing a portion of unbonded fiber. The factor S can be interpreted as the area under the $\mathrm{d}W/\mathrm{d}V$ versus r curve and holds in general for any material and defect configuration. Unlike the stress quantity, the $1/r$ relation for $\mathrm{d}W/\mathrm{d}V$ remains unchanged for large and finite[9] deformation theories, materials undergo plastic[10] deformation, and all defect shapes.[11]

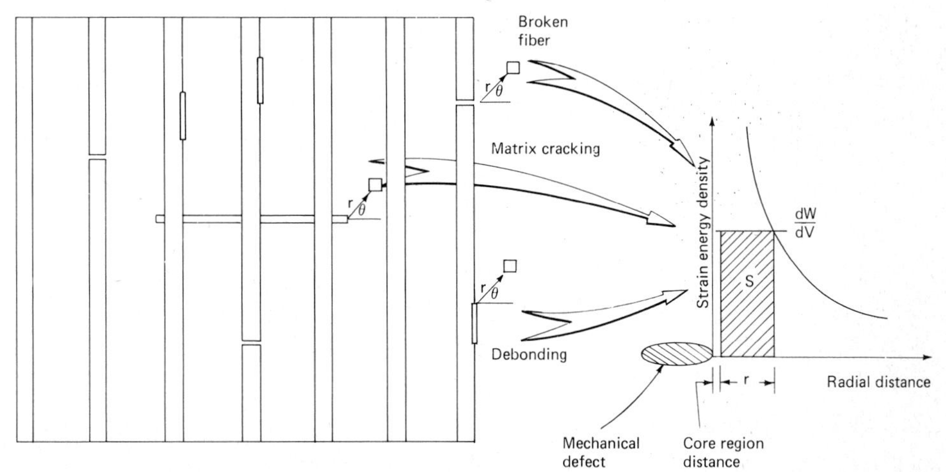

FIG. 23. Strain energy density decay near mechanical defect: broken fiber, matrix cracking and debonding.

[9] In finite elasticity, each stress component attains a different order of singularity near the crack tip depending on the constitutive relation.

[10] The order of the stress singularity depends on the strain hardening exponent.

[11] The order of the stress singularity changes on the border of a surface crack.

Failure criteria based on stress or strain quantities are limited in use because their basic character depends on material and defect type and can lead to inconsistencies when applied to describe physical phenomena. A case in point is the maximum normal stress criterion that assumes a crack to spread in the direction normal to the largest local tensile stress component. This contradicts the dynamic stress solution of a running crack where the maximum local stress component is parallel to the crack.

5.1.2. Failure direction and rate

Failure modes in composites are more complex than those in homogeneous materials. Cracks initiating from defects are not likely to grow in a self-similar manner nor will they follow paths that are known as an a priori. Moreover, subcritical crack growth or material damage becomes the rule in composites and may not be ignored. The strain energy density criterion is particularly suited to include these failure behaviors. Stated briefly are the following basic hypotheses:

Hypothesis 1. Failure by yielding and/or fracture is assumed to initiate at sites corresponding, respectively, to the maximum values of the relative maximum and minimum of the strain energy density function, i.e., $\partial(\mathrm{d}W/\mathrm{d}V)/\partial\theta = 0$ for $\theta = \theta_{\max}$ and $\theta_{\min}$:

$$\left(\frac{\mathrm{d}W}{\mathrm{d}V}\right)_{\max}^{\max} \quad \text{at } \theta = \theta_{\max} \text{ (yielding)} \tag{107}$$

and

$$\left(\frac{\mathrm{d}W}{\mathrm{d}V}\right)_{\min}^{\max} \quad \text{at } \theta = \theta_{\min} \text{ (fracture)}. \tag{108}$$

Hypothesis 2. Onset of yielding and/or fracture is assumed to occur when $(\mathrm{d}W/\mathrm{d}V)_{\max}^{\max}$ and $(\mathrm{d}W/\mathrm{d}V)_{\min}^{\max}$ reach their respective critical values:

$$\left(\frac{\mathrm{d}W}{\mathrm{d}V}\right)_{\max}^{\max} = \left(\frac{\mathrm{d}W}{\mathrm{d}V}\right)_{\mathrm{p}} \tag{109}$$

and

$$\left(\frac{\mathrm{d}W}{\mathrm{d}V}\right)_{\min}^{\max} = \left(\frac{\mathrm{d}W}{\mathrm{d}V}\right)_{\mathrm{c}} \quad \text{or} \quad \left(\frac{\mathrm{d}W}{\mathrm{d}V}\right)_{\mathrm{c}}^{*}. \tag{110}$$

Hypothesis 3. The failure rate by yielding and/or fracture is assumed to occur incrementally in accordance with the following conditions

$$\left(\frac{\mathrm{d}W}{\mathrm{d}V}\right)_{\mathrm{p}} = \frac{S_1}{r_1} = \frac{S_2}{r_2} = \cdots = \frac{S_j}{r_j} = \cdots = \frac{S_{\mathrm{p}}}{r_{\mathrm{p}}} \tag{111}$$

and

$$\left(\frac{\mathrm{d}W}{\mathrm{d}V}\right)_{\mathrm{c}} \quad \text{or} \quad \left(\frac{\mathrm{d}W}{\mathrm{d}V}\right)_{\mathrm{c}}^{*} = \frac{S_1}{r_1} = \frac{S_2}{r_2} = \cdots = \frac{S_j}{r_j} = \cdots = \frac{S_{\mathrm{c}}}{r_{\mathrm{c}}} \tag{112}$$

such that for unstable yielding[12]

$$\begin{aligned} & r_1 < r_2 < \cdots < r_j < \cdots < r_{\mathrm{p}}\,, \\ & S_1 < S_2 < \cdots < S_j < \cdots < S_{\mathrm{p}} \end{aligned} \tag{113}$$

and for unstable fracture[13]

$$\begin{aligned} & r_1 < r_2 < \cdots < r_j < \cdots < r_{\mathrm{c}}\,, \\ & S_1 < S_2 < \cdots < S_j < \cdots < S_{\mathrm{c}}\,. \end{aligned} \tag{114}$$

The above three hypotheses when used simultaneously can describe all failure modes ranging from plastic collapse to brittle fracture. There still remains the fundamental problem of how composite damage should be modeled appropriately by the theory of continuum mechanics. It is the scale of the continuum element relative to defect size and microstructure construction of the composite that determines whether homogeneous anisotropy or nonhomogeneous isotropy would best simulate the composite behavior.

5.2. Modeling of composite damage: matrix cracking

Composite damage cannot always be modeled arbitrarily without appropriate scaling of the defect dimension with reference to the composite microstructure and the mode of failure. Several different types of failure modes may occur depending on the scale of the continuum element $\Delta x \Delta y$ as illustrated in Figs. 24(a) and (b). For a defect whose opening, say 2δ, in Fig. 24(a), is large in comparison with the fiber spacing $2h$ and diameter d, then the stress or energy state can be approximated by invoking the assumption of homogeneous anisotropy. This implies that the failure of an element involves the breaking of both fibers and matrix. When the defect size is of the same order as the fiber spacing and diameter, fiber and matrix cracking should be considered separately. Fig. 24(b) illustrates the nonhomogeneity of the system where the element is embedded either entirely in the matrix or fiber. The latter model is more realistic for unidirectional composites that failure by matrix cracking which will be discussed subsequently.

[12] If yielding is self-arresting, then the inequality in Equations (111) would be reversed and r_{p} and S_{p} correspond to their respective values at arrest.

[13] Equations (114) would be modified by reversing the inequalities and replacing r_{c} and S_{c} by r_0 and S_0, respectively.

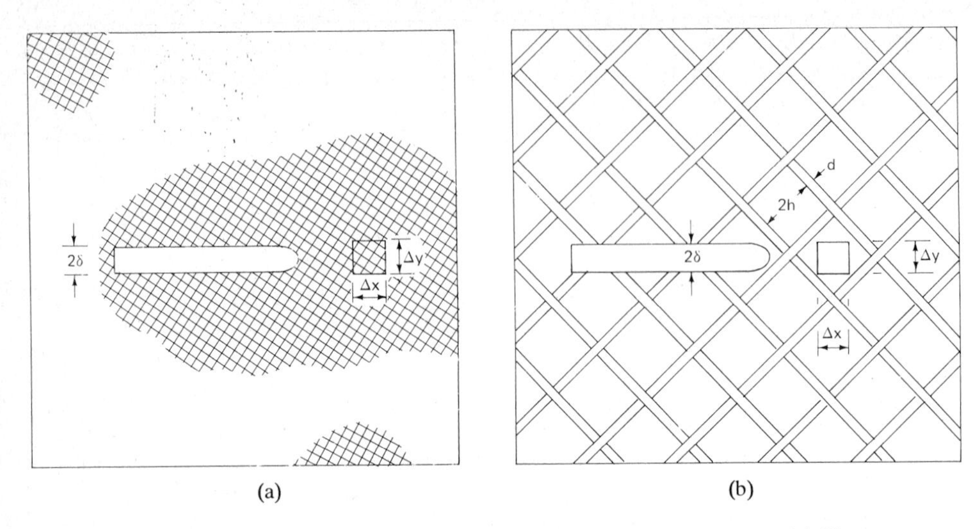

FIG. 24. Scaling of defect size with reference to composite microstructure. (a) Homogeneous anisotropic system ($\delta \gg h$ or d). (b) Nonhomogeneous isotropic system ($\delta \simeq h$ or d).

5.2.1. Matrix cracking by impact

Consider the geometry of a crack parallel to the fibers as shown in Fig. 1(a). The crack experiences both normal and shear impact of magnitude σ_0 and τ_0, respectively. This results in a mixed mode loading situation where both stress intensity factors $k_1(t)$ will vary as a function of time. Refer to Figs. 3 and 4 for their numerical values with crack length to fiber spacing ratios of $a/h = 0.5$, 1.0 and 2.0. Once $k_1(t)$ and $k_2(t)$ are known from Equations (16) and (20), the strain energy density factor S in Equation (106) for a linear elastic material can be found from

$$S = a_{11}k_1^2(t) + 2a_{12}k_1(t)k_2(t) + a_{22}k_2^2(t) \tag{115}$$

in which the coefficients a_{ij} $(i, j = 1, 2)$ are given by

$$\begin{aligned} a_{11} &= \frac{1}{16G_m}(3 - 4\nu_m \cos\theta)(1 + \cos\theta), \\ a_{12} &= \frac{1}{8G_m}\sin\theta(\cos\theta - 1 + 2\nu_m), \\ a_{22} &= \frac{1}{16G_m}[4(1 - \nu_m)(1 - \cos\theta) + (3\cos\theta - 1)(1 + \cos\theta)]. \end{aligned} \tag{116}$$

The local polar coordinates r and θ in Equations (115) and (116) are defined in Fig. 1(b).

Because of the antisymmetric nature of impact, crack may not initiate in a direction parallel to the fibers, i.e., $\theta_0 = 0°$ in Fig. 1(b). The combination of normal and shear impact may be related to a dynamic load σ applied at an angle β_0 with reference to the x-axis in Fig. 1(a):

$$\sigma_0 = \sigma \sin^2 \beta_0 , \qquad \tau_0 = \sigma \sin \beta_0 \cos \beta_0 . \tag{117}$$

For a fixed distance r that is small as compared to the crack length and fiber spacing, equation (108) may be employed to determine the direction of crack initiation[14] θ_0 as a function of β_0 by taking $\partial S/\partial\theta = 0$ for θ_0 corresponding to S_{min}. Substituting Equations (117) into (16) and (20) and making use of Equations (115) and (108) for r = const., a relation between $-\theta_0$ and β_0 is established. The results for $G_c/G_m = 10.0$, $\nu_c = \nu_m = 0.29$, $\rho_c = \rho_m$ and $a/h = 1.0$ are presented numerically in Fig. 25. As it is to be expected, the time dependency effect becomes important only for small time or $(c_2)_m t/a$ values. This can be observed more readily in Fig. 26 which displays the variations of $-\theta_0$ with $(c_2)_m t/a$.

The critical stress at failure depends on the crack position β_0 and the elapsed time at which S_{min} first reach S_c being characteristic of the material. A plot of $16\, G_m S_{min}/\sigma^2 a$ versus $(c_2)_m t/a$ is given in Fig. 27 for $\beta_0 = 30°, 45°, \ldots, 90°$. The curves rise to a peak and then decrease in magnitude. This effect is not

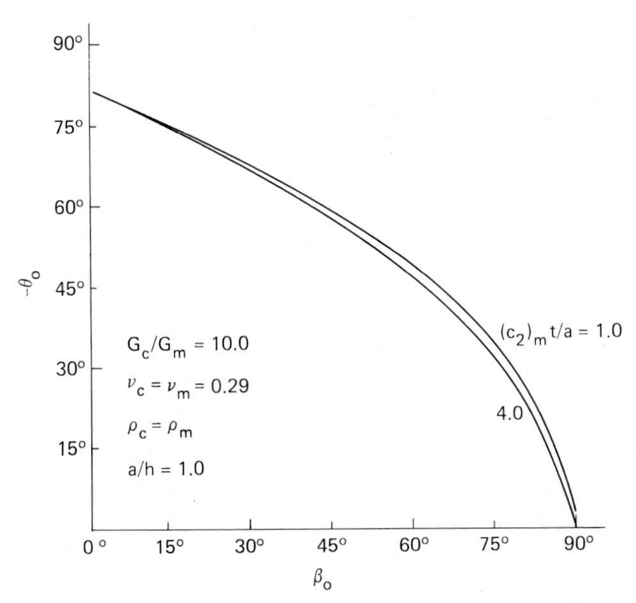

FIG. 25. Fracture angle versus crack position for different time.

[14] Although matrix cracking may appear to occur parallel to the fibers, the direction of crack initiation in the immediate vicinity of the crack tip may be oriented at angle θ_0 different from zero degree.

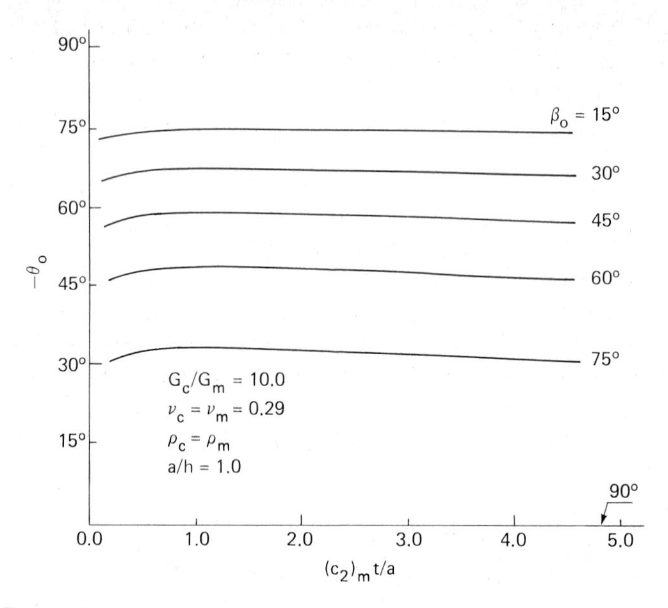

FIG. 26. Direction of crack initiation as a function of time for different β_0.

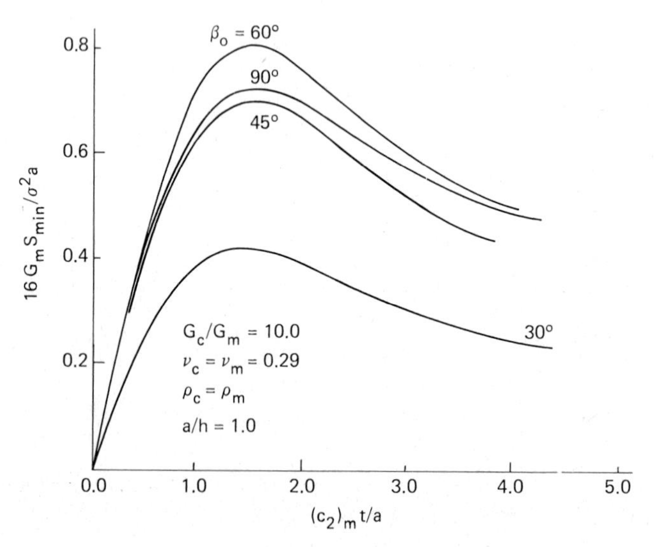

FIG. 27. Normalized minimum strain energy density factor versus time for different crack position β_0.

monotonic in terms of β_0 as the maximum peak of S_{min} is seen to occur at approximately $\beta_0 = 60°$. Consider the ratio of the dynamic to static failure stress σ_m/σ_s that can be expressed as

$$\frac{\sigma_m}{\sigma_s} = \sqrt{\frac{1-2\nu_m}{4G_m S_{min}/\sigma^2 a}} \tag{118}$$

which is plotted against $(c_2)_m t/a$ in Fig. 28. Failure by unstable crack growth is assumed when $S_{min} = S_c$ regardless of whether the load is applied statically or dynamically. From Fig. 28, the lowest failure stress σ_m is found to occur at $(c_2)_m t/a \simeq 1.5$ and $\beta_0 \simeq 60°$. The most vulnerable position of the crack is not the one that corresponds to $\beta_0 = 90°$ for a homogeneous material.

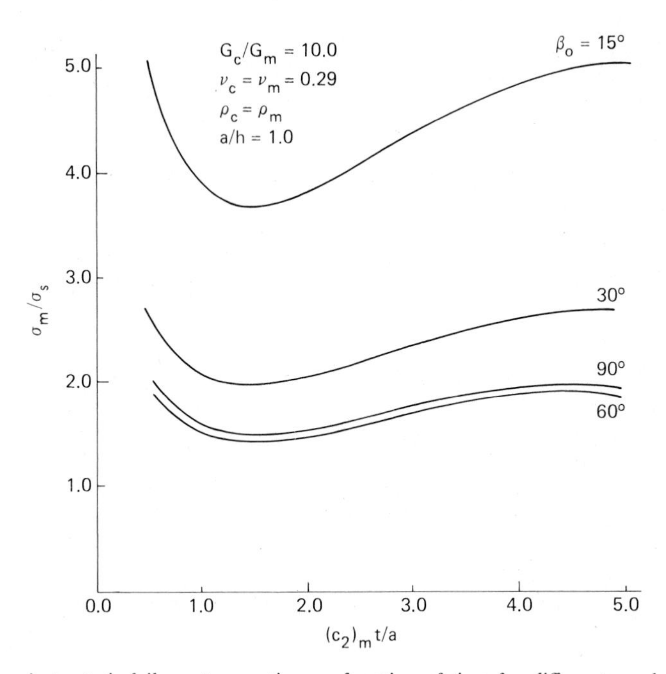

FIG. 28. Dynamic to static failure stress ratio as a function of time for different crack position β_0.

5.2.2. *Running crack in matrix*

Matrix involves the stage of crack propagation. This problem was considered in Section 3.2 with a crack running at a constant velocity c in the matrix material, Fig. 14. The near field stresses in Equations (53) may be used to compute dW/dV whose $1/r$ coefficient gives

$$\frac{8G_m S}{\sigma^2 a} = \frac{R_1^2[c/(c_2)_m]}{H[(\lambda_1)_m, (\lambda_2)_m]} [[1 + (\lambda_2)_m^2]^2\{2(1 - \nu_m)[1 + (\lambda_2)_m^2]^2 - (1 - 2\nu_m)[2(\lambda_1)_m^2 + 1 - (\lambda_2)_m^2][1 + (\lambda_2)_m^2]\}f^2[(\lambda_1)_m] + 32(\lambda_1)_m^2(\lambda_2)_m^2 f^2[(\lambda_2)_m] - 16(\lambda_1)_m(\lambda_2)_m[1 + (\lambda_1)_m^2] \times [1 + (\lambda_2)_m^2]f[(\lambda_1)_m]f[(\lambda_2)_m] + 8(\lambda_1)_m^2 \times [1 + (\lambda_2)_m^2]^2\{g[(\lambda_1)_m] - g[(\lambda_2)_m]\}], \tag{119}$$

where $H[(\lambda_1)_m, (\lambda_2)_m]$ is defined in Equation (54) and the numerical values of $R_1[c/(c_2)_m]$ can be found in Fig. 15. The parameters λ_1 and λ_2 are given by Equations (45).

Values of θ_0 corresponding to S_{min} may be obtained from Equation (119) by taking $\partial S/\partial\theta = 0$ as in the case of a stationary crack. Table 1 summarizes the

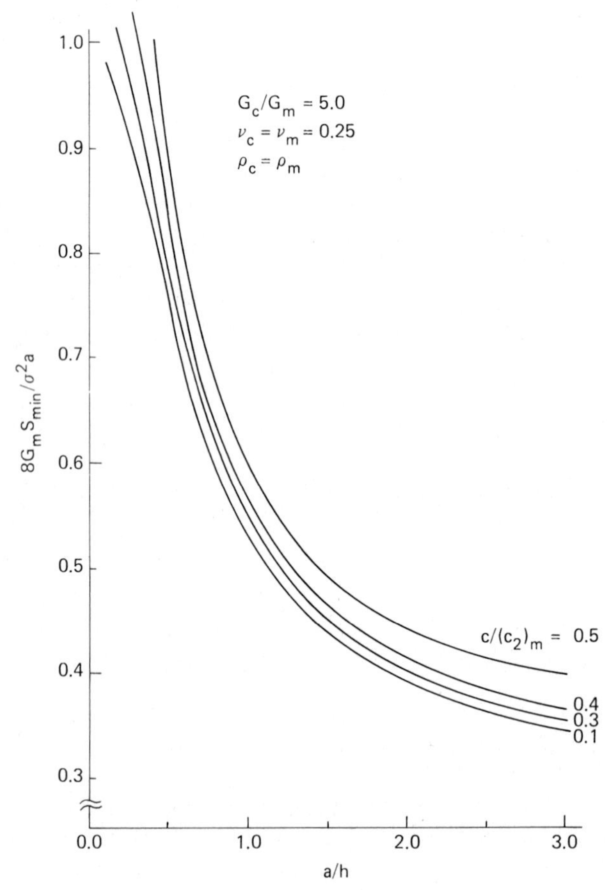

FIG. 29. Normalized minimum strain energy density factor versus crack length to fiber spacing ratio for different crack velocities.

TABLE 1. Crack angles and strain energy density factors for a unidirectional composite with $\nu_c = \nu_m = 0.25$, $\rho_c = \rho_m$, $G_c/G_m = 5.0$.

	$c/(c_2)_m$						
	0.1	0.2	0.3	0.4	0.5	0.6	0.8
$\pm\theta_0$	0°	0°	0°	0°	15.1°	48.6°	65.2°
$\dfrac{8G_m S_{min}}{\sigma^2 a}$	1.008	1.036	1.090	1.185	1.361	1.631	3.892

results for $G_c/G_m = 5.0$, $\nu_c = \nu_m = 0.25$ and $\rho_c = \rho_m$. The minimum value of S or S_{min} occurred at $\theta_0 = 0°$ for crack velocity c less than $0.4(c_2)_m$. Bifurcation is predicted for c within the range of $0.4(c_2)_m$ and $0.5(c_2)_m$ where S_{min} tends to occur to the sides of the moving crack at approximately $\theta_0 = \pm 15.1°$. At these sites, the energy density factor associated with dilatation or volume change S_v can be shown to be larger than that associated with distortion or shape change S_d, i.e., $S_v > S_d$ where

$$S = S_v + S_d\,. \tag{120}$$

The case of $S_d > S_v$ corresponds to locations where yielding tends to dominate. This is a brief physical interpretation of the strain energy density criterion.

The influence of fiber spacing is illustrated in Fig. 29 with $8G_m S_{min}/\sigma^2 a$ plotted as a function of a/h for $c/(c_2)_m$ varying from 0.1 to 0.5. For large values of a/h, a slight variation of the applied stress σ can lead to large changes in the crack velocity c.

5.2.3. *Crack propagation in orthotropic materials*

The strain energy density factor S will be calculated for a crack propagating in a unidirectional composite modeled as a homogeneous orthotropic medium as indicated in Fig. 17. For a linear elastic material, Equation (103) reduces to

$$\frac{dW}{dV} = \tfrac{1}{2}\sigma_{ij}e_{ij}\,. \tag{121}$$

The stress and strain relations in Equations (63) may be used to express dW/dV in terms of the stress components σ_x, σ_y and τ_{xy} alone:

$$\frac{dW}{dV} = \frac{1}{2}\left[\frac{c_{22}\sigma_x^2 + c_{11}\sigma_y^2 - 2c_{12}\sigma_x\sigma_y}{c_{11}c_{22} - c_{12}^2} + \frac{\tau_{xy}^2}{G_{12}}\right]. \tag{122}$$

Substituting the local stress components in Equations (73) into (121) and applying the relation in Equation (106), it is found that

$$\frac{8G_{12}S}{k_1^2} = \frac{1}{N_5^2}\left\{\frac{N_3 - N_1}{\mu_1^2}(\mu_1 + \cos\theta) + \frac{N_3 - N_2}{\mu_2^2}(\mu_2 + \cos\theta) + \frac{2N_4}{\mu_1\mu_2}\sqrt{\mu_1 + \cos\theta}\sqrt{\mu_2 + \cos\theta} + N_3\left[\frac{\sqrt{\mu_1 - \cos\theta}}{\mu_1} - \frac{\sqrt{\mu_2 - \cos\theta}}{\mu_2}\right]^2\right\} \tag{123}$$

in which k_1 is given by Equation (76) and N_j ($j = 1, 2, \ldots, 5$) stand for

$$N_1 = (M_2^2 - 1)\frac{\rho c^2}{G_{12}}, \quad N_2 = M_1^2(M_3^2 - 1)\frac{\rho c^2}{G_{12}}, \quad N_3 = M_1^2(\eta_2 + M_3)^2,$$
$$N_4 = \frac{M_1}{G_{12}}[2c_{11} - c_{12}(\eta_1 M_2 + \eta_2 M_3) - \rho c^2], \quad N_5 = \frac{M_4}{G_{12}}. \tag{124}$$

Refer to Equations (75) for the definitions of the quantities M_j ($j = 1, 2, \ldots, 4$).

The influence of material orthotropy on crack propagation may now be investigated by calculating for the angles θ_0 that are assumed to correspond with the crack path or S_{min} in the strain energy density criterion. For the E-glass epoxy composite, Equations (78) show that the elastic modulus E_1 in the crack growth direction is much larger than E_2 being normal to the crack, i.e., $E_1 \gg E_2$. Table 2 gives the normalized values $4\,G_{12}S_{min}/k_1^2$, the crack growth directions $\pm\theta_0$ and the dimensionless crack velocity $c/\sqrt{G_{12}/\rho}$. Because of the high degree of orthotropy, bifurcation is predicted for all crack velocities. This result is likely to be unrealistic for the actual E-glass epoxy composite which when pulled in a direction normal to the fibers will produce a running crack parallel to the fibers rather than branching. The effect of *nonhomogeneity* is significant and cannot be neglected for running cracks with openings less than or approximately equal to the fiber spacings. This is an example which demonstrates the inadequacy of the assumption of *homogeneous orthotropy*.

Unlike the E-glass epoxy composite, the combination of stainless steel fibers with an aluminum matrix yields a much weaker orthotropy. For a 40% fiber volume fraction, the elastic constants are

$$E_1 = 18.30 \times 10^3\ \text{ksi}, \quad \nu_{12} = \nu_{13} = 0.32,$$
$$E_2 = 15.51 \times 10^3\ \text{ksi}, \quad \nu_{23} = 0.36, \tag{125}$$
$$G_{12} = 5.87 \times 10^3\ \text{ksi},$$

and

$$c_{11}/G_{12} = 4.278, \quad c_{12}/G_{12} = 1.813, \quad c_{22}/G_{12} = 3.804. \tag{126}$$

Table 3 shows that the crack did grow in a straight line for low crack velocities. Crack branching occurred at approximately $c = 0.50\sqrt{G_{12}/\rho}$ with $\theta_0 = 7.0°$. The

TABLE 2. Crack angles and strain energy density factors for E-glass epoxy unidirectional fibrous composite (56.5% fiber volume fraction).

	$c/\sqrt{G_{12}/\rho}$								
	0.0	0.1	0.2	0.3	0.4	0.5	0.6	0.7	0.8
$\pm\theta_0$	36.6°	36.6°	36.8°	37.2°	37.9°	39.1°	41.0°	44.1°	49.20°
$\frac{4G_{12}S_{min}}{k_1^2}$	0.9113	0.915	0.927	0.947	0.981	1.034	1.121	1.277	1.620

TABLE 3. Crack angles and strain energy density factors for stainless steel aluminum unidirectional fibrous composite (40% fiber volume fraction).

	$c/\sqrt{G_{12}/\rho}$								
	0.0	0.1	0.2	0.3	0.4	0.5	0.6	0.7	0.8
$\pm\theta_0$	0.0°	0.0°	0.0°	0.0°	0.0°	7.0°	36.0°	50.4°	60.7°
$\frac{4G_{12}S_{min}}{k_1^2}$	0.726	0.731	0.748	0.781	0.840	0.946	1.118	1.416	2.185

larger branching angles correspond to higher crack velocities. Hence, the assumption of homogeneous anisotropy becomes more realistic for composites whose fiber and matrix moduli do not differ significantly. This is to be expected on intuitive grounds.

6. Concluding remarks

Composite materials possess distinctive crack growth characteristics that often cannot be adequately explained by fracture mechanics based on the assumptions of isotropy and homogeneity. The anisotropy and nonhomogeneity from the standpoint of fiber versus matrix or multiple laminae of different orientations are the principal cause of concern. Modeling of the local failure modes has to be justified by proper scaling of the defect size relative to the microstructure dimensions of the composite. Inappropriate stress analyses can lead to unrealistic predictions regardless of the failure criteria. This was clearly demonstrated by the example of a running crack in a unidirectional fiber reinforced composite modeled by a homogeneous orthotropic system.

There is also the more fundamental question on the philosophy of specimen testing for composite materials. Since anisotropy and nonhomogeneity can interact with specimen size, loading rate, material microstructure, etc., in a complex manner, it is not clear how the tests data on the smaller specimens could be used in the design of larger size composite structures. There is strong evidence in past investigations that composite specimens may have to be analyzed as structures due to their high degree of nonhomogeneity. There remains much to be done in terms of developing appropriate stress and failure analyses so that more reliable design procedures for high performance composites may be established in the future.

References

ARCISZ, M. and G.C. SIH (1982), Moving Cracks in Orthotropic Media, Institute of Fracture and Solid Mechanics Technical Report IFSM 82-113, Lehigh University, Bethlehem PA, U.S.A.

ASTM STP 568 (1973), Foreign Object Impact Damage of Composites, American Society for Testing and Materials, Philadelphia, PA, U.S.A.

SIH, G.C., R.S. RAVERA and G.T. EMBLEY (1972), Impact response of a finite crack in plane extension, *Internat. J. Solids and Structures* **8**, 977–993.

SIH, G.C. (1971), A review of the three-dimensional stress problem for a cracked plate, *Internat. J. Fracture Mechanics* **7**, 39–61.

SIH, G.C. (1973), A Special Theory of Crack Propagation, in: G.C. SIH, ed., *Mechanics of Fracture, Vol. I: Methods of Analysis and Solution of Crack Problems* (Noordhoff, The Netherlands), 21–45.

SIH, G.C., E.P. CHEN, S.L. HUANG and E.J. MCQUILLEN (1975), Material characterization on the fracture of filament-reinforced composites, *J. Composite Materials* **9**, 167–186.

SIH, G.C. (1979), Fracture Mechanics of Composite Materials, in: G.C. SIH and V.P. TAMUSZ, eds., *Fracture of Composite Materials* (Sijthoff and Noordhoff, The Netherlands), 111–130.

SIH, G.C. and E.P. CHEN (1980), Effect of material nonhomogeneity on crack propagation characteristics, *Internat. J. Engineering Fracture Mechanics* **13**, 431–438.

SIH, G.C. and E.P. CHEN (1981a), *Cracks in Composite Materials* (Martinus Nijhoff, The Netherlands), 15–81.

SIH, G.C. and E.P. CHEN (1981b), Sudden bending of a cracked laminate, *Internat. J. Engineering Sci.* **19**, 979–991.

SIH, G.C. and E.P. CHEN (1981c), Sudden stretching of a four-layered composite plate, *J. Engineering Fracture Mechanics* **15**(1, 2), 243–252.

SIH, G.C. and E.P. CHEN (1982), Moving cracks in layered composites, *Internat. J. Engineering Sci.* **20**, 1181–1192.

SIH, G.C. and E. MADENCI (1983), Crack growth resistance characterized by the strain energy density function, *Internat. J. Engineering Fracture Mechanics.* (In press).

CHAPTER IV

The Influence of Time and Temperature on the Reinforced Plastic Strength

J.V. Suvorova

Mechanical Engineering Research Institute
Academy of Sciences of the U.S.S.R.
Moscow
U.S.S.R.

Contents

HANDBOOK OF COMPOSITES, VOL. 3 – Failure Mechanics of Composites
Edited by G.C. SIH and A.M. SKUDRA

List of Symbols

σ	Stress
σ_0	Stress connected with the instantaneous deformation curve
σ^+	Failure stress
σ_0^+	Ultimate stress of the instantaneous deformation curve
σ_*	Stress achieved at the moment of unloading
σ_f	Stress in fibers
σ_f^+	Failure stress in fibers
τ	Shear stress
τ^u	Ultimate shear stress
τ_0^u	Ultimate shear stress of the instantaneous deformation curve
σ_e	Component of the stress connected with the reversible creep deformation
σ_p	Component of the stress connected with the irreversible failure deformation
$\dot{\sigma}$	Stress rate
$\dot{\sigma}_l$	Rate of loading
$\dot{\sigma}_{un}$	Rate of unloading
e	Strain
e_u	Ultimate strain
e_*	Strain achieved at the moment of unloading (according to deformation)
e_f	Strain of fibers
e_{fu}	Ultimate strain of fibers
$\dot{e}$	Strain rate
$\varphi(e)$	Curve of instantaneous deformation
t	Time
t_u	Time to failure
t_*	Time at the moment of unloading (according to deformation)
$K^*\sigma$	Hereditary operator connected with the inclination of the real stress-strain diagram from the instantaneous one
$K(t-\tau)$	Kernel of the operator $K^*\sigma$
$L^*\sigma$	Hereditary operator describing the inclination of the real stress-strain diagram connected with the viscous properties of the material
$L(t-\tau)$	Kernel of the operator
$M^*\sigma$	Hereditary operator describing the inclination of the real diagram connected with the damage accumulation process
$M(t-\tau)$	Kernel of the operator
T	Temperature

$f(T)$ Function of the temperature influence
E_f Elastic modulus of the fibers
E_m Elastic modulus of the matrix
ω Parameter of the damage accumulation
n Number of the cycles to failure
a Temperature shift factor
V_f Fiber volume content

1. Introduction

The studies of time-dependent failure of composites have been rather scarce until the present time. Still there exists a strong dependence of failure properties upon the loading time.

Some experimental stress-rupture curves can be found in literature, but, for example, information about the strain rate and temperature dependence of strength is very scanty, and there are practically no sufficiently general mechanical models.

The difficulty consists in the impossibility of the investigation of failure process without analysing the properties of components and their mutual work in composite structure. But the calculation of the whole number of necessary factors leads to the construction of the mathematical models which are in poor correlation with experimental results and can be used only for the qualitative analysis.

Therefore we must prefer the formulation of the models which of course must be based on the information about microstructural peculiarities of deformation, in the terms of those magnitudes, which can be measured in the macroscopic experiments and which can be used in structural design and optimization.

The purpose of this chapter is not in the enumeration of the experimental results, which describe the strength of composites in the dependences of time and loading conditions (this was done, for example, in the survey of LIFSHITZ (1974) and SKUDRA et al. (1971)) but in the discussion of possible approaches to the investigation of long-time failure and models corresponding to the properties of the kind.

It is necessary to note that the statistical models constructed to describe long-time failure have not been inserted in this survey, they must be a subject of an independent investigation. It will be a phenomenological model that will attract our principal attention. This model is based on the hereditary presentations about processes of deformation and failure of composites. If we have the possibility we shall add a description of possible microstructural deformational processes of a composite structure to our phenomenological analysis. It will help us to settle the bounds of applicability of mathematical formulations having been made.

2. Influence of the structure of a composite on creep rupture

2.1. Failure modes

It is impossible to construct the creep rupture models without the previous analysis of the failure modes corresponding to the given loading conditions. The various modes of failure are known to take place, the most common of them being the following:

(1) Damage accumulation. It is considered that the certain concentration of the damage accumulation leads to the rupture of the material.

(2) Propagation of a crack or a system of cracks which have appeared in the material beforehand.

(3) Thorough and practically instant rupture due to the break of the weakest element.

The task is to learn the conditions under which that or other mode of failure can occur and to formulate the corresponding mathematical model.

We have no intention to investigate the failure modes of composites. There are a lot of works concerning this subject, for example, the works of MULLIN et al. (1968), TSAI and HAHN (1978), CRAIG and COURTNEY (1975), DHARAN (1978) and others. But we obtain a valuable information, if the determination of the strength properties is related with the determination of the failure mode, because it allows us to observe the tendency in the changing of the time-dependent strength properties and in consequence of that to predict the behaviour of the construction under the given conditions.

The attention is usually attracted to the investigation of some mode of failure and not to the transition from one mode to another, though the latter task is of great importance for obtaining the concrete information about the behaviour of a composite and for constructing the mechanical model. Because of the change of the failure mode under the change of surrounding conditions or a micro-structure of composites one can obtain the nonmonotonical dependence of strength on the parameter under variation. Probably MILEIKO (1973, 1976, 1979) was the first who paid attention to the transferring from one mode to another. These works considered the fiber volume content to be a variable parameter. Mileiko formulated the appropriate model which gave the logical completeness to the theory of failure of composites. This model allows to understand the conditions of occurring some mode of failure in the material and as a consequence to recognize the character of its behaviour under load. It also allows to make some certain conclusions about the possibility and necessity of variation in the microstructure of a composite and in its technology to secure the work of a construction in the most favourable conditions.

Change in the failure character can be a consequence not only of the fiber volume content variation but also of surrounding conditions, i.e. strain rate and temperature. It was shown by SUVOROVA et al. (1980a) and their results are given in Fig. 1. It describes the loading rate dependence of strength of two types of carbon fiber reinforced plastics (cfrp) distinguished in the adhesion

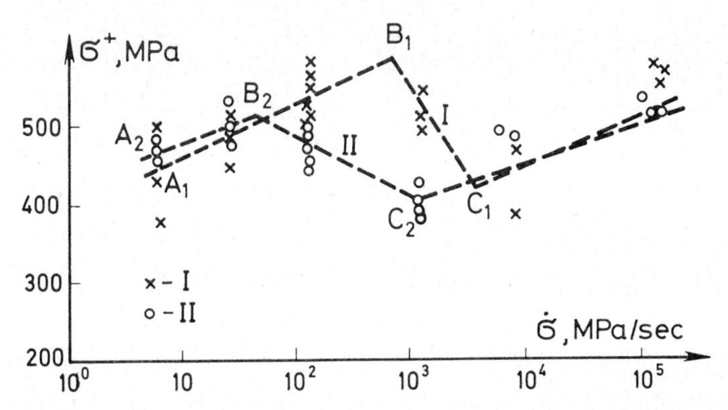

FIG. 1. Dependence of strength on the loading rate for two types of orthogonal-reinforced carbon fiber plastics with epoxy-phenol resin 5211-B. Material I: fibers with normal surface; Material II: fibers with surface treated by HNO_3.

properties between fibers and epoxy-phenol resin. One can see that there are three parts in the curve $\sigma \sim \lg \dot{\sigma}$. The first of them is connected with the damage accumulation failure, strength increases with increasing loading rate. The less the loading time, the less the amount of damage inside the material, the higher is it's strength. The second part of the $\sigma \sim \lg \dot{\sigma}$ curve shows that there the failure character is not the same as the first one and strength decreases with increasing loading rate. According to the Mileiko's model, it must be explained by the growth of the defects or cracks having occurred in the material and failure of the sample is the consequence of the macrocrack propagation. At last the third part of the curve is the instantaneous failure which occurs because of the weakest element break.

Fig. 2 illustrates this supposition. It shows the surfaces of the samples have failed at different rates of loading.

Sample 1 represents the material with weak adhesion strength between fibers and resin, and Sample 2 represents the material in which the adhesion properties have been improved by the fiber surface treating by HNO_3. Both samples were ruptured at the stress-rate $\dot{\sigma} = 6$ MPa/sec ($\lg \dot{\sigma} = 0.222$). One can see that in both cases the surfaces are rough, the fibers are pulled out from the resin and failure is the consequence of the damage accumulation. Samples 3 and 4 were ruptured at the stress-rate $\dot{\sigma} = 125$ MPa/sec ($\lg \dot{\sigma} = 2.097$). Under these conditions the material with weaker adhesion strength (Sample 3; Curve 1; Fig. 1) shows the same failure character as at the lower stress rate and its surface is the same as at the stress-rate $\dot{\sigma} = 6$ MPa/sec. Sample 4 shows another situation, its failure surface is smoothed, and it means that the sample has been ruptured by the macrocrack propagation. Fig. 1 affirms this, one can see that failure occurring in Material II at the stress-rate $\dot{\sigma} = 1.25$ MPa/sec is of another mode than at the stress-rate $\dot{\sigma} = 6$ MPa/sec. At last both Figs. 1 and 2 show that at the stress-rate $\dot{\sigma} = 125$ MPa/sec ($\lg \dot{\sigma} = 3.097$) Samples 5 (Material I) and 6 (Material II) have been broken identically and their surfaces are smooth.

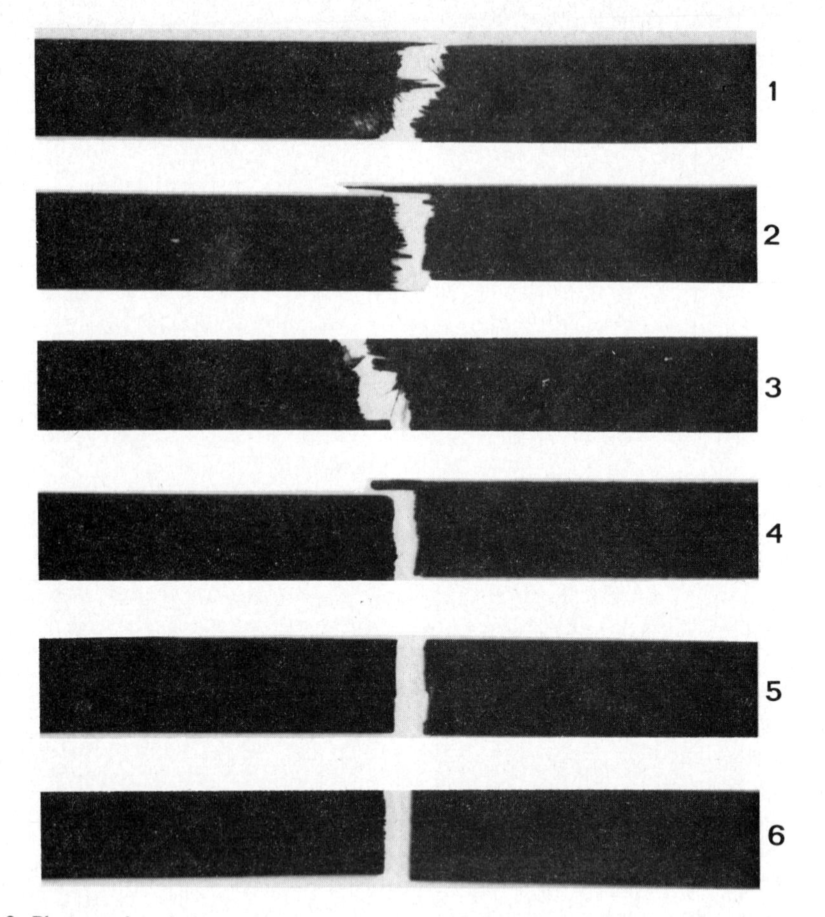

FIG. 2. Photographs of the surfaces of specimens failed at various rates of loading. 1, 3, 5: specimens made of material I; 2, 4, 6: specimens made of material II; 1 and 2: $\dot{\sigma} = 6$ MPa/s ($\lg \dot{\sigma} = 0.222$); 3 and 4: $\dot{\sigma} = 125$ MPa/s ($\lg \dot{\sigma} = 2.097$); 5 and 6: $\sigma = 1250$ MPa/s ($\lg \sigma = 3.097$).

The experimental results of Fig. 1 show that the adhesion characteristics of a composite are of great importance in the locations of the extreme points of the curve $\sigma \sim \dot{\sigma}$. The weak adhesion strength results in the debonding of the material and in the dividing fibers from resin. And this process, in its turn, prevents the propagation of the macrocrack, so the point of transition from failure caused by the accumulation of damage to failure caused by the propagation of the crack for the material with low adhesion strength would correspond to higher loading rates. Knowing this we come to a conclusion, that we should not aim at creating a very firm bond between fibers and resin in composites intended for short-time use in constructions under instant overload, as debonding will extinguish the impact energy and the value of ultimate stress will increase by 30% for Material I, at a rate increasing by two orders of magnitude. However, the material with low adhesion strength will be less affective under the repeated cyclic load or the long-time load as the process of damage accumulation is more intensive in it, and the probability of damage is higher.

Similarly the material strength changes with the change of the fiber volume content. And it is clear that loading rates, corresponding to the point of transition from mode 1 to mode 2 of failure, will be different for composites with various fiber volume contents. And vice versa, if we construct dependences of strength σ on V_f at different rates of loading, we will see that the change of the mode of failure also takes place at various volume contents. For example, SIERAKOVSKI et al. (1971) showed the dependences of strength on the fiber volume content at different rates of loading for the epoxy resin reinforced by steel fibers. At a low rate of loading ($\dot{e} \approx 6.5 \cdot 10^{-5}\ \text{sec}^{-1}$) strength is continuously increasing with the increase in the fiber volume content (the data is given for the composite containing up to a 40% volume fraction of fibers), in this case the change of fracture mode is not observed. At high loading rates ($\dot{e} = 1 \cdot 10^3\ \text{sec}^{-1}$), even at $V_f = 25\%$ a transition to failure is observed caused by the propagation of the crack. Thus, for every rate of loading, we can establish some bounds according to V_f, in the limits of which some or other mode of failure is observed and vice versa for every value of the fiber volume content corresponding values of loading rates can be found. Apparently the same tendency of the shift of the point of transition towards the zone of less values of V_f will be observed at temperature decreasing, though such experimental data are unknown to the author.

Thus, the analysis of the working conditions of composite constructions and the peculiarities of its structure is of great importance for understanding this behaviour under the load and for estimating the possibilities of failure. The higher is the fiber volume content, the lower is the number of fibers, the more homogeneous fiber properties, the less non-effective fiber length, the more probable is the failure caused by the propagation of the crack. The analysis of the influence of these factors on the mode of fracture is given, for example, in the works of SKUDRA et al. (1971), MULLIN et al. (1974), CRAIG and COURTNEY (1975), DHARAN (1978). By changing the mentioned characteristics of a material we can optimize its properties, if the conditions under which the material works are known. And vice versa, if there is any concrete material and it is impossible to change the technology of its manufacture, then we can say under what conditions we can use it more effectively.

It is to be stressed also, that when we speak about the fracture mechanics, based on the processes of cracks propagation, in application to composites, we should learn the conditions (rate and temperature intervals) under which the cracks existing in the material and defects can increase and lead to the complete fracture. Only then the obtained estimates would be of great value.

2.2. Role of viscous properties of components in creep rupture processes

Viscous properties of composite components are of essential importance in the creep rupture processes, as, for example, the viscous properties of the resin determine the character of stress state in the broken fiber ends and its dependence on time, and viscous properties of fibers influence the accumulation of damage process, which is related to the failure of the material. Rather a detailed review of experimental works concerning the influence of various components properties on creep rupture, the influence of the volume content, viscous properties of the resin and temperature, is given in the works of SKUDRA et al. (1971) and LIFSHITZ (1974). Apparently, it is not necessary to enumerate all of them here. We shall give only some of the last publications, not included in the mentioned above reviews, which help to establish the most general regularities of the creep rupture process.

GUNYAEV et al. (1978) tested the influence of the modification of the matrix properties (epoxy resin) of the carbon plastic by reinforcing it with whisker crystals TiO_2. It has been established that the penetration of these crystals increases long-time strength, the best results are achieved at 2.5–3 V% of whisker content. The decrease in strength of such a material after 100 hrs is 11%, while the decrease in strength of the material with homogeneous matrix without whiskers is 35%, i.e. three times as much. But further increasing whisker content leads to the strength decrease of the composite material. The obtained results show clearly the role of matrix failure, if we change its properties, it leads to changing of stress conditions of the fiber-resin system, the existence of whiskers leads to the redistribution of stresses and decreases stress concentration.

The role of viscous properties, i.e. of the matrix creep compliance was investigated by LIFSHITZ (1974) on glass-reinforced plastics containing the same volume fraction of fibers and based on different resins (polyester and epoxy). The results show that the decrease in strength of the glass-reinforced epoxy after 10^5 minutes is 12%, while the decrease in strength of the glass-reinforced polyester, the matrix of which has the higher creep compliance, after the same period of time, is 29%. It points to the fact, that the higher creep compliance of the matrix at the same regime of loading leads to the more intensive failure of fibers, therefore the accumulation of damage, which determines creep rupture, will be more intensive too.

Another matter is the viscous properties of fibers. SUVOROVA et al. (1980b) described the experimental data on creep rupture of the epoxy resin reinforced by different fibers – carbon and organic. While the decrease in strength of the carbon reinforced plastic after 1000 hrs is for about 30%, for the organic plastic – it is only 11%, and it is quite natural that its ultimate elongation is essentially greater.

As regards the influence of the adhesion properties of the material on the long-time strength, apparently, it is not considerable. It can be seen in the first section of Fig. 1, which shows that the slopes of dependences $\sigma \sim \lg \dot{\sigma}$ for materials with different adhesion properties (shear strength of one is twice as much as of the other) are nearly the same and in Fig. 3, which gives data on creep rupture for the same two materials. The decrease in strength for them after the same period of time is seen to be approximately equal. In that way the role of adhesion strength comes in the main to the determination of the failure

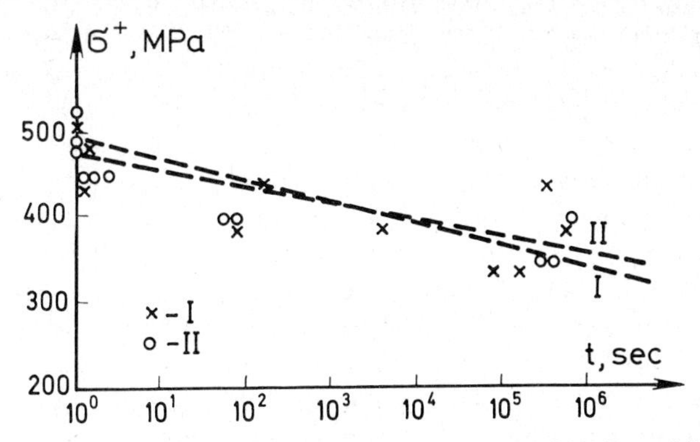

FIG. 3. Long-time strength of carbon fiber reinforced plastics I and II. Dotted lines: calculations according to Equation (17) at σ = constant.

mode, and the change of adhesion properties affects first of all the possibility of the crack propagation, points B_1 and B_2 in Fig. 1, which designate the failure mode and the moment of transition from the ductile failure, which occurs by an accumulation of damage to the brittle fracture caused by the crack propagation, differ by an order of rates of loading.

2.3. *Time-dependent failure models*

The construction of time-dependent models, based on the analysis of composite structure peculiarities, i.e. on studying the processes of mutual deformation and failure of the matrix (resin) and fibers, appears to be a rather complicated problem. The difficulty consists in the fact that this analysis requires some idealization of both the composite structure and the component properties, also idealizing the character of bond between them. It may be the reason why there are so few works concerning this subject, and in those of them which give the comparison with experiments, the parameters of the equations are determined, as a rule, according to macroscopic tests.

Rather a detailed analysis of the process of interaction between stiff short fibers and viscous resin which leads to creep of a composite and its failure is given by MILEIKO (1971). In this work an idealized scheme of the material structure is assumed. It is supposed that the fibers are of equal length, hexagonal section and are distributed regularly. First of all the author finds out what is meant by the term 'short fibers'. He introduces for this two critical values of ρ^* and ρ^{**} ($\rho = L/h$, L fiber length, h fiber radius) and the fibers are considered short, if $\rho < \min(\rho^*, \rho^{**})$. The introduction of critical values is caused by the conditions of mutual deformation of fibers and resin. The first value means that, if the fiber length is more than that defined by the critical value, the fibers are overloaded and can fail. So the value of ρ^* is naturally related to the fibers strength and also to other parameters, determining the condition of transferring of the load on fibers through the resin, and it depends on the level of the applied load too. The introduction of the second critical value means that if fiber length is more than that defined by this value, the fiber's creep rate is like the resin's one, that is why such a composite will behave as the composite reinforced with continuous fibers. This value also depends on the fiber and resin properties and on the level of applied load. The existence of the two limitations to the value of ρ, related to the fact, that beforehand it is unknown which of the values ρ^* or ρ^{**} will be less. Further for the investigation of the failure process of a composite, for which $\rho < \min(\rho^*, \rho^{**})$, it is supposed an ideal bond between the fibers and resin up to some critical shear strain value of γ along the fiber-matrix interface, at $\gamma = \gamma^*$ the bond is broken and this fiber carries no more load. It is dropped out the whole chain of fibers, the load on which is determined by the value of mutual overlapping of pairs of adjacent fibers in the axial section.

The complete failure occurs, when all chains are broken. The conducted analysis (in supposition of creep power laws of the fiber and resin materials) allows to calculate the time to failure:

$$t_u = \frac{\gamma^*}{\rho \dot{e}} \frac{I - V_f^{(m-1)/2}}{m - I} (I + m\psi(m)) . \tag{1}$$

Here V_f is the fiber volume content, $\dot{e}$ the creep rate of a composite, determined by the given level of load and resin properties, m the exponent in the creep law, and:

$$\psi(m) = \int_0^1 x^{m-I} \left[I - \left(\frac{2x}{I + x} \right)^2 \right]^m \mathrm{d}x .$$

As the author has made strong suppositions, related to the idealization of the composite properties and structure, formula (1) can hardly give the reliable quantitative estimates, but, however, the qualitative picture corresponds to reality. Besides that, by using the results of the analysis, the author manages to establish, that the initial moment of the failure process of a composite, i.e., the moment, when the first weakest chains begin to break, corresponds to the strain value which is half of the ultimate. This fact is affirmed by a lot of experiments, particularly, the experiments made on the composites with continuous fibers, for example, SUVOROVA et al. (1979, 1980b), though the conducted analysis is unapplicable in this case.

The construction of the creep rupture models for the composites with continuous fibers is based usually on the statistical character of the rupture process. It has already been mentioned, that such models are not considered in this chapter, it is a subject of an independent discussion; however, we should mention the original work of LIFSHITZ and ROTEM (1970) which gives the modification of Rosen's model with regard for viscoelastic properties of resin; it allows to establish the dependence of the ultimate stress on the time of loading.

When the matrix material is viscoelastic the shear stress along the fiber-matrix interface relaxes in time causing a decrease of the axial stress in the fiber near its broken end. It turns out, that the ineffective length δ (determined, for example, as a distance along the fiber over which the stress value changes from zero to some value of $\varphi\delta$) is an increasing function of time. It means that failed material volume is increasing all the time even without additional breaks of fibres. This effect of increasing ineffective length of fibres is called 'penetration'. LIFSHITZ and ROTEM (1970) suppose that penetration can be calculated by the raise of the number of fibers which are considered broken. The relative number of these additional breaks of fibers in a layer can be represented by the value of $R(t)$, dependent on time:

$$R(t) = F[\sigma_f(0)]\left[\frac{\delta(t)}{\delta(0)} - 1\right] + \int_{F[\sigma_f(0)]}^{F[\sigma_f(t)]} \left[\frac{\delta(t-\tau)}{\delta(0)} - 1\right] \mathrm{d}t . \qquad (2)$$

Here $\sigma_f(t)$ is the stress in an effective fiber at time t, $F(\sigma) = I - \exp(-\alpha\delta\sigma\beta)$ is a strength distribution of the Weibull type and α and β its parameters.

In formula (2) the contribution from fibers that were broken at time zero is considered as well as the contribution from fibers that failed at a later time.

The analysis of the ineffective length change δ due to the viscoelastic properties of resin allows the authors to establish the following dependence of δ on time:

$$\delta(t) = B[J(t)]^{1/2} \qquad (3)$$

where $J(t)$ is the matrix creep compliance in shear;

$$B = [E_f ra/2]^{1/2} \ln[I/(I - \varphi)]$$

where E_f the Young's modulus of fiber, r the fiber radius, and a the thickness of the adhesive layer between fibers and resin.

If we consider, that the stress $\tilde{\sigma}$ applied to the composite at time zero $t = 0$ is a step function and is applied only to unpenetrated fibers, then the stress value of $\sigma(t)$, applied to the unpenetrated part of the composite can be defined as:

$$\sigma(t) = \tilde{\sigma}/(I - R(t)) .$$

The failure criterion according to the general notions of Rosen's statistical model is then given as:

$$\frac{\tilde{\sigma}}{I - R(t)} = V_f(\alpha\delta\beta \, \mathrm{e})^{-I/\beta} \qquad (4)$$

(e the base of natural logarithms).

Equations (2)–(4) allow to determine the dependence of the ultimate stress $\tilde{\sigma}$ on the time to failure. They are used by LIFSHITZ (1974) for the estimation of the time-dependent strength of two types of polymeric (epoxy and polyester) composites, which distinguish in dependence $\delta(t)$ defined by formula (3).

The results demonstrate that the decrease in strength of the reinforced epoxy after 10^5 min is 12%, while the decrease in strength of the reinforced polyester, the creep compliance of which is much more, after the same period of time is 29%. The theoretical results have been compared with experimental ones. It is to be noted, that the obtained conformity is rather of the qualitative than of quantitative character. The author thinks that not all necessary data have been included in the analysis, for example, the change of the fiber strength is to be considered as well as some others factors.

The model suggested by LIFSHITZ and ROTEM (1970) undoubtedly let us see clearly the role of matrix in creep rupture processes of composites. But like Rosen's model it only explains what happens to the composite at failure, but cannot give any quantitative estimates according to the macroscopic analysis.

There are some other, at first sight simpler, attempts to calculate analytically the contributions of resin and fibers in the creep rupture process. For example, this was done in the works of SKUDRA et al. (1971), LIFSHITZ (1974), CRAIG and COURTNEY (1975), ROGINSKY and KANOVICH (1979), CHIAO et al. (1977), AKAY and SAIBEL (1978). Most of them are based on the generalization of the known relation:

$$\sigma_c^+ = V_f E_f + (I - V_f) E_m e_{u_f} \tag{5}$$

by the introduction of the time-dependent moduli: $E_f(t)$ and $E_m(t)$ either as empirically suggested formulae or as dependences determined by both differential and integral visco-elastic models.

Probably the most general approaches are represented by relation suggested in the works of SKUDRA et al. (1971), ROGINSKY and KANOVICH (1979), there hereditary presentations are used for the description of the change of moduli in time. But in this, as well as in other models of the kind, agreement with the experiment can be achieved only by the introduction of special parameters to which some sense is attributed.

But the determination of these parameters and investigation of time-dependent characteristics of fibers and resin is not always possible, for example, we are unable to manufacture high quality testing specimens of some polymeric resin. Then we are to regard the introduced dependences as some empirical formulae, determined in macroexperiment. That is why some investigators consider it quite reasonable to treat the composite as a homogeneous material and to use the models received for homogeneous materials (for example, polymers and metals at high temperatures) and apply available formulae, certainly with necessary modifications, to the description of the creep rupture curves. For example, CHAIO et al. (1977) take exponential dependence of the time to failure t_u on the value of the applied stress σ^+ for the creep rupture curve description:

$$t_u = t_0 \exp[(U_0 - \alpha\sigma^+)/kT]\,. \qquad (6)$$

The authors attribute to constants t_0, U_0, α and k the usual meaning as in equations for homogeneous materials, for example, t_0 the period of thermal oscillation of bound atoms, U_0 the energy of activation, but such an interpretation is hardly possible for nonhomogeneous materials or composites. Probably that is why the above mentioned constants were determined with the aid of the creep rupture curves of the composites as some parameters of macroexperiment.

AKAY and SAIBEL (1978) made an attempt to modify somehow the type of equation (6) taking into consideration the fact that a composite is a material consisting of minimally two different materials having different creep rupture characteristics. For each of the materials the equation is considered to be right:

$$\frac{dN_i}{dt} = -N_i \frac{kT}{h} \exp\left(-\frac{F_i}{RT}\right) \exp\left(\frac{W_i}{kt}\right), \qquad (7)$$

where N is the number of bonds per unit cross-sectional area of the composite, F the energy of activation, T temperature, k the Bolzman constant, h the Plank constant, R universal gas constant, and W the work done by the stress applied to the material for breaking a bond. Further it is considered that a composite consists of elastic and viscoelastic elements. For the elastic material:

$$W = \sigma_0\lambda/2N$$

(where λ is the mean separation between the equilibrium positions of minimum force potential), for a viscoelastic material:

$$W = \frac{\lambda\sigma_0}{EN} I - \exp\left(-\frac{E}{\eta}t\right)$$

(E Young's modulus, η material toughness).

Equations are solved numerically and in that way observe the process of bonds breaking in time for each of the materials. The composite is considered failured when all bonds in it are broken.

AKAY and SAIBEL (1978) give the results of the numerical calculations of the time to failure at various magnitudes of included constants.

But we can consider the obtained data as qualitative ones, because it is hardly possible to determine, for example, N, the number of bonds per unit area, and also some other parameters included in (7), in a reliable way, for the calculations made according to the suggested models should be in good agreement with experimental data.

In that way, the above listed attempts to describe the process of creep rupture of composites, give sometimes the opportunity to reveal the reasons for observed regularities, but they do not represent quantitative data, so they are not used for the prediction of the composite's behaviour and its life-time under conditions not similar to those taken for the determination of parameters of equations.

For the investigation of the time-dependent strength properties under different and rather complicated loading conditions, it is necessary to construct the phenomenological models, which must allow to formulate the constitutive equations and strength criteria in the terms of values which can be determined in macroexperiment. The construction of such models and the interpretation of results must undoubtedly correspond to the analysis of the inner processes taking place in the material at deformation and failure, the main task of this analysis is to establish the bounds in which we can use the suggested mathematical formulations.

3. Phenomenology of time-dependent failure

3.1. Strength criteria

There are several types of strength criteria known in mechanics of solids. Criteria, which establish some definite relationship between components of the strain and stress tensors, are widely used. These criteria give satisfactory results for elastic materials. But, however, they cannot predict the materials behaviour under repeated loading conditions and so, naturally, they are unapplicable for materials with viscous properties.

The other type of strength criteria is based on the presentation of damage accumulation. As a rule, it is introduced the measure of damage accumulation of the material, i.e., some value characterizing the time-dependent failure process. The sample will fail when the measure of damage achieves some definite value.

At last, one more approach to characterize strength properties of a material is the analysis of the crack propagation based on linear (or nonlinear) fracture mechanics. In this case the material's strength properties are estimated according to some characteristic values of stress concentration factors. The development of this approach is under great attention, but the calculations are rather difficult, especially for anisotropic and three-dimensional analysis.

In application to composites every approach has its peculiarities. Composites are inhomogeneous, constitutive materials, which can have different characteristic failure mechanisms, even one and the same material can fail differently under different conditions. However, it is not always easy to determine the mechanism of failure. So the construction of strength criteria for the composites is carried out usually either in the way of generalizing the known relations for isotropic materials, or by simplifying the analysis according to microstructural presentations.

Phenomenological strength criteria for composites are highly developing now, for example, in the works of GOLDENBLATT and KOPNOV (1965), MALMEISTER (1966), TSAI and WU (1971). They have rather complicated equations, containing a great number of experimental constants, and cannot be applied without computers, not even for rather simple problems, but they do give the opportunity to take into consideration the influence of anisotropic properties, which is of great importance for such inhomogeneous materials as composites. This approach facilitates some simplifications according to microstructural presentations. This was done in the works of HASHIN and ROTEM (1973), RABOTNOV and POLILOV (1977), UEMURA (1979). The works investigate only two mechanisms of failure: under normal and shear stresses in the direction of fibers. This approach considers the failure mode and allows the authors to establish one general criterion including both possible mechanisms. The simplest variant of this criterion is the achievement of some value (constant) by linear or quadratic combination of normal and shear stresses. More simplications were assumed in Uemura's work (1979). In this work it is investigated that the fracture of helically-filamentwound composite cylinders made of two unidirectionally layers with fibers in adjacent layers, plied in a definite angle. The author considers different loading conditions and finds out four corresponding possible fracture mechanisms:

(1) under the fiber tensile stress;
(2) under compressive stress;
(3) under the stress normal to fibers which leads to debonding;
(4) under shear stresses along the fiber-matrix interface.

For every case there are accepted very simple failure criteria, consisting in the fact that corresponding components achieve some critical value.

A large number of works is devoted to the investigation of the damage accumulation processes in composites and to construction of corresponding failure conditions. As a rule, as well as for isotropic materials, it is introduced some value of ω characterizing the degree of damage accumulation of the material, i.e., the part of axial section area which cannot carry any load. For the failure conditions the relation $\omega = \omega^*$ (in a particular case $\omega^* = I$) is usually taken. These are the well-known KACHANOV's (1974) and RABOTNOV's (1966) hypotheses. A lot of works deal with the application of these hypotheses to composites, TAMUZH and KUKSENKO (1978) give a rather detailed review of these works. The development of these presentations is conducted either in statistical analysis or in constructing hypothetical kinetic relations, describing the damage accumulation process and correlating well with microstructural analysis or macroexperiment.

And, at last, the third approach to constructing strength criteria, based on the application of linear fracture mechanics to composites, is also developing, but there are some limits to its use, connected with structural inhomogeneity of materials, so the main attention of the investigators is directed at ascertaining the possibilities of the development of such methods. SMITH (1978) gives the review of these works.

Unfortunately, though it has been found out that just one and the same material can fail differently under different conditions, most works, connected with investigation of failure and construction of strength criteria, do not establish bounds in which the suggested relations can be used. If we again return to Fig. 1, we notice that the approaches to constructing strength criteria, described in the present chapter, correspond to different areas and so to conformable failure modes.

Phenomenological models, based on constructing critical relations for stress components, can be succesfully used in Zone III. Linear fracture mechanics, based on the analysis of crack propagation is used in Zone II. And by studying material behaviour in Zone I we can make investigations of damage accumulation processes. In 2.1 it was pointed out, that transitions from one zone to another, i.e., from one failure mode to another, depend on composite structure, loading rate and temperature. So, before we choose this or that strength criteria to be used in calculations, we must learn the loading conditions for a given material and determine in which zone it works under such conditions. Most of modern composites, used in real structures, work in Zone I, where loading rates are not very high and temperatures are not very low. The necessity for the damage accumulation process analysis appears to be especially great under the conditions of long-time loading.

3.2. Creep process and accumulation of damage

Analysis of damage accumulation is impossible without the investigation of the process of material deformation. For the so-called 'elastic' composites, such as glass-, carbon-, or boron-reinforced plastics, which do not creep at room temperature, the study of stress-strain diagrams, obtained under different loading conditions, allows to find out the kinetics of the damage accumulation process. However for just the same materials at elevated temperatures, and for materials which creep at room temperatures, it is necessary, together with accumulation of damage, to investigate also the process of the viscous flow of a material in order to separate these two processes and find out the failure deformation at any moment.

For describing creep behaviour of materials various models exist, the most general of them are hereditary type models. It is convenient to use these models as they have wide possibilities and allow to use one set of parameters for describing different regimes of loading in a large scope of times and rates. This was shown, for example, in the work of SUVOROVA (1979).

For achieving some uniformity in describing creep and damage accumulation, which is convenient for models construction and calculations, we can suppose, that damage accumulation process can also be described by some operator of the hereditary type. (The hereditary character of the failure process is accepted also, for example, by ILYUSHIN (1967).)

Gradual failure of material leads inevitably to appearance of residual or irreversible deformation on unloading, which is very often connected with nonlinear behaviour. So it is more convenient to conduct this analysis within the framework of the nonlinear model, which in some particular cases can easily be turned into linear.

For the governing relation we shall take Rabotnov's equation:

$$\varphi(e) = \sigma + K^*\sigma \tag{8}$$

where $\varphi(e)$ is an instantaneous stress-strain curve, which gives the upper limit to possible deformation of the material.

$$K^*\sigma = \int_0^t K(t-\tau)\sigma(\tau)\,\mathrm{d}\tau . \tag{9}$$

It is the integral operator characterizing the creeping down from an instantaneous curve. $K(t-\tau)$ is the kernel of the integral operator, which can vary subject to material properties and behaviour. The choice of kernels is a question of an independent interest and we shall not dwell on it here. The works of OSOKIN and SUVOROVA (1978) and SUVOROVA (1979) show that for practical purposes it is convenient to choose the simplest Abel type kernel $K(t-\tau) = K(I-\alpha)/(t-\tau)^{\alpha}$ which satisfactorily reproduces the behaviour of the material under various loading conditions, and at the same time is very simple in calculations.

The integral operator $K^*\sigma$ characterizes, as has already been mentioned, the degrees of declination of real stress-strain curve from the instantaneous one, which takes place according to time effects.

This declination can be caused by two factors operating either separately (for materials of different types) or mutually. The first factor is viscous flow of the material. Let us represent the decrease in the value of stress subjected to creeping as $\Delta\sigma_e$. The second factor is gradual failure of the material, i.e., damage accumulation. And the decrease in the value of stress subjected to the failure process we represent as $\Delta\sigma_p$. As a result, it turns out that the value of average stress can be expressed by the following relation:

$$\sigma = \sigma_0 - \Delta\sigma_e - \Delta\sigma_p \tag{10}$$

Here σ_0 is the stress value corresponding to the instantaneous deformation. It can be connected with strain both linearily (then it will be a linear-constitutive equation) and nonlinearily.

Declination from an instantaneous curve, as we have already mentioned, can be represented by some hereditary type operator:

$$\Delta\sigma_e = L^*\sigma . \tag{11}$$

Suppose that the process of accumulation of damage can be described by some operator, and that the value of $\Delta\sigma_p$ can be represented in the following form:

$$\Delta\sigma_p = M^*\sigma . \tag{12}$$

The value of $\Delta\sigma_p$ is related to the irreversible deformation of the material.

On loading both components $\Delta\sigma_e$ and $\Delta\sigma_p$ are summed and cannot be distinguished. On unloading they do not behave similarly. The elastic-hereditary deformation is reversible, when the load is taken away the strain relaxes to zero, and Equation (11) is valid both for loading and unloading conditions. As far as the irreversible deformation is concerned, it cannot decrease. (It is to be noted that for the moment of beginning of unloading we take the moment when deformation achieves its maximum value, and not the moment when stress changes sign.) Therefore, considering (8), (10)–(12), we get the following constitutive equation:

$$\begin{aligned} \varphi(e) &= \sigma + \int_0^t [L(t-\tau) + M(t-\tau)]\sigma(\tau)\,\mathrm{d}\tau , \quad 0 \leqslant t \leqslant t_* , \\ \varphi(e) &= \sigma + \int_0^t L(t-\tau)\sigma(\tau)\,\mathrm{d}\tau + \int_0^{t_*} M(t_* - \tau)\sigma(\tau)\,\mathrm{d}\tau , \quad t > t_* . \end{aligned} \tag{13}$$

These equations can be rewritten in the following way:

$$\varphi(e) = \sigma + \int_0^t K_1(t-\tau)\sigma(\tau)\,\mathrm{d}\tau , \quad 0 \leqslant t \leqslant t_* ,$$

$$\varphi(e) = \sigma + \int_0^t K_{11}(t-\tau)\sigma(\tau)\,\mathrm{d}\tau + \sigma_p^* , \quad t > t_* ,$$

$$\sigma_p^* = \int_0^{t_*} M(t_* - \tau)\sigma(\tau)\,\mathrm{d}\tau .$$

Here $K_1(t) = L(t) + M(t)$; $K_{11}(t) = L(t)$.

Actually, it turns out that such an approach gives a description of the process of loading which does not differ from the traditional one and which takes place in accordance with an equation of type (8). The process of unloading also takes place in accordance with Equation (8) but with another kernel.

These equations have been used in the work of SUVOROVA et al. (1979) for interpreting experimental data. In this work the simplest way was chosen, i.e., components $\Delta\sigma_e$ and $\Delta\sigma_p$ were described by the same Abel's hereditary operators with equal parameters of singularity:

$$L(t-\tau)=\frac{l(1-\alpha)}{(t-\tau)^{\alpha}}; \qquad M(t-\tau)=\frac{m(1-\alpha)}{(t-\tau)^{\alpha}}.$$

Even such a simple case allows for good agreement between theory and experimental data. It was observed however, that the values of residual strain obtained in calculations were overestimated in comparison to the ones obtained experimentally. This suggests that on unloading some defects can be cured, which naturally causes the decrease in the value of strain. At this, if only small defects appear under load, they can be cured completely when load is removed, the appearance of large or oriented defects is an irreversible process. In accordance with this hypothesis Equation (13) was rewritten in the following way, if the size of defects is determined by the value of strain achieved at the moment of unloading:

$$\begin{aligned}
\varphi(e) &= \sigma+\int_0^t L(t-\tau)\sigma(\tau)\,\mathrm{d}\tau+\int_0^t M(t-\tau)\sigma(\tau)\,\mathrm{d}\tau, \quad 0\leq t\leq t_*, \\
\varphi(e) &= \sigma+\int_0^t L(t-\tau)\sigma(\tau)\,\mathrm{d}\tau+\Phi(e_*)\int_0^{t_*} M(t_*-\tau)\sigma(\tau)\,\mathrm{d}\tau, \quad t>t_*.
\end{aligned} \tag{14}$$

Here the function $\Phi(e_*)$ is introduced, which characterizes recovery and depends only on ultimate strain e_*. Experimental data obtained at various regimes of loading allow to find the value of this function for different values of e_*. For example, in Table 1 data obtained for organoplastic 7t are given. It represents creep regimes at constant stress and regimes of loading (unloading) according to the law $\dot{\sigma}$ = constant.

Dependences $\Phi(e_*)$ at various temperatures are plotted in Fig. 4. It shows the existence of some value of strain at which the small defects transfer into larger ones. These large defects cannot be cured at unloading. Micromechanical investigations of the failure process made by TAMUZH and KUKSENKO (1978) affirm this result obtained by macroanalysis.

Supposition about the possibility of partial strain recovery allows to explain material behaviour under fatigue conditions. If fatigue deformation takes place in that zone, where $\Phi(e_*)=0$, i.e., where complete recovery occurs, then every cycle of loading is reversible (according to strain). If deformation takes place in a transition zone $0<\Phi(e_*)<1$, then at some value of strain the steady cycle establishes the number of defects having appeared correspond to the number of disappeared. And at last, at $\Phi(e_*)=1$ after several cycles fracture necessarily occurs.

The given analysis shows that the value of strain e_* at which the quality changing of defects takes place, is a very important materials characteristic. It is determined, in the first place, by viscous properties of resin, and if we change these properties in some definite way, we can move the value of e_* in the zone of higher values of strain.

TABLE 1. Experimental values of function $\Phi(e_*)$ obtained at various regimes of loading.

Regime of loading	Time t_* (min)	Ultimate strain e_* (%)	$\Phi(e_*)$
σ = constant σ = 200 MPa	360	0.91	0.06
σ = constant σ = 300 MPa	360	1.5	0.4
$\dot{\sigma}_l = 0.1$ MPa/s $\dot{\sigma}_{un} = -25$ MPa/s $\sigma_* = 350$ MPa	58.3	1.64	0.9
$\dot{\sigma}_l = 25$ MPa/s $\dot{\sigma}_{un} = -0.1$ MPa/s $\sigma_* = 400$ MPa	2.267	1.86	1.0
$\dot{\sigma}_l = 25$ MPa/s $\dot{\sigma}_{un} = -25$ MPa/s $\sigma_* = 450$ MPa	0.3	1.9	1.0
σ = constant σ = 400 MPa	360	2.16	1.0

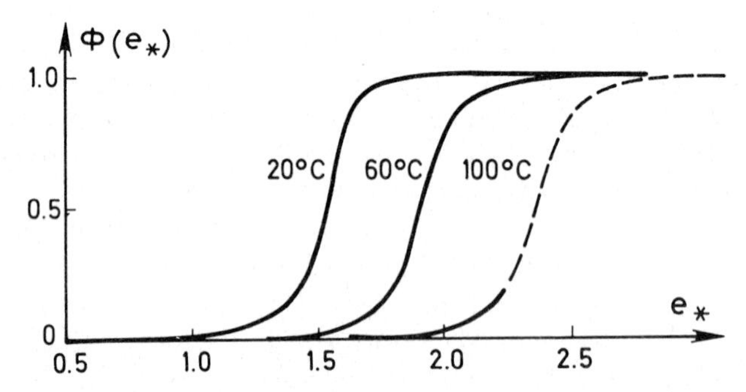

FIG. 4. Dependence of the function characterising recovery of defects on the value of ultimate strain achieved at the moment of unloading. Hypothetical section at 100°C is marked by a dotted line.

3.3. *Criterion of time-dependent failure*

Apparently, the most reasonable approach to the formulation of failure conditions is considering that it is the consequence of the process of deformation, being its natural continuation. Besides that, it is desirable to determine the parameters included in the strength criterion by independent experiments (particularly, for example, according to stress-strain curves).

Let us consider firstly the composites without viscous properties (for example, glass-, carbon-, boron-reinforced plastics). In these composites under long-time load only the gradual failure occurs. This means that Equations (13) and (14) contain only the operator M^* describing the failure process.

Let σ_0 represent true stress in a material; if no failure occurs in a material, it is this stress that should govern its stress-strain curve $\sigma_0 = \varphi(e)$. Because of the damage accumulation the measured stress σ, in accordance with the aforementioned, turns to be smaller by the value of $\Delta\sigma$:

$$\sigma = \sigma_0 - \Delta\sigma \tag{15}$$

So the stress-strain curve, obtained in experiment $\sigma = \varphi(e) - \Delta\sigma$ will be lower than the curve corresponding to undamaged material. The value of $\Delta\sigma$ can be related to the usually introduced value of ω $(\sigma_0 = \sigma/(1-\omega))$ in the following way:

$$\Delta\sigma = \sigma_0 - \sigma = \varphi(e)\omega$$

In accordance with (12) gradual failure or accumulation of damage, i.e., the change of the value $\Delta\sigma$ can be represented and described by some hereditary type operator $\Delta\sigma = M^*\sigma$, connected with the kinetics of accumulation of damage in the following way: $\omega = M^*\sigma/\varphi(e)$.

From Equations (13) or (15) it follows that

$$\sigma + M^*\sigma = \sigma_0 = \varphi(e)\,. \tag{16}$$

This equation describes the process of material deformation, i.e. the diagram $\sigma \sim e$ under the condition of accumulation of damage. Now we can interpret the failure criterion as the achievement of its maximum value by true stress, i.e.

$$\sigma_0 = \sigma_0^+ \quad \text{or} \quad \sigma + M^*\sigma = \sigma_0^+\,. \tag{17}$$

In this form the criterion takes into consideration the known apporaches to the construction of strength criteria: on the one hand the measure of damage is introduced, determined by the operator $M^*\sigma$, on the other hand we interpret the failure criterion as the achievement of its maximum value by stress. In particular cases the equation can be reduced to the usual phenomenological failure criterion and to condition $\omega = \omega$.

Suppose in Equation (17) $\sigma = \text{const} = \sigma^+$. In this case the dependence is established between the value of applied stress and the time to failure, which allows to describe creep rupture curves.

From Equation (17) it follows:

$$\sigma^+ = \frac{\sigma_0^+}{1 + M^*} = \sigma_0^+(1 - N^*)\,.$$

Here N^* is the resolvent operator. We get different equations for description of creep rupture curves subject to the type of chosen operator M^* (or N^*) including exponential dependence.

If we choose M^* as an integral operator of hereditary type and its kernel as an exponent $\exp[-\beta(t-\tau)]$, we have according to (17) for the creep rupture curve the following equation:

$$\sigma^+ = \sigma_0^+/[1+(1-e^{-\beta t_u})/\beta]\,. \tag{18}$$

Equation (18) has two limiting values:

$$\frac{\sigma_0^+}{1+1/\beta} \leqslant \sigma^+ \leqslant \sigma_0^+\,,$$

corresponding to the instant loading, at which theoretical strength is practically utilized, and to small levels of applied load, to which the material can resist infinitely long.

It is convenient, apparently, to choose the operator in such a way that it allows to describe also the deformation process of a material. As previously for the investigation of deformation Abel's kernel has been chosen, and supposing

$$M(t-\tau) = m(1-\alpha)(t-\tau)^{-\alpha}$$

we have

$$\sigma^+ = \sigma_0^+(1+mt_u^{1-\alpha})\,. \tag{19}$$

The lower limit of strength in (19) makes zero, and using this kernel we see that even the smallest, non-zero, stress σ can lead to failure.

The failure criterion in the form of (17) can be used for any material with damage accumulation under loading, and it gives the opportunity to relate the process of deformation of the material with its ultimate characteristics, i.e., with failure stress, not only in the case when the stress is constant (long-time strength) but at any loading law.

Let us observe how we can explain some results of the application of the criterion to composites.

As a rule, for composites with polymeric matrix the high-strength fibers are used – they behave as an elastic material and have a linear stress-strain curve $\varphi(e) = E_f e$. When a tensile load is applied in the direction of the fibers, it is by the fibers that practically all the applied load is carried. So for the strength of the material the following simple formula is usually suggested:

$$\sigma_c^+ = V_f E_f e_{u_f}\,. \tag{20}$$

In fact the experimental values show lower values of strength than the theoretical prediction. This can be explained by the fact, that the work of filament is not taken into consideration, i.e., the effect of bond or interface between fibers and matrix, which can be partly broken. Besides that, the fibers break in weak sections, due to their strength distribution, and portions of fibers near broken ends become ineffective. These processes lead to the fact, that failure occurs at stresses lower than the theoretical prediction. So one usually introduces the value of the fiber's effective strength

$$\sigma_{\text{eff}} = \sigma_f^+ \Omega , \qquad \Omega < 1$$

and considers that the composite strength is $\sigma_c^+ = \sigma_{\text{eff}} V_f$. Methods of determining σ_{eff} may vary – for example, they can be based on the investigation of material structure. Attempts have been made to account for the redistribution of stresses in a broken fiber, based on the determination of 'ineffective length', but, however, the values of composite strength predicted according to this value, are, as a rule, essentially higher than experimental data. Apparently, this can be connected with the fact that actually the value of 'ineffective length' is much higher, due to debonding, which takes place along the interface of fiber and matrix at broken ends. The situation is the same as at estimating processes of deformation and failure of metals. Despite the fact that dislocation mechanisms which determine the appearance and development of plastic strain have been studied well, we are not able to use them in the quantitative analysis of the processes of deformation and failure. Also the micromechanics of fibrous composites does not give the exact quantitative prediction of their macroscopic behaviour, although it allows a qualitative understanding of it.

The above described model results in criterion (17), allows to use macroscopic analysis for explaining and gives a quantitative description of the decrease in the strength value in comparison with values predicted by formula (20), not basing this on the introduction of the fiber effective strength, but on supposing that on loading the value of V_f changes.

Let us take for example the widely used type of a composite viscoelastic (polymeric) resin and high-strength but brittle fibers. Fibers are considered as load-carrying elements, and the role of resin consists in transferring stresses from broken to unbroken fibers. A single fiber behaves as an elastic material all the way to fracture. The break of it means the complete loss of its load carrying capacity. But when it is embedded in resin, the break of a fiber does not mean the fracture of a specimen, simply portions of fibers near broken ends do not carry load. If we construct the stress-strain curve of the whole specimen, we obtain a nonlinear curve with the initial slope corresponding to fiber modulus E_f. Because of the breaks of the fibers it is seen that the value of V_f during the deformational process gradually decreases and the deviation from the linear law in the diagram $\sigma \sim e$ can be observed. According to degree and character of nonlinearity we can judge to what extent the material has been damaged. It is to be noted that with the accumulation of damage, i.e., with the increase in the number of breaks, the role of the matrix, transferring load between the fibers and depending considerably on time, is growing.

For the true stress in a composite it is natural to take the stress in a fiber. Then the process of deformation can be written in the following way:

$$\sigma + M^*\sigma = E_f e_f V_f = \sigma_f V_f$$

and the failure criterion as:

$$\sigma + M^*\sigma = \sigma_f^+ V_f\,.$$

(In a general case in the right-hand side of this equation the relation $(E_f V_f + E_m V_m)\, e_{u_f}$ can stand.) The equation shows that the ultimate stress corresponding to failure is proportional to the given fiber volume content in a composite. According to the formula, the description of the dependence of ultimate stress on time (using Abel's kernel) will be:

$$\sigma^+ = \frac{\sigma_f V_f}{1 + m t_u^{1-\alpha}}\,.$$

Experiments on glass-reinforced plastics show that at the same fiber volume content, a small change of the value of stress (a decrease of ~20%) leads to a considerable increase in life-time t_u (by 70 times approximately). This proves that the index of rate at t_u must be small. The same result was obtained in the work of SUVOROVA et al. (1979) on organoplastics. The conclusion is in good agreement with the results of quasistatical deformation.

If in Equation (17) we suppose $\dot{\sigma}$ = constant, we will get the following dependence between the ultimate stress σ^+ and the loading rate:

$$\dot{\sigma}^{(1-\alpha)} = \frac{m(\sigma^+)^{(2-\alpha)}}{\sigma_f V_f - \sigma^+}$$

or simply the dependence of σ^+ on the time to failure $t_u = \sigma^+/\dot{\sigma}$:

$$\sigma^+ = \frac{\sigma_f^+ V_f}{1 + \frac{m}{2-\alpha} t_u^{1-\alpha}}\,. \tag{21}$$

If we suppose that ultimate strain is a constant, depending neither on time to failure nor on fiber volume content, it follows from (21) that the increasing of the rate of loading (the less t_u) is connected with the increasing of failure stress σ^+. This is in good correlation with experimental data (Fig. 1, Zone 1).

Macroanalysis of the failure process allows not only to predict the strength of the whole composite, but also allows to make some qualitative conclusions concerning the necessary properties of its components. Thus, for example, fibers with a larger coefficient of variation are broken at very low stresses, so the degree of nonlinearity of the initial section of the diagram is higher than for fibers with a small coefficient of variation, although the ultimate values of strength can be equal. The description of the failure process with the help of hereditary presentations naturally relates this process with nonlinearity of the stress-strain curve. The kernel parameters as well as the position of the stress-strain instantaneous curve can serve as characteristics of this process. Thus, for an ideal material $\varphi(e) = Ee$, the nonlinearity of the instantaneous curve allows to estimate the quality of material fabrication. It is to be stressed that it is not always necessary to choose fibers with higher modulus for manufacturing of composites. This is required only when instantaneous diagrams are linear, in which case the increase in modulus leads to the increase in strength. Nonlinear material with initial damage shows more intensive accumulation of defects and can fail at lower stresses as material with lower initial modulus.

Criterion (17) can be developed further and generalized by considering the strength properties of resin and by statistical analysis of the failure process.

3.4. Experimental results

3.4.1. Strength at various regimes of loading

Usually for quantitative estimating the processes of time-dependent failure, for example, the values of creep rupture strength, the authors construct some relationships describing satisfactorily the obtained experimental data, but not general enough for describing experiments on other types of loading.

So one of the main problems consists in constructing such a strength criterion, that will allow to compare various loading conditions and to describe time dependences of strength under different laws of changing $\sigma(t)$ by one and the same equation with definite parameters.

We shall take, for example, experiments of POTTER (1974), who describes the shear experiments on unidirectional carbon-fiber reinforced epoxy. It has been obtained:

(1) creep rupture curve;
(2) dependence of strength on loading rate;
(3) dependence of strength on the number of cycles under repeated loading conditions.

To determine the parameters of Equation (17), rewritten in this case in the following way:

$$\tau + M^*\tau = \tau_0^{\mathrm{u}}, \tag{22}$$

it is necessary to use one of the given experimental curves. The creep rupture curve plotted in Fig. 5, has been used. The following values: $\alpha = 0.94$, $m = 1.47\ \mathrm{min}^{-(1-\alpha)}$, $\tau_0^{\mathrm{u}} = 208.2\ \mathrm{MPa}$ have been obtained. The creep rupture curve calculated with these parameters according to Equation (22), is also given in Fig. 5. Further, the same values have been used for calculating the dependence of strength on the loading rates, Fig. 6, and on the number of cycles to failure, Fig. 7. The results of calculations correlate well with experimental data. This

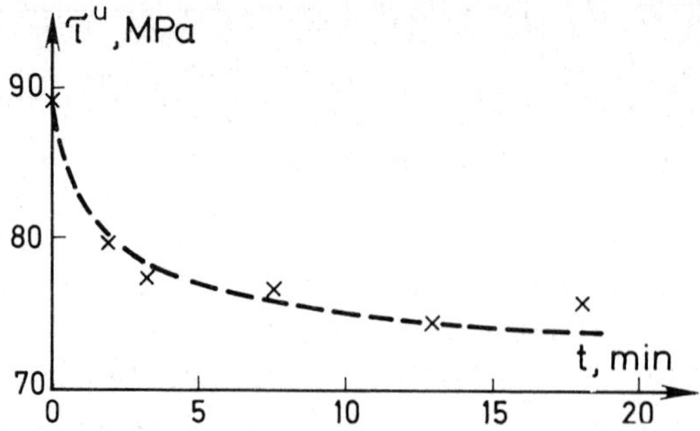

FIG. 5. Long-time strength at shear for orthogonal reinforced carbon fiber plastics; ×: experimental data of POTTER (1974). Dotted line: calculations according to Equation (22) at τ = constant.

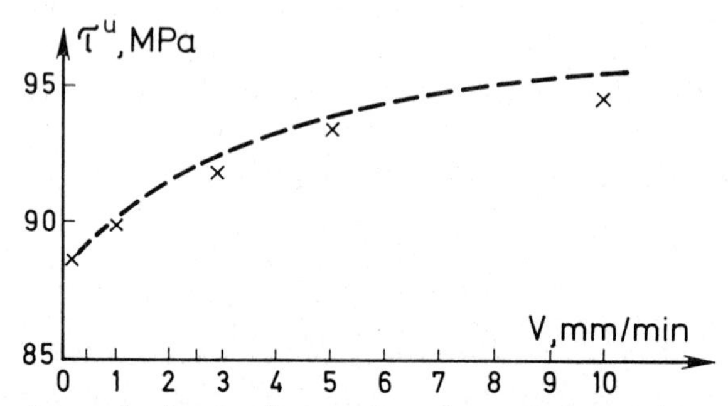

FIG. 6. Dependence of strength τ^u on loading rate for orthogonal reinforced carbon fiber plastic; ×: experimental data of POTTER (1974). Dotted line: calculations according to Equation (22) at $\dot{\tau} =$ τ t.

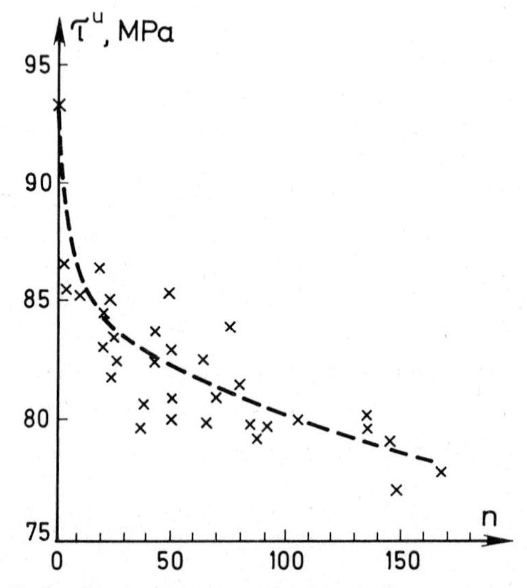

FIG. 7. Influence of cyclic loading on shear strength of orthogonal carbon reinforced plastic at a loading rate of 5 mm/min. ×: experimental data of POTTER (1974). Dotted lines: calculations according to (22) at $\tau = \tau^* \cos \omega t$.

affirms the fact that in all mentioned cases of loading, failure was caused by the accumulation of damage in the material. This process is governed by the conditions of load application and can be described by the operator M^*, characterized by two parameters (α and m).

Let us give another example. The works of SUVOROVA et al. (1980 a, b) show the results of experiments and calculations for carbon-fiber reinforced plastics and for organoplastics. The calculations of creep rupture curves and dependences of ultimate stress on loading rate were made with the same set of parameters. Some of the obtained results are plotted in Figs. 1 and 3. These figures show the results of experiments on two types of orthogonal reinforced carbon fiber plastics with epoxyphenol resin, which differ in their adhesive characteristics. Parameters of the failure criterion have been determined by sections A_1B_1, A_2B_2, (Fig. 1), and it has been assumed that $M(t-\tau) = m(1-\alpha)/(t-\tau)^\alpha$. The obtained values are given in Table 2.

The results of calculations, made in accordance with Equation (17), considering the loading rate to be constant $\dot{\sigma}$ = constant, are marked in Fig. 1 by solid lines. The same parameters as given in Table 2, have been used for calculation of creep rupture curves, Fig. 3.

TABLE 2. Values of parameters of failure criterion for carbon-reinforced plastics.

	α	$m(s^{-(1-\alpha)})$	σ_0^+(PMa)
Material I	0.92	0.29	650
Material II	0.96	0.946	983

3.4.2. *Failure of materials with viscous properties*

If a material has viscous properties, then two processes take place under loading – creep rupture (reversible) and accumulation of damage. The corresponding model is described in detail in Section 3.2. The parameters of kernels of operators describing each of the processes can be determined by the stress-strain curve under loading and unloading conditions. However, as only one of the processes, namely the accumulation of damage described by the operator $M^*\sigma$, is connected with the failure of a material, in the formulation of fracture conditions we only use this operator. So the failure criterion for both viscous and nonviscous materials is written in the same way.

The constitutive equation, describing the deformation process of the material with viscous properties, is as follows:

$$\sigma + L^*\sigma + M^*\sigma = \varphi(e) \tag{23}$$

and for nonviscous material:

$$\sigma + M^*\sigma = \varphi(e) \tag{24}$$

and the failure criterion for both of them is:

$$\sigma + M^*\sigma = \sigma_0^+ . \tag{25}$$

Comparing Equations (25) and (24) we observe that at the moment of failure $\varphi(e_u) = \sigma_0^+$, i.e. strain is constant and independent of time. For viscous materials the comparison of Equations (25) and (23) gives

$$\varphi(e_u) = \sigma_0^+ + L^*\sigma|_{t_u} .$$

If, for example, operator L^* is chosen as an operator of Abel's type, $L(t) = l(1-\alpha)t^{-\alpha}$, and loading occurs according to the law $\sigma = \dot{\sigma}t$, then:

$$\varphi(e_u) = \sigma_0^+ + \frac{l}{2+\alpha}\,\sigma^+ t_u^{1-\alpha} . \tag{26}$$

Formula (26) indicates that in this case ultimate strain efficiently depends on time of load application t_u.

In Figs. 8 and 9 the experimental data of SUVOROVA et al. (1980 a, b) are plotted. Represented are the values of ultimate strain at various loading rates for carbon-reinforced plastics, i.e., nonviscous materials (Fig. 8) and for efficiently viscous organoplastics (Fig. 9). While for carbon-reinforced plastics we fail to observe the dependence of e_u on loading rates within the scatter of data, for organoplastic it is rather considerable. The results of calculations according to formula (26) are plotted in Fig. 9 by solid lines (the value of parameters will be given below).

The determination of ultimate values of stress corresponding to the given loading conditions for a material with viscous properties have been carried out by Equation (25) as well as in 3.4.1.

The procedure of determining parameters by stress-strain curves at loading and unloading is described in the mentioned works. The diagrams chosen for calculations are plotted in Fig. 10. The data presented are sufficient for determining parameters of kernels of both operators L^* and M^*, and for

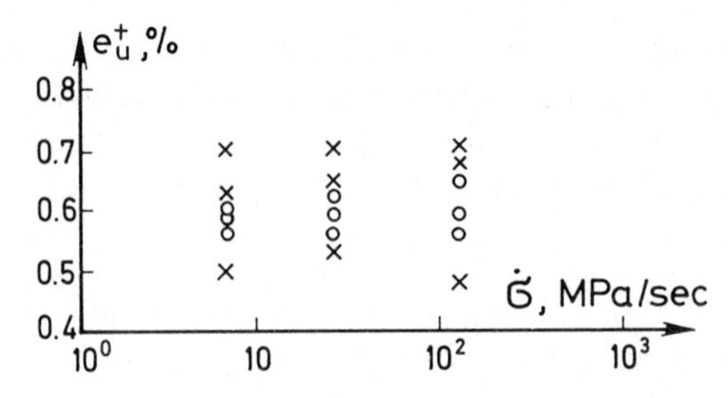

FIG. 8. Dependence of ultimate strain on loading rate for two types of carbon fiber reinforced plastics (see Fig. 1). ×: material I; O: material II.

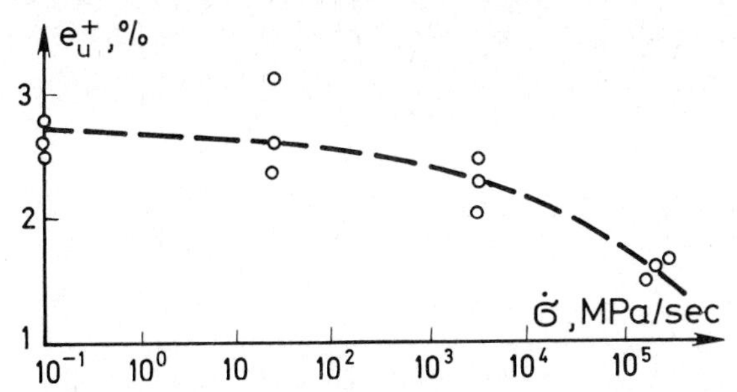

FIG. 9. Dependence of ultimate strain on loading rate for organoplastic; Dotted line: calculations according to (26).

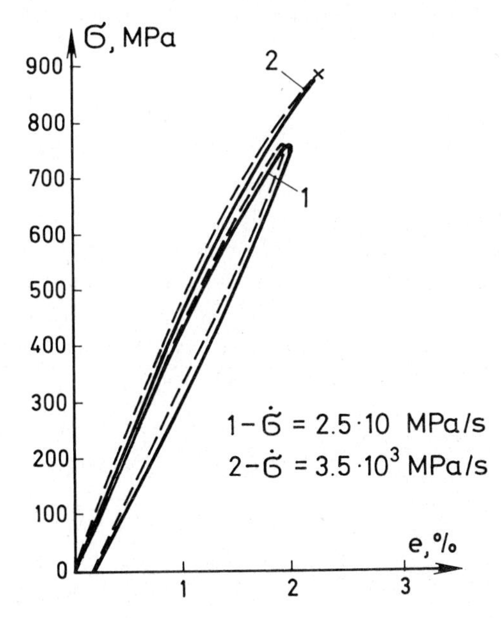

FIG. 10. Stress-strain curves obtained for organoplastic at different rates of loading; solid lines: experimental curves, dotted lines: calculations according to Equation (13).

constructing a curve $\varphi(e)$. The following values have been obtained: $\alpha = 0.97$; $l = 1.5$ min; $m = 0.2\ \text{min}^{-(1-\alpha)}$. The stress-strain curves, constructed according to these parameters are also given in Fig. 10. For using the failure criterion, it is also necessary to know σ_0^+. In the aforementioned work this value was determined as a parameter of macroexperiment by a singular test led to fracture (it is marked by a cross in Fig. 10). A value of $\sigma_0^+ = 107$ MPa has been obtained. Further, all these parameters have been used for determining the dependence of strength on loading rate, Fig. 11, and for constructing creep rupture curves, Fig. 12.

It should be emphasized that criterion (25) is to be used only when failure is

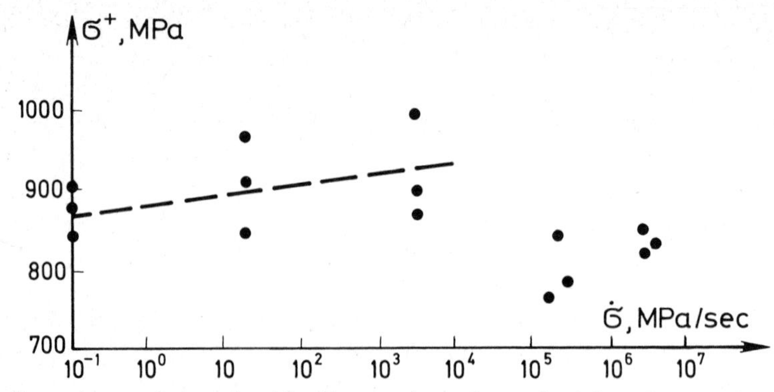

FIG. 11. Dependences of strength on loading rate for orthogonal reinforced organoplastic; dotted lines: calculations according to Equation (17) at $\sigma = \dot{\sigma}t$.

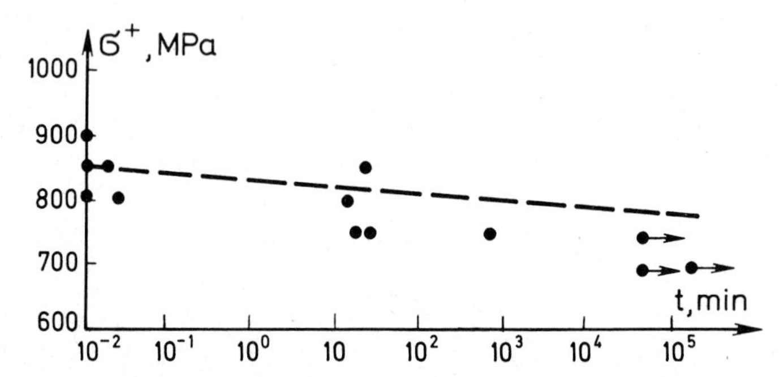

FIG. 12. Long-time strength of orthogonal reinforced organoplastic, dotted lines: Calculations according to (17) at σ = constant.

a consequence of the accumulation of damage in a material, but not of crack propagation.

The change of failure character occurs at a considerable increase in loading rates, it was mentioned in Section 2, while discussing experiments on carbon-reinforced plastics (Fig. 1). The same picture is observed for organoplastics, for them the decrease in strength occurs at $\dot{\sigma} \sim 10^5$ MPa/sec, approximately (Fig. 11). Some suppositions concerning the modification of criterion (25) in order to describe the change of failure mode, are given in the work of SUVOROVA et al. (1980a).

4. Phenomenology of temperature influence

4.1. *Constitutive equation*

The main advantage of composite materials consists in the fact that the designer choosing the resin and fiber materials in the proper way, can manufacture the specimen having all necessary properties under given loading conditions, particularly at high and low temperatures. For example, carbon-reinforced plastics can endure very high temperature retaining their resistance. However, temperature efficiently effects the processes of deformation and failure of a material. It is connected with the temperature sensitivity of the resin properties, temperature influences the possibility to transfer load and reduce stress concentrations in broken ends, etc. If we return to Fig. 1, we can say that if temperature increases, the point of transition from Zone I to Zone II, i.e. from failure connected with the accumulation of damage to failure caused by crack propagation, will shift to the right in the zone of higher volume contents and loading rates. On the whole, the problem of temperature influence on composite properties has not been investigated properly. All works in this field consist generally in constructing stress-strain curves at various temperatures and in attempts to describe temperature effect by artificial introduction of parameters dependent on temperature into the constitutive equations. Some empirical or semi-empirical relationships allowing to analyze the influence of temperature in a given interval have been suggested, but they are not general enough for a systematic study.

Now the principle of time-temperature analogy formulized for polymers by FERRY (1961) is of wide extension. It suggests to introduce the quantity of 'reduced time'

$$\tilde{t} = \int_0^t a[T(t)]\, \mathrm{d}t \, .$$

For isothermal process $\tilde{t} = a(t)t$. The convenience of this approach consists in the fact, that inserting temperature in constitutive equation one only changes the scale of time, and owing to that we can consider thermoreologically identical processes. The relationship between stress, strain and reduced time, is the same as the relationship between values of σ, e, and the usual time t at a constant temperature.

A large amount of works is devoted to the problem of application of time-temperature analogy to investigation of polymers properties. Attempts have been made to use this principle for composites, for example, in the works of KOVRIGA et al. (1972), KOLTUNOV and TROJANOVSKY (1970), URZHUMCEV and MAXIMOV (1975), DAUGSTE (1979). However, the results in this case are somewhat worse than for pure polymers. It turns out that in many cases the temperature shift factor a for composites depends on temperature and stress in a rather complicated manner. One has to insert in a constitutive equation additional parameters determined by additional experiments. It limits the use of this principle especially for nonisothermal processes.

This situation arises because the principle of time-temperature analogy is not a mechanical model, but represents only a convenient way of interpreting experimental data. It was very useful at first stages of investigation of temperature dependencies for inelastic materials, now it also gives the opportunity to systematize experimental curves by arranging them in a proper sequence, and it also allows to compare various materials according to their thermal sensitivity and investigate material properties at different regimes and duration of loading.

But, however, it is necessary to construct also mechanical models which can describe the influence of temperature on creep and failure in a sufficiently general manner. One of the possible approaches to constructing these models within the framework of the hereditary principle was proposed by SUVOROVA (1977) and suggests that the instantaneous stress-strain curve $\varphi(e)$ is independent of temperature. This assumption seems to be quite natural, because this curve is the upper limit to the possible area of material deformation, irrespective of the conditions under which the experiment is carried out (rate, temperature, moisture, etc.). It leads to the idea, that elevated temperature causes the inclination from the instantaneous curve, and $\varphi(e)$ for itself corresponds to the process of deformation at 0°K (which is of course some mathematical abstraction). No matter at what rate the material is loaded at 0°K, its stress-strain curve will correspond to the instantaneous one. The same diagram can be obtained at any other temperature, if the process occurs at infinitely high rate. This supposition allows to write the constitutive equation with regard to temperature in the following way:

$$\varphi(e) = \sigma + \int_0^t K(t-\tau)\sigma(\tau) f[T(t), T(\tau)]\, d\tau \tag{27}$$

Here f is some function which characterises the influence of temperature, $f \equiv 0$ at $T = 0$°K. The function naturally must depend on the actual temperature and on the history of its change, i.e., on τ. For its definition it is necessary to conduct special experiments. For example, the data of URZHUMCEV and MAXIMOV (1975), SUVOROVA (1977, 1980c), MAXIMOV (1974) indicate that in a number of cases we can neglect the temperature history and consider that the function depends only on temperature at the present moment. Such an approach greatly simplifies the calculations. The form of the function must be determined according to conveniency and to its compliance with experimental data.

The above mentioned works allow us to choose for the function f the following form:

$$f[T(t)] = T^{\gamma}(t)\,. \tag{28}$$

Thus for the complete ascertaining of temperature influence on material properties, it is necessary to know one additional parameter determined by an additional experiment. It is to be stressed, that in Equation (27) the kernel does not depend on temperature.

Experimental verification of the model (27) and (28) concerning its applicability for describing processes of deformation and failure for various materials, both for isothermal and nonisothermal processes, was carried out by SUVOROVA et al. (1980c) and in some other works. We shall not dwell on the question of temperature influence in the case of isothermal loading; it is clear that the increase in temperature leads to a decrease in diagrams to greater or less extent, which is determined by temperature sensitivity of fibers and resin. Let us represent two examples of nonisothermal loading showing that the function of temperature influence can be chosen in shape (28) at least for the tested materials. Isothermal and nonisothermal creep curves for organoplastic are given by SUVOROVA et al. (1980c) and shown in Fig. 13. (Nonisothermal creep occurs according to the following conditions: the material was heated from 20 to 100°C in 4 h at a constant rate, then it was kept at 100°C during two hours. $K(t-\tau) = K(1-\alpha)(t-\tau)^{-\alpha}$, $f[T(t)] = T^{\gamma}(t)/273$, $k = 2.06\ \text{min}^{-(1-\alpha)}$, $\alpha = 0.96$, $\gamma = 2.04$.)

A nonisothermal curve is seen to cross isothermal creep curves in the points corresponding to adequate temperatures. The same result was obtained by MAXIMOV et al. (1974) by analysing experiments on epoxyphenol glass-rein-

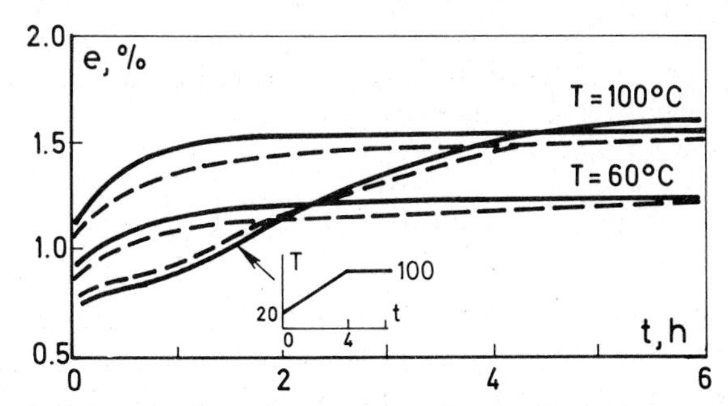

FIG. 13. Isothermal and nonisothermal creep curves for organoplastic 7t at $\sigma = 200$ MPa; solid lines: experimental curves; dotted lines: calculations according to Equations (27) and (28).

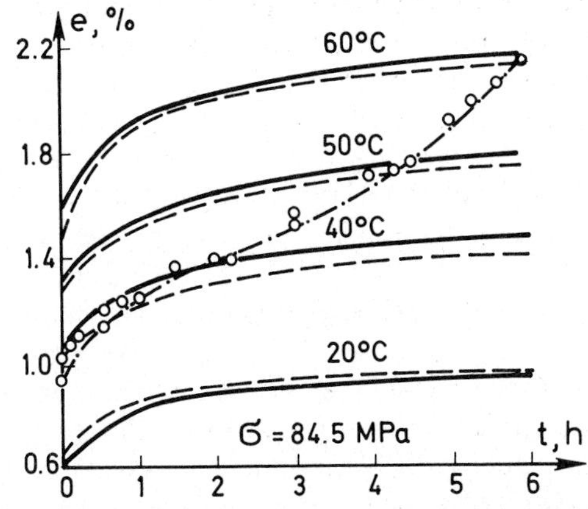

FIG. 14. Isothermal and nonisothermal creep curves for glass-reinforced plastic; solid lines: experimental curve of MAXIMOV et al. (1974); dotted lines: calculations according to Equations (27) and (28).

forced plastics (Fig. 14). The results of the calculations of these experiments made by SUVOROVA (1977) according to Equations (27) and (28) are shown in Fig. 14 by dotted lines.

In general it is necessary to solve the question of temperature heredity for every material used, though the organization of a nonisothermal experiment appears not to be an easy task. The difficulty consists in carrying out the process in such a way that no temperature gradients arise, otherwise for the analysis of experimental data one must solve a different boundary problem. It is to be stressed that the suggested model is correct only at temperatures not causing any microstructural changes of the material.

4.2. The influence of temperature on damage accumulation

It is within the framework of the model, as it allows to divide the processes of viscous flow and damage accumulation described in Section 3.2, to represent the constitutive equation for loading and unloading conditions in the following way:

$$
\begin{aligned}
\varphi(e) &= \sigma + \int_0^t L(t-\tau) f_l(T)\sigma(\tau)\,\mathrm{d}\tau + \int_0^t M(t-\tau) f_m(T)\sigma(\tau)\,\mathrm{d}\tau, \\
& 0 \leqslant t \leqslant t_*, \\
\varphi(e) &= \sigma + \int_0^t L(t-\tau) f_l(T)\sigma(\tau)\,\mathrm{d}\tau + \Phi(e_*, T) \int_0^{t_*} M(t_*-\tau) f_m(T)\sigma(\tau)\,\mathrm{d}\tau, \\
& t > t_* .
\end{aligned}
\tag{29}
$$

Temperature functions $f_l(T)$ and $f_m(T)$ may be different for processes of viscous flow and damage accumulation.

It was shown by SUVOROVA et al. (1980c) that kernels L and M can be of the same Abel's type with the same values of singularity α. Corresponding parameters of l and m have been determined by short-time creep curves. Values of $l_T = l(T/273)^{\gamma_1}$ and $m_T = m(T/273)^{\gamma_2}$ at various temperatures are given in Table 3.

Table 3 shows, that ratios l_T/m_T for various temperatures are approximately the same ($l_{20}/m_{20} = 4.5$; $l_{60}/m_{60} = 4.66$; $l_{100}/m_{100} = 4.5$).

This means that, at least for the material tested, we can assume $\gamma_1 = \gamma_2 = 2.04$, i.e., $f_l(T) = cf_m(T)$; c constant. This conclusion effeciently simplifies the analysis of processes occurring at various regimes of loading.

The function $\Phi(e_*, T)$ has been determined experimentally at elevated temperatures (Table 4). It characterizes the damage accumulation process. The same data for 20°C is given in Section 3.2, Table 1.

Corresponding diagrams of the change of $\Phi(e_*)$ in dependence of the value of strain are given in Fig. 4. Only the initial portion of $\Phi(e_*)$ at 100°C has been obtained for lack of sufficient number of experimental data. Its further hypothetical position is represented by a dotted line in Fig. 4.

TABLE 3. Parameters of Equations (29).

T (°C)	$l_T(\text{min}^{-(1-\alpha)})$	$m_T(\text{min}^{-(1-\alpha)})$
20	2.025	0.45
60	2.8	0.6
100	3.33	0.74

TABLE 4. The values of function $\Phi(e_*)$ at $T = 60$°C obtained at various regimes of loading.

Regime of loading	Time of loading t_*	Ultimate strain e_* (%)	$\Phi(e_*)$
σ = constant $\sigma = 200$ MPa	360	1.26	0.04
σ = constant $\sigma = 300$ MPa	360	2.03	0.1
$\dot{\sigma}_l = 0.1$ MPa/s $\dot{\sigma}_{un} = -25$ MPa/s $\sigma_* = 400$ MPa	66.7	2.08	0.9
$\dot{\sigma}_l = 0.1$ MPa/s $\dot{\sigma}_{un} = -0.1$ MPa/s $\sigma_* = 400$ MPa	66.7	2.08	1.0
$\dot{\sigma}_l = 25$ MPa/s $\dot{\sigma}_{un} = -0.1$ MPa/s $\sigma_* = 400$ MPa	4.2	2.45	1.0

Fig. 4 indicates that the value of ultimate strain, at which small defects transfer in larger ones not recovering under unloading conditions, increases with the increase in temperature. It also affirms the supposition that the point of transition from failure caused by the accumulation of damage to failure caused by crack propagation, with increase in temperature, will move to the right in the zone of larger volume contents and higher loading rates.

5. Conclusions

In this chapter an attempt has been made to analyze the time-dependent failure processes of composites. There are not as yet enough experiments investigating the problem. The influence of composite structure and components properties on failure mode is still to be investigated in detail. Of great importance also is the comparison of data obtained at different loading regimes and temperatures.

The present chapter is primarily concerned with accumulation of damage, though it has been stressed, that the change of failure mode transition to crack propagation is possible under special conditions.

The proposed mathematical model allows to investigate the accumulation of damage process in terms of true stresses. It gives the opportunity to analyze also the process of crack propagation, an analysis that can be based on the fracture mechanics presentations.

Of course, it is necessary to clear up the reasons causing the transition from one mode of failure to another for establishing optimal regimes of specimen work. Besides that, investigation of failure processes must be accompanied by statistical analysis, which can be introduced in the model.

All this is the task of future investigations.

References

AKAY, A. and E. SAIBEL (1978), *J. Composite Materials* **12**, 262.

CHIAO, C.C., R.J. SHERRY and N.W. HETHERINGTON (1977), *J. Composite Materials* **11**, 79.

CRAIG, W.H. and T.H. COURTNEY (1975), *J. Material Sci.* **10**, 1119.

DAUGSTE, CH.L. (1974), *Mech. Polymers* **3**, 427 (in Russian).

DERGUNOV, N.N., L.H. PAPERNIK and YU.N. RABOTNOV (1971), *Appl. Mech. Tech. Phys.* **2**, 276 (in Russian).

DHARAN, C.K.H. (1978), *J. Engrg. Material Technology* **100**, 233.

FERRY, J.D. (1963), *Viscoelastic Properties of Polymers* (Foreign Literature Press, Moscow), 236.

GOLDENBLATT, I.I. and V.A. KOPNOV (1965), *Mech. Polymers* **2**, 70 (in Russian).

GUNYAEV, G.M., I.M. MACHMUTOV, T.G. SORINA, E.I. STEPANYCHEV, A.I. SURGUCHIEVA and G.N. PHINOGENOV (1978), *Problems of Strength* **9**, 70 (in Russian).

HASHIN, Z. and A.A. ROTEM (1973), *J. Composite Materials* **7**, 448.

ILYUSHIN, A.A. (1967), *Mech. Solids* **3**, 21 (in Russian).

KATCHANOV, L.M. (1974), *Foundations of Fracture Mechanics* (Nauka, Moscow), 140 (in Russian).

KOLTUNOV, M.A. and I.E. TROYANOVSKY (1970), *Mech. Polymers* **2**, 217 (in Russian).

KOVRIGA, V.V., E.S. OSIPOVA, I.I. FARBEROVA and K.YA. ARTANOVA (1972), *Mech. Polymers* **2**, 360 (in Russian).

LIFSHITZ, J.M. (1974), Time-dependent fracture of fibrous composites, in: L.J. BROUTMAN, ed., *Composite Materials*, Vol. 5, Fracture and Fatigue (Academic Press, New York and London), Ch. 6, 249.

LIFSHITZ, J.M. and A. ROTEM (1970), *Fibre Sci. Technology* **3**, 1.

MAXIMOV, R.D., CH.L. DAUGSTE and E.A. SOKOLOV (1974), *Mech. Polymers* **3**, 415 (in Russian).

MALMEISTER, A.K. (1966), *Mech. Polymers* **4**, 519 (in Russian).

MILEIKO, S.T. (1971), *Problems of Strength* **7**, 3 (in Russian).

MILEIKO, S.T. (1979), Fracture mechanics of composites, in: E.M. SOKOLOVSKAYA, ed., *Composite Materials* (Moscow University Press, Moscow), 274.

MILEIKO, S.T., N.M. SOROKIN and A.M. CIRLIN (1973), *Mech. Polymers* **5**, 840 (in Russian).

MILEIKO, S.T., N.M. SOROKIN and A.M. CIRLIN (1976), *Mech. Polymers* **6**, 1010 (in Russian).

MITO, S. and K. MACHIDA (1979), *Trans. Japan. Soc. Mech. Engrg. A* **45**, **394**, 559.

MULLIN, J., J.M. BERRY and A. GATTI (1968), *J. Composite Materials* **2**(1), 82.

OSOKIN, A.E. and J.V. SUVOROVA (1978), *Apppl. Mech. Math.* **42**(6), 1107 (in Russian).

POTTER, R.T. (1974), *Composites* **5**(6), 261.

RABOTNOV, YU.N. (1969), *Creep Problems in Structural Members* (North-Holland, Amsterdam), 379.

RABOTNOV, YU.N. (1977), *Elements of Hereditary Mechanics of Solids* (Nauka, Moscow), 54 (in Russian).

RABOTNOV, YU.N. and A.N. POLILOV (1977), Strength criteria for fibre-reinforced plastics, in: D.M.R. TAPLIN, ed., *Proc. 4th Int. Conf. Fracture*, Waterloo, 1977 (University of Waterloo Press, Canada), 1059.

ROGINSKY, S.L. and M.Z. KANOVICH (1979), *Rep. Acad. Sci. USSR* **248**(3), 565 (in Russian).

ROSEN, B.W. and N.F. DOW (1976), Mechanics of Failure of Fibrous Composites, in: H. LIEBOWITZ, ed., *Fracture*, Vol. 7, Part I (Mir, Moscow), Ch. 5, 300 (in Russian).

SIERAKOWSKI, R.L., G.E. NEVILL, C.A. ROSS and E.R. JONES (1971), *J. Composite Materials* **5**, 363.

SKUDRA, A.M., F.YA. BULAVS and K.A. ROCENS (1971), *Creep and Static Fatigue of Reinforced Plastics* (Zinatne, Riga), 135 (in Russian).

SMITH, K. (1978), Limitations of application of fracture mechanics to composites, in: C.T. HERAKOVICH, ed., *Inelastic Behaviour of Composite Materials* (Mir, Moscow), Ch. 6, 221 (in Russian).

SUVOROVA, J.V. (1977), *Problems of Strength* **2**, 43 (in Russian).

SUVOROVA, J.V. (1979), Creep of fibrous composites, in: E.M. SOKOLOVSKAYA, ed., *Composite Materials* (Moscow University Press, Moscow), 403.

SUVOROVA, J.V., I.V. VIKTOROVA and G.P. MASHINSKAYA (1979), *Mech. Composite Materials* **5**, 794 (in Russian).

SUVOROVA, J.V., T.G. SORINA, I.V. VIKTOROVA and V.V. MICHAILOV (1980a), *Mech. Composite Materials* **5**, 847 (in Russian).

SUVOROVA, J.V., I.V. VIKTOROVA and G.P. MASHINSKAYA (1980b), *Mech. Composite Materials* **6**, 1010 (in Russian).

SUVOROVA, J.V., I.V. VIKTOROVA, G.P. MASHINSKAYA, G.N. PHINOGENOV, A.E. VASILJEV (1980c), *Mech. Engineering* **2**, 67 (in Russian).

TAMUZH, V.P. and V.S. KUKSENKO (1978), *Micromechanics of Failure of Polymer Materials* (Zinatne, Riga), 129 (in Russian).

TSAI, S.W. and E.M. WU (1971), *J. Composite Materials* **5**(1), 58.

TSAI, S.W. and H.T. HAHN (1978), Analysis of failure of composites, in: C.T. HERAKOVICH, ed., *Inelastic Behaviour of Composite Materials* (Mir, Moscow), Ch. 3, 104 (in Russian).

UEMURA, M. (1979), Strength of helically filament-wound composite cylinders, in: E.M. SOKOLOVSKAYA, ed., *Composite Materials* (Moscow University Press, Moscow), 364.

URZHUMCEV, YU.S. and R.D. MAXIMOV (1975), *Prediction of Deformation Behaviour of Polymer Materials* (Zinatne, Riga), 96 (in Russian).

CHAPTER V

Methods of Static Testing for Composites

YU.M. TARNOPOL'SKII and T. KINCIS

Institute of Polymer Mechanics
Academy of Sciences of the Latvian SSR
Riga 226006
Latvian S.S.R.
U.S.S.R.

Contents

HANDBOOK OF COMPOSITES, VOL. 3 – Failure Mechanics of Composites
Edited by G.C. SIH and A.M. SKUDRA

List of Symbols

x_1, x_2, x_3 Axes of elastic symmetry in the rectangular system of coordinates
x'_1, x'_2, x'_3 An arbitrary rectangular system of coordinates
x_θ, x_z, x_r Axes of elastic symmetry in the cylindrical system of coordinates
σ_i, e_i Components of stress and strain matrices in the rectangular system of coordinates; $i = 1, \ldots, 6$
$\sigma_{\theta\theta}, \sigma_{zz}, \sigma_{rr}, \sigma_{zr}, \sigma_{\theta r}, \sigma_{\theta z}$ Components of the stress matrix in the cylindrical system of coordinates
$e_{\theta\theta}, e_{zz}, e_{rr}, e_{zr}, e_{\theta r}, e_{\theta z}$ Components of the strain matrix in the cylindrical system of coordinates
E_1, E_2, E_3 Moduli of elasticity in the directions x_1, x_2, x_3
$E_{\theta\theta}, E_{zz}, E_{rr}$ Moduli of elasticity in the directions x_θ, x_z, x_r
$E^{\pm 45°}$ Modulus of elasticity of material reinforcement oriented $\pm 45°$ to x_1-axis
E_m Matrix modulus of elasticity
G_{23}, G_{13}, G_{12} Shear moduli in the planes x_2x_3, x_1x_3, x_1x_2
$G_{zr}, G_{\theta r}, G_{\theta z}$ Shear moduli in the planes $x_zx_r, x_\theta x_r, x_\theta x_z$
ν_{ij} Poisson's ratio for transverse strain in the j-direction, when strained in the i-direction; $i, j = 1, 2, 3$.
v_f Volume content of reinforcement
v_V Volume content of voids
u_1, u_2, u_3 Displacements in the directions x_1, x_2, x_3 or x_θ, x_z, x_r
P Concentrated load
P_u Concentrated load at the failure of the specimen
p Uniform pressure
p_u Uniform pressure at the failure of the specimen
M Bending moment
M_u Bending moment at the failure of the specimen
M_T Torque
M_T^u Torque at the failure of the specimen
$\sigma_1^+, \sigma_2^+, \sigma_3^+$ Tensile strength in the directions x_1, x_2, x_3
$\sigma_{\theta\theta}^+, \sigma_{zz}^+, \sigma_{rr}^+$ Tensile strength in the directions x_θ, x_z, x_r
$\sigma_1^-, \sigma_2^-, \sigma_3^-$ Compression strength in the directions x_1, x_2, x_3
$\sigma_{\theta\theta}^-, \sigma_{zz}^-, \sigma_{rr}^-$ Compression strength in the directions x_θ, x_z, x_r
τ_{ij}^u Shear strength; $i, j = 1, 2, 3$ or $i, j = \theta, z, r$
b, h, l Characteristic linear dimensions of the specimen, explained in the summary tables of text
R, R_i, R_0 Mean, inner and outer radii of ring or tube specimens

1. Introduction

For the purpose of estimating the specific strength and stiffness the structural materials are subjected to mechanical testing. Historically, mechanical testing subsequently followed by practical application of structural materials, may be traced back to July 4, 1662 (GORDON (1968)). Strengths of tows made from Riga and Holland yarns were compared; the thinner Riga yarn turned out to be more advantageous. Since then, the test procedures for materials, mainly metals, have attained a comparatively high degree of perfection. The advent of composites and their widespread application in primary structures have forced turning to the subject of mechanical testing anew. New test methods are continuously being developed, the already existing techniques verified and reexamined. Research practice has far exceeded the development of test methods, specified by standards. So far there are only several standards written in the terms of anisotropy the world over. In practice, very diverse specimen shapes, sizes, manufacture techniques and experimental methods are used. This leads to incomparable ambivalent judgements on the potential capabilities of the materials. The fact still further enforces the need for a critical analysis of the existing methods, their estimation and generalization.

In the survey an attempt at selecting the most promising methods of tensile, compression, shear and bending tests for composites is made. In the main, the test methods for unidirectional materials are treated. In cases when the loading schemes and ways of data processing are applicable to orthogonally reinforced materials (with reinforcement orientation at 0°, 90° and ±45° and the number of layers being sufficient for transition to continuous medium), the necessary explanations are given in the text and summary tables. In selecting the test methods the loading along the axis of elastic symmetry of the material is preferred; exceptions are noted especially.

The basic structural unit of composites is a lamina which is a flat or curved arrangement of unidirectional or woven fibers in a matrix. A uniform throughout the thickness laminate may also be considered as a lamina.

Testing of a lamina (Fig. 1) is essential not only for specification of the engineering material. The methods of evaluation of the properties of hybrid materials and those of various fiber orientation along the thickness are also based on the concept. As seen from Fig. 1, in the testing of a lamina, plates uniform along the thickness, rings and cylinders (tubes) are used. Plates may be subdivided into strips and bars, tubes into rings. Selection of the loading scheme depends on the aim of testing, i.e. on the characteristics to be determined.

The widespread use of filament wound structures has served as an impetus to development of numerous tests, performed on ring specimens; in this field of materials testing the advances have been the most impressive. The fact allows to consider here from a common standpoint – according to loading type – the test methods for flat and ring specimens in tension, compression, shear and bending.

The material for each loading type is presented in summary tables, contain-

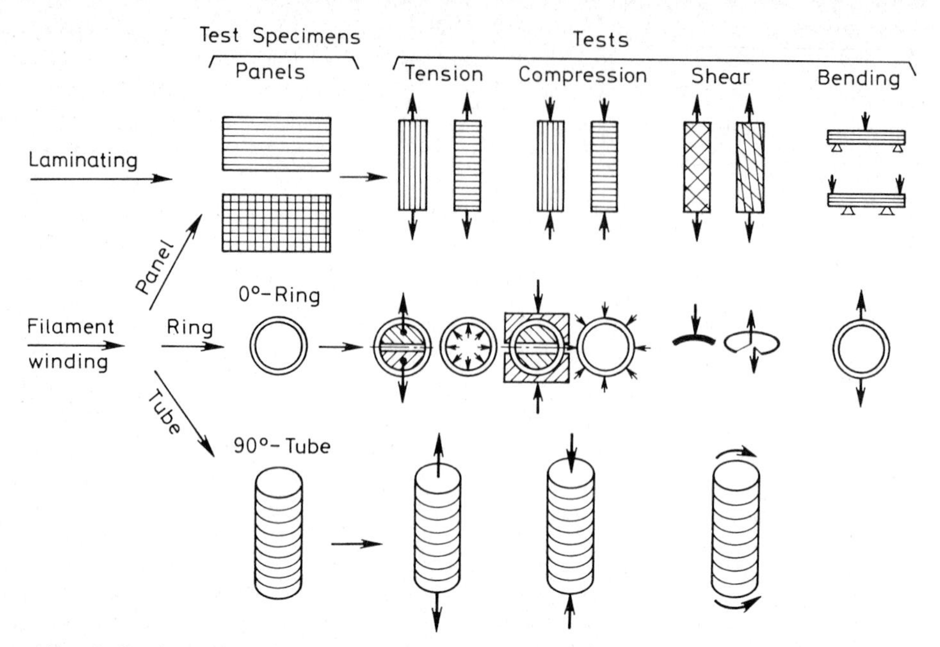

FIG. 1. Specimen shapes and test techniques, based on Fig. 7 of CHIAO and HAMSTAD (1976).

ing the loading scheme, determinable characteristics, measured experimental values, calculation relations and structural, physical and geometrical limitations. There are two types of tables; for convenience, the summary tables are designated by Arabic numerals, the rest by Roman numerals. The number of the loading scheme is in bold type, for example **2-4**; the designation is further used for description of the loading schemes. In many cases, the presence of initial stresses is characteristic for composites. The survey of the methods for evaluation of the initial stresses on ring specimens is presented at the end of the chapter.

In spite of the progress, the degree of mastering and standardization is different for each type of testing. As before, the shear stiffness and, especially, shear strength estimation presents some difficulties. With the accumulation of experience, a number of methods for estimation of strength and stiffness is in need of further corrections. The survey is based on the third revised and supplemented edition of authors' book (TARNOPOL'SKII and KINCIS (1981)), presenting itself a generalization of worldwide experience in the field of static testing of composites.

2. Basic features

Advanced fibrous composite materials of unidirectional, laminated and spatial reinforcement are nonhomogeneous, essentially anisotropic materials. The customary terms – tension, compression, shear and bending – become meaningless without specification of the direction between the load and the axes of elastic symmetry of the material. Therefore two systems of coordinates must be employed: material symmetry axes and loading axes. Their coincidence is preferred.

The majority of fibrous composites exhibit low interlaminar shearing and transverse resistance. Shear resistance is characterized by the relations E_1/G_{13} and $\sigma_1^{+(-)}/\tau_{13}^{u}$, transverse tension and compression resistance by E_1/E_2, $\sigma_1^{+}/\sigma_3^{+}$, $\sigma_1^{+}/\sigma_3^{-}$. Here E_1 and E_3 are elastic moduli in the directions x_1 and x_3; G_{13} is the interlaminar shear modulus; $\sigma_1^{+(-)}$ and $\sigma_3^{+(-)}$ are the strengths in the x_1 and x_3 directions; τ_{13}^{u} is the shear strength in the plane x_1x_3; the x_1- and x_2-axes are located in the reinforcement plane, and the x_3-axis is perpendicular to this plane; the (+) and (−) designate tension and compression, respectively. The range of these parameters for three classes of unidirectional composites are presented in Table I.

The material anisotropy and the structural peculiarities cause a number of serious difficulties. The first difficulty is connected with determination of the number of determinable strength and elastic characteristics, necessary for a sufficiently rigorous identification of the material. The number of determinable characteristics depends on the type of anisotropy (LEKHNITSKII (1977), TARNOPOL'SKII and KINCIS (1981)). Characteristics of an orthotropic body and methods of their experimental evaluation are presented in Table II. For a transversely isotropic lamina it is necessary to determine the following characteristics in a plane state of stress: the moduli of elasticity in tension and compression $E_1^{+(-)}$ and $E_2^{+(-)}$, shear modulus G_{12}, Poisson's ratio ν_{12}, strengths in tension and compression $\sigma_1^{+(-)}$ and $\sigma_2^{+(-)}$ and shear strength τ_{ij}^{u} in two planes: in the reinforcement plane and in the plane of isotropy.

Selection of the loading type, for which the material characteristics are most simply related to the experimentally determinable values, selection of the calculation relations for processing experimental data and estimation of the range of their validity are of principal significance. Since all the relations for calculation in the chapter are based on the theory of elasticity for a homogeneous anisotropic body, it is necessary to accomplish transition from a

TABLE I. Typical parameters, characterizing strengths of unidirectional composites in interlaminar shear, transverse tension and compression (ZHIGUN and POLYAKOV (1978)).

Parameters of anisotropy	GFRP	CFRP	BFRP	OFRP
E_1/G_{13}	20–35	40–80	30–60	25–40
$\sigma_1^{+}/\tau_{13}^{u}$	30–40	20–40	20–50	10
E_1/E_3	5–8	20–30	8–12	12–18
$\sigma_1^{+}/\sigma_3^{+}$	25	25–50	15–30	50
$\sigma_1^{+}/\sigma_3^{-}$	6–10	6–10	10	15–20

TABLE II. Characteristics of an orthotropic body and their experimental determination.

	General state of stress		Plane state of stress		Techniques of experimental determination
	Rectangular coordinates	Cylindrical coordinates	Rectangular coordinates	Cylindrical coordinates	
Elastic constants	E_1, E_2, E_3	$E_{\theta\theta}, E_{zz}, E_{rr}$	E_1, E_2	$E_{\theta\theta}, E_{zz}$	Tension, compression, bending
	G_{23}, G_{13}, G_{12}	$G_{zr}, G_{\theta r}, G_{\theta z}$	G_{12}	$G_{\theta z}$	Tension, compression, bending, torsion, panel shear
	$\nu_{23}, \nu_{13}, \nu_{12}$	$\nu_{zr}, \nu_{\theta r}, \nu_{\theta z}$	ν_{12}	$\nu_{\theta z}$	Tension, compression, bending
Strength (See Note 1)	$\sigma_1^*, \sigma_2^*, \sigma_3^*$	$\sigma_{\theta\theta}^*, \sigma_{zz}^*, \sigma_{rr}^*$	σ_1^*, σ_2^*	$\sigma_{\theta\theta}^*, \sigma_{zz}^*$	Tension, compression
	$\tau_{23}^u, \tau_{13}^u, \tau_{12}^u$	$\tau_{zr}^u, \tau_{\theta r}^u, \tau_{\theta z}^u$	τ_{12}^u	$\tau_{\theta z}^u$	Tension, compression, bending, torsion, panel shear

Note 1. The following upper indices (*) denote the type of loading at failure due to normal stresses: b = bending, − = compression, + = tension.

Note 2. Direction of axes: in the rectangular system of coordinates x_1-axis coincides with the longitudinal specimen axis; in the cylindrical system of coordinates the x_θ-axis coincides with a tangent to the circumference having the radius R. In coordinate systems the left ones.

non-homogeneous laminated material to continuous medium and to evaluate the error arising from the transition. Such transition is possible when the number of structural members (fibers, prepregs, etc.) is sufficiently large; the problem is treated in detail in the works (BOLOTIN et al. (1972), LOMAKIN (1978)). For instance, in the case of bending the minimal necessary number of layers, for which the transition is possible, depends on the degree of anisotropy of the material, evaluated in terms of the parameter

$$\kappa_b = \frac{\pi}{2}\frac{h}{l}\sqrt{\frac{E_1^b}{G_{13}^b}}.$$

For a simply supported bar under the sinusoidal load the minimal necessary number of layers n is as follows:

κ_b	1	3	5	7
n	5	15	17	20

The majority of errors in the testing of thin laminates in bending are due to the condition being not satisfied.

In the case of fibrous composites the principal difficulties lie in the establishing of a uniform stress[1] in the representative volume of the material, even for the simplest types of tests. The difficulties increase with the increase in the degree of material anisotropy, i.e., for materials reinforced with high modulus and high strength fibers (boron, carbon and organic fiber reinforced composites). In composite testing, the measurable strain essentially depends on the boundary conditions, i.e., on the method of fastening and loading the specimen. The phenomenon is a characteristic of highly anisotropic materials and represents itself a specific manifestation of St. Venant's principle. According to St. Venant's principle, for an isotropic, elastic medium, disturbances introduced by a balanced system of forces rapidly diminish at distances from the source of the disturbances greater than the characteristic size of the specimen H, i.e. at $\Lambda \sim H$ (Fig. 2). In the case of an anisotropic medium the disturbances are damped differently in different directions. In the direction of the greatest stiffness, they are damped more slowly and, in the direction of the least stiffness, more rapidly. As a result, the region of noticeable disturbances is elongated in the direction of the greatest stiffness. A characteristic size of the disturbance region in this direction is on the order of

$$\Lambda \sim H\left(\frac{E_i}{G_{ij}}\right)^{1/2} \tag{1}$$

where E_i, G_{ij} are the moduli of elasticity and shear, respectively; $i, j = 1, 2, 3$.

Anisotropy of elastic properties enhances the requirements imposed on the

[1] In this chapter, mean stresses and strains are dealt with.

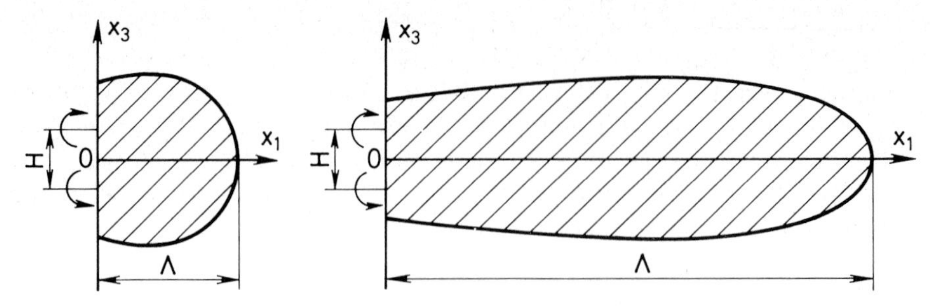

FIG. 2. Damping of perturbations in isotropic (a) and anisotropic (b) materials (BOLOTIN et al. (1972)).

specimen shape and size, the elimination of edge effects the proper choice of the distance from the grips to reference lines on a specimen, the type of load transmission and the fiber orientation. The strength anisotropy in the case of improper selection of the type of loading and specimen fastening leads to the alteration of the failure mode, for instance, to delamination or failure of the specimen in grips.

A special problem is the selection of the specimen width. It is important to avoid edge effect, i.e., the source of critical interlaminar stresses. Therefore, in the description of methods the physical and geometrical limitations, necessary for obtaining reliable material characteristics, are outlined in tables.

The approach to technology is also novel. The material and the structural designs are now being performed simultaneously and the composites are extremely sensitive to their force-temperature prehistory. The specimen technology should precisely simulate the operational conditions of the material in actual structures, but the shape should allow to easily vary the technological parameters. Technology imperfections require special attention; in particular, porosity, fiber waviness and misalignment. Porosity manifests itself in evaluation of the matrix-determined properties (for instance, shear strength).

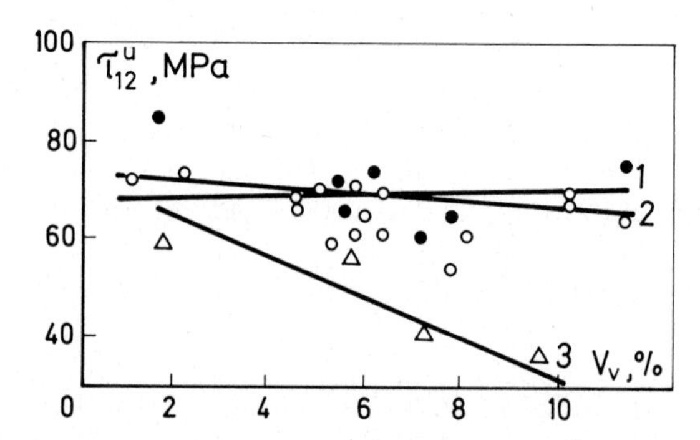

FIG. 3. Shear strength dependence of unidirectional carbon composites on the void content v_V and test techniques (PURSLOW (1977)). (1) tension of a strip (○); (2) transverse compression (●); (3) three-point bending (△).

However, not all the test methods are equally sensitive in this respect. During testing of unidirectional CFRP the most sensitive to porosity was the method of three-point bending, least sensitive was the method of tension of a strip (see Fig. 3).

Fibre waviness manifests itself in evaluation of the properties in the direction of reinforcement (Fig. 4). For ring specimens the measured elastic modulus $E^*_{\theta\theta}$ is equal to (TARNOPOL'SKII et al. (1973))

$$E^*_{\theta\theta} = \frac{E_{\theta\theta}}{1 + \frac{E_{\theta\theta}}{G_{\theta r}} \frac{f^2}{2}} \tag{2}$$

where $E_{\theta\theta}$ is the circumferential modulus of elasticity of the material under tension along the 'straight' fibers; $G_{\theta r}$ is the interlaminar shear modulus; f is a parameter, characterizing the fiber waviness (for the case of a sinusoidal waviness $f = \pi Ak/l$, A is the amplitude of the sinusoid, k is the number of half-waves in base l). Formula (2) makes allowance for the low shear resistance only. For composites, reinforced with anisotropic fibers, the ratio of moduli of elasticity along and across the fibers $E_{\theta\theta}/E_{rr}$ and all the terms incorporating f^2 must also be taken into account:

$$E^*_{\theta\theta} = \frac{E_{\theta\theta}}{\frac{2+f^2}{2(1+f^2)^{3/2}} + \frac{E_{\theta\theta}}{G_{\theta r}} \frac{f^2}{2(1+f^2)^{3/2}} + \frac{E_{\theta\theta}}{E_{rr}} \left[1 + \frac{2+3f^2}{2(1+f^2)^{3/2}}\right]} \tag{3}$$

Even with small waviness ($f^2 \sim 0.01$) the modulus of elasticity $E^*_{\theta\theta}$ and strength $\sigma^*_{\theta\theta}$ of actual materials can be considerably lower than those of the

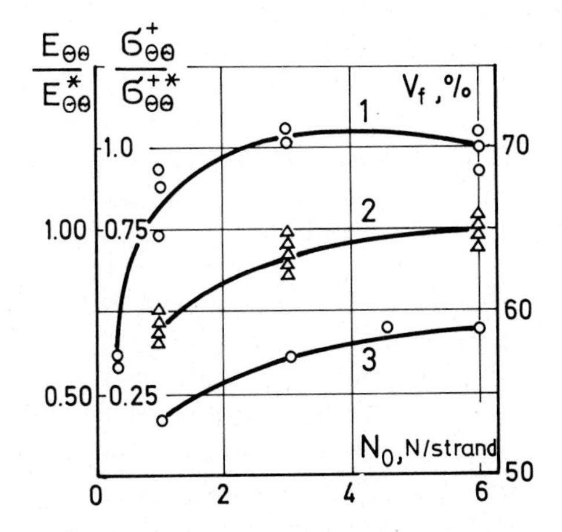

FIG. 4. The effect of tensioning of a strand N_0 on the variation in strength (1), modulus of elasticity (2) and fiber volume fraction (3) in the reinforcing direction under tension; $E_{\theta\theta}$ and $\sigma^+_{\theta\theta}$ are related to their values at $N_0 = 6N$/strand (designated by an asterisk), TARNOPOL'SKII et al. (1973).

material with ideally 'straight' fibers. The elastic modulus perpendicular to the direction of the reinforcement depends little on the reinforcement waviness

$$E^*_{zz} = \frac{E_{zz}}{1 + \dfrac{E_{zz}}{G_{\theta z}} \dfrac{f^2}{2}} \tag{4}$$

since E_{zz} and $G_{\theta z}$ are quantities of the same order. Analogously, it can be shown that the shear modulus $G^*_{\theta z}$ of a medium, with small initial irregularities, depends little on the degree of waviness of the reinforcement.

3. Tension and compression

3.1. Tension of flat specimens

The data for flat specimens are summarized in Table 1. Regardless of the analogy of loading for flat composite specimens in tension and compression, only the calculation relations, taking into account the sign of stresses and strains, are common in both cases. Requirements imposed on the specimen shape and size, gripping methods, loading types and failure modes in both cases essentially differ. Both the methods have been standardized. The standards cover determination of the tensile and compressive properties (strength $\sigma_1^{+(-)}$, modulus of elasticity $E_1^{+(-)}$, Poisson's ratio $\nu_{12}^{+(-)}, \nu_{13}^{+(-)}$) of symmetric unidirectionally and orthogonally reinforced glass, boron and carbon resin-matrix composites. Load is applied in the directions of the principal axes of elastic symmetry of the material.

The standards are commonly meant for specimens of constant cross section (strips) only, but in practice also waisted specimens ('dog bones') are often used for measurement of the tensile and compression strengths. The advantages of waisted specimens are as follows: a characteristic fracture section, lower ultimate load (for the equal size of gripped cross sections), facilitating the specimen gripping and load transmission, lower sensitivity to inaccurate specimen installation in the testing machine; the main disadvantages are nonuniform state of stress and time-consuming fabrication. In separate cases, the specimens of circular cross section and sandwich beams have been used.

In the specimen to be tested under uniaxial tension three functionally different zones can be distinguished: the gage length, two transitional zones and two loaded zones. Strain measurements are made in the gage length of the specimen and stresses are computed according to its geometrical dimensions and external load. Transition zones are meant for the localization of stress-strain state disturbances, associated with the specimen fastening and loading. Loaded zones are meant for the installation of the specimen into the testing machine, they carry and transmit the external tensile load to the gage section.

The dimensions of the gage zone should answer the following requirements:

TABLE 1. Tension and compression of flat specimens.

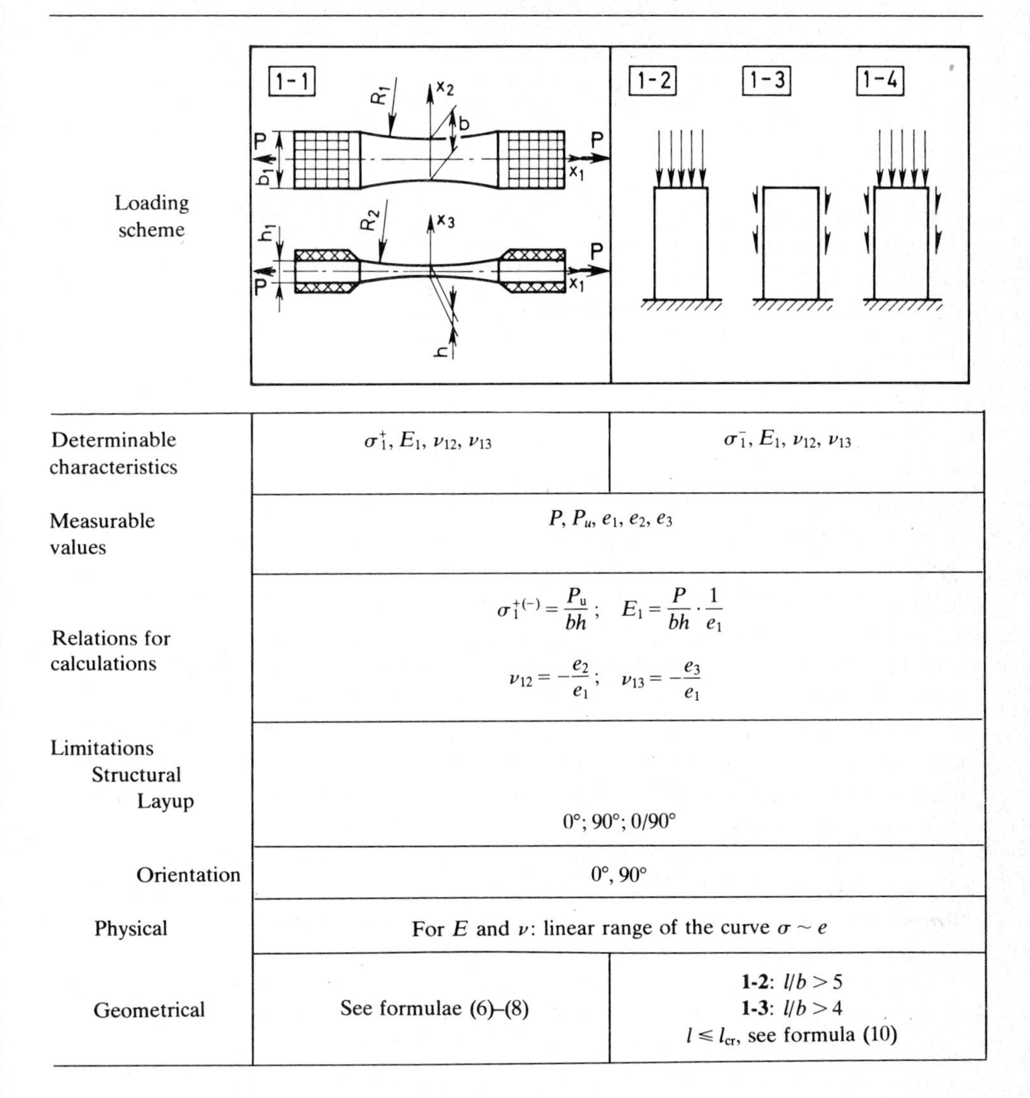

	1-1	1-2, 1-3, 1-4
Loading scheme		
Determinable characteristics	$\sigma_1^+, E_1, \nu_{12}, \nu_{13}$	$\sigma_1^-, E_1, \nu_{12}, \nu_{13}$
Measurable values	P, P_u, e_1, e_2, e_3	
Relations for calculations	$\sigma_1^{+(-)} = \dfrac{P_u}{bh}$; $E_1 = \dfrac{P}{bh} \cdot \dfrac{1}{e_1}$ $\nu_{12} = -\dfrac{e_2}{e_1}$; $\nu_{13} = -\dfrac{e_3}{e_1}$	
Limitations Structural Layup	0°; 90°; 0/90°	
Orientation	0°, 90°	
Physical	For E and ν: linear range of the curve $\sigma \sim e$	
Geometrical	See formulae (6)–(8)	**1-2**: $l/b > 5$ **1-3**: $l/b > 4$ $l \leq l_{cr}$, see formula (10)

in the gage zone the state of stress should be uniform; the measurable values must be independent of the shape and size of specimen cross section; reliable instrumentation must be ensured. In order to satisfy the first requirement the overall specimen length for a given gage length must be chosen with the allowance for the end effect. It has been theoretically and experimentally (POLYAKOV and ZHIGUN (1978, 1979)) that the length of end effect zone depends on the relations of elastic constants E_1/E_3 and $(E_1/G_{13} - 2\nu_{13})$, on the relative geometrical dimensions of specimen gripping zone (c/h, a/h, b/a; see Fig. 5) and on the tensile load transmission (a ratio q/σ_1). Calculations show that even

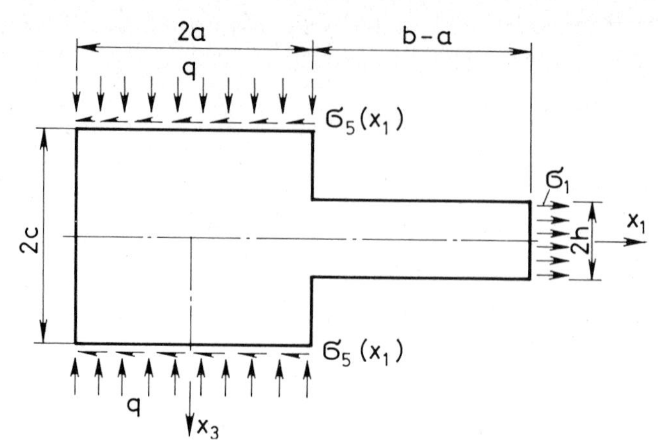

FIG. 5. The loading scheme under tension (POLYAKOV and ZHIGUN (1978)); specimen width b = const.

for highly anisotropic materials the length of end effect zone does not exceed 8.5 h. The experimental investigations of strain distribution along the specimen length and height (see Table III) are in good agreement with the theoretical derivations and they show that strain distribution depends on the degree of anisotropy and the type of reinforcement. In the region of specimen fastening the outer layers are overloaded, the inner layers underloaded. In strip specimens ($c/h = 1$) the perturbation of stress-strain state is levelled out slower than in waisted specimens ($c/h > 1$). Increase in the length of loading region, characterized by the ratio a/h, leads to the decrease in strains at the side of the cross section. Variation of loading conditions, i.e. the ratio q/σ_1, inessentially affects the distribution of normal stress σ_1. Magnitude of stresses σ_3 and σ_5 in the end effect zone are two or three orders lower than the stress σ_1.

In general, the specimen dimensions, recommended in the existing standards, answer the requirements. Length, width and thickness are adopted in the standards, depending on the type of reinforcement.

The basic technical difficulty of tension test for composite materials, especially unidirectional ones, concerns the safe tensile force transmission (ZHIGUN and MIKHAILOV (1978)). Tensile forces are transmitted to the specimen by friction; special specimen designs are also employed, in which the side surfaces of specimen heads are at the same time the supporting surfaces. Frictional forces are induced by wedge grips or fastening bolts. Reliable installation of the specimen in wedge grips is ensured by fulfilling the requirement

$$\frac{f}{2\,\mathrm{tg}(\alpha+\varphi)} > 1 \tag{5}$$

where f is the friction coefficient of the material of wedge surfaces; α is the angle of slope of grip jaws (Fig. 6); φ is the reduced angle of rolling friction on the inclined surface of grips.

TABLE III. Strain distribution along the length and height of the specimen in the testing of composites in tension (POLYAKOV and ZHIGUN (1979)).

Materials and layup	Specimen dimensions			Coordinates of strain measurements and their relative values							
				$x_3/h = 1$				$x_3/h = 0$			
	c/h	a/h	b/a	x_1/a	e_1/e_n	x_1/a	e_1/e_n	x_1/a	e_2/e_n	x_1/a	e_2/e_n
GFRP, AG-4S, $0^\circ_5/90^\circ$	1.7	5.7	11.2	1.8	1.07	3.3	1.00	1.8	0.96	3.3	1.00
	1.0	5.7	11.2	1.8	1.07	3.3	1.01	1.8	0.92	3.3	0.98
CFRP, 0°	1.7	5.6	4.8	2.4	1.12	3.1	1.02	2.4	0.88	3.1	1.00
	1.0	5.6	4.8	2.4	1.23	3.1	1.04	2.4	0.80	3.1	0.95
CFRP, 0°/90°	1.5	5.7	4.6	1.8	1.05	3.2	1.00	1.8	0.95	3.2	1.00
	1.0	5.7	4.6	1.8	1.06	3.2	1.00	1.8	0.92	3.2	0.98
	2.5	4.0	3.8	1.2	1.12	2.0	1.00	1.2	0.81	3.0	1.00
Spatially reinforced GFRP:											
base	1.0	4.0	3.8	1.2	1.14	2.0	1.03	1.3	0.81	3.0	1.00
weft	2.5	6.0	3.8	1.2	1.10	2.0	1.04	1.2	0.79	3.0	1.02

Note. The values e_n were measured in the middle of the specimen gage section.

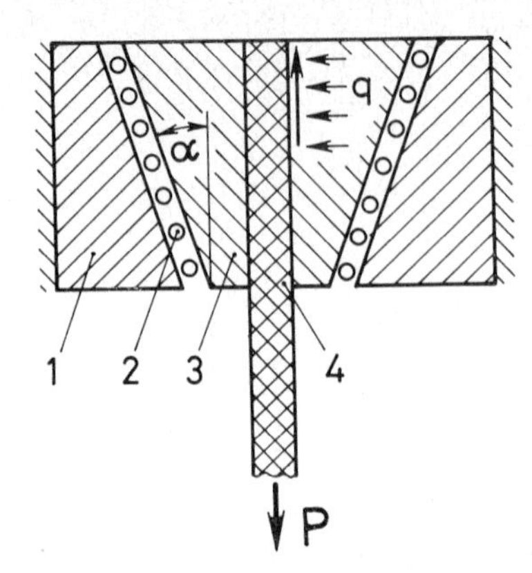

FIG. 6. The loading scheme for a specimen, installed in a self-closing wedge grip (ZHIGUN and MIKHAILOV (1978)). (1) a body of the gripping device; (2) roller race bearings; (3) movable grip jaws; (4) a specimen.

In the selection of the angle α the low transverse resistance of composites must be taken into account by fulfilling the condition

$$l_1 > \frac{1}{2\,\mathrm{tg}(\alpha+\varphi)} \frac{\sigma_1^+}{\sigma_3^-} \frac{b}{b_1} h \tag{6}$$

where l_1 is the length of the specimen section sustaining the pressure q; σ_1^+, σ_3^- are the strengths under tension and transverse compression, respectively; b, b_1 are the specimen widths in the gage section and at the load input, respectively; h is the thickness of the gage section.

In the case of waisted specimens the shear fracture of their heads should be prevented by a proper choice of width ratio b/b_1

$$(b_1 - b)\frac{b}{b_1} < 2l_1 \frac{\tau_{13}^u}{\sigma_1^+} \frac{h_1}{h} \tag{7}$$

where τ_{13}^u is the interlaminar shear strength of the material of the specimen; h_1 is the thickness of the specimen zone under pressure q.

The waisting profile of specimens must be chosen so that the tensile load were transmitted to the gage section without loss of specimen integrity. The evaluation of the regularities of change of the longitudinal section along the specimen length $\mathrm{d}x_3/\mathrm{d}x_1 = f(x_1)$ is based on the selected strength criterion (see, for instance, EWINS (1974)). The most effective way of gage section reduction involves the change in specimen width, but the fulfillment of the tensile load transmission conditions in the cases makes the specimen too long. Therefore, often double-waisted specimens are used, in which the reduction in cross section area is achieved by reducing both the specimen thickness and width. The reduction in the thickness of a gage section, i.e., flexural stiffness in the plane perpendicular to the reinforcement plane is also favoured because in the plane the inaccuracy of the installation of the specimen in grips and, consequently, the effect of bending is the highest. However, the reduction in specimen thickness is permissible only for materials having a uniform reinforcement through the thickness.

The load transmission can be considerably improved by means of tabs bonded to the end sections of the specimen. The tabs are made of the material, the modulus of elasticity of which is significantly lower and relative elongation at failure greater than those of the specimen material. Tabs are commonly made of glass fiber reinforced plastics, although good results have also been obtained with veneer and soft aluminium tabs. Thickness of tabs is equal to $(1.5–4)h$, where h is the specimen thickness.

Dimensions of tabs are determined by the shear strength τ_{13}^{u} of the adhesive, the specimen or the tabs (whichever is lower), thickness h and width b of the specimen gage section and the estimated strength σ_1^+ of the material being tested

$$\sigma_1^+ bh < 2\tau_{13}^{u} b_{tab} l_{tab} . \tag{8}$$

The failure mode under tension depends on the direction of external load relative to the fiber direction and on the type of reinforcement. This problem far exceeds the scope of the survey.

3.2. *Compression of flat specimens*

In compression testing, the dimensions of specimens with constant cross section are to be chosen by taking into account the concentration of the normal σ_2 and tangential stresses σ_6. Investigations (ZHIGUN et al. (1979)) have shown that the minimum value of the relation l/b depends on the type of load input; the respective values of l/b for a specimen of orthotropic material are presented in Table 1.

The main technical problem in compression testing of flat specimens is the load input and ensuring that the specimen failure occurred by compression. In the case of loading with normal forces (Scheme **1-2**), it is practically impossible to achieve close contact between the end faces of the specimen and the platens of the testing machine. As a result a premature failure of the specimen occurs. In the case of loading with tangential forces (Scheme **1-3**), the compression force transmission is not perfect either, especially when flat wedges instead of split collet-type grips are used. The most efficient is the combined loading: with normal forces applied at the end faces and tangential forces – along side faces of the specimens (Scheme **1-4**).

Under combined loading the angle of slope of grip jaws must be chosen by taking into account the load distribution between the end and side faces of the specimen. It has been experimentally established that crushing of the specimen is out of question, if the load at the end faces amounts to 45–50% of the load at failure (without any side load). In the existing test devices the angle of slope is equal to 14–17°. The specimen end sections are protected by tabs.

Under compression of unidirectional composites in the fiber direction the three basic failure modes have been observed (GRESZCZUK (1973)):

– buckling of reinforcing fibers (matrices of low stiffness, $E_m = 15\text{–}25$ MPa),

– transverse breakage due to Poisson's ratios difference of the material constituents and nonuniform transverse strain distribution along the specimen length (matrices of average stiffness, $E_m = 200\text{–}700$ MPa) and

– shearing of reinforcing fibers at an angle near 45° without local buckling (matrices of high stiffness, $E_m > 2000$ MPa).

Materials, reinforced at an angle to the longitudinal axis, fail due to shearing without collapse at the ends; all of the shearing load is taken by the polymer matrix.

The aforementioned basic failure modes may be accompanied by a series of other phenomena: inelastic and nonlinear behavior of the reinforcing fibers and the matrix, interlaminar stresses, peeling, overall and local loss of stability, destruction of the end faces, brooming, splitting across the layer. Various combinations of the phenomena can significantly interfere with the establishment of the actual failure mode.

It should be noted that in the absence of overall delamination the invisible buckling of the side face may cause erroneous strain measurements. The formula of critical stress σ^*_{cr}, at which local loss of stability occurs, accompanied

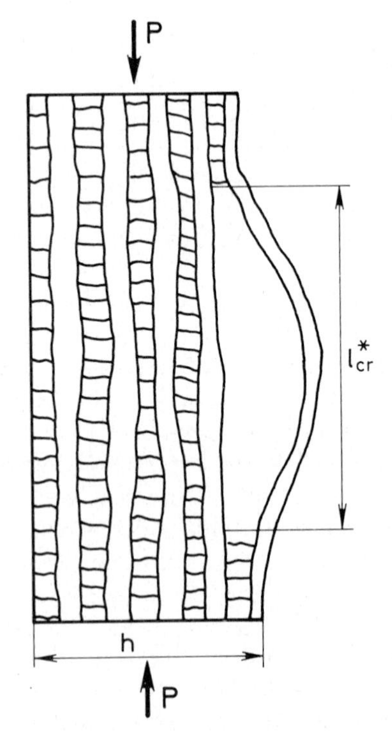

FIG. 7. A model of composite peeling under compression.

by matrix failure (Fig. 7), is as follows (KACHANOV (1976), TARNOPOL'SKII and SKUDRA (1979)):

$$\sigma_{cr}^{*} = \frac{2\pi^2}{3} E_{1}^{-} \left(\frac{k_c^2}{2}\right)^{1/5} \tag{9}$$

where $k_c = 36\gamma/(\pi^4 E_1^- l_{cr}^*)$; γ is the specific surface energy at failure; l_{cr}^* is the peeled length.

In compression testing, the overall loss of stability of the specimen and, especially, of its gage section should be prevented. In order to prevent the loss of stability of the specimen such test devices are used, in which the specimen side surfaces are in contact with prisms, preventing the overall specimen buckling, but not restraining in-plane deformations.

In order to prevent the loss of stability in the specimen gage section, the free length l should be less than the critical length

$$l_{cr} = 0.907\, h \sqrt{E_1^- \left(\frac{1}{\sigma_{cr}^-} - \frac{1.2}{G_{13}}\right)} \tag{10}$$

where E_1^- and G_{13} are the elastic and shear moduli of the material being tested; σ_{cr}^- is the critical compression stress.

Tension and compression test procedures are meant for flat specimens, provided the fibers are oriented parallel or perpendicular to the specimen axis. The problem of testing the specimen with fiber orientation at an angle to the specimen axis, is unsolved as yet. There is no theory for experimentators, allowing to take into account all the effects of off-axis ply interaction, to evaluate the end and edge effects, to determine the width of specimen gage section and ensure a uniform state of stress.

3.3. *Tension of rings*

The most widespread methods for tensile testing of rings are listed in Table 2. The tensile test by means of a split-disk loading device (Scheme **2-1**) is widely used. Testing by uniform internal pressure, introduced by a compliant ring (Scheme **2-2**) or hydraulically (Scheme **2-3**), are not so popular. Tension tests conducted on rings under split-disk loading have been standardized (ASTM D 2290-76). The standard is intended for thin-walled rings (D_i = 146.05 mm, h = 1.52 or 3.18 mm). Initially, the method was developed for evaluation of the effect of glass fiber roving chemical treated methods on the strength of the composite, but later the field of its use has been unwarrantedly enlarged, which led to the appearance of a large number of erroneous results. Only systematic researches allowed one to state the capabilites of the method and specimen dimensions. Tension test by split-disk loading is a simple technique as regards its utilization (manufacturing of specimens and their installation into the testing fixture present no difficulties, it is even possible to use half-disks as mandrels in the fabrication of specimens) and data processing. However, the method has several essential disadvantages, reducing the value of the information obtained: distribution of the strain around the circumference is

TABLE 2. Tension of ring specimens.

Loading scheme	2-1, 2-2	2-3	2-4
Determinable characteristics	$E^+_{\theta\theta}, \sigma^+_{\theta\theta}$		
Measurable values	$P, P_u, e_{\theta\theta}$	$p, p_u, e_{\theta\theta}$	
Relations for calculations	$E_{\theta\theta} = \frac{P}{2bh} \cdot \frac{1}{e_{\theta\theta}}$ $\sigma^+_{\theta\theta} = \frac{P_u}{2bh}$	$E_{\theta\theta} = \frac{pD_i}{2he_{\theta\theta}}, \quad \sigma^+_{\theta\theta} = \frac{p_u D_i}{2h}$	
Limitations Structural Layup	0°; 90°; 0/90°		
Orientation	0°, 90°		
Physical	For $E_{\theta\theta}$: linear range of the curve $P \sim e_{\theta\theta}$ or $p \sim e_{\theta\theta}$		
Geometrical	For $\sigma^+_{\theta\theta}$: $0.08 \leqslant h/R \leqslant 0.18$ (for GFRP) $L \geqslant (2/3)R$	For $\sigma^+_{\theta\theta}$: $0.08 \leqslant h/R \leqslant 0.18$ (for GFRP)	

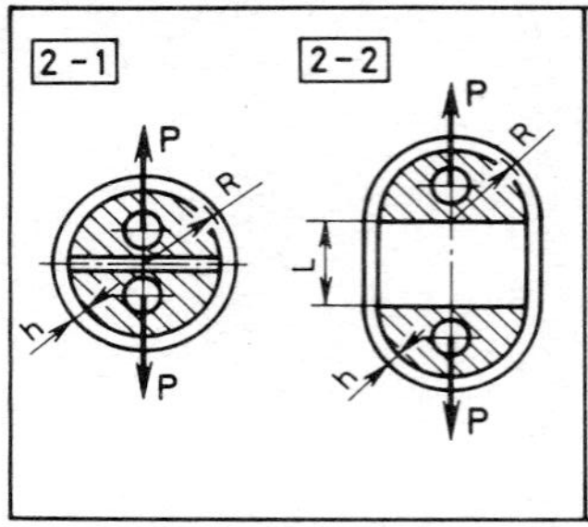

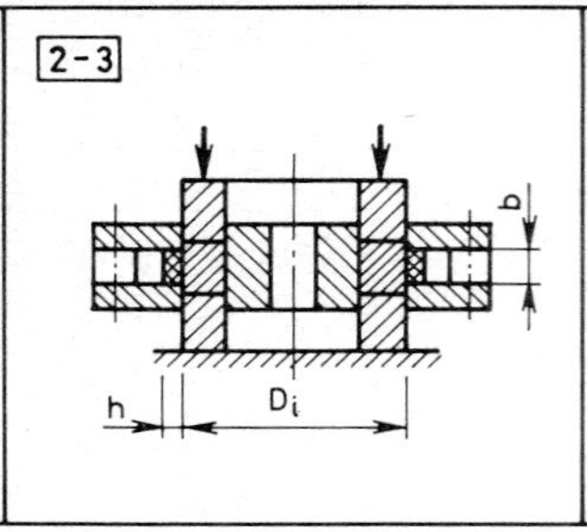

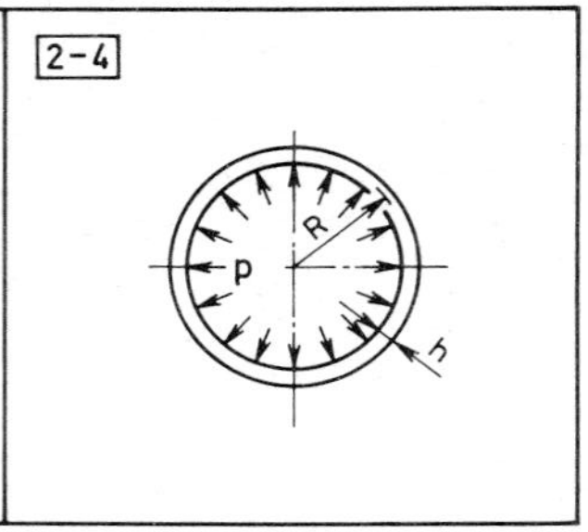

nonuniform (Fig. 8) and, mainly, in the gaps between the two half-disks the change in radius of curvature introduces stress concentration in the specimen. The concentration of radial tensile stresses σ_{rr} exerts small effect in thin-walled rings, but interlaminar tangential stresses $\sigma_{\theta r}$ may exceed the ultimate value before the specimen failure under normal circumferential stresses $\sigma_{\theta\theta}$ (KNIGHT (1977)) occurs. All the phenomena are enhanced with the increase in the relative specimen thickness h/R, degree of anisotropy and the magnitude of ultimate strain of the material. As a result of stress concentration, the strength measured in split-disk tests is lower and may serve merely for a qualitative comparison of the composites. The coefficients of correction have not been used in practice. Attempts at reducing the effect of stress concentration by employing elongated ring specimens, in which both halfrings are joined by straight and parallel sections (Scheme **2-2**), due to technological difficulties (to ensure the uniformity of material structure, over straight specimen sections the application of extra pressure is necessary), turned out to be ineffective as well as the use of multisectional fixtures (Fig. 9), the complicated production and exploitation of which is not always worth the data gain (PARTSEVSKII and GOL'DMAN (1970), TARNOPOL'SKII and KINCIS (1981)). For measuring the circumferential modulus of elasticity $E_{\theta\theta}$ the strain gages are located in the specimen sections having the most uniform strain distribution – at an angle 30–45° to the midplane of disks. During testing of an elongated ring specimen the strain gages are located in the middle of the two straight specimen sections, where the deformation field is uniform.

Stress concentration in the specimen is eliminated during tests under uniform internal pressure by means of a compliant ring (Scheme **2-3**) or hydraulically (Scheme **2-4**). The main disadvantage of the test method by means of a

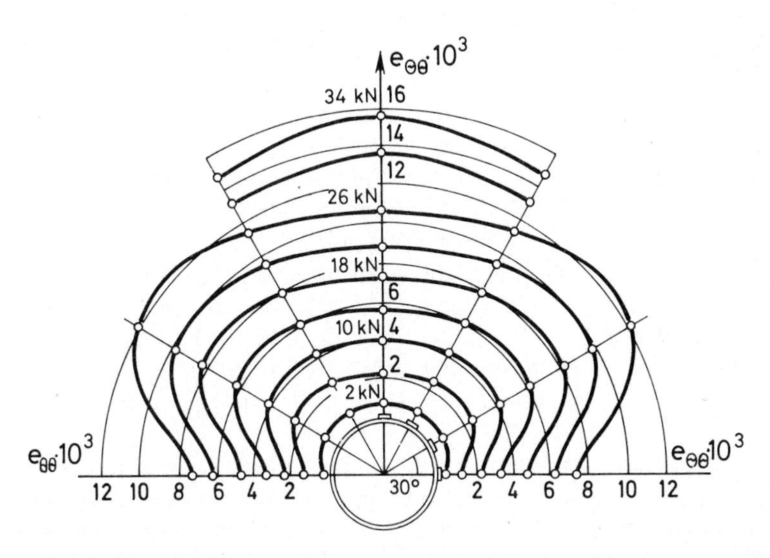

FIG. 8. Distribution of composite hoop strain $e_{\theta\theta}$ of rings, installed on rigid half-disks (KLEINER and LÜßMANN (1969)).

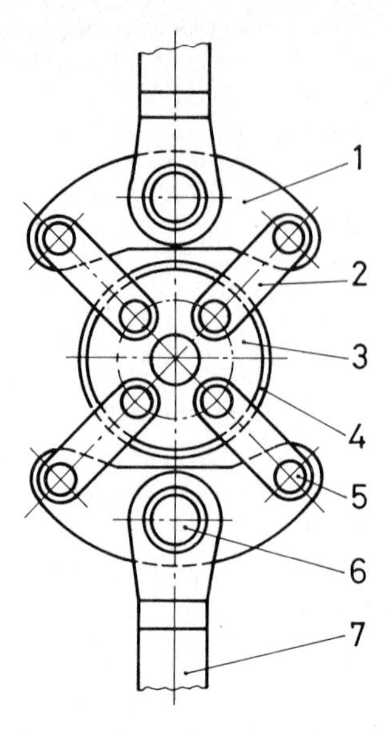

FIG. 9. A four-link fixture for tension of circular specimens (PARTSEVSKII and GOL'DMAN (1970)). (1) a balancier; (2) pull rods; (3) rigid sectors; (4) a ring under test; (5) and (6) pins; (7) forks.

compliant (rubber) ring is the necessity of preliminary and systematic calibration of the loading element and very careful specimen surface treatment, of hydraulic tests – the necessity of a special, comparatively complicated and expensive equipment.

3.4. *Compression of rings*

In-plane compression of rings is accomplished under external pressure; the loading schemes, used in practice, and calculation relations are presented in Table 3. The compression testing in the split-disk fixture (Scheme **3-1**) differs from the tensile tests in the same fixture in that in the former case it is possible to reduce the effect of stress concentration in the gap between the two half-disks. The best results have been obtained in the testing of rings in fixtures with half-frames and girder locks, excluding the possibility of increasing the horizontal diameter of the specimen (SBOROVSKII et al. (1971)).

Compression tests under external pressure, produced by means of a compliant ring (Scheme **3-2**) and by hydraulics (Scheme **3-3**) are conducted in the same way as loading of rings under internal pressure by respective methods. The compliant ring is an elastic foundation for the specimen, and, to a certain extent, it increases the critical pressure, at which the ring specimen loses stability.

External pressure can be produced by means of mechanical devices, the load in which is produced by a large number of identical loading levers (usually 72), which are connected to the test machine piston through the loading plunger. The advantage of the multipin device is the simple (although laborious to make) design and comparative accuracy of the experimental results. However,

TABLE 3. Compression of ring specimens.

Loading scheme	3-1 (P, R, h)	3-2 (b, h, D_0)	3-3 (p, R, h)
Determinable characteristics	$E_{\theta\theta}$, $\sigma^{-}_{\theta\theta}$		
Measurable values	P, P_u, $e_{\theta\theta}$	p, p_u, $e_{\theta\theta}$	
Relations for calculations	$E_{\theta\theta} = \frac{P}{2bh} \cdot \frac{1}{e_{\theta\theta}}$ $\sigma^{-}_{\theta\theta} = \frac{P_u}{2bh}$	$E_{\theta\theta} = \frac{pD_0}{2he_{\theta\theta}}$ $\sigma^{-}_{\theta\theta} = \frac{p_u D_0}{2h}$	
Limitations Structural Layup	0°; 90°; 0/90°		
Orientation	0°, 90°		
Physical	For $E_{\theta\theta}$: linear range of the curve $P \sim e_{\theta\theta}$ or $p \sim e_{\theta\theta}$		
Geometrical	For $\sigma^{-}_{\theta\theta}$: $0.08 \leqslant h/R \leqslant 0.18$ (for GFRP)		

even with high precision of production, the irregularity in pressure of a ring specimen can reach 10%. In operation, the irregularity increases, which requires periodic adjustment and calibration of the device. The power of these devices is limited.

The main difficulty in compression tests by external pressure is selection of the relative thickness of the specimen, at which the specimen fails only in compression. Depending on the relative thickness of the specimen h/R and the degree of anisotropy of the material $E_{\theta\theta}/G_{\theta r}$ three types of failure of ring specimens under compression may be observed: the loss of stability (thin-walled rings), compression (rings of average relative thickness) and destruction of thick-walled rings. In the latter case, in analyzing the strength not only circumferential stress $\sigma_{\theta\theta}$, but also radial stress σ_{rr} must be taken into account.

The critical pressure, at which the loss of stability of a ring occurs, can be estimated by the formula

$$p_{cr} = \frac{1}{1 + x_r^2} p_{cr}^* \tag{11}$$

where $p_{cr}^* = 3E_{\theta\theta}J/R^3$ is the critical pressure per unit length of a ring axis, calculated without consideration of shear; R is the average radius of the ring;

$$x_r = \frac{h}{R}\sqrt{\frac{E_{\theta\theta}}{G_{\theta r}}}.$$

Upon loading of rings with external pressure it is often impossible to correctly estimate the compression strength due to peeling of the inner layer (Fig. 10). Peeling of the inner layer and the subsequent loss of stability occur, when the difference in energy of the inner layer, preserving a circular configuration, and the same layer after peeling and stability loss is higher than the bond energy of the inner layer. At the layer thickness h_0, determined by the reinforcement, the critical stress, at which the peeling occurs, is equal to (TARNOPOL'SKII (1979)):

$$\sigma_{cr} = 0.916\, E_{\theta\theta}\left[\left(\frac{h_0}{R_i}\right)^2 + k_r\left(\frac{h_0}{R_i}\right)^{-1}\right]^{1/2} \tag{12}$$

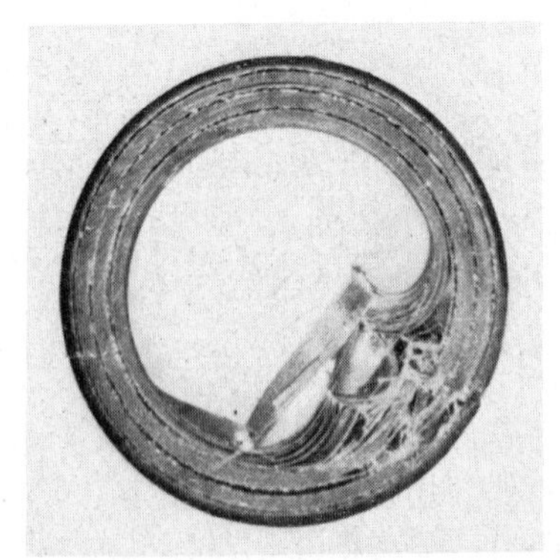

FIG. 10. Peeling of the inner ply of a ring under compression.

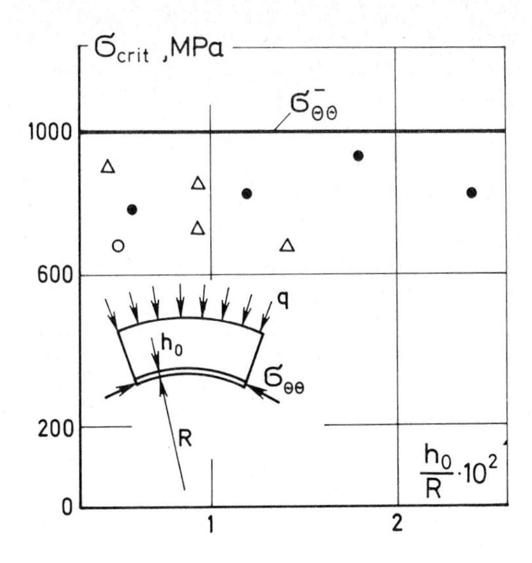

FIG. 11. The effect of peeling and relative thickness h/R on the compression strength of a ring (TARNOPOL'SKII (1979)). $h/R = 0.11$ (△); 0.21 (○); 0.34 (●).

where $k_r = 4.77\gamma/E_{\theta\theta}R_i$; γ is the specific fracture energy according to Griffith; R_i is the inner radius of a ring.

As is seen from Fig. 11, in the testing it is impossible to reach the strength of the material due to peeling of the inner layer; the deviation depends also on the relative thickness of a ring. The delamination as a result of peeling can further proceed in two ways: at helical winding it may initiate unwinding, at circular reinforcement a subsequent layer-by-layer peeling can occur.

4. Shear

Methods of evaluating the shear resistance of composites on flat and ring specimens have experienced unprecedented development of late years (TARNOPOL'SKII and KINCIS (1981)). The advantages and disadvantages of a series of loading modes for variegated specimen shapes have been investigated. The most widely used in research practice shear test methods, performed on flat specimens, are listed in Tables 4 and 5, on ring specimens in Table 6, on tubes in Table 5. Methods of estimation the shear characteristics in bending of flat and ring specimens are treated in Section 5. In the summary tables the measurable values, calculation relations and structural, physical and geometrical limitations are presented. The number of loading types, having been investigated and described in literature (see, for instance, TARNOPOL'SKII and KINCIS (1981)) is more numerous.

TABLE 4. Methods of investigation in-plane shear.

Loading scheme	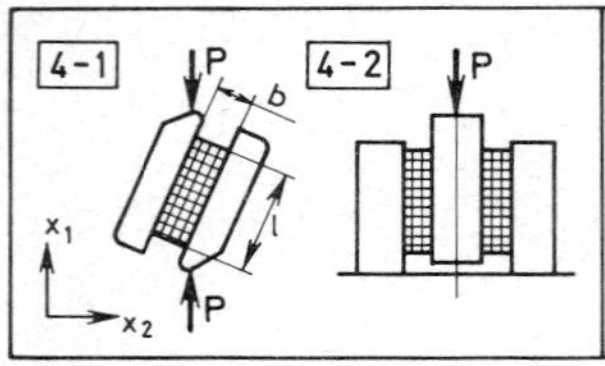 4-1, 4-2	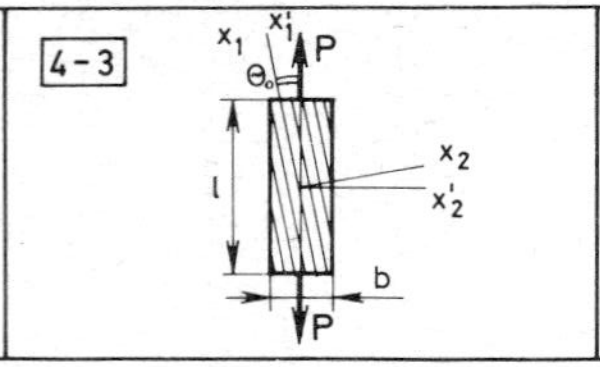 4-3	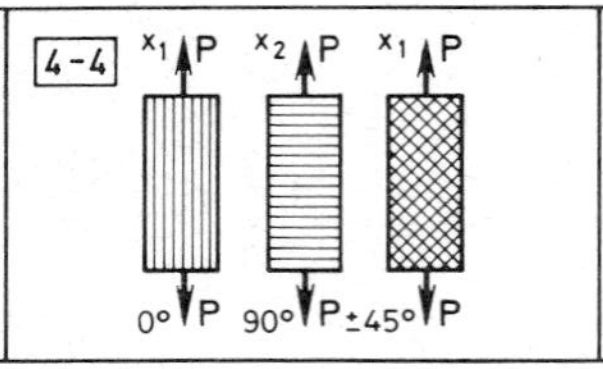 4-4	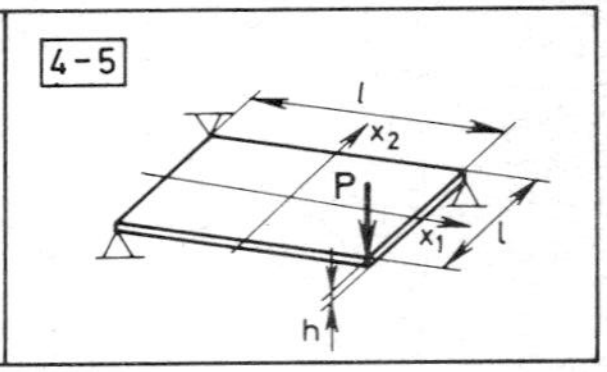 4-5
Determinable characteristics	G_{12}, τ_{12}^{u}	G_{12}, σ_6^{u} (?)	G_{12}	G_{12}
Measurable values	P, P_u, e_{45°	P, e_{0°, e_{45°, e_{90° or e_{0°, e_{120°, e_{240°	P, e_1, e_2, $e^{\pm 45^\circ}$	P, u_3
Relations for calculations	$G_{12} = P/2Ae_{45^\circ}$; $\tau_{12}^{u} = P_u/A$; **4-1**: $A = lh$; **4-2**: $A = 2lh$	See formulae (13)–(19)	See formulae (20)–(23)	$G_{12} = \dfrac{3Pl^2}{h^3 u_3}$
Limitations Structural Layup	0°; 90°; 0/90°	0°	0°; 90°; 0/90°	0°; 90°; 0/90°
Orientation	0°, 90°	10°–15°	0°, 90°, 45°	0°, 90°
Physical	–	–	–	Linear range of the curve $P \sim u_3$; $u_3 \leq 0.5h$
Geometrical	$l/b > 10$	$l/b \simeq 14$	–	For BFRP: $1/25 > h/l > 1/100$

Loading scheme

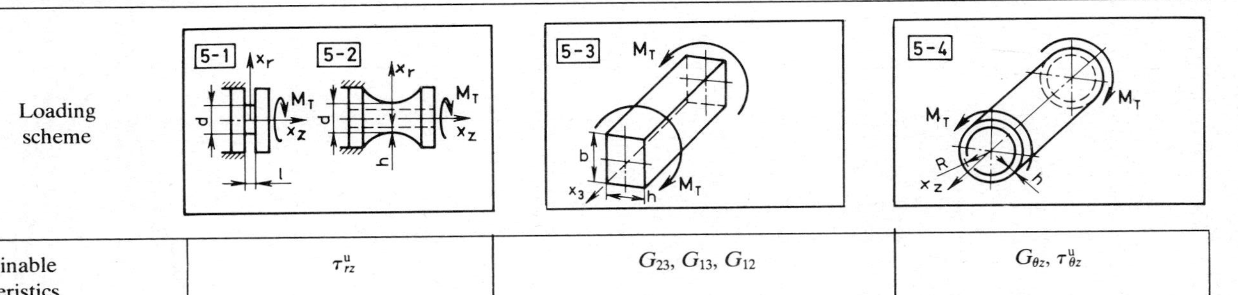

	5-1, 5-2	5-3	5-4
Determinable characteristics	τ^{u}_{rz}	G_{23}, G_{13}, G_{12}	$G_{\theta z}, \tau^{u}_{\theta z}$
Measurable values	M^{u}_{T}	$M_T, \varphi_1, \varphi_2$	$M_T, M^{u}_{T}, \varphi_1, \varphi_2, e^{+45°}, e^{-45°}$
Relations for calculations	**5-1**: $\tau^{u}_{rz} = \frac{16}{\pi}\frac{M_T}{d^3}$ **5-2**: $\tau^{u}_{12} = \frac{2}{\pi}\frac{M_T}{(d-h)^2 h}$	$C_3 = G_{13} b h^3 F(\eta)$, where $F(\eta) = \frac{32\eta^2}{\pi^4} \sum_{k=1,3,5...}^{\infty} \frac{1}{k^4}\left(1 - \frac{2\eta}{k\pi} th \frac{k\pi}{2\eta}\right)$; $\eta = \alpha\beta;\quad \alpha = \frac{b}{h};\quad \beta = \sqrt{\frac{G_{23}}{G_{13}}}$	$G_{\theta z} = \frac{2M_T}{\pi(R_0^4 - R_i^4)} \frac{l}{(\varphi_1 - \varphi_2)}$ or $G_{\theta z} = \frac{2M_T}{\pi(R_0^4 - R_i^4)} \frac{1}{e^{+45°} - e^{-45°}}$ $\tau^{u}_{\theta z} = \frac{2M_T R}{\pi(R_0^4 - R_i^4)}$
Limitations — Structural — Layup	Symmetrical with respect to x_2-axis	0°; 90°; 0/90°	
Orientation	–	0°, 90°	
Physical	–	Linear range of the curve $M_T \sim \varphi$	For $G_{\theta z}$; linear range of the curve $M_T \sim \varphi$
Geometrical	**5-1**: $d < 15$ mm; $l/d = 0.2$–1.0 **5-2**: $h \simeq 2$ mm (for GFRP)	–	Approximately: for GFRP $h/R < 1/10$ for CFRP $h/R < 1/40$

TABLE 6. Torsion of ring specimens.

Loading scheme	6-1	6-2	6-3
Determinable characteristics	$G_{\theta r}$, $G_{\theta z}$		$G_{\theta z}$, $G_{\theta r}$ (?)
Measurable values	P, u_3		
Relations for calculations	$C = \frac{PR^3}{u_3}\theta$	$C = \frac{3}{\frac{u_3}{\pi PR^3} - \frac{12}{E_{\theta\theta}hb^3}}$	$C = \frac{0.037\,885}{\frac{2u_3}{PR^3} - \frac{5.605\,08}{E_{\theta\theta}hb^3}}$
Limitations: Structural: Layup	0°; 90°; 0/90°		
Orientation	0°, 90°		
Physical	–		
Geometrical	–	$12/E_{\theta\theta}hb^3 \to 0$ at $b/h > 3$	$R/h > 12$ at $b/h = 0, 5$–10

Notes. b and h are width and thickness, respectively, (h is directed along the radius); u_3 is the deflection at the point of loading **6-3** or the mutual displacement of load application points (**6-1** and **6-2**).

4.1. In-plane shear

Historically, the first loading type was that of panel shear or picture-frame loading fixture, borrowed from the testing of plywood. However, the method has essential disadvantages – nonuniform strain in the specimen gage section, large specimen dimensions, the possibility of buckling of thin specimens and strong edge squeezing. The kind of shear load input is a special problem. In standard test fixtures, having a rigid four-link frame, the axes of rotation of pin-joints, as a rule, do not coincide with the corner points of the specimen test area. This is the reason of nonuniform strain distribution and concentration in the specimen test area. Diagonals of the four-link rig are elongated and shortened per equal amount and as a result in the testing of unbalanced materials, the axes of elastic symmetry of which coincide with rig diagonals, it is impossible to ensure the state of pure shear. In order to eliminate the disadvantages, the fixtures have been suggested, in which the axes of rotation of pins are extending over the specimen surface and are linked with the corner points of the specimen test area (Fig. 12): fixtures, in which the tensile load is applied through the system of mutually disconnected levers at a distance from the gage section of the specimen (Fig. 13); fixtures, in which shearing forces are

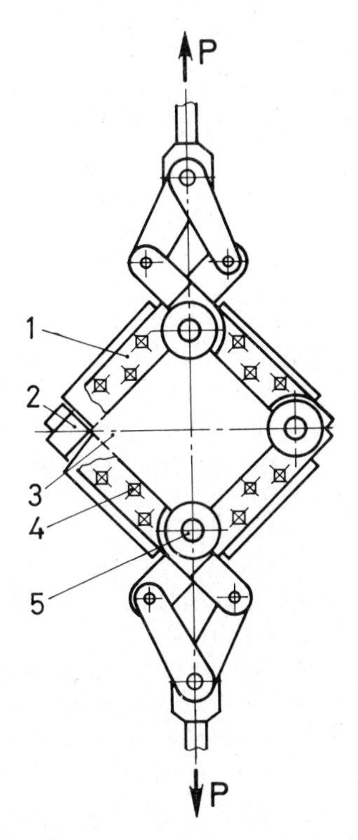

FIG. 12. A scheme of improved four-link test rig (PANSHIN et al. (1970)). (1) and (2) pinned frames; (3) a specimen; (4) a fastener; (5) a pin.

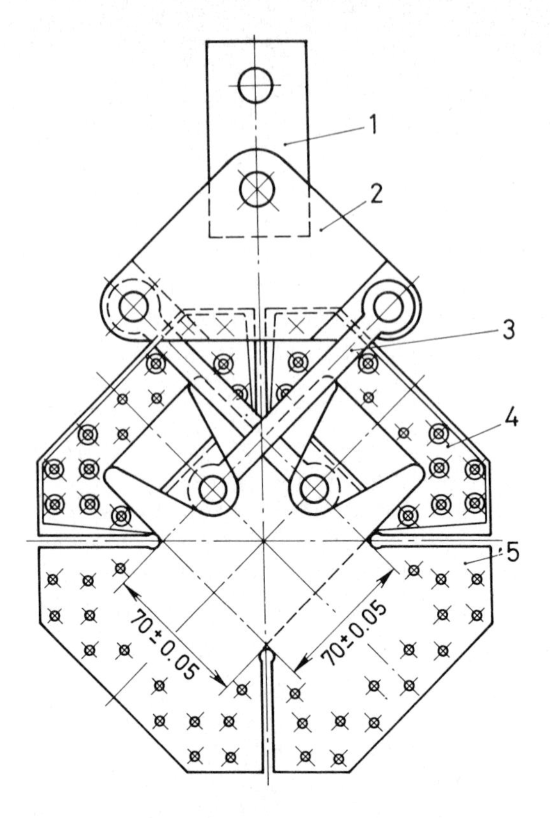

FIG. 13. A modified four-link test rig (SKUDRA et al. (1975)). (1) a specimen holder; (2) a traverse; (3) a pull rod; (4) a link of the test rig; (5) a specimen.

transmitted through the rigid rig frame and adjustable links with pin-joints (Fig. 14). The mentioned disadvantages and technical difficulties have led to controversial estimation of the method and it has gradually been given up, although lately there is a renewed interest into the picture-frame type test.

The picture-frame shear test was followed by the rail shear test (Schemes **4-1** and **4-2**). The rail shear test is a simple and economical technique, however some limitations are characteristic of the method. At the free specimen edges the state of stress differs from the state of pure shear. Fixed specimen edges undergo clamping in rails. The effect of end zones and uniformity of tangential stress distribution across the specimen width depend on the length to width ratio of the specimen gage section l/b and on the relation of elastic constants G_{12}/E_2 of the material being tested. For composites, the effect of edge zones is negligible at $l/b > 10$, except for materials characterized by high Poisson's ratio (for instance, with reinforcement at an angle ±45°). Data for elastic constants obtained by the rail shear test are less sensitive to the relative ratio l/b, since the measurements are being taken in the center of the specimen gage section, where the state of stress is the most uniform. In measuring the strength τ_{12}^{u}, the

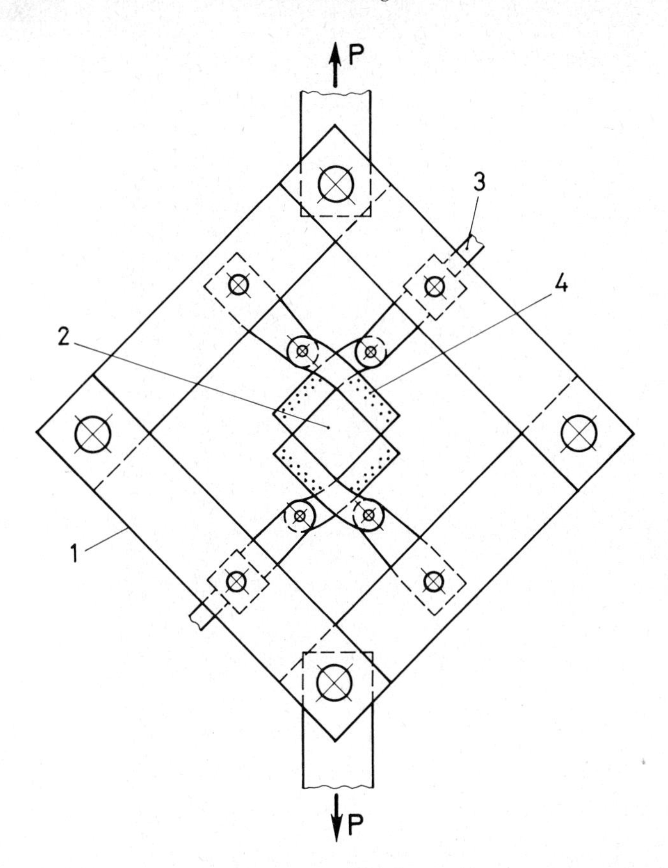

FIG. 14. A loading scheme in a four-link test rig (TERRY (1979)). (1) a pinned steel frame; (2) a specimen; (3) calibrated adjustable loading links; (4) steel edge members for fastening the specimen.

clamped edges and the loading mode exert considerable effect. In simple fixtures (Scheme **4-1**), the specimen is in a biaxial state of stress and this may cause a premature specimen failure. That is why the double fixtures (Scheme **4-2**) are more reasonable. In comparison to the picture-frame shear test the rail shear test yields somewhat higher values of shear modulus and lower strength values (see Table IV).

The method of tension of an anisotropic strip with various fiber orientation (Schemes **4-3** and **4-4**) is famous for its apparent simplicity. However, the anisotropic strips are not employed for determination of the in-plane shear strength, because the method yields lower values. In tension of a strip of the fiber layup ±45° the normal stresses are also acting on shear planes and the state of pure shear is not created. This results in the composite shear curve $\sigma_6 \sim e_6$, having a lower ultimate stress than the shear curve determined in pure shear.

TABLE IV. In-plane shear characteristics, determined by two different techniques (AUZUKALNS et al. (1976)).

Material	G_{12}, GPa		τ_{12}^{u}, MPa	
	Four-link rig	Strip	Four-link rig	Strip
GFRP, 0°	5.25	5.48	48.0	44.4
GFRP, 0°/90°	5.47	5.53	50.2	48.2
BFRP, 0°/90°	10.40	11.00	49.5	48.2
CFRP, 0°/90°	3.92	4.10	44.2	43.5
Epoxy binder	0.15	0.17	42.0	40.3

In the tension of a strip of the unidirectional material, reinforced at an angle θ_0, the value of the angle, at which the relative shear strains e_6/e_1' reach their maxima and the tangential stress σ_6 reaches its ultimate value, depends on the anisotropy of elastic and strength properties of the material (CHAMIS and SINCLAIR (1977)).

The relation e_6/e_1' reaches its maximum at an angle equal to

$$\theta_0 = \pm \operatorname{arctg}\left[(B + D^{1/2})^{1/3} + (B - D^{1/2})^{1/3} - \frac{\delta_1}{3}\right]^{1/2} \tag{13}$$

where

$$B = \frac{1}{2}\frac{E_2}{E_1} - \frac{\delta_1^3}{27} - \frac{\delta_1\delta_2}{6},$$

$$D = \frac{1}{4}\frac{E_2^2}{E_1^2} - \frac{\delta_1}{3}\frac{E_2}{E_1}\left(\frac{\delta_2}{2} + \frac{\delta_1^2}{9}\right) - \frac{\delta_2}{27}\left(\frac{\delta_1^2}{4} + \delta_2\right),$$

$$\delta_1 = 3 - \frac{E_2}{G_{12}} + 2\nu_{12}, \qquad \delta_2 = 3\frac{E_2}{E_1} - \frac{E_2}{G_{12}} + 2\nu_{12}.$$

For advanced composites, the optimal value of the angle θ_0 falls within a narrow range and it is equal to 10–15° (for isotropic materials $\theta_0 = 45°$). It has been stated (CHAMIS and SINCLAIR (1977)) that the stress relations σ_1/σ_1', σ_2/σ_1' and σ_6/σ_1' are more sensitive to the variation of the angle θ_0 (Fig. 15) than the strain relations e_1/e_1', e_2/e_1' and e_6/e_1' (Fig. 16). As a result the allowances for fiber orientation, strain gage positioning and load alignment have been specified as equal to ±1°. In order to ensure the uniformity of stress state, relatively narrow specimens have been used ($l/b \approx 14$).

In the rail shear test and tension of an anisotropic strip a curve $\sigma_6 \sim e_6$ (in a general case, nonlinear) is plotted, from which the shear modulus (tangential or secant) can be determined. Therefore, in the testing according to Schemes **4-1** to **4-4** physical limitations are not imposed on the curve $\sigma_6 \sim e_6$.

Upon loading according to Schemes **4-1** and **4-2** the tangential stresses are

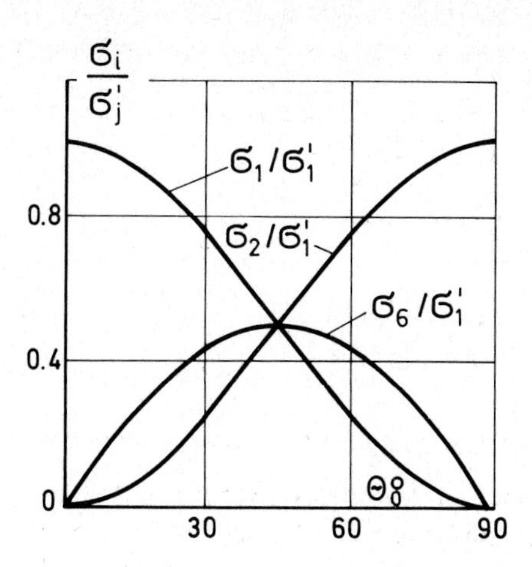

FIG. 15. Change in relative stresses σ_1/σ_1', σ_2/σ_1' and σ_6/σ_1', depending on the angle θ_0 (CHAMIS and SINCLAIR (1977)) for unidirectional carbon composite Modmor I/ERLA 4617.

obtained from the formula

$$\sigma_6 = \frac{P}{A} \tag{14}$$

where $A = lh$ (Scheme **4-1**) and $A = 2lh$ (Scheme **4-2**); the shear strain is equal to

$$e_6 = 2e_{45°} . \tag{15}$$

In the case of loading according to Scheme **4-3** the tangential stress is equal to

$$\sigma_6 = \frac{P}{2bh} \sin 2\theta_0 \tag{16}$$

the shear strain to

$$e_6 = (e_2' - e_1') \sin 2\theta_0 + e_6' \cos 2\theta_0 . \tag{17}$$

When employing a 60-degrees delta rosette (the angle is registered anti-clockwise) the relative strains are equal to

$$e_1' = e_{0°} , \qquad e_2' = \frac{1}{3}(2e_{120°} + 2e_{240°} - e_{0°}) ,$$
$$e_6' = \frac{2}{\sqrt{3}}(e_{240°} - e_{120°}) \tag{18}$$

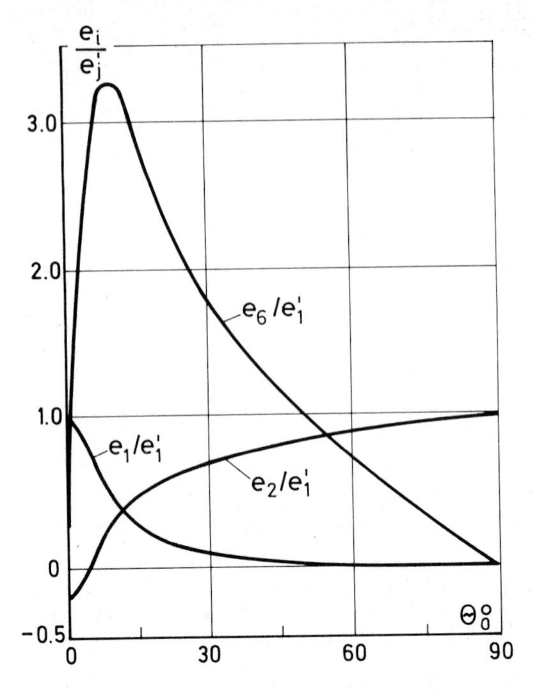

FIG. 16. Change in relative strains e_1/e_1', e_2/e_1' and e_6/e_1', depending on the angle θ_0 (CHAMIS and SINCLAIR (1977)) for unidirectional carbon composite Modmor I/ERLA 4617.

for the rectangular rosette

$$e_1' = e_{0^\circ}, \qquad e_2' = e_{90^\circ}, \\ e_6' = -e_{0^\circ} + 2e_{45^\circ} - e_{90^\circ}. \tag{19}$$

Formulae (16) and (17) are also used in the case of tension of a strip with fiber layup at an angle $\pm 45^\circ$ with the aim of estimation of the in-plane shear modulus of the constituent unidirectional laminae (HAHN (1973), ROSEN (1972)); in the case $\sigma_6 = \frac{1}{2}\sigma_1'$ and $e_6 = e_2' - e_1'$.

Upon loading according to Scheme **4-4** (PETIT (1969)) the characteristics of a unidirectional material (E_1, E_2, ν_{12} and ν_{21}) and the tangential modulus of elasticity $E^{\pm 45^\circ}$ of a strip having a layup $\pm 45^\circ$ (for all the three series of specimens $V_f = item$) are determined from independent tests; thereby the stress-strain curve $\sigma^{\pm 45^\circ} \sim e^{\pm 45^\circ}$ is broken down into increments, for which the relation $\sigma \sim l$ is practically linear. Furthermore, for each increment of the curve $\sigma^{\pm 45^\circ} \sim e^{\pm 45^\circ}$ the shear strain is calculated from the experimentally determined values of $e^{\pm 45^\circ}$ and $\nu_{12}^{\pm 45^\circ}$

$$\Delta e_6 = (1 + \nu_{12}^{\pm 45^\circ}) e^{\pm 45^\circ} \tag{20}$$

and tangential stresses

$$\Delta\sigma_6 = G_{12}\,\Delta e_6 \tag{21}$$

where the tangential shear modulus is equal to

$$G_{12} = \frac{L_1 E^{\pm 45^\circ}}{4(L_1 - E^{\pm 45^\circ})}, \tag{22}$$

$$L_1 = (E_1 + E_2 + 2\nu_{21}E_1)/(1 - \nu_{12}\nu_{21}) \tag{23}$$

From the incremental data $\Delta\sigma_6$ and Δe_6 the laminae stress-strain curve is plotted.

The twist of a square plate has been extensively used for measuring the in-plane shear modulus of composites. The loading pattern for composites was first substantiated by S.W. Tsai (TSAI (1965), TSAI and SPRINGER (1963)) (Scheme **4-5**). Its wide acceptance may be attributed to the simplicity of its calculation relation. However, the experiments should be performed with utmost care. The formula on Table 4 is applicable only at small deflections ($u_3 \leq 0.1\,h$) of plates from uniform across the thickness and orthotropic materials in the specimen axes. In the work of CHANDRA (1976) another formula is given with due allowance for large deflections and gravity of the plate

$$m_1\bar{u}_3^3 + m_2\bar{u}_3 - \bar{P} - \bar{\rho} = 0\,, \tag{24}$$

where

$$m_1 = \frac{128}{\pi^8}\left(1 + \frac{E_1}{G_{12}} + \frac{E_1}{E_2} - 2\nu_{12}\right)^{-1}, \qquad m_2 = \frac{G_{12}}{3E_1},$$

$$\bar{P} = \frac{Pl^2}{E_1 h^4}, \qquad \bar{u}_3 = \frac{u_3}{4}, \qquad \bar{\rho} = \frac{\rho l^4}{4E_1 h^4}.$$

By setting the first and the last terms in the formula (24) equal to zero, the formula of Table 4 for calculation of the shear modulus G_{12} is obtained. Results of the experimental verification of the formula (24) are not known. Calculations according to refined formulae (CHANDRA (1976), FOYE (1967)) for various materials show that the relation $P \sim u_3$ remains linear up to $u_3/h \approx 1$ and higher (see Fig. 17). However, due to possible loss of stability of the plate, one must confine to $u_3 \leq 0.5\,h$ (PURSLOW (1977)). The relative plate thickness h/l is determined by two conditions: the effect of transverse shear on deflection (at large relations h/l) and the possibility of stability loss (at small relations h/l). The boundaries of the relation h/l, given in Table 4, are meant for BFRP (LUBIN (1969)). However, it has been stated as a result of tests performed on GFRP, CFRP and BFRP with various types of reinforcement that reliable data

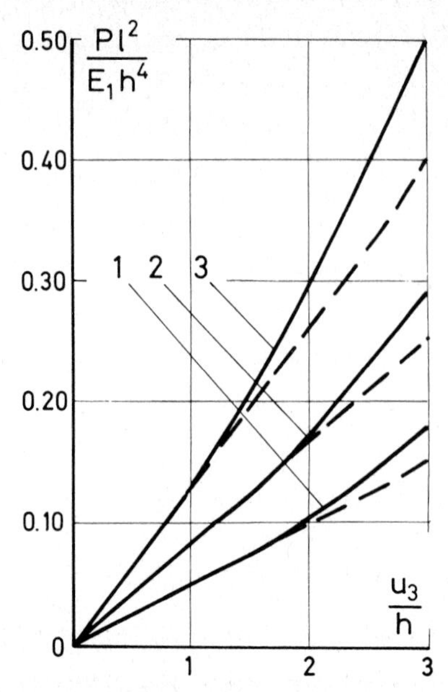

FIG. 17. Dependence of the relation 'deflection-load' in square plate twist of glass-epoxy (1), boron-epoxy (2) and carbon-epoxy (3) composites (CHANDRA (1976)). —— nonlinear theory; ----- linear theory.

can already be taken at $h/l < 1/15$ (see Fig. 18) (ZHIGUN et al. (1979)); at higher relations h/l the measured shear modulus is decreasing. The specimen should be strictly planar, without any initial deflections or curvatures, its thickness constant, since the value h^3 enters the calculation formula. The distance from

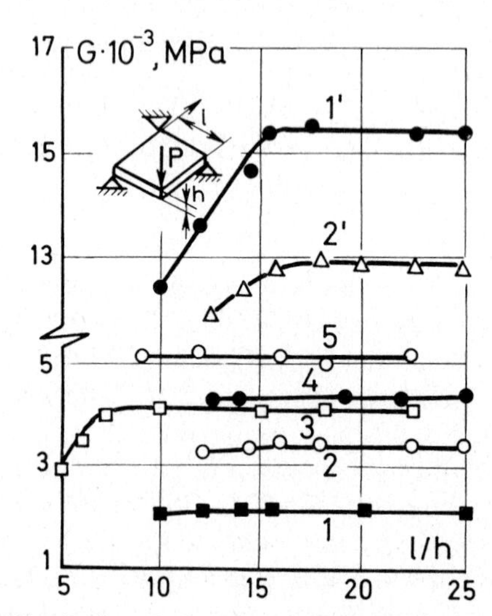

FIG. 18. Dependence of G_{12} (1–5) and G_{45} (1′ and 2′) of carbon-epoxy (1–4) and glass-epoxy composites (5) on the ratio l/h of a square plate in torsion according to the three-point scheme (ZHIGUN et al. (1976b)). Fiber layup: 1 and 1′: 0/90°; 2 and 2′: ±30°; 3: $0_2/90^\circ_2$; 4: 0°; 5: 0/90°.

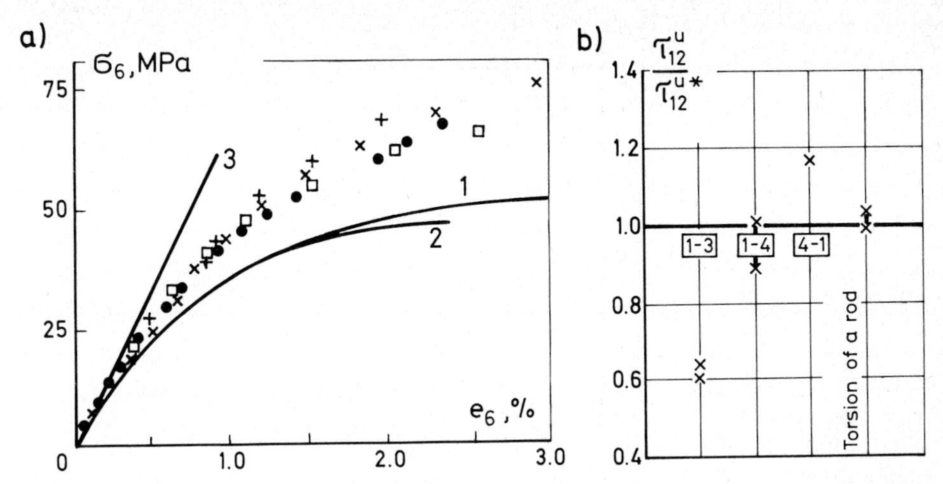

FIG. 19. (a) Stress-strain curves for carbon-epoxy composite in shear. Methods and layups: panel shear, a four-link rig: ● (0°), □ (0/90°); rail shear, + (0/90°); tension of a strip, × (±45°). Torsion: (1) a tube of circular and (2) quadratic cross section with a layup (0°/90°$_2$/0°/90°$_2$/0°/90°$_2$/0°); (3) a square plate.

(b) A comparative evaluation of different methods for determination of shear resistance: τ_{12}^{u*} is the shear strength, determined by torsion test of a thin-walled tube. ○ CFRP, × OFRP.

the point of support or load application to the apex of angles of the plate should not exceed $2h$; at larger distances of the points of support from the corner of the plate, the measured shear modulus increases.

Experimental evaluation of the panel shear test, rail shear test, tension of an anisotropic strip and square plate twist shows, that all these methods yield comparable values of the in-plane shear modulus (see the initial section of strain curves in Fig. 19). In determination of the shear strength, the panel shear test and tension of an anisotropic strip are quantitatively comparable, but the strengths, obtained in the rail shear test and twist test stand out against other strength values.

4.2. *Interlaminar shear*

Possibilities of estimating the interlaminar shear characteristics in comparison to in-plane shear characteristics are more scarce. This is attributed to the composite structure and to the difficulty of producing the necessary stress-strain state.

The three-point bending test of straight bars and ring segments is a rather widespread test method for estimation of the interlaminar shear characteristics. However, the method has significant limitations. Capabilities and disadvantages of the method of three-point bending are outlined in detail in Section 5.

Owing to the simplicity of realization, the method of determining the interlaminar shear strength in tension or compression of prismatic or ring specimens with notches (Fig. 20) is attractive. Still the method has essential

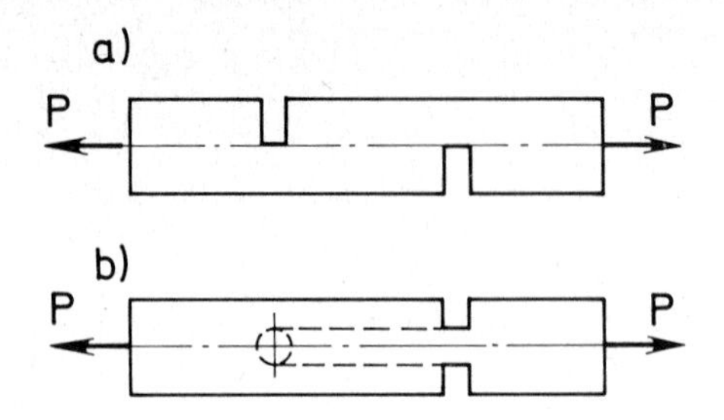

FIG. 20. A loading scheme for a specimen with nonsymmetrically (a) and symmetrically (b) cut notches.

disadvantages (CHIAO et al. (1977), MARKHAM and DAWSON (1975), MENGES and KLEINHOLZ (1969)): for testing samples with nonsymmetrically placed notches (Fig. 20(a)), the fixtures for prevention of specimen bending are necessary; time-consuming and invalid estimation of stress concentrations, high sensitivity to the accuracy of cutting the notches. In practice, due to unqualified notches an inadmissibly wide scatter of results is observed (see Table V) and the method may not be considered reliable.

Restrictions imposed on the range of application of the three-point bending test have led to the quest for other methods; in determination of the shear strength, good results, particularly for spatially reinforced materials, have been obtained by the torsion of waisted specimens. Specimens of two types have been used: hollowed (Scheme **5-2**) and non-hollowed (Scheme **5-1**). Selection of geometrical parameters of the specimen, its relative width l/d, diameter d, and wall thickness h is of significance. Investigations (ZHIGUN et al. (1976b)) show that within the range of relation $l/d = 0.2$–1.0 the length of the specimen gage section l does not affect the value of the strength being measured τ^{u}_{rz} (Fig. 21). The increase in the diameter of the specimen gage section d from 5 to 15 mm does not affect the strength τ^{u}_{rz} either, but by further increasing the diameter a sharp fall in the measured strength τ^{u}_{rz} is observed (Fig. 21). The wall thickness h of GFRP specimens equals 2 mm (MCKENNA et al. (1974)). Structural and physical restrictions are not imposed.

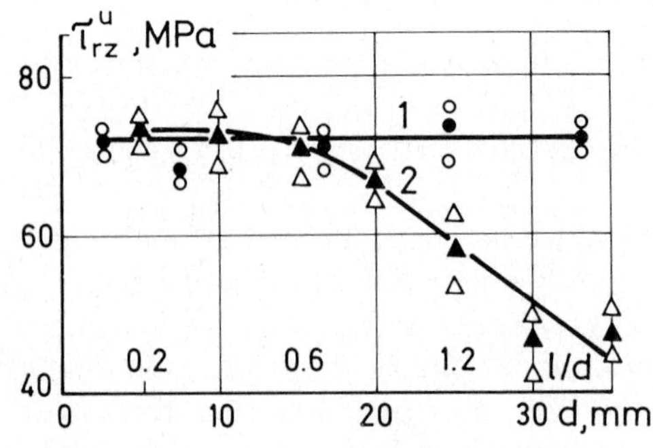

FIG. 21. Dependence of the strength τ^{u}_{13} on the relative width l/d on a circular notch (1) and the diameter d of the gage section of a specimen (2) (ZHIGUN et al. (1976b)).

TABLE V. Shear characteristics of unidirectional OFRP, obtained from various tests (data taken from CHIAO et al. (1977)).

Methods	Strength			Secant shear modulus at $e_6 = 0.5\%$		
	$(\tau_{12}^u)_i$ (MPa)	v (%)	$(\tau_{12}^u)_i/(\tau_{12}^u)_{11}$	G_i (GPa)	v (%)	G_i/G_{11}
Tension of 10° off-axis laminates:						
(1) coated edges	19.4	3.6	0.620	2.082	5.7	1.194
(2) bare edges	19.1	9.4	0.610	1.903	6.3	1.091
Tension of ±45° off-axis laminates:						
(3) midplane symmetry (7-layer)	29.4	2.0	0.939	1.923	4.7	1.103
(4) midplane symmetry (7 layers, repeated)	27.9	2.4	0.891	1.889	0.8	1.083
(5) nonsymmetrical (8-layer)	31.7	3.6	1.013	1.875	3.9	1.075
Three-point bending						
(6) prismatic bars	36.8	5.1	1.176	–	–	–
(7) ring segments	38.4	4.1	1.227	–	–	–
(8) Tension of notched bars	7.7–23.5	2.8–22.0	0.246–0.750	–	–	–
Torsion of a 0° wound composite rod:						
(9) strain measured with gages	31.3	7.2	1.000	1.965	3.6	1.127
(10) strain from angle twist	32.5	7.5	1.038	1.758	1.0	1.008
(11) Torsion of 90° wound thin tube	31.3	6.0	1.000	1.744	2.1	1.000

4.3. Torsion

The testing of straight bars by torsion (Scheme **5-3**) has been experimentally well established and, regardless of the relatively high expenses and high requirements as regards the accuracy of the experiment and data processing, the method is successfully used for measuring the shear moduli of composites (NIKOLAEV and NOVICHKOV (1968), STÖFFLER (1980), SUMSION and RAJAPAKSE (1978)). For measurement of the shear strength the method is as yet insufficiently developed.

Shear moduli are calculated in terms of the experimentally determined torsional stiffness

$$C = M_T \frac{l}{\varphi_1 - \varphi_2} \tag{25}$$

where M_T is the torque, l is the gage length (equal to approximately 1/3 the free length of a bar); φ_1 and φ_2 are twist angles at both ends of the gage length. The readings are taken, as a rule, over the linear section of the torsion diagram $M_T \sim \varphi$.

To obtain the shear moduli from experimentally measured torsional stiffness C the analytical relations, relating the two shear moduli and the geometrical dimensions of the cross section of an orthotropic bar are used (see Table 5).

Calculation of the shear moduli from experiments on bars of circular cross section offers no difficulties. However, the method is applicable only to monotropic or cylindrically orthotropic materials. To determine the shear moduli of specimens of rectangular cross section the tests of several sets of specimens, having different ratios b/h, have been conducted.

In the processing of test data for specimens of nearly square cross section from the experimentally obtained 4–5 values of torsional stiffness C_i by the technique of least squares a curve (Fig. 22)

$$\frac{C_i}{b_i h_i^3} = f\left(\frac{b_i}{h_i}\right)$$

is plotted, by the help of which two values of torsional stiffness C_1 and C_2 are selected and the shear moduli G_{12} and G_{13} calculated.

The scheme of shear moduli calculation by means of a computer is as follows:

– from the known values of b_1, h_1, C_1 and b_2, h_2, C_2 the following relation is obtained

$$F(\beta) = \frac{b_2}{b_1} \frac{h_2^3}{h_1^3} \frac{C_1}{C_2}; \tag{26}$$

– from the known values of b_1, h_1, b_2, h_2 and the given relations $\beta =$

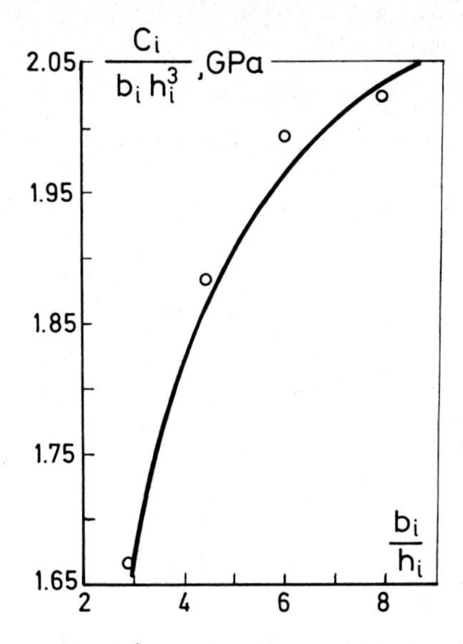

FIG. 22. The curve $C_i/b_ih_i^3 = f(b_i/h_i)$ (SUMSION and RAJAPAKSE (1978)).

$\sqrt{G_{13}/G_{12}}$ (the range of β depends on the material being tested; usually $\beta \leqslant 1$) with the help of a computer and the torsional function (see Table 5, Scheme **5-3**) the following relation is determined:

$$F(\beta) = \frac{f(\alpha_1\beta)}{f(\alpha_2\beta)} \tag{27}$$

where $\alpha_1 = b_1/h_1$ and $\alpha_2 = b_2/h_2$;

– from the equality of numerical values of relations, obtained according to the formulae (26) and (27), β, the function $f(\alpha_1\beta)$ and the shear moduli are calculated

$$G_{12} = \frac{C_1}{b_1h_1f(\alpha_1\beta)} \tag{28}$$

and

$$G_{13} = \beta^2 G_{12}\,. \tag{29}$$

In case there is no computer available, the graphs or tables of torsional function must be consulted (see NIKOLAJEV and NOVICHKOV (1968), TARNOPOL'SKII and KINCIS (1981)) or the method of successive approximations used (STÖFFLER (1980)).

The difficulties in the computation, associated with the application of the torsional function $F(\eta)$ might be eliminated by using a strip instead of bars, for which $b \gg h$ or $b \ll h$.

Under the condition when

$$\frac{h}{b}\sqrt{\frac{G_{23}}{G_{13}}}<\frac{\pi}{4} \tag{30}$$

or

$$\frac{h}{b}\sqrt{\frac{G_{23}}{G_{13}}}>4\pi\,, \tag{31}$$

the torsional stiffness C may be expressed through the following relationships, respectively (LEMPRIERE et al. (1969))

$$\frac{3}{bh^3}C_z = G_{23} - 0.630\,25\frac{h}{b}G_{23}\sqrt{\frac{G_{23}}{G_{13}}} \tag{32}$$

or

$$\frac{3}{hb^3}C_z = G_{13} - 0.630\,25\frac{b}{h}G_{13}\sqrt{\frac{G_{13}}{G_{23}}}\,. \tag{33}$$

In order to determine the shear moduli G_{13} and G_{23} the torsional stiffness C_z is measured experimentally at various h/b ratios and a graph of "the reduced torsional stiffness $(3/bh^3)C_z$ – ratio h/b" is plotted (Fig. 23). By extrapolating, at $h/b = 0$ one obtains

$$G_{23} = \frac{3}{bh^3}C_z \tag{34}$$

and further

$$G_{13} = 0.3972\frac{G_{23}^3}{k_1^2} \tag{35}$$

where k_1 is the slope of the straight line $(3/bh^3)C_z$.

The approximate method requires a very high accuracy of determining the shear modulus G_{23} and the slope of the straight line k_1, since the factor G_{23} triples any error in the intercept, while the error in the slope is doubled in the factor k_1. The high requirements imposed on the accuracy of determining the specimen cross section dimensions and the first determined value of the shear modulus, are disadvantageous for the data processing during torsional tests of bars of noncircular cross section.

For evaluation of the in-plane shear modulus and shear strength, the torsional test of thin-walled tubes (Scheme **5-4**) has often been adopted as a standard. In the torsion of thin-walled tubes, the tangential stresses along the

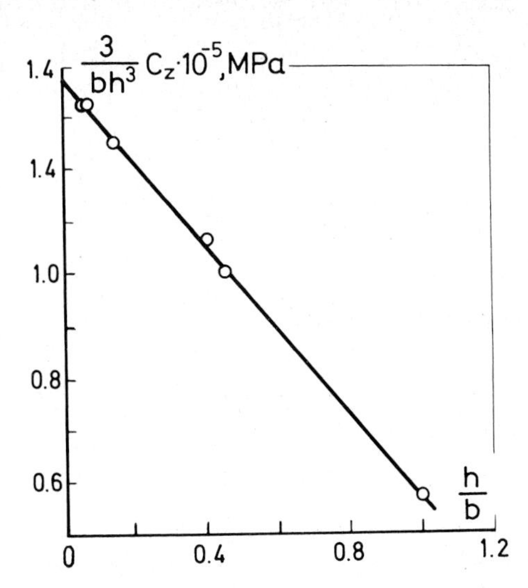

FIG. 23. Determination of the shear modulus by means of a diagram "reduced stiffness – h/b" (LEMPRIERE et al. (1969)).

circumference and along the specimen length are distributed uniformly; shear strains along the specimen wall thickness are practically constant. In torsion the concept of "a thin-walled tube" depends on the anisotropy of the material $E_{zz}/E_{\theta\theta}$ (PAGANO and WHITNEY (1970)) and the relative thickness h/R may vary within a wide range (see Table 5). The disadvantages of the method are: the method is applicable only to filament wound materials or specimens of special configurations (for example, with the fiber layup parallel to the specimen axis), rather large specimen dimensions and the need of special test fixtures. For evaluation of the in-plane shear strength $\tau^{u}_{\theta r}$ the buckling of the specimen is in admissible.

In Tables V and VI the shear characteristics of composites, based on various materials and having various reinforcement patterns, are listed. Test results for flat specimens of unidirectional OFRP have been compared with the characteristics, obtained in the torsion test of a thin-walled tube (Table V). It follows from the comparison that all the methods of Table V yield shear modulus values within the range of experimental data scatter. In evaluation of the shear strength the methods of tension of an anisotropic strip, the three-point bending and tension-compression of a notched bar must be sharply distinguished. There are several reasons of such distinction.

The method of tension of an anisotropic strip may appear to be invalid due to the shearing of the strip or due to incorrectly chosen angle $\theta_0 = 10°$. In the three-point bending test the disadvantages of the method as well as the unique features of the material (the behavior of OFRP under compression is often nonlinear; in such cases, the formulae of the technical theory of bending are invalid) may affect the test results. The disadvantages of tests conducted on notched specimens have been pointed out earlier.

TABLE VI. In-plane shear moduli (in GPa), determined by different techniques (data taken from ZHIGUN et al. (1979)).

Material, layup and fiber volume fraction	Square plate twist $(G_{12})_1$	Tension of a strip		Bar flexure		Torsion of a rod	
		$(G_{12})_2$	$\frac{(G_{12})_2}{(G_{12})_1}$	$(G_{12})_3$	$\frac{(G_{12})_3}{(G_{12})_1}$	$(G_{12})_4$	$\frac{(G_{12})_4}{(G_{12})_1}$
Glass textolite ($v_f = 0.45$)	3.50	3.60	1.200	2.60	0.743	3.20	0.914
HM GFRP, 0/90° ($v_f = 0.54$)	5.10	5.50	1.078	4.35	0.853	5.20	1.020
CFRP, 0° ($v_f = 0.50$)	2.90	3.20	1.103	2.10	0.724	2.70	0.931
CFRP, 0/90° ($v_f = 0.34$)	2.30	2.15	0.935	5.00	2.174	2.30	1.000
CFRP, ±15° ($v_f = 0.35$)	8.70	–	–	6.00	0.690	9.00	1.034
CFRP, ±30° ($v_f = 0.39$)	15.90	12.30	0.774	8.00	0.503	15.90	1.000
BFRP, 0° ($v_f = 0.58$)	4.00	3.95	0.988	2.80	0.700	4.60	1.150

TABLE VII. Shear moduli $G_{\theta r}$ and $G_{\theta z}$, determined by various techniques (KINCIS and SHLITSA (1978)).

Materials[a]	$G_{\theta z}$ (GPa) v%				$G_{\theta r}$ (GPa) v%		
	Torsion of split rings **(6-1)**	Torsion of prismatic bars **(5-2)**	Square plate twist **(4-5)**	Torsion of a cylinder **(5-3)**	Torsion of split rings **(6-1)**	Bending of rings **(8-1)**	Torsion of prismatic bars **(5-2)**
CFRP, UP-I	5.30	4.10	5.10		2.65	2.70	3.20
(v_f = 56.1–58.6%)	19	7	12		22	26	25
CFRP, UP-II	4.50	4.60	4.30		3.80		3.80
(v_f = 64.0–64.7%)	18	7	5		13		21
BFRP	7.80				7.50	7.10	
(v_f = 57.0–58.0%)	15				19	18	
HM GFRP	10.80			8.20	10.50	9.20	
(v_f = 67.0–69.0%)	14			16	15	19	

[a] Matrix is a modified epoxy resin. The parameter of anisotropy of BFRP and OFRP $E_{\theta\theta}/G_{\theta r}$ = 20–30, of HM GFRP $E_{\theta\theta}/G_{\theta r}$ = 8.

The analysis of the data of Table VI shows that the most reliable readings – in reference to the method of twist of a square plate, being 'the purest' in its relization – are obtained by the method of tension of an anisotropic strip, and torsion of a bar of rectangular cross section, the widest, practically inadmissible data scatter is observed in the three-point bending test.

In recent years, the information obtained from the testing of rings has been greatly enriched. Several loading types for estimation of the shear moduli from torsional tests of rings have been developed; the most effective ones are presented in Table 6. As regards the uniformity of the stress state, the best loading type is **6-1** (NIKOLAEV (1971)), because the effect of bending and transverse shear is negligible. Technically, the loading type **6-2** (GRESZCZUK (1969)) is easier to accomplish, yet in this case the bending effect is negligibly small only under the conditions, indicated in Table 6. The use of whole rings (Scheme **6-3**, KINCIS and SHLITSA (1978)) for evaluation of the shear moduli due to strict geometrical limitations is reasonable only in the cases when the elastic constants and strengths are estimated on the same specimens. Shear moduli, in terms of the torsional stiffness C, are calculated analogously to the case of torsion of straight bars.

There are relatively few comparable experimental data for evaluation of the torsional test on rings (GRESZCZUK (1969), KINCIS and SHLITSA (1978), NIKOLAEV (1971)). Table VII lists the shear moduli $G_{\theta z}$ (in the reinforcement plane and $G_{\theta r}$ (interlaminar) of glass-, boron- and carbon-epoxy composites. All the methods of Table VII, except twist of a square plate, are characterized by a vast data scatter but, in general, the methods yield qualitatively comparable results.

The interlaminar shear modulus $G_{\theta r}$ and interlaminar shear strength $\tau^{u}_{\theta r}$ can be estimated also by testing whole or split rings in bending; for these methods see Section 5.

5. Bending

The most popular type of loading in the bending tests of prismatic composite bars is the three-point bending (Schemes **7-1** and **7-2**). 'Pure' or four-point bending (Scheme **7-3**) and five-point bending (Scheme **7-4**) tests are less popular, although these loading types have several advantages versus the three-point bending.

5.1. Three-point bending

The three-point bending tests allow to determine the modulus of elasticity E^{b}_{1}, interlaminar shear modulus G^{b}_{13}, strength due to normal stresses σ^{b}_{1} and interlaminar shear strength τ^{bu}_{13}.

Theoretically, the moduli of elasticity in tension, compression and bending are equal ($E^{+} = E^{-} = E^{b}$). However, due to technological defects and peculiarities of the state of stress in bending, the modulus of elasticity E^b may somewhat differ from E^+ and E^-. The peculiarity of bending test is designated by index *b*.

In order to experimentally estimate the elastic constants in bending E_1^b and G_{13}^b the refined formulae for deflection of a bar at the midspan, taking into account the effect of shear, are used. In bending tests, the modulus of elasticity E_1^b can be estimated in two ways: together with the shear modulus G_{13}^b or without taking into account the effect of shear. Simultaneously, the values of E_1^b and G_{13}^b from the experimentally measured load P_i and deflection u_{3i} in the middle of the span l on several sets of specimens, having different relations (h/l), can be obtained by two techniques: graphically and analytically.

In the case of a graphical solution in the system of coordinates $(h/l)_i^2$, $1/E_{f_i}$ from the experimentally obtained data for the 4–5 sets of specimens, having different relative thicknesses (h/l), a straight line is plotted

$$\frac{1}{E_{f_i}} = \frac{1}{E_1^b} + \frac{1.2}{G_{13}^b}\left(\frac{h}{l}\right)_i^2 \tag{36}$$

whereof

$$E_{f_i} = -\frac{P_i}{4b_i u_{3i}}\left(\frac{l}{h}\right)_i^3$$

In the system of coordinates $(h/l)_i^2$, $1/E_{f_i}$ a tangent of the slope angle of the straight line (36) to the abscissa is equal to $1.2/G_{13}^b$, and it crosses the ordinate at the point $1/E_1^b$ (Fig. 24).

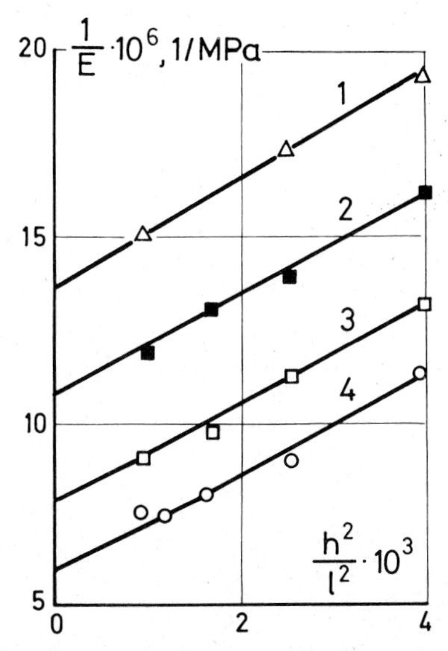

FIG. 24. An example of the data processing in estimating the modulus of elasticity E_1^b and shear modulus G_{13}^b of boron composites. Layups: (1) 1:1:1; (2) 1:1; (3) 2:1; (4) 1:0.

Analytically, the moduli E_1^b and G_{13}^b are obtained by solving the system of two equations (36), when the points $i = 1$ and $i = 2$ had been chosen from the experimentally plotted straight line (36). When employing the technique, it must be taken into account that the deflection due to transverse shear $u_{3\tau}$ only partly contributes to the total specimen deflection u_3. In order to reliably determine the shear modulus by properly selecting the relative span l/h, depending on the anisotropy degree of the material E_1^b/G_{13}^b, a sufficiently high ratio of deflections $k_\tau = u_{3\tau}/u_3$ (no less than 0.3) must be ensured. In order to state the necessary relative span length l/h, depending on the given ratios of E_1^b/G_{13}^b and $k_\tau = u_{3\tau}/u_3$, the following relation can be used:

$$\frac{l}{h} \leqslant 1.095 \sqrt{\frac{1-k_\tau}{k_\tau} \frac{E_1^b}{G_{13}^b}} . \tag{37}$$

The minimal allowable ratio must be $(l/h)_{\min} > 8$.

To determine the modulus of elasticity by neglecting the shear effects, the relative span length l/h must be chosen so that the shear effect was negligible. In Fig. 25 the dependence of the necessary relative span length l/h on the degree of anisotropy E_1^b/G_{13}^b for determination of the modulus of elasticity E_1^b by neglecting the shear effect with prescribed error δ is illustrated. As is seen from the figure, the necessary relative span lengths are rather large for highly anisotropic materials, as a rule, more than 40. This somewhat complicates the measurement of the deflections and loads and the change in a span length l and deflection u_3 must be taken into account in the case of supporting the specimen on cylindrical supports:

$$l_1 = l - 2\,\Delta l , \tag{38}$$

$$|u_3'| = |u_3| - |\Delta u_3| \tag{39}$$

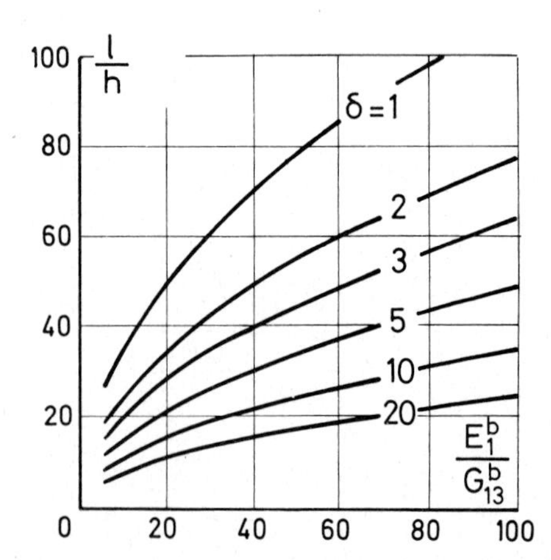

FIG. 25. Dependence of the value of minimum ratio l/h on the relation E_1^b/G_{13}^b in estimation of the modulus of elasticity, without taking account of shears with prescribed error δ (%).

where

$$\Delta l = \frac{R_2}{\frac{1}{4}\frac{l}{u_3} + \frac{u_3}{l}}; \qquad \Delta u_3 = \frac{2R_2}{1 + \frac{l^2}{4u_3}},$$

R_2 is the radius of cylindrical supports.

In bending of composites, the strengths computed on the basis of normal σ_1^b and tangential stresses τ_{13}^{bu} should be strictly differentiated, since unlike isotropic materials, the values may differ per order of magnitude or more and the tangential stresses may essentially affect the failure mode. If the failure occurs due to normal stresses, the fracture of the extreme extended or compressed plies is observed, due to tangential stresses – the delamination approximately on the level of the specimen midplane. In very short beams, the third type of failure is observed – due to bearing and shearing, accompanied by an apparent increase in the material resistance in shear. Redistribution of stresses in the specimen and the change of failure modes with the relative span l/h are shown in Fig. 26. Laminates may have one more failure mode – peeling of the compressed outer ply. The critical stress, at which the failure mode is likely to occur, is equal to (TARNOPOL'SKII (1979))

$$\sigma_{cr} = 16\pi^2 E_1^b \left(\frac{h_0}{l}\right)^2 \left(1 + \sqrt{1 + \frac{3\gamma l^4}{64\pi^4 E_1^b h_0^5}}\right) \tag{40}$$

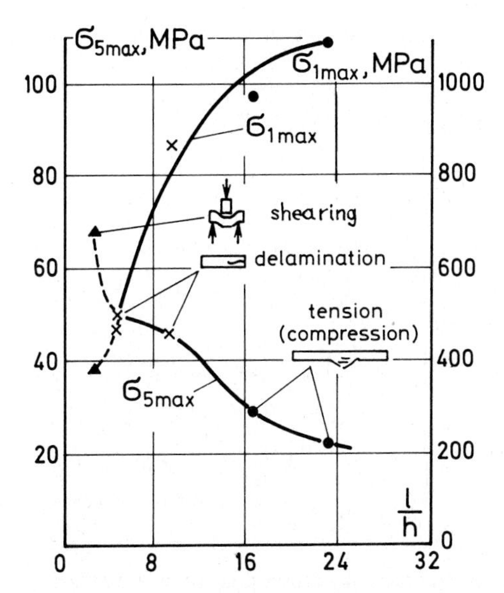

FIG. 26. Dependence of maximum normal and tangential stresses and the failure mode in bending on the relative span l/h (MENGES and KLEINHOLZ (1969)). Material: fiberglass–polyester, specimen thickness 4 mm.

where h_0 is the peeled-off thickness, γ is the specific surface energy at failure.

Failure in bending of spatially reinforced composites differs essentially from that of laminates (ZHIGUN and POLYAKOV (1978)).

Owing to the peculiarities of estimating the bending strength, the failure mode should be indicated, otherwise the experimental data will be incomparable.

The expected failure mode – due to normal or tangential stresses – can approximately be estimated from the graph l/h, σ_5 (Fig. 27). The magnitude of tangential stresses in a specimen at the moment of failure due to normal stresses is equal to

$$\sigma_{5\max} = \frac{1}{2}\frac{h}{l}\sigma_1^b . \tag{41}$$

Formula (41), in the system of coordinates l/h, σ_5 represents a rectangular hyperbola, the asymptotes of which coincide with the coordinate axes (Fig. 27). If the interlaminar shear strength τ_{13}^{bu} is plotted on the graph, the point of intersection of the straight line τ_{13}^{bu} with the curve $\sigma_{5\max}$ will determine the value of the relative span l/h, at which the failure mode changes due to normal or tangential stresses. For real composites, owing to technological defects, it is impossible to find the distinctive intersection locus of the curve $\sigma_{5\max}$ and the straight line τ_{13}^{bu} and near the theoretical intersection locus there is always some transitional region, where the failure may be caused by normal, as well as

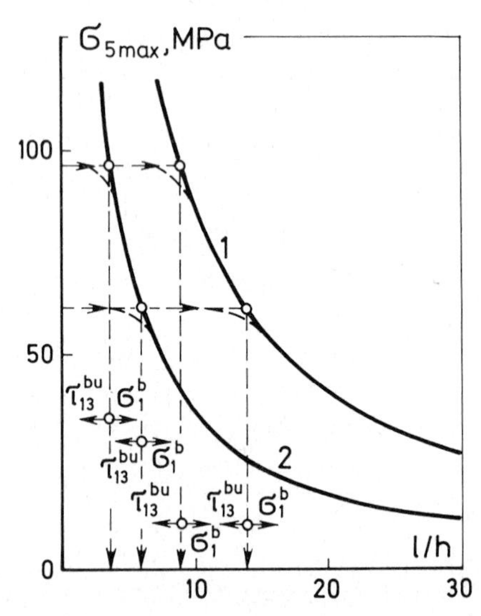

FIG. 27. Change in maximum tangential stresses (at a given strength σ_1^b) with the relative span l/h in three-point bending (MULLIN and KNOELL (1970)). Material: unidirectional boron composite.
(1) $\sigma_{5\max} = 840/(l/h)$; $\sigma_1^b = 1680$ MPa; $\tau_{13}^{bu} = 96$ MPa; $v_f = 70\%$; $v_V = 0$.
(2) $\sigma_{5\max} = 360/(l/h)$; $\sigma_1^b = 720$ MPa; $\tau_{13}^{bu} = 62$ MPa; $v_f = 30\%$; $v_V = 6\%$.

tangential stresses. Also owing to reasons, outlined below, the strength τ_{13}^{bu} is not of a constant value, but it falls with the increase in l/h, as a result the abscissa of the intersection point is shifted toward higher values of l/h.

The flexure formula for computation of the maximum normal stresses in the three-point bending test is a simple one, the effect of shear on the normal stress distribution for parameters

$$x_{\text{b}} = \frac{\pi}{2}\frac{h}{l}\sqrt{\frac{E_1^{\text{b}}}{G_{13}^{\text{b}}}} < 1.2$$

being negligible, but at larger values of x_{b} the effect is essentially weaker than that on deflection (TARNOPOL'SKII and KINCIS (1981)).

In experimental research practice, the three-point bending test, conducted on short beams or ring segments (Table 7, Schemes **7-1** and **7-2**), is the most widely used method for measuring the interlaminar shear strength τ_{13}^{bu}. Initially, when the resistance to interlaminar shear of composites was low, the method yielded good results. As a result of the development of composites technology and the increase in the interlaminar shear strength there is a need for specimens of shorter relative span length l/h. However, the refined solution of the problem of bending a short anisotropic bar (BERG et al. (1972), TARNOPOL'SKII et al. (1977)) has shown that the state of stress essentially differs from the state of stress, assumed in the technical bending theory. Distribution of tangential stresses along the height of a short anisotropic bar only in the middle of a halfspan corresponds approximately to the quadratic parabola of the technical bending theory: in the vicinity of concentrated load application the distribution of tangential stresses along the height of a bar reaches well defined maxima near the loaded surface (Fig. 28(a)). Relatively short anisotropic bars lack the increment with a constant ordinate of maximum tangential stresses (Fig. 28(b)). Besides, along the entire length of a relatively short bar the compressive transverse stresses σ_3 become active and high compressive contact stresses arise in the contact regions. Due to the deviations from the elementary theory, the experimentally obtained interlaminar shear strength τ_{13}^{bu} decreases with the increase in the relative span (Fig. 29) and, therefore, the results of the three-point bending test of a relatively short bar may serve only for a qualitative comparison of the mechanical properties of various composites.

For determination of the interlaminar shear strength of ring segments the formula for prismatic bars was used. In this case, it should be taken into account that in ring segments, in contrast to prismatic bars, the normal interlaminar stresses σ_{rr} are acting along the entire span and their direction depends on the loading pattern. In the case of segment loading with convexity upwards (Scheme **7-2**) the stresses are tensile, with convexity downwards they are compressive; in the latter case the stresses prevent the opening of a delamination crack due to tangential stresses $\sigma_{\theta r}$ and thereby they 'increase' the interlaminar shear resistance of the material. It should be taken into account

TABLE 7. Bending of bars.

Loading scheme	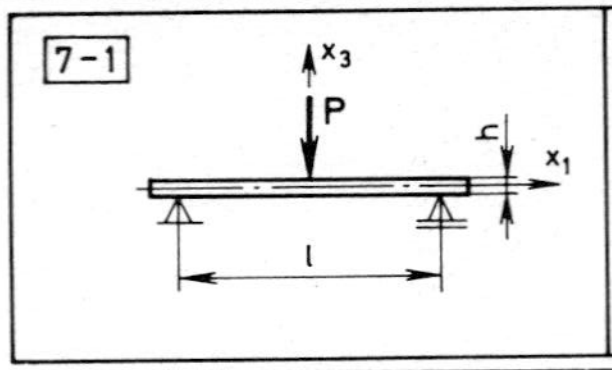 7-1	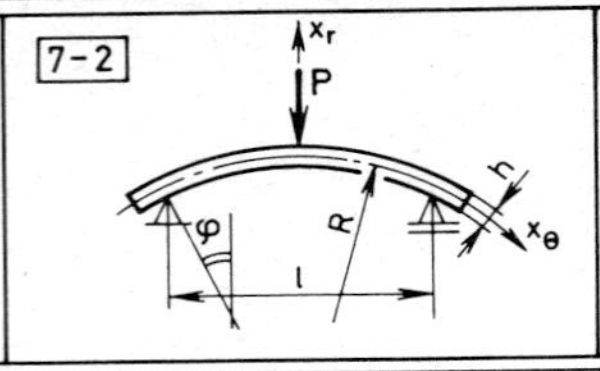 7-2	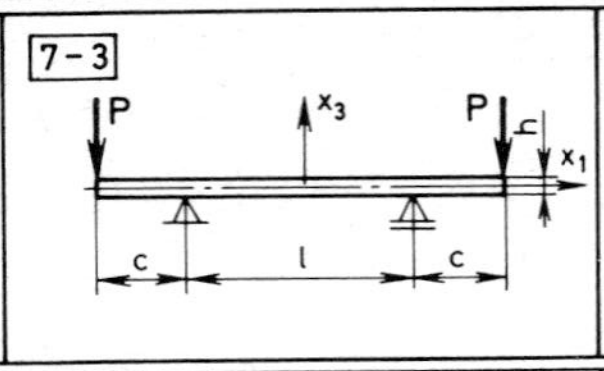 7-3	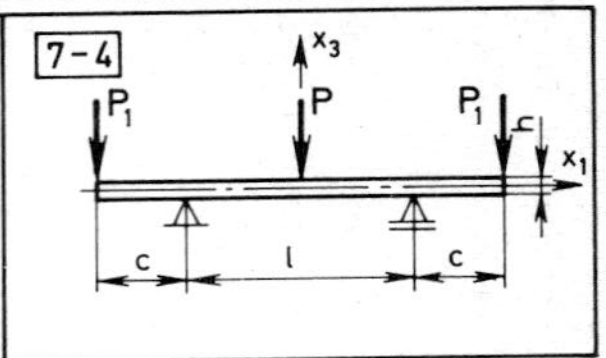 7-4
Determinable characteristics	E_1^b, G_{13}^b, σ_1^b, τ_{13}^{bu}	$\tau_{\theta r}^{bu}$	E_1^b, σ_1^b	E_1^b, G_{13}^b, E_1^b/G_{13}^b
Measurable values	P_1, P_u, u_3	P_u	P, P_u, u_3, e_t, e_c, φ	P, P_1, u_3
Relations for calculations	E_1^b and G_{13}^b: See formula (36) $E_1^b = -\dfrac{Pl^3}{4bh^3u_3}$ (without shear effect) $\sigma_1^b = \dfrac{3}{2}\dfrac{P_u l}{bh^2}$; $\tau_{13}^{bu} = \dfrac{3}{2}\dfrac{P_u}{bh}$	$\tau_{\theta r}^{bu} = \dfrac{3}{2}\dfrac{Q_u}{bh}$ For Q_u see, formulae (42)	For E_1^b see formulae (44)–(47) $\sigma_1^b = \dfrac{6P_c}{bh^2}$	See formulae (48)–(52)
Limitations Structural Layup	0°; 90°; 0/90°			
Orientation	0°, 90°			
Physical	E_1^b and G_{13}^b: linear range of the curve $P \sim u_3$	–	E_1^b: linear range of the curve $P \sim u_3$	Linear range of the curve $P \sim u_3$
Geometrical	For E_1^b: $l/h \geqslant 40$ For E_1^b and G_{13}^b: $8 \leqslant \dfrac{l}{h} \leqslant 1.095\sqrt{\dfrac{1-k_\tau}{k_\tau}\dfrac{E_1^b}{G_{13}^b}}$	$\dfrac{l}{2R} < \dfrac{\sigma_{rn}}{\tau_{\theta r}^{bu}}$; $\dfrac{h}{2l} > \dfrac{\tau_{\theta r}^{bu}}{\sigma_{\theta\theta}^b}$	For σ_1^b: $\dfrac{c}{h} > \dfrac{\sigma_1^b}{4\tau_{13}^{bu}}$	$\dfrac{c}{h} > \dfrac{\sigma_1^b}{4\tau_{13}^{bu}}$; $P_{max} = P_{1max} < \dfrac{bh^2}{6c}\sigma_1^b$

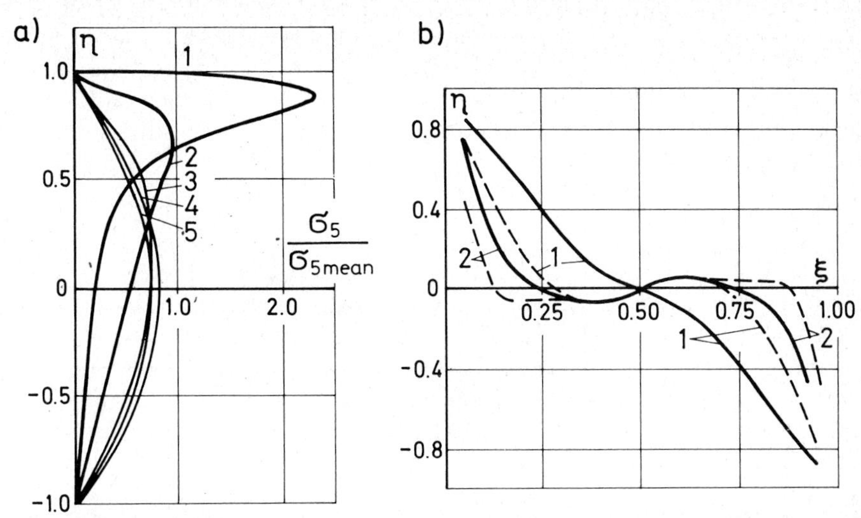

FIG. 28. (a) Variation of the ratio $\sigma_5/\sigma_{5\text{mean}}$ along the ordinate $\eta = x_3/h$ with the parameters $\xi = x/l$ and $\bar{l} = l/h$. (TARNOPOL'SKII et al. (1977)). $\bar{l} = 4$; $\xi = 0.05$ (Curve 1); 0.15 (2); 0.25 (3); 0.35 (4); 0.50 (5). (b) Variation of coordinates of maximum of tangential stresses for anisotropic (——) and isotropic (---) materials (TARNOPOL'SKII et al. (1977)). $\bar{l} = 4$ (Curve 1); 10 (2); ξ is recorded at the midspan.

that the value of shearing force Q depends on the specimen supporting mode: in case of supporting on a plate or sharp edges of a prism

$$Q_u = \tfrac{1}{2} P_u \tag{42}$$

in case of supporting on cylindrical supports

$$Q_u = \frac{P_u}{2} \frac{1}{\cos \varphi}\,. \tag{43}$$

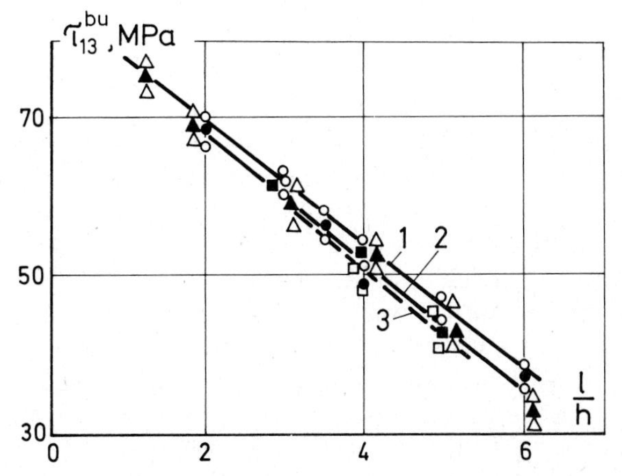

FIG. 29. Dependence of strength τ_{13}^{bu} on the relative span l/h in three-point bending (ZHIGUN et al. (1976a)). $b = 10$ mm; $h = 10$ (1); 15 (2); 20 mm (3).

For measurement of the interlaminar shear strength $\tau^{u}_{\theta r}$ the segment dimensions must be selected so that the normal circumferential stresses $\sigma_{\theta\theta}$ and normal radial stresses σ_{rr} were negligible compared to the tangential stresses $\sigma_{\theta r}$. The necessary conditions in selecting the relation of dimensions l/R and h/l are presented in Table 7.

5.2. Pure or four-point bending

The merits of pure bending are as follows: a uniform state of stress along the entire specimen's length, lack of transverse shear and contact stresses due to concentrated forces. Readings can be taken over the entire specimen's length. The main disadvantage – loading of the specimen according to the design scheme, the bending moment on its end faces – is practically impossible.

Modulus of elasticity E^{b}_{1} in pure bending tests can be evaluated by several techniques: by measuring the relative strain of extended or compressed outer plies (e_t or e_c), deflection at mid-span u_3 or the slope of the end faces of a specimen φ. The respective formulae for computation have the form

$$E^{bt}_{1} = \frac{6M}{bh^2}\frac{e_t + e_c}{2e_t^2} \tag{44}$$

or

$$E^{bc}_{1} = \frac{6M}{bh^2}\frac{e_t + e_c}{2e_c^2}, \tag{45}$$

$$E^{b}_{1} = \frac{3}{2}\frac{Ml^2}{bh^3 u_3} \tag{46}$$

or

$$E^{b}_{1} = \frac{6Ml}{bh^3\varphi}. \tag{47}$$

In practice, pure bending is generated under four-point loading (Scheme **7-3**). Four-point bending differs from pure bending by the loading type – the bending moment is due to concentrated forces, applied on a span or outside the span length l. Consequently, over the sections of the specimen from the supports to the point of load application, the shearing force, i.e., transverse shear is acting. In the case of load application outside the span length l, as shown in Table 7, the phenomenon does not affect the uniformity of the state of stress over the entire specimen length between the supports, i.e. this part of the specimen has all the advantages of pure bending. The loading type with concentrated forces on a span l, being less 'pure', is not treated here. In determination of the strength σ^{b}_{1} the cantilever length c must be chosen with the allowance for the bending moment M and shearing force Q.

The modulus of elasticity E_1^b in the four-point bending is computed according to formulae (44) to (47) by substituting $M = Pc$.

5.3. Five-point bending

Five-point bending (Scheme **7-4**) incorporates three- and four-point bendings with all their advantages and disadvantages. Under five-point loading the state of stress varies along the entire specimen length and this must be kept in mind when selecting the specimen dimensions. In contrast to the two foregoing loading types, the five-point bending allows to estimate the interlaminar shear modulus by two different techniques. The five-point bending test is inapplicable for measurement of the strengths σ_1^b and τ_{13}^{bu} and it has the disadvantage of having two independent loading systems.

There are three variants of employing the five-point load test in bending:

(1) in the case of force ratio

$$\frac{P_1}{P} = \frac{l}{6c} \tag{48}$$

the interlaminar shear modulus is equal to

$$G_{13}^b = 0.3 \frac{Pl}{bhu_3} \tag{49}$$

where u_3 is the deflection at midspan l;

(2) in the testing of a beam at different values of P, P_1 and c (span length l remains constant) in the system of coordinates $Pl/4P_1c$, u_3/P_1c by the technique of least squares a straight line is plotted

$$u_3 = \frac{Pl^3}{4E_1^b bh^3} - \frac{3P_1cl^2}{2E_1^b bh^3} + \frac{0.3Pl}{bhG_{13}^b} \tag{50}$$

which intersects the ordinate at the point $-3l^2/2E_1^b bh^3$, but for an abscissa $Pl/4P_1c = 3/2$ has the ordinate $1.8/bhG_{13}^b$;

(3) under the loading of the specimen, so that deflection at midspan $u_3 = 0$, the relation is

$$\beta^2 = \frac{E_1^b}{G_{13}^b} = \frac{1}{1.2}\left(\frac{l}{h}\right)^2\left(\frac{6c}{l}\frac{P_1}{P} - 1\right) \tag{51}$$

and furthermore, according to the known β^2 and G_{13}^b the modulus of elasticity

$$E_1^b = \beta^2 G_{13}^b \tag{52}$$

can be computed.

For limitations imposed on loads P and P_1, see Table 7.

5.4. Bending of rings

Loading schemes for whole and split rings in bending are present in Table 8. Bending of whole rings with concentrated forces (Scheme **8-1**) has been experimentally verified and, provided the relative thickness of the specimen has been selected correctly, the method yields quite reliable results. Circumferential modulus of elasticity $E_{\theta\theta}$ and the interlaminar shear modulus $G_{\theta r}$ in terms of the experimentally measured load P and the change in a vertical diameter u_3 are estimated in the same way as in the three-point bending of prismatic bars: according to the experimentally obtained data $[u_{3i}, (h/R)_i]$ in the system of coordinates $(h/R)_i^2$, $1/E_{\theta\theta f_i}$, a straight line is plotted

$$\frac{1}{E_{\theta\theta f_i}} = \frac{1}{E_{\theta\theta}} + \frac{0.528}{G_{\theta r}}\left(\frac{h}{R}\right)_i^2 \tag{53}$$

where

$$E_{\theta\theta f_i} = 1.785\frac{P_i}{b_i u_{3i}}\left(\frac{R}{h}\right)_i^3 .$$

The straight line intersection with the ordinate yields the value $1/E_{\theta\theta}$, which with the abscissa makes an angle, the tangent of which equals $0.528/G_{\theta r}$.

In order to ensure reliable values of the interlaminar shear modulus $G_{\theta r}$ by the method of loading rings with concentrated forces, the relative specimen thickness h/R must be selected with the allowance for the degree of anisotropy of the material $E_{\theta\theta}/G_{\theta r}$ and the partial contribution of tangential stresses to the change in a vertical diameter $k_\tau = u_{3\tau}/u_3$, which must be no less than 0.25–0.30.

In evaluation of the interlaminar shear strength $\tau_{\theta r}^{u}$ by the technique of bending whole rings, the selection of the relative specimen thickness must ensure the failure by shear – delamination in the midplane. The method is not used for determination of the strength according to normal circumferential stresses $\sigma_{\theta\theta}^{+}$.

When evaluating the interlaminar shear strength on split rings, the specimens are loaded with concentrated forces P, applied to a rigid cantilever so that the line of their action passed through the ring center (Scheme **8-2**). The width of gage section AB is approximately equal to a half of the specimen width. In order to prevent the specimen failure in bending, the ring section along the horizontal diameter is enforced by a steel clamp.

During bending test of split rings one can estimate the transverse tension strength σ_{rr}^{+} of the filament wound material (Scheme **8-3**). The bending moment $M = Pl$ is introduced by two centilevers, the force P being applied at the ends of the cantilevers. The specimen failure in the case is likely to have resulted from the joint effect of normal stresses σ_{rr} and tangential stresses $\sigma_{\theta r}$. The effect of tangential stresses

$$\sigma_{\theta r} = \frac{\sin\varphi}{l/R + \cos\varphi}\,\sigma_{rr\,\max} \tag{54}$$

TABLE 8. Bending of ring specimens.

Loading scheme	8-1	8-2	8-3
Determinable characteristics	$E_{\theta\theta}$, $G_{\theta r}$, $\tau^{u}_{\theta r}$	$\tau^{u}_{\theta r}$	σ^{+}_{rr}
Measurable values	P, P_u, u_3	P_u	P_u
Relations for calculations	For $G_{\theta r}$: see formula (53) $\tau_{\theta r}=\frac{3}{4}\frac{P_u}{bh}$	$\tau^{u}_{\theta r}=\frac{3}{2}\frac{P_u}{bh}$	$\sigma^{+}_{rr}=\frac{3}{2}\frac{P_u l}{bhR}\left(1+\frac{R}{l}\cos\varphi\right)$
Limitations Structural Layup	0°; 90°; 0/90°		
Orientation	0°, 90°		
Physical	For $G_{\theta r}$: linear range of the curve $P \sim u_3$	–	–
Geometrical	For $G_{\theta r}$: $\frac{R}{h}\leqslant 0.727\sqrt{\frac{1-k_\tau}{k_\tau}\frac{E_{\theta\theta}}{G_{\theta r}}}$ For $\tau^{u}_{\theta r}$: $h/R = 0.08$–0.18 (for GFRP)	–	See formulae (54) and (55)

is governed by the selection of the cantilever and gage section AB lengths.

The specimen failure can also be attributed to normal circumferential stresses $\sigma_{\theta\theta}$. In order to prevent such failure, the specimen dimensions should satisfy the inequality

$$\frac{h}{R} > 4\frac{\sigma_{rr}^{+}}{\sigma_{\theta\theta}^{+}} \tag{55}$$

where $\sigma_{\theta\theta}^{+}$ is the strength according to normal circumferential stresses.

6. Initial stresses and their experimental determination

(BLAGONADJOZHIN et al. (1976))

Composites are characterized by initial (or technological) stresses, induced in the fabrication process. There are initial stresses of two kinds: microstresses and macrostresses. Initial microstresses are caused mainly by the difference in coefficients of thermal expansion of the material constituents. Initial microstresses are completely maintained in specimens cut from the material. Therefore, mechanical characteristics, obtained in the testing of specimens, incorporate the effect of initial microstresses.

Initial macrostresses are built up as thermoelastic stresses in the post-cure cooling process or as a result of fiber tensioning or material inhomogeneity. Upon cutting the specimens from the material, initial macrostresses are partly or completely obliterated. Therefore, mechanical tests provide strength characteristics, devoid of the effect of initial macrostresses. Consequently, the initial macrostresses should be determined separately and in the process of structural design they must be summated with the stresses under external load.

The most widely used experimental technique of determinining the initial macrostresses are the generalized Sachs' and Davidenkov's methods; both methods are exclusively meant for ring specimens. In order to study the process of initial macrostresses generation, allowing to evaluate the effect of various technological factors, tensiometric mandrels and embedded strain gages are used.

6.1. A generalized Sachs' method

The method involves a subsequent removal of thin inner or outer cylindrical layers and recording of circumferential strain on the opposite side of the ring. For wound cylinders and rings from preimpregnated tape of fabric, Sachs' method is accomplished by layer-by-layer unwinding of the inner or outer layers of the specimen. For specimens, fabricated by filament or strand winding, the layer unwinding is inapplicable and the cylindrical layers are removed by cutting, for example, machining of specimens. Practically, the outer layer unwinding or machining is more convenient. Sachs' method allows one to determine not only the absolute values of radial (σ_n^0) and circumferential ($\sigma_{\theta\theta}^0$) initial stresses, but also the distribution of the initial stresses along the specimen radii $\sigma_{rr}^0 = f(r)$ and $\sigma_{\theta\theta} = \varphi(r)$. By applying a tensiometer on the inner surface of rings from homogeneous material, the radial and circumferential normal stresses are calculated according to the formulae:

$$\sigma_{rr}^0 = -\frac{kE_{rr}}{2}\left[\left(\frac{r}{R_i}\right)^{k-1} - \left(\frac{r}{R_i}\right)^{-(k+1)}\right]e_{\theta\theta_i}(r), \tag{56}$$

$$\sigma_{\theta\theta}^0(r) = r\frac{\mathrm{d}\sigma_{rr}^0}{\mathrm{d}r} + \sigma_{rr}^0(r) \tag{57}$$

where E_{rr} and $E_{\theta\theta}$ are the moduli of elasticity of the material in the radial and circumferential directions, respectively; $k = \sqrt{E_{\theta\theta}/E_{rr}}$; r is the actual radius; $e_{\theta\theta_i}$ is the relative circumferential strain, measured on the inner surface of the ring ($r = R_i$).

In the testing of cylinders, the moduli of elasticity E_{rr} and $E_{\theta\theta}$ are substituted by the moduli for a plane strain

$$E_{rr}^* = \frac{E_{rr}}{1-\nu_{\theta r}\nu_{r\theta}}, \qquad E_{\theta\theta}^* = \frac{E_{\theta\theta}}{1-\nu_{\theta r}\nu_{r\theta}}.$$

According to formula (56) and the moduli of elasticity E_{rr} and $E_{\theta\theta}$, radii R_i and r and experimentally evaluated circumferential strain $e_{\theta\theta_i}$ the curve $\sigma_{rr}^0 = f(r)$ is plotted, subsequently used for calculation of the stresses $\sigma_{\theta\theta}^0 = \varphi(r)$ from formula (57).

Calculation relations for thick-walled elements ($R/h > 4$) of inhomogeneous materials are more complex and computer-performed (BLAGONADJOZHIN et al. (1976)).

6.2. *A generalized Davidenkov's method*

Davidenkov's method differs from Sachs' method in that before the removal cylindrical layers the ring or cylinder is cut through the radius. In other words, tension (or compression) of a ring according to Sachs is substituted by the study of a curved bar in flexure.

The circumferential modulus of elasticity of thin rings is assumed to be constant through the ring thickness and the formulae for calculating the initial stresses take the form:

$$\sigma_{\theta\theta}^0(a) = E_{\theta\theta}\left[\left(1-\frac{2a}{h}\right)e_{\theta\theta_i} - \frac{1}{3}(h-a)\frac{\mathrm{d}e_{\theta\theta_i}}{\mathrm{d}a} + \frac{2}{3}\int_0^a \frac{\mathrm{d}e_{\theta\theta_i}}{\mathrm{d}\eta}\left(\frac{2h-3a+\eta}{h-\eta}\right)\mathrm{d}\eta\right], \tag{58}$$

$$\sigma_{rr}^0(a) = \frac{1}{r}\int_{R_i}^{r} \sigma_{\theta\theta}^0(a)\,\mathrm{d}r \tag{59}$$

where a is the layer coordinate, for which the stress is calculated; h is the thickness of the specimen; η is the thickness of the removed layer.

More complex cases of the distribution of circumferential modulus of elasticity $E_{\theta\theta}$ through the thickness of the ring (linear, hyperbolic) as well as the general case of testing thick-walled elements have been treated elsewhere (BLAGONADJOZHIN et al. (1976)).

When evaluating and making choice between Sachs' and Davidenkov's methods it should be taken into account that the circumferential strain $e_{\theta\theta_i}$ according to Sachs' method is due to extension of an annular specimen, according to Davidenkov's method – due to bending of a curved bar. Therefore, when employing Sachs' method there is a need of high precision of recording the strain $e_{\theta\theta_i}$. This is of particular importance in the testing of thin-walled specimens.

The formulae of Sachs' method assume that the stresses σ^0_{rr} are proportional to radial modulus of elasticity E_{rr} or its function, the accuracy of their determination is considerably lower than that of measuring the circumferential modulus of elasticity $E_{\theta\theta}$, which is the basic in the formulae of Davidenkov's method. Circumferential stresses $\sigma^0_{\theta\theta}$ are calculated according to Sachs' method by differentiation of the relation $\sigma_{rr} = f(r)$, resulting in additional errors. According to Davidenkov's method, the circumferential stress $\sigma_{\theta\theta}$ is first determined in the processing of experimental data, i.e., the value is determined with more accuracy, and the radial stresses are estimated by integrating the relation $\sigma^0_{\theta\theta} = \varphi(r)$.

In view of the distinction, Davidenkov's method is more accurate, but also more time-consuming. Besides, the scatter of experimental data is unlikely to justify the gain in accuracy of initial data estimates.

6.3. *Investigation of initial macrostresses generation*

The process of initial macrostresses generation has been investigated by means of tensiometric mandrels and by application of the embedded strain gages. Strain gaging of the mandrel is a subsidiary method, by which the information about initiation and building up of the initial stresses in the process of winding and subsequent heat treatment due to various technological factors, is obtained by measuring the pressure on mandrel from specimen's side. For this sake, heat resistance strain gages are glued on the inner surface of the metallic mandrel: operation gages – in circumferential direction, compensational – in axial direction. Simultaneously, the mandrel temperature and various technological parameters (for instance, tensioning of glass tape being wound) are measured.

The method of embedded strain gages is used for investigation of the generation of circumferential initial strains and stresses in the process of heat treatment of thick-walled cylinders, wound from preimpregnated fabric. For this reason, in the winding of a cylinder, the tensiometric tapes, made of the same fabric as the cylinder, with heat resistant microgages glued on them are placed between fabric layers. In the vicinity of operating strain gages the compensating strain gages and thermopairs are placed. The method of embedded strain gages is treated in detail in the works (DANIEL et al. (1972), VARUSHKIN (1971)).

7. Concluding remarks

The activities aimed at standardization of the test methods for composites have been started quite recently. At the beginning of the 1960s there was practically no standard, pertaining to static testing of composites.

At present, the standards for static testing of composites are either already available or under development in the majority of technically developed countries. However, research practice has already outstripped – not only as regards the quantity, but also quality of the obtained information – the development of the test methods, specified by standards.

There is an urgent need to unite efforts over the generalization and practical evaluation of the accumulated experience, selection of the most reliable test methods, their theoretical analysis and experimental characterization – with the objective to conduct researches on the level of international standards for this class of materials. It is these reliable test methods – for the stage of material development as well as for structural parts – which will allow one to better understand and estimate the structural capabilities of polymeric, metallic or ceramic matrix-based advanced composites.

The methods of static testing of composites are based on the experience gained for fibrous composites of 'the first generation', based on a polymeric matrix, reinforced with conventional and high-modulus fibers having unidirectional, laminated and three-dimensional reinforcement layup. The practice has shown that the methods have proved to be beneficial and successfully applicable to fibrous composites of 'the following generations' with carbon, metallic and ceramic matrices. However, the experience gained for the first generation of composites cannot be directly applied to the advanced composites without due regard for the specific properties of matrices.

References

AUZUKALNS, YA.V., A.N. BIRZE and F.YA. BULAVS (1976), Strength properties of reinforced plastics in compression at an angle to the fiber direction, in: *Methods of Nondestructive Testing for Building Materials* **2** (Riga Polytechnic, Riga), 86 (in Russian).

BERG, C.A., J. TIROSH and M. ISRAELI (1972), Analysis of short beam of fiber reinforced composites, in: *ASTM STP* **460** (American Society for Testing and Materials, Philadelphia, PA), 206.

BLAGONADJOZHIN, V.L., G.H. MURZAHANOV and V.P. NIKOLAEV (1976), *Test Methods for Composites and Structures* (MEI, Moscow) (in Russian).

BOLOTIN, V.V., I.I. GOL'DENBLAT and A.F. SMIRNOV (1972), *Structural Mechanics. State-of-Art and Future Perspectives* (Stroyizdat, Moscow) (in Russian).

CHAMIS, C.C. and J.H. SINCLAIR (1977), *Experimental Mech.* **17**, 339.

CHANDRA, R. (1976), *AIAA J.* **14**, 1130.
CHIAO, C.C., R.L. MOORE and T.T. CHIAO (1977), *Composites* **8**, 161.
CHIAO, T.T. and M.A. HAMSTAD (1976), Testing of fiber composite materials, in: *Proc. 1975 Internat. Conference Composite Materials*, Geneva, Switzerland and Boston, MA, USA, 1975 (The Metallurgical Society of AIME, New York) **2**, 884.
CHRISTIANSEN, A.W., J. LILLEY and J.B. SHORTALL (1974), *Fibre Sci. Tech.* **7**, 1.
DANIEL, J.M., J.L. MULLINEAUX, F.J. AHIMAZ and T. LIBER (1972), The embedded strain gage technique for testing boron/epoxy composites, in: *ASTM STP* **497** (American Society for Testing and Materials, Philadelphia, PA) 257.
EWINS, P.D. (1974), Techniques for measuring the mechanical properties of composite materials, in: *Composites – Standards, Testing and Design, Conference Proc.*, Nat. Phys. Labor., 1974 (IPC Science and Technology Press, Guildford), 144.
FOYE, R.L. (1967), *J. Composite Materials*, 194.
GORDON, J. (1968), *The New Science of Strong Materials or Why You Don't Fall Through the Floor* (Penguin Books, Harmondsworth).
GRESZCZUK, L.B. (1969), Shear-modulus determination of isotropic and composite materials, in: *ASTM STP* **460** (American Society for Testing and Materials, Philadelphia, PA), 140.
GRESZCZUK, L.B. (1973), Compressive strength and failure modes of unidirectional composites, in: *ASTM STP* **521** (American Society for Testing and Materials, Philadelphia, PA), 192.
HAHN, H.T. (1973), *J. Composite Materials* **7**, 383.
KACHANOV, L.M. (1976), *Polymer Mech.* **12**, 812.
KINCIS, T.YA. and R.P. SHLITSA (1978), *Polymer Mech.* **14**, 764.
KLEINER, W. and W. LÜSSMANN (1969), *Kunststoffe* **59**, 941.
KNIGHT, C.E., Jr. (1977), Failure analysis of the split-D test method, in: *ASTM STP* **617** (American Society for Testing and Materials, Philadelphia, PA), 201.
LEKHNITSKII, S.G. (1977), *Theory of Elasticity for an Anisotropic Body* (Nauka Press, Moscow) (in Russian).
LEMPRIERE, B.M., R.W. FENN, Jr., D.D. CROOKS and W.C. KINDER (1969), *AIAA J.* **7**, 2341.
LOMAKIN, V.A. (1978), *Mech. Solid Bodies*, 45 (in Russian).
LUBIN, G., ed. (1969), *Handbook of Fiberglass and Advanced Plastics Composites* (Van Nostrand Reinhold, New York).
MARKHAM, M.F. and D. DAWSON (1975), *Composites* **6**, 173.
MCKENNA, G.B., J.F. MANDELL and F.J. MCGARRY (1974), Interlaminar strength and toughness of fiberglass laminates, in: *Proc. 29th Annual Conf. SPI Reinforced Plastics/Composites Insitute*, Washington, D.C., 1974 (SPI, New York) 13-C.
MENGES, G. and R. KLEINHOLZ (1969), *Kunststoffe* **59**, 959.
MULLIN, J.V. and A.C. KNOELL (1970), *Materials Res. Standards* **10**, 16.
NIKOLAEV, V.P. (1971), *Polymer Mech.* **7**, 984.
NIKOLAEV, V.P. (1973), *Polymer Mech.* **9**, 675.
NIKOLAEV, V.P. and YU.N. NOVICHKOV (1968), Experimental determination of GFRP shear modulus, in: *Strength Calculations* **13** (Mashinostroyeniye, Moscow), 355 (in Russian).
PAGANO, N.J. and J.M. WHITNEY (1970), *J. Composite Materials* **4**, 360.
PANSHIN, B.I., L.P. KOTOVA and D.V. KOLCHEV (1970), *Factory Laboratory* **36**, 1371 (in Russian).
PARTSEVSKII, V.V. and A.YA. GOL'DMAN (1970) Mechanical testing of GFRP ring specimens, in: *Works of MEI, Dynamics and Machine Strength* **74** (MEI, Moscow), 125 (in Russian).
PETIT, P.H. (1969), A simplified method of determining inplane shear stress-strain response of unidirectional composites, in: *ASTM STP* **460** (American Society for Testing and Materials, Philadelphia, PA), 83.
POLYAKOV, V.A. and I.G. ZHIGUN (1978), *Polymer Mech.* **14**, 883.
POLYAKOV, V.A. and I.G. ZHIGUN (1979), *Mech. Composite Materials* **15**, 109.
PURSLOW, D. (1977), Aeronautical Research Current Paper No. 1381 (HMSO, London).
ROSEN, B.W. (1972), *J. Composite Materials* **6**, 552.
SBOROVSKII, A.K., V.D. POPOV, N.F. SAVELJEVA and A.V. LAVROV (1971), *Tech. Ship Building* **25**, 116 (in Russian).

SKUDRA, A.M., F.J. BULAVS and K.A. ROCENS (1975), *Kriechen und Zeitstandverhalten von verstärkten Plasten* (VEB Deutscher Verlag für Grundstoffindustrie, Leipzig).

STÖFFLER, G. (1980), *J. Composite Materials* **14**, 95.

SUMSION, H.T. and Y.D.S. RAJAPAKSE (1978), Simple torsion test for shear moduli determination of orthotropic composites, in: *Proc. 1978 Internat. Conference Composite Materials*, Toronto, Canada, 1978 (The Metallurgical Society of AIME, New York), 994.

TARNOPOL'SKII, YU.M. (1979), *Mech. Composite Materials*, 331 (in Russian).

TARNOPOL'SKII, YU.M. and T.YA. KINCIS (1981), *Methods for Static Testing of Reinforced Plastics* (Khimiya Press, Moscow) (in Russian).

TARNOPOL'SKII, YU.M., G.G. PORTNOV, YU.B. SPRIDZANS and V.V. BULMANIS (1973), *Polymer Mech.* **9**, 592.

TARNOPOL'SKII, YU.M. and A.M. SKUDRA (1979), Advanced test methods for composites, in: *Proc. USSR–Japan Symp. Composite Materials*, Moscow, 1979 (Moscow State University, Moscow), 326.

TARNOPOL'SKII, YU.M., I.G. ZHIGUN and V.A. POLYAKOV (1977), *Polymer Mech.* **13**, 52.

TERRY, G.A. (1979), *Composites* **10**, 233.

TSAI, S.W. (1965), *Trans. ASME* **B 87**, 315.

TSAI, S.W. and G.S. SPRINGER (1963), *Trans. ASME* **E 30**, 467.

VARUSHKIN, E.M. (1971), *Polymer Mech.* **7**, 925.

ZHIGUN, I.G. and V.V. MIKHAILOV (1978), *Polymer Mech.* **14**, 586.

ZHIGUN, I.G. and V.A. POLYAKOV (1978), *Properties of Spatially Reinforced Plastics* (Zinātne Press, Riga) (in Russian).

ZHIGUN, I.G., V.A. POLYAKOV and V.V. MIKHAILOV (1979), *Mech. Composite Materials*, 1111 (in Russian).

ZHIGUN, I.G., V.A. YAKUSHIN and YU.N. IVONIN (1976a), *Polymer Mech.* **12**, 640.

ZHIGUN, I.G., V.A. YAKUSHIN, V.V. TANEVSKII and V.V. MIKHAILOV (1976b), *Polymer Mech.* **12**, 112.

CHAPTER VI

Methods of Testing Composite Materials

I.M. Daniel

Illinois Institute of Technology
Chicago, IL 60616
U.S.A.

Contents

HANDBOOK OF COMPOSITES, VOL. 3 – Failure Mechanics of Composites
Edited by G.C. SIH and A.M. SKUDRA

1. Introduction

To design a composite structure and evaluate its safety and durability, one must know the structure geometry, the loading and environment to which the structure will be subjected, conduct a stress analysis, and apply relevant failure criteria. The complexity of loading conditions, structure geometry, and the inhomogeneity, anisotropy, and inelasticity of the material make the use of experimental methods indispensable for stress and failure analysis. Even when the stress analysis can be handled analytically it is necessary to characterize experimentally the basic material to provide the data needed in the analysis. Existing failure theories are either inadequate or cumbersome to apply in predicting static failure and fatigue life.

Specific applications of experimental methods in composite materials include the following:

(1) Characterization of constituent materials, i.e., fiber and matrix, for use in micromechanics analyses. Knowing these properties one can predict, in principle, the behavior of the lamina and hence of laminates and structures.

(2) Verification of micromechanics analyses, especially in the case of nonlinear, inelastic behavior, including effects of curing stresses, temperature, and moisture.

(3) Qualification and characterization of the basic unidirectional lamina which forms the building block of all laminated structures. This characterization, especially the determination of irreversible and ultimate properties, is essential to provide data for subsequent stress analysis using lamination theory.

(4) Strain (stress) analysis. Strain distributions are determined experimentally and stress distributions are obtained from these using the appropriate constitutive relations for the material.

(5) Fracture characterization, including identification of fracture mechanisms and modes, initiation, and propagation.

(6) Assessment of structural integrity and reliability. This assessment is done in service by continual monitoring using nondestructive techniques.

(7) Life prediction through accelerated testing.

A variety of experimental methods have been used for the various applications above. Most of these deal with measurement of deformations or strains in composite materials. Some employ modeling procedures and others are applied directly to the prototype composite. Some give point-per-point information and others give full-field representations. The most commonly used experimental methods are:

(1) Photoelastic methods. These include two- and three-dimensional,

microphotoelastic, and dynamic photoelastic modeling techniques, anisotropic photoelasticity using birefringent fibrous composites, and birefringent coatings applied directly to the prototype.

(2) Strain gages.

(3) Moiré methods.

(4) Holographic and interferometric methods.

(5) Nondestructive evaluation methods, including ultrasonics, acoustic emission, X-ray radiography, and thermography.

Reviews of various experimental methods applicable to composites have been given in the literature (DANIEL and ROWLANDS (1972), BERT (1974), DANIEL (1974, 1975 a, b), WHITNEY et al. (1982)). Brief reviews of these methods are given in the following section along with specific applications to testing of composite materials.

2. Experimental methods

2.1. *The photoelastic method*

The method makes use of transparent models viewed in the polariscope. In the case of two-dimensional models, stresses are related to the fringe order or retardation in wavelengths by the 'stress-optic' law:

$$\sigma_1 - \sigma_2 = 2nf/t \tag{1}$$

where σ_1, σ_2 = principal stresses in plane of the model; n = fringe order; f = material fringe value (constant for material); t = specimen thickness.

In three-dimensional applications, a three-dimensional scaled model of the prototype is machined out of certain polymeric materials having the desired 'stress-freezing' properties. The model is loaded inside an oven and slowly heated to a 'critical' temperature, held there for some time and finally, slowly cooled down to room temperature. This process 'freezes' or 'locks-in' the deformations at the elevated temperature. This state of stress is not disturbed by careful slicing of the model. Slices are analyzed in the same manner as two-dimensional models.

To resolve the stress components individually from photoelastic data, auxiliary methods are used. One of these, the shear difference method, is based on numerical integration of the equations of equilibrium (FROCHT (1941)).

Another three-dimensional method is the scattered-light method. This method is nondestructive and the experiments can be conducted at room temperature. The specimen is illuminated by an intense polarized beam, such as that emitted by a laser, and the scattered light fringe pattern is viewed in a direction perpendicular to the primary beam. This fringe pattern is related to the state of stress in the interior of the three-dimensional model.

2.2. *Anisotropic photoelasticity*

This method combines the photoelastic approach with more realistic modeling of the prototype fibrous composites (PIH and KNIGHT (1969), SAMPSON (1970), DALLY and PRABHAKARAN (1971), BERT (1972), PIPES and ROSE (1974), KNIGHT and PIH (1976), DANIEL (1975b), DANIEL et al. (1981)). Macromechanical stress analysis can be conducted by means of anisotropic photoelasticity using transparent birefringent composites. These materials are glass-fiber reinforced plastics with the matrix and the fibers having the same index of refraction. Transparent fibrous composites can be made to simulate the anisotropy of opaque fibrous composites such as boron/epoxy, graphite/epoxy, etc. These transparent composites are treated as homogeneous materials with anisotropic elastic and optical properties.

The stress-optic law for an orthotropic material takes the form (SAMPSON (1970)):

$$N = \sqrt{\left(\frac{\sigma_x}{2f_x} - \frac{\sigma_y}{2f_y}\right)^2 + \left(\frac{2\tau_{xy}}{f_{xy}}\right)^2} \tag{2}$$

where N = birefringence per unit thickness; f_x, f_y, f_{xy} = three principal material fringe values; and the x- and y-directions are those of the principal material axes. The components of birefringence

$$N_x = \frac{\sigma_x}{2f_x}, \qquad N_y = \frac{\sigma_y}{2f_y}, \qquad N_{xy} = \frac{\tau_{xy}}{f_{xy}} \tag{3}$$

follow the same transformation relations as the corresponding stress components.

2.3. *Photoelastic coating method*

This method consists of bonding a thin sheet of birefringent material to the surface of the composite specimen, so that the bonding interface is reflective. When the specimen is loaded, the surface strains are transmitted to the coating and produce a fringe pattern which is recorded and analyzed by means of a reflection polariscope. For perfect strain transmission

$$\varepsilon_1^c = \varepsilon_1^s, \qquad \varepsilon_2^c = \varepsilon_2^s \tag{4}$$

where the superscripts c and s refer to the coating and specimen, respectively. The difference in principal strains and the coating birefringence are related by the strain-optic law:

$$\varepsilon_1^c - \varepsilon_2^c = \varepsilon_1^s - \varepsilon_2^s = \frac{Nf_\varepsilon}{2h} = NF_\varepsilon \tag{5}$$

where f_ε is the strain fringe value, F_ε the coating fringe value, h the coating thickness, and N the fringe order.

The application of coatings to composite materials is based on the same principles. The surface strain field of the anisotropic composite produces a photoelastic response in the isotropic coating. The birefringence in the coating is related to the difference in principal strains by the strain-optic law above. The stresses in the composite are computed from the strains using the appropriate anisotropic stress-strain relations.

When the composite material is orthotropic with the elastic axes of symmetry 1, 2 coinciding with the axes of material, geometric and loading symmetry, the analysis is simplified. Then, the coating birefringence is related to the principal stresses in the composite by

$$\frac{NF_\varepsilon}{2h} = \frac{\sigma_{11}^{s}}{E_{11}}(1+\nu_{12}) - \frac{\sigma_{22}^{s}}{E_{22}}(1+\nu_{21}) \tag{6}$$

where $E_{11}, E_{22}, \nu_{12}, \nu_{21}$ are the elastic constants referred to the axes of elastic symmetry.

One of the major limitations in the application of birefringent coatings is the effect of Poisson's ratio mismatch (POST and ZANDMAN (1961)). If Poisson's ratio of the coating is different from that of the substrate a distortion in the displacement field is produced in the coating which is especially pronounced at the boundaries.

For a specially orthotropic laminate, it can be assumed that the maximum principal strain is parallel to the boundary and is transmitted to the coating ($\varepsilon_1^c = \varepsilon_1^s$) and that the minimum principal strain is given by $\varepsilon_{22}^s = -\nu_{12}^s\varepsilon_{11}^s$. The transverse strain in the coating at the boundary varies between

$$\varepsilon_{22}^c = \varepsilon_{22}^s = -\nu_{12}^s\varepsilon_{11}^s \tag{7}$$

at the interface, and

$$\varepsilon_{22}^c = -\nu^c\varepsilon_{11}^c = -\nu^c\varepsilon_{11}^s \tag{8}$$

at the free surface. It has been shown experimentally that Equation 8 above represents fairly accurately the average transverse strain in the coating (DALLY and ALFIREVICH (1969)). Thus, the principal strain along the boundary can be obtained from the relation

$$\varepsilon_{11}^s = \frac{Nf_\varepsilon}{2h}\frac{1}{1+\nu^c} \tag{9}$$

and the nonzero principal stress along the boundary is

$$\sigma_{11}^s = \frac{E_{11}}{1+\nu^c}\frac{Nf_\varepsilon}{2h}. \tag{10}$$

Difficulties arise also in a small transition region near the boundary, where the photoelastic response is a function of both Poisson's ratios, those of the coating and the composite. For a uniaxial state of stress the strain can be expressed (DALLY and ALFIREVICH (1969)) as

$$\varepsilon_{11}^{s} = \frac{Nf_{\varepsilon}}{2h}\left[\frac{1}{1+\nu_{12}^{s}+C_{\nu}(\nu^{c}-\nu^{s})}\right] \tag{11}$$

where C_{ν} is a correction factor ranging in value from 1 at the boundary to 0 away from it. The transition region extends for approximately four coating thicknesses from the boundary. The transition region, of course, is reduced by decreasing the coating thickness and is completely eliminated when Poisson's ratio of the composite equals that of the coating.

2.4. *Electrical resistance strain gages*

Electrical resistance strain gages are very sensitive for measuring deformations in composite materials. Strain measurement is based on the electrical resistance change of the gage bonded to the part undergoing deformation. The strain is given by

$$\varepsilon = \frac{1}{S_{g}}\left(\frac{\Delta R}{R}\right) \tag{12}$$

where S_{g} is the gage factor, a calibration constant which is a function of the gage alloy and the backing materials (DALLY and RILEY (1965)). Commercially available foil gages come in a variety of sizes, configurations and combinations (rosettes), and are very suitable for composite materials.

In many strain gage applications, test conditions are not isothermal. Hence, techniques must be employed to compensate for free thermal strains the material exhibits due to a change in temperature. Although this phenomenon is also encountered in isotropic materials, the anisotropy of composite materials requires that temperature compensation of these materials receive special considerations. A change of ΔT in ambient temperature produces the following resistance change:

$$\left(\frac{\Delta R}{R}\right)_{\Delta T} = (\alpha_{\theta}-\beta)S_{g}\Delta T + \gamma\Delta T \tag{13}$$

where α_{θ} is the coefficient of thermal expansion of the composite in the direction parallel to the gage axis, β is the thermal coefficient of expansion of the gage, and γ is the temperature coefficient of resistivity of the gage material. For experiments where the temperature variation exceeds $\pm 10°C$, it is necessary to take into account the temperature dependence of the gage factor S_{g}, i.e., the change in strain sensitivity of the gage alloy.

A practical approach for temperature compensation of strain gages on composite materials relies on the use of a reference material of known thermal expansion. For every type of gage applied to a composite material a similar gage is applied to the reference material, which undergoes the same temperature history as the composite. Then, the true thermal strain in the composite is given by

$$\varepsilon_t = \varepsilon_a - \varepsilon_r + \varepsilon_{tr} \tag{14}$$

where ε_t is the true strain, ε_a the apparent (uncorrected) strain in the composite, ε_r the apparent strain in the reference material, and ε_{tr} the known $(\alpha_r \Delta T)$ thermal expansion of the reference material. Reference materials used normally are aluminum oxide, fuzed quartz, and titanium silicate with coefficients of thermal expansion of $6.8 \times 10^{-6}\,K^{-1}$ (3.8 μin./in./°F), $0.7 \times 10^{-6}\,K^{-1}$ (0.4 μin./in./°F), and $0.03 \times 10^{-6}\,K^{-1}$ (0.017 μin./in./°F), respectively. Fig. 1 shows the thermal response of strain gages, temperature-compensated for steel,

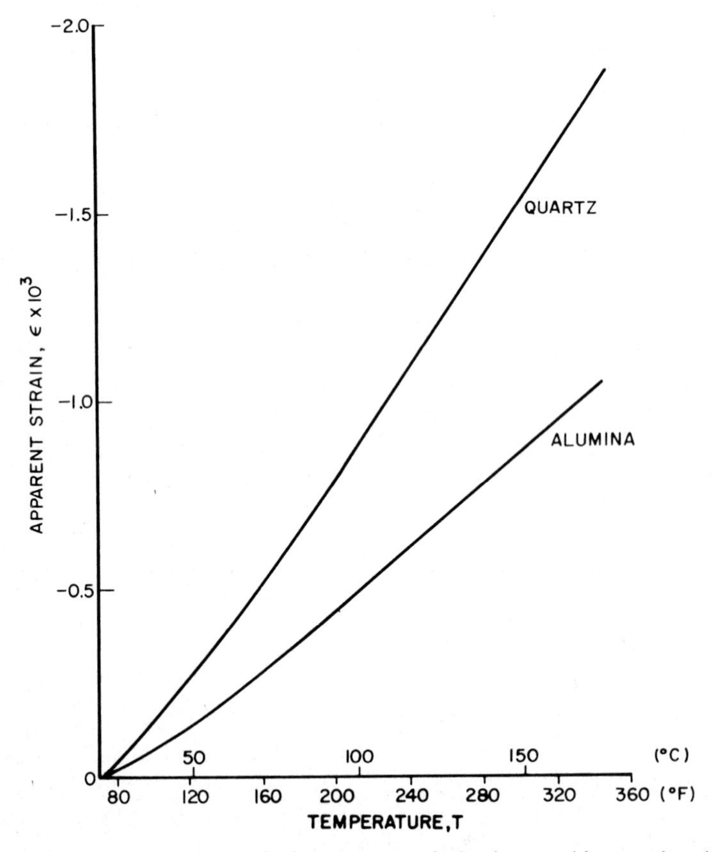

FIG. 1. Outputs of strain gages attached to quartz and aluminum oxide as a function of temperature (coefficients of thermal expansion for quartz and alumina are $0.7 \times 10^{-6}\,K^{-1}$ [0.4×10^{-6} in./in./°F] and $6.8 \times 10^{-6}\,K^{-1}$ [3.8×10^{-6} in./in./°F], respectively).

mounted on aluminum oxide and fuzed quartz. To minimize the purely thermal response of the gage it is preferable to use gages with zero-expansion temperature compensation ($\beta \cong 0$). The thermal response of such a gage mounted on a titanium silicate specimen is illustrated in Fig. 2.

The gage factor given by the manufacturer is defined as

$$S_g = \frac{\Delta R/R}{\varepsilon_{xx}} \quad \text{when } \varepsilon_{yy} = -0.285\,\varepsilon_{xx}\,. \tag{15}$$

It is apparent that if the gage factor above is used to determine strain in any strain field other than the one under which S_g was determined, an error will result. The magnitude of this error can vary from insignificant to appreciable, depending on the case. It is possible to correct the error by taking the cross-sensitivity factor of the gage into consideration. This factor is defined as

$$K = \frac{S_A}{S_T} \tag{16}$$

where S_A and S_T are the axial and transverse strain sensitivities defined as

$$S_A = \frac{\Delta R/R}{\varepsilon_{yy}} \quad \text{when } \varepsilon_{xx} = 0 \tag{17}$$

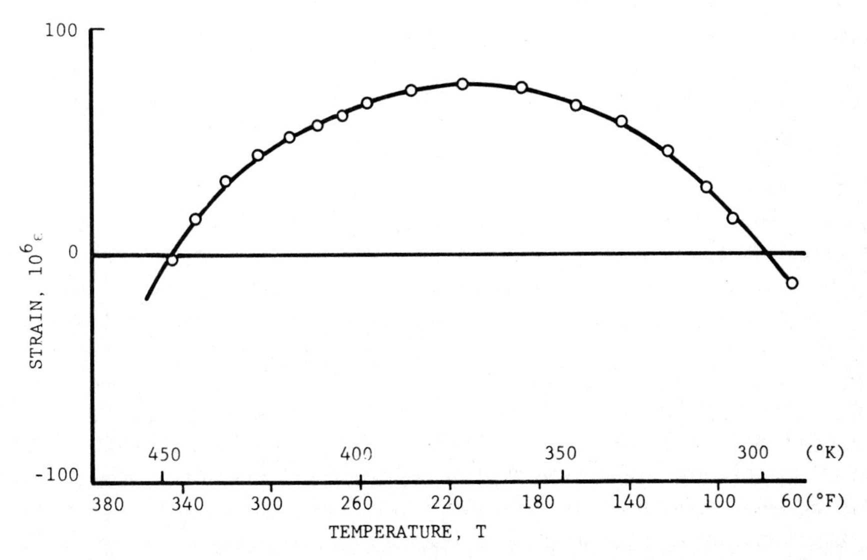

FIG. 2. Apparent strain as a function of temperature of WK-00-125TM-350 gage bonded on titanium silicate.

and

$$S_{\mathrm{T}} = \frac{\Delta R/R}{\varepsilon_{xx}} \quad \text{when } \varepsilon_{yy} = 0\,. \tag{18}$$

For commonly used gages the cross-sensitivity factor K ranges between ± 0.01 and ± 0.02.

Taking the cross-sensitivity factor into consideration, the true strains at a point along two mutually perpendicular directions are obtained from the apparent strains as follows:

$$\begin{aligned} \varepsilon_{xx} &= \frac{1-\nu_0 K}{1-K^2}\,(\varepsilon'_{xx} - K\varepsilon'_{yy})\,, \\ \varepsilon_{yy} &= \frac{1-\nu_0 K}{1-K^2}\,(\varepsilon'_{yy} - K\varepsilon'_{xx}) \end{aligned} \tag{19}$$

where ε_{xx}, ε_{yy} are the actual strains, ε'_{xx} and ε'_{yy} are the apparent strains and $\nu_0 = 0.285$ is Poisson's ratio of the gage calibration material. The transverse sensitivity correction becomes appreciable whenever the strain to be measured is small compared to the strain in the perpendicular direction. This situation is encountered frequently in composites.

In the case of Poisson's ratio determination the true value is related to the uncorrected value as follows:

$$\nu = -\frac{\varepsilon_{xx}}{\varepsilon_{yy}} = -\frac{\varepsilon'_{xx} - K\varepsilon'_{yy}}{\varepsilon'_{yy} - K\varepsilon'_{xx}} = \frac{\nu' + K}{1 + \nu' K} \cong \nu' + (1-\nu'^2)K \cong \nu' \pm 0.02 \tag{20}$$

where ν = actual Poisson's ratio; ν' = uncorrected Poisson's ratio.

The correction above is meaningful for the determination of the minor Poisson's ratio, ν_{21}, in unidirectional composites.

Composite laminates are amenable to embedment of strain gages between various plies during the fabrication process. This allows for strain measurements in the interior of composite laminates, monitoring of strains during curing and at elevated temperatures. Techniques have been developed and applied to the measurement of subsurface strains in boron/epoxy, boron/polyimide, graphite/epoxy, graphite/polyimide and glass/epoxy laminates (DANIEL et al. (1972, 1975), DANIEL and LIBER (1975, 1976 a, b, 1977 a, b)). It was demonstrated that conventional foil gages when embedded do not produce local thickening of the specimen, do not reduce the strength of the material, and do not affect the average mechanical properties. Valid strain readings were obtained up to failure of the material (DANIEL et al. (1972)). Fig. 3 shows a cross-section through a $[0/\pm 45/\bar{0}]_s$ boron/epoxy specimen with embedded gages. Fig. 4 shows strain gages with ribbon leads attached located on an interior ply of the same laminate.

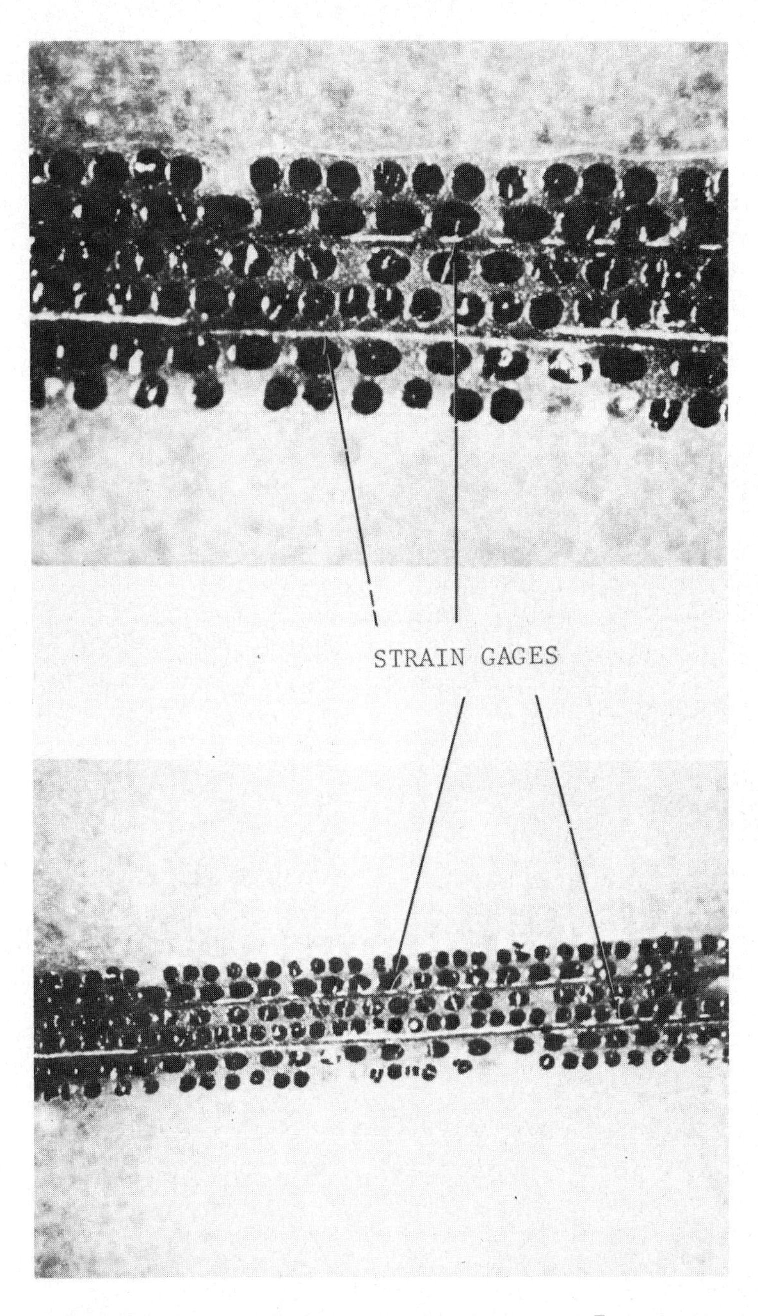

FIG. 3. Section through a gage location of an instrumented $[0/\pm45/\bar{0}]_s$ boron/epoxy specimen showing embedded 3-gage rosettes on the second and fourth plies. (Note the separate sections of the gage grid exposed below the second ply.)

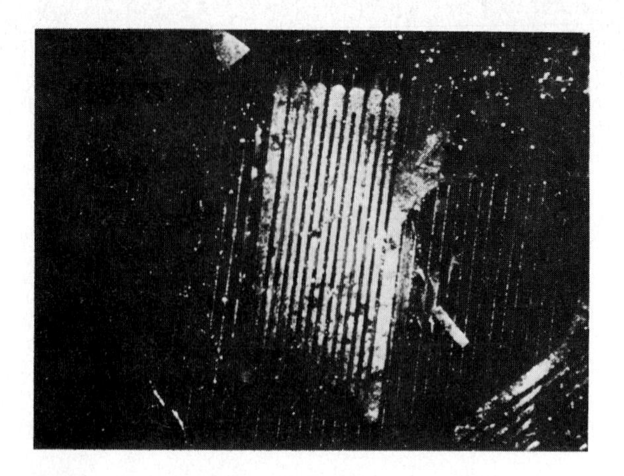

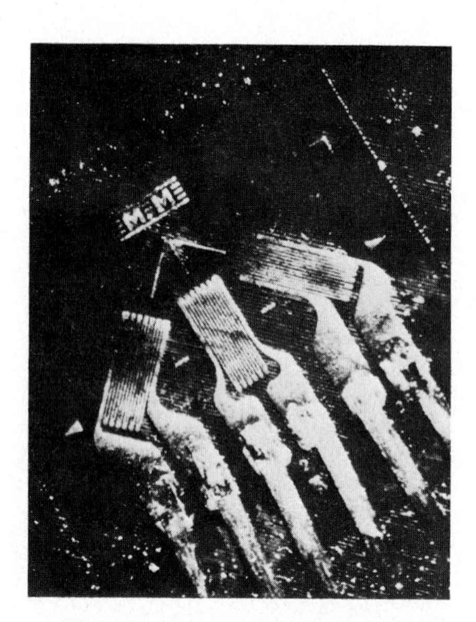

FIG. 4. Strain gages with ribbon lead wire located on boron/epoxy composite.

2.5. *Moiré methods*

The moiré effect is an optical phenomenon observed when two closely spaced arrays of lines are superimposed and viewed with either transmitted or reflected light. If the two arrays consist of opaque parallel lines which are not identical in either spacing (pitch) or orientation, then interference between the two arrays occurs and moiré fringes are produced. The method yields full-field information of the in-plane surface displacements from which strain and stress fields can be derived. Being strictly geometric in nature, the technique is independent of the anisotropy and inelasticity of the material.

The sensitivity of the moiré method depends on the density of the rulings used and on the magnitude of the deformation. Glass/epoxy composites in general, and graphite/epoxy and boron/epoxy laminates with a large number of ±45-deg plies, undergo sufficient deformation to yield analyzable moiré fringes using model arrays of 40 lines/mm (1000 lpi) and superimposing a similar master array during the test. However, this line density represents a practical upper limit for model arrays and, in many cases, the deformations to be measured are small. Therefore, the sensitivity of the moiré method must be enhanced by other means, such as fringe interpolation and multiplication techniques.

The application of moiré techniques in general consists of three basic tasks:

(1) application of rulings or grids to the specimen,

(2) photographic recording of information in the form of fringe patterns or deformed grids, and

(3) analysis of moiré data.

In conventional practice one array or a grid of lines is applied to the specimen surface by photographic, etching, transfer, or bonding techniques. A practical upper limit of line density is 40 lines/mm (1000 lpi). In some applications to translucent glass/epoxy laminates, film replicas of these arrays can be bonded directly to the specimen with the emulsion side outward, without any special preparation of the specimen surface. In the case of opaque materials, such as most composites, the surface is prepared with a smooth and partially reflective substrate and the line array or grid is either bonded or photoprinted onto this surface by the photoresist process. A transparent reference array (master grid) is then placed near or in contact with the model. As the specimen is loaded, its array of lines deforms to follow the surface displacements and a moiré interference pattern is formed. This pattern which is a measure of the displacement field is recorded photographically. Sometimes it is more suitable to photograph the specimen grid in the no-load and load conditions and form the moiré patterns on the bench subsequently. Moiré patterns are related to the displacement field since they represent loci of points having the same component of displacement normal to the line array. Strains are obtained from the displacements by graphical, optical (mechanical), or numerical differentiation techniques. Moiré techniques, including that of fringe multiplication, and their application to composites have been discussed previously (DANIEL and ROWLANDS (1972), DANIEL et al. (1973)).

2.6. Holographic and interferometric methods

The holographic method is based on the optical interference produced by superposition of coherent light waves reflected from the object under consideration (object beam) and those of a coherent reference beam. The laser is an ideal source of coherent monochromatic light. The normal procedure is to record the interference pattern on film (hologram) and subsequently view it with a similar laser beam. The emerging light from the hologram contains the wave patterns characteristic of the original object.

One of the important applications of holography to structural and stress analysis is interferometry, i.e., the measurement of small surface displacements in a body produced by mechanical or thermal loadings. In applying holographic interferometry, two holograms of the test object are recorded on the same photographic plate with an alteration of the object surface occurring between the two recordings. The surface of the test object can be altered in several ways, including mechanical loading, pressurization, thermal loading, and acoustic vibration. Upon reconstruction, the two images will interfere with each other and will produce a set of fringes representing contours of displacement in the direction of the viewing axis. Each fringe represents a relative displacement with respect to its neighboring fringe of approximately one-half the wavelength of the light used in reconstruction. For a helium-neon laser the sensitivity of holographic interferometry is approximately 0.3 μm. In addition to the full-field nature and the high sensitivity, the method has the added advantage that it does not require any special surface preparation and can be applied to surfaces of nonoptical quality such as those of composites (ROWLANDS and DANIEL (1972), ROWLANDS et al. (1974b), ERF (1974), SANDECKYJ et al. (1978), MADDUX and SENDECKYJ (1979)).

An improved holographic technique which has many advantages is image-plane holography. In this approach the light waves emanating from the object are collected by an imaging lens and the film plate is placed near the image plane. This technique allows the use of white light for reconstruction and results in sharper fringe patterns. It can also be applied to larger specimens than those used in conventional holography.

Holographic techniques commonly used in experimental mechanics fall into four basic categories:

(1) Double-exposure holographic interferometry;
(2) Real-time holographic interferomery;
(3) Time-average or vibration holographic interferometry;
(4) Dynamic double-exposure pulsed holographic interferometry.

Since holographic techniques are best suited for measuring out-of-plane displacements most applications to composites deal with flexure and vibration of plates. An important application of holography is NDE of material integrity. The technique relies on the fact that if the state of stress in the component is changed, the surface displacements will be altered locally around surface or near surface defects. This alteration is manifested as a local anomaly in the overall fringe pattern. This nondestructive detection is performed either by double-exposure or real-time holography. Cracking and delamination in composite laminates and bond defects in composite sandwich panels or adhesive joints can be detected. The type of loading applied to reveal the presence of a defect depends on the type of defect sought, the material properties, and the component geometry.

Another interferometric technique applicable to composites is speckle interferometry. It makes use of the speckle pattern produced on the surface of an object illuminated by coherent light. It has many characteristics complementary to those of holographic interferometry. It is less sensitive than the latter and is primarily indicative of in-plane displacements.

The most important development in this area is that of speckle-shearing interferometry introduced by HUNG and TAYLOR (1973). This technique allows direct determination of derivatives of surface displacements. It overcomes or alleviates many of the limitations of conventional interferometry, namely:

(1) the setup is relatively simple and does not require laborious alignment of optics,

(2) mechanical and environmental stability are not critical,

(3) coherent length of light is minimized,

(4) sensitivity can be controlled over a wider range,

(5) films of much lower resolution can be used,

(6) the fringes always localize on the specimen surface, and

(7) strains can be recorded directly without the need of differentiating displacements.

More recent developments introduced by HUNG et al. (1975, 1978) allow the simultaneous determination of derivatives of surface displacements along any direction and with variable sensitivity using a single photographic record (shearing-specklegram).

2.7. Nondestructive evaluation methods

Flaws can be introduced in composite laminates during processing and fabrication. These include contaminants, voids, unpolymerized resin, nonuniform matrix distribution, incorrect fiber orientation, inclusions, ply gaps, delaminations, and crazing. Sources of stress concentration and damage growth, such as holes or cutouts of various sizes and shapes, are also introduced for design functions. Defects are also induced or enhanced in service with age, loading, and environmental conditions. These cases include matrix aging and degradation, moisture absorption, ultraviolet radiation, damage zones around initial stress concentrations, and surface gouges and scratches.

A variety of nondestructive evaluation (NDE) techniques are used for evaluating the integrity of composite materials. They include radiographic (X-ray and neutron), optical (moiré, birefringent coatings, holography, and interferometry), thermographic, acoustic (acoustic wave, acoustic emission), embedded sensor, and ultrasonic techniques. These various methods are best suited for detecting different types of flaws. Reviews of the state of the art of NDE of composites have been given by DANIEL and LIBER (1979) and MATZKANIN et al. (1979).

The most widely used method of flaw detection and characterization is the ultrasonic one (HAGEMAIER (1974), HARRIGAN (1978), KULKARNI et al. (1978), LIBER et al. (1979b), MOOL and STEPHENSON (1971), PETTIT and LAURAITIS (1980), ROSE et al. (1973), SENDECKYJ et al. (1978), SHELDON (1978), WHITESIDE et al. (1973)). It is based on the attenuation of high-frequency sound passing through the specimen. The attenuation results from three sources, i.e., viscoelastic effects in the resin matrix, geometric dispersion caused by material heterogeneity, and geometric attenuation caused by internal defects such as delaminations and cracks. The effects of the latter are maximized by proper selection of the sound wave frequency.

The transducers and the specimen are immersed in a tank of water to provide a uniform coupling medium for transmission of the ultrasonic waves. In some applications to large components in the aerospace industry the parts are not immersed. Coupling medium is provided by a steady jet of water bridging the gap between the transducer and part.

Two modes of scanning are normally used. In the through transmission mode two transducers are used: a transmitting one in front of the specimen and a receiving one in the back of the specimen (Fig. 5). The presence of defects is related to the attenuation of the received signal. In the pulse-echo mode the ultrasound transmitted by the single transducer is received by the same transducer after reflection from the back surface of the specimen (Fig. 6). The attenuation of the reflected pulse is related to internal defects in the laminate. A variation of the two techniques above is the pitch-catch method which employs two transducers on the same side of the specimen. Pulses emitted by the transmitting transducer at an angle to the specimen surface are received by the receiving transducer after reflection from the back surface of the specimen.

Ultrasonic inspection records can be obtained in many forms such as an amplitude-time display at a specific point on the specimen (A-scan), cross-

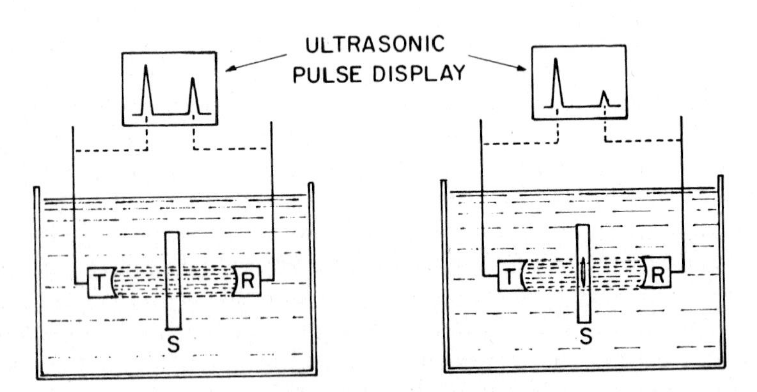

FIG. 5. Ultrasonic through transmission method.

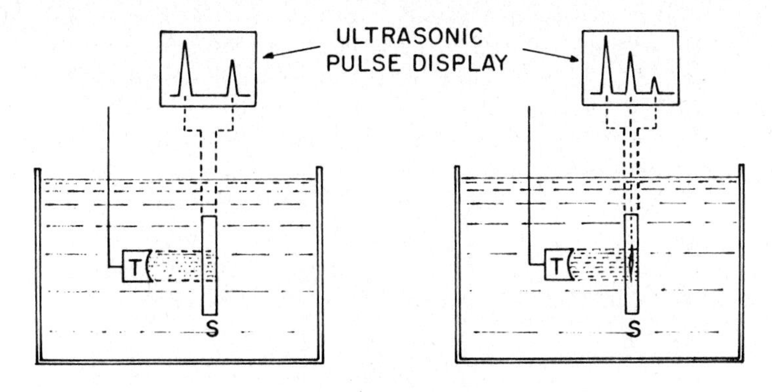

FIG. 6. Ultrasonic pulse echo method.

sectional view of the specimen along a scan line (B-scan), and a series of scans covering the surface of the specimen and giving a plan view image (C-scan). Examples of A-scans and C-scans are shown in Figs. 7 and 8. Two modes of recording C-scans are shown. In the first mode, the pen-lift mode, an alarm circuit with a limit is used. The limit is adjusted to lift the pen of the X-Y recorder whenever the gated peak voltage of the received pulse becomes smaller than a preset value corresponding to known flaws of a standard specimen (Fig. 8a). In the second mode, the analog mode, the gated peak voltage of the received pulse is recorded as a deflection of the pen normal to the scanning direction. The variations of this deflection are related to the presence of flaws. A typical analog record of the same specimen shown in Fig. 8(a) is shown in Fig. 8(b). Fig. 8(c) shows an offset analog scan, where a component of the X-axis signal is fed into the Y-axis signal resulting in a perspective view of the specimen.

Computer-aided procedures have been developed for recording and for more sophisticated processing of ultrasonic data (ELSLEY (1978)). Plotting of the data in any desired form can be done at any time after inspection using different discriminator levels to detect smaller or larger flaws.

X-ray radiography is also a widely used method for inspecting composite laminates (CHANG et al. (1975), DOMANUS and LILHOLT (1978), SENDECKYJ et al. (1978)). Radiographic methods were applied to the study of fatigue damage around holes in angle-ply boron-reinforced epoxy laminates (RODERICK and WHITCOMB (1976)). X-ray radiographs clearly show the sites of fiber breaks in thin laminates. The most effective means of detecting damage consisting of cracks and delaminations in graphite/epoxy and hybrid composites is by enhancing the X-ray method with opaque penetrants such as tetrabromoethane (TBE), di-iodobutane (DIB), and zinc iodide solution.

Acoustic emission is also classified as a nondestructive method. It is based on

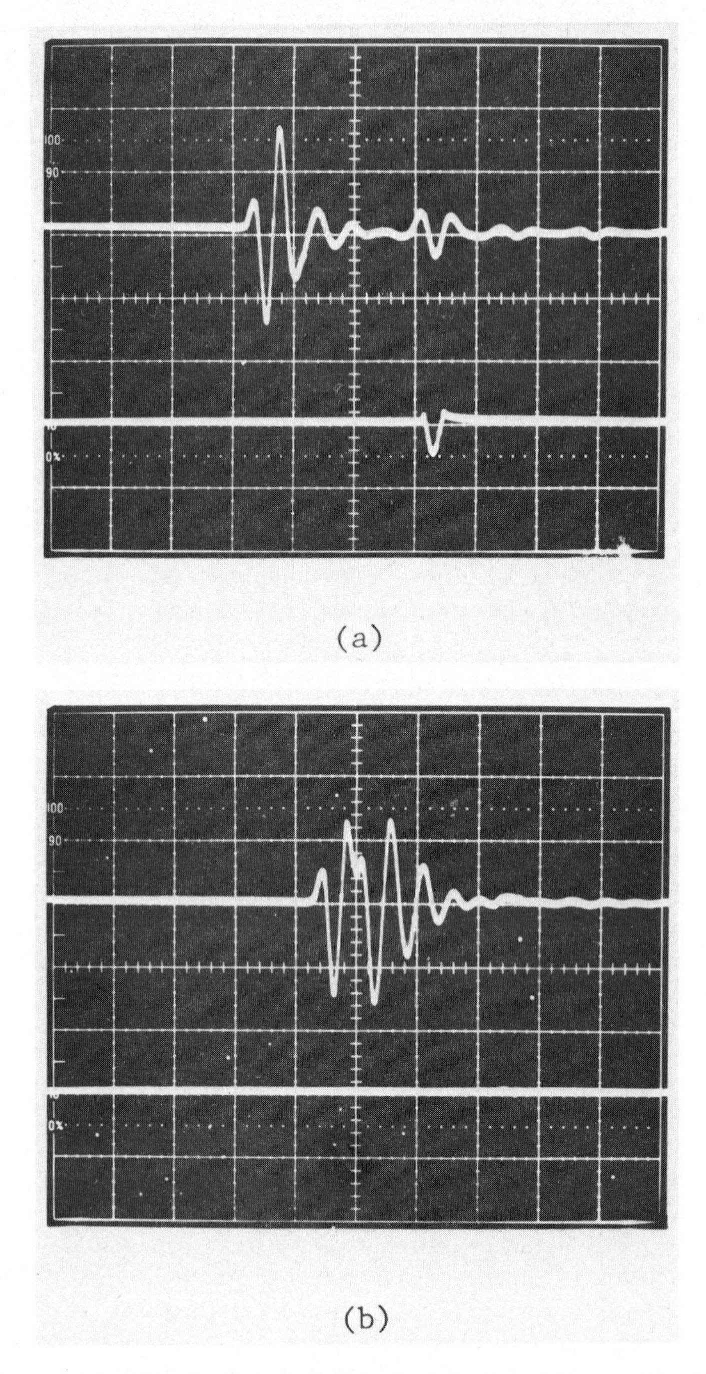

FIG. 7. Records of reflected ultrasonic pulses through the thickness of a $[(0/\pm45/90)_s]_2$ graphite/epoxy laminate. (a) Unflawed section, (b) at point with embedded film. (Upper traces: full pulse; lower traces: gated portion.)

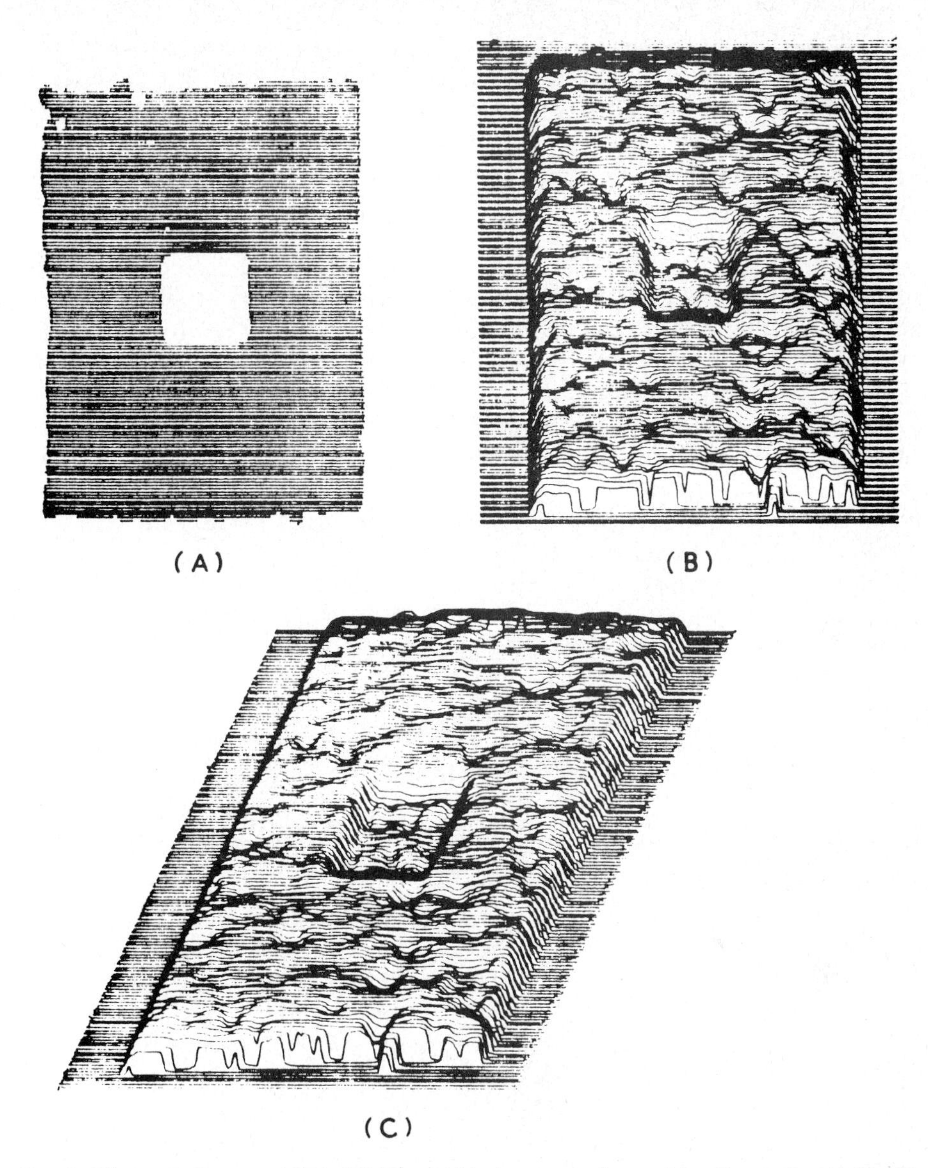

FIG. 8. Ultrasonic C-scans of $[(0/\pm45/90)_s]_2$ graphite/epoxy specimen with a film patch. (A) Pen-lift scan, (B) Analog scan (normal), (C) Analog scan (perspective).

the phenomenon that the sudden release of energy inside a material results in emission of acoustic pulses. Energy release occurs as a result of deformations or failure processes caused by loading. The primary failure mechanisms in fiber composites are matrix cracking, fiber debonding, and fiber breakage. The acoustic signals are detected by piezoelectric transducers in contact with the specimen through a coupling medium, electronically processed, and recorded.

The usual procedure is to count the number of pulses above a preset amplitude threshold. The result can be recorded and presented in terms of a cumulative number of counts, which indicates the extent of damage, or rate of counts, which is related to the rate of damage growth. The various failure mechanisms in composites produce signals of different amplitudes. Thus, fiber breakage produces a higher acoustic emission activity than fiber debonding, which in turn produces more measurable counts than matrix cracking.

To be able to detect and discriminate defects and fracture modes, it is not sufficient to measure number and rate of acoustic emission counts. Attempts have been made to discriminate and identify the various microfracture modes by amplitude distribution and frequency analysis of the acoustic pulses (ROTEM (1977)). Computer pattern recognition techniques have been used to analyze acoustic emission signals and to correlate frequency and amplitude distributions with physical and mechanical properties of the composite materials, such as moisture content and strength (GRAHAM and ELSLEY (1978)). To date, however, no quantitative failure criteria exist in terms of easily measurable acoustic emission parameters.

Other less frequently used techniques include thermographic techniques using liquid crystals and infrared cameras, and electromagnetic techniques using microwaves.

3. Micromechanics

3.1. Introduction

The behavior of composite structures subjected to loading and environmental fluctuations is intimately related to the micromechanics of load transfer between the constituent parts of the composite, i.e., fiber and matrix.

Matrix stresses on a transverse plane of a unidirectional composite arise due to matrix shrinkage during curing, differential thermal expansion, moisture absorption, and external loading. For relatively low fiber volume ratios, resin shrinkage produces compressive radial stresses and longitudinal shear stresses around the fibers in the matrix. The fibers themselves are subjected to radial and longitudinal compression. Shrinkage stresses are usually studied by means of two-dimensional photoelastic models of the transverse cross section of the composite.

Transverse tensile loading of a unidirectional composite results in high strain concentrations in the matrix. In this case the matrix stresses and strains are the governing criteria of failure. The transverse tensile behavior of unidirectional composites is greatly influenced by the residual stresses produced by matrix shrinkage, temperature, and moisture absorption. The extent to which these parameters influence the behavior of the composite is very important and merits careful experimental investigation.

A great deal of analytical work has been reported on the micromechanics of unidirectional composites (KIES (1962), HILL (1963, 1964), HASHIN and ROSEN (1964), ADAMS and DONER (1967), HASHIN (1972), CHRISTENSEN (1979)). Related experimental work has consisted primarily of two-dimensional photoelastic studies. Shrinkage stresses around inclusions have been studied by DANIEL and DURELLI (1961, 1962), and KOUFOPOULOS and THEOCARIS (1969) by means of two-dimensional models. The effects of external loading were studied by SAMPSON (1962) and DANIEL (1966, 1969). An extensive three-dimensional study of the effects of shrinkage and external loading was conducted by MARLOFF and DANIEL (1969) using a realistic three-dimensional fiber-reinforced composite model.

3.2. Two-dimensional model studies

Stresses on a transverse plane of a unidirectional composite, such as shrinkage or loading stresses, are usually studied by means of two-dimensional photoelastic models. Models simulating the transverse cross section of a unidirectional composite are prepared by casting epoxy around an array of disk inclusions. Upon curing, the matrix shrinks around the inclusions producing a state of stress manifested in the form of isochromatic fringe patterns.

The state of residual (shrinkage) stress in the model and prototype composite specimens can be determined photoelastically and expressed in dimensionless form by dividing the actual stresses by the nominal interface pressure around a single isolated inclusion. This dimensionless stress distribution is independent of size of inclusion, matrix material properties, and amount of effective shrinkage.

The same two-dimensional models used for determination of shrinkage stresses are used for evaluation of stress distributions due to external loading. The specimens are loaded in uniaxial compression in the plane of the model as shown in Fig. 9. Birefringence readings and photographs of fringe patterns for a model of a composite of 0.50 fiber volume ratio are shown in Fig. 10. The data of interest in testing such models are the maximum fringe order n_1 at the boundary of one of the central inclusions of the array along the vertical diameter and the fringe order n_0 at a similarly located point around the single isolated inclusion. The variation of these two fringe orders with applied stress is shown in Fig. 11. The horizontal intercepts on the fringe order axis represent initial birefringence due to residual stress.

The state of stress in the matrix of the two-dimensional specimens is a complex three-dimensional one due to the out-of-plane restraint introduced by the inclusions at their interface. Away from the interface, the state of stress tends to approach the plane stress condition. In two-dimensional photoelastic analyses, the birefringence measured is related to the in-plane stresses averaged over the thickness (optical path) of the specimen. Stress-strain relations can be expressed in terms of these values by integrating them through the thickness of the specimen.

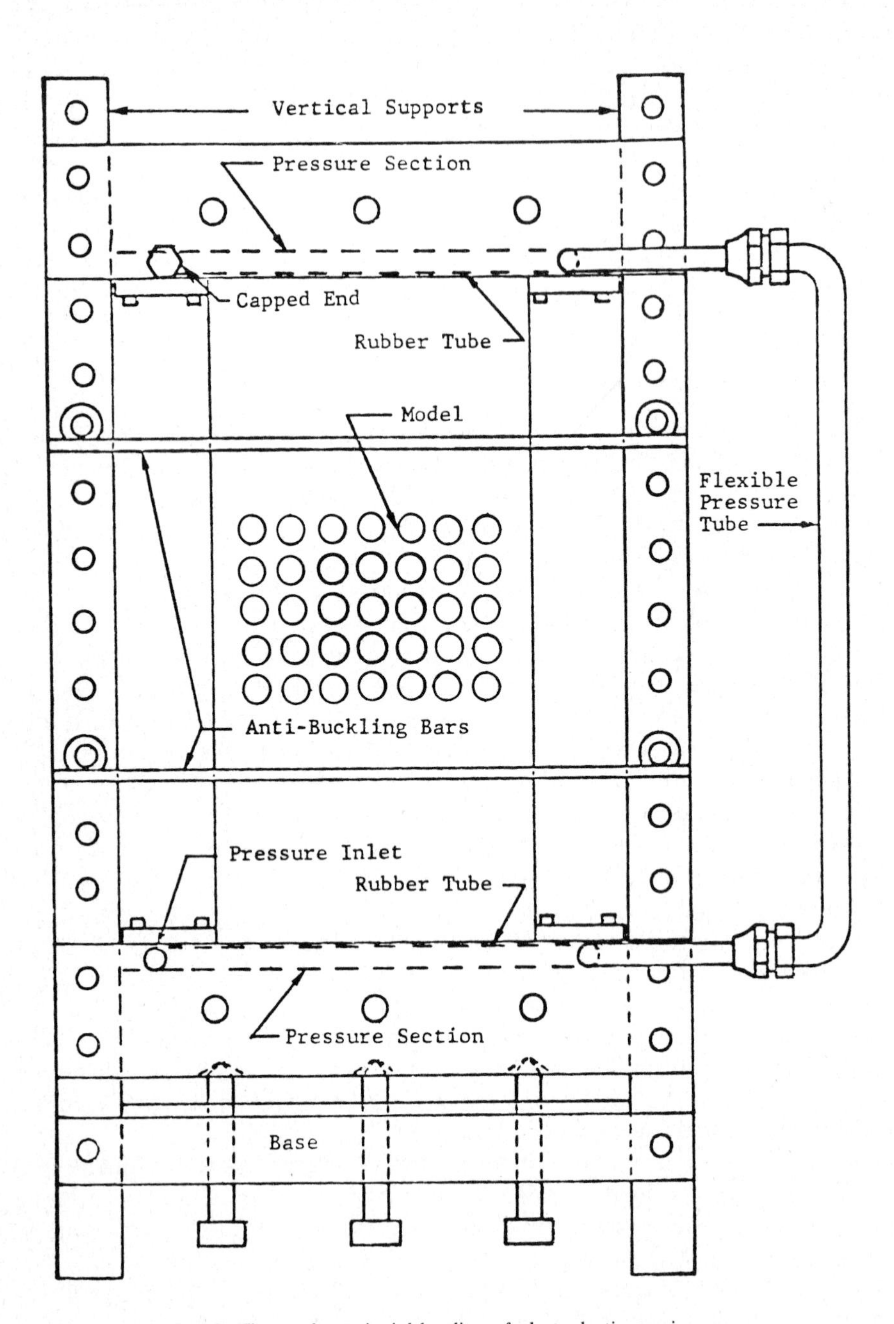

FIG. 9. Fixture for uniaxial loading of photoelastic specimens.

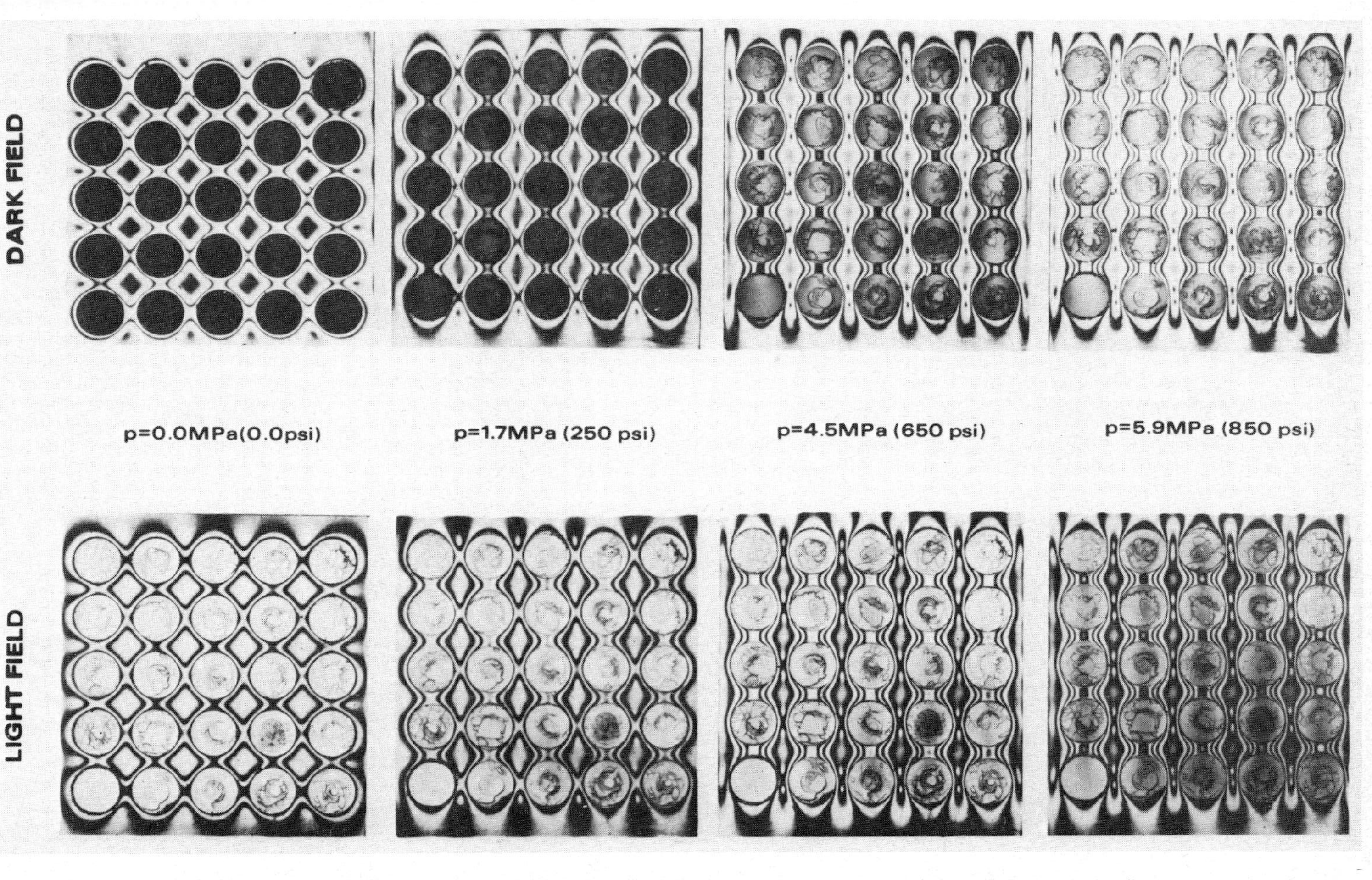

FIG. 10. Isochromatic fringe patterns for composite model with square array of inclusions under uniaxial loading.

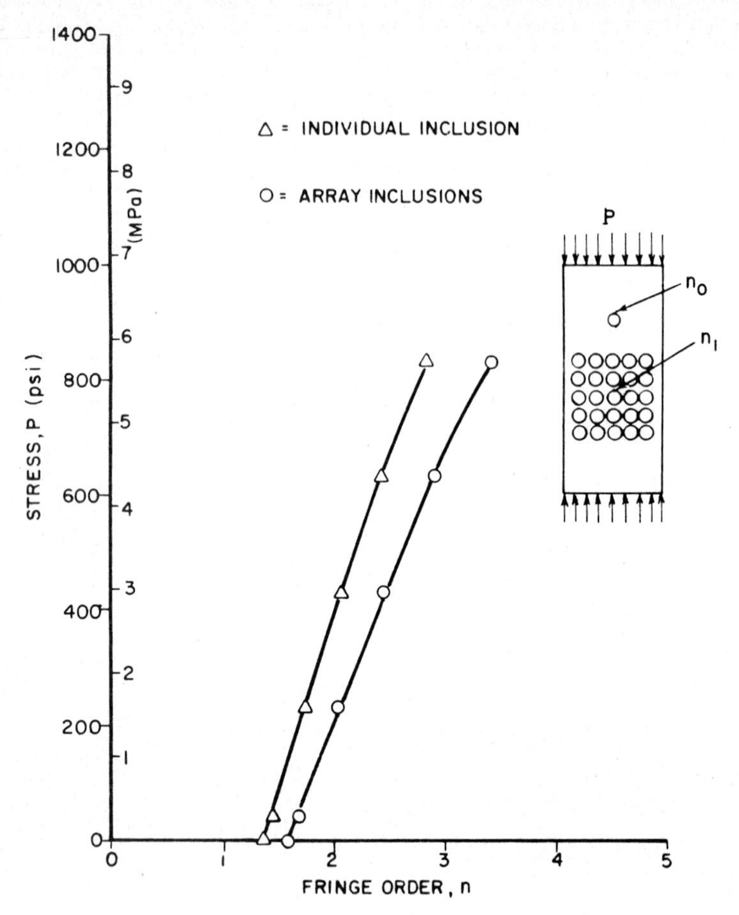

FIG. 11. Stress-birefringence curves for photoelastic specimen with 1.27 cm (0.50 in.) inclusion array.

$$\bar{\sigma}_x = \frac{E}{(1+\nu)(1-2\nu)}[(1-\nu)\bar{\varepsilon}_x + \nu(\bar{\varepsilon}_y + \bar{\varepsilon}_z)] ,$$

$$\bar{\sigma}_y = \frac{E}{(1+\nu)(1-2\nu)}[(1-\nu)\bar{\varepsilon}_y + \nu(\bar{\varepsilon}_x + \bar{\varepsilon}_z)] , \tag{21}$$

$$\bar{\sigma}_z = \frac{E}{(1+\nu)(1-2\nu)}[(1-\nu)\bar{\varepsilon}_z + \nu(\bar{\varepsilon}_x + \bar{\varepsilon}_y)]$$

where $\bar{\sigma}_x$, $\bar{\sigma}_y$, $\bar{\varepsilon}_x$, $\bar{\varepsilon}_y$, are in-plane stresses and strains transverse and parallel to loading direction and averaged through the thickness of the specimen: $\bar{\sigma}_z$ and $\bar{\varepsilon}_z$ are the out-of-plane stress and strain averaged through the thickness.

The inclusions can be considered rigid compared with the matrix due to the high ratio of moduli. This fact introduces the following condition on the boundary of the inclusion on the vertical axis of symmetry where the maximum stress occurs:

$$\bar{\varepsilon}_x = \bar{\varepsilon}_z = 0\,. \tag{22}$$

From the Equations (21), it follows that

$$\bar{\sigma}_x = \bar{\sigma}_z = \frac{\nu}{1-\nu}\bar{\sigma}_y\,. \tag{23}$$

Substituting in the stress-optic law of photoelasticity

$$\bar{\sigma}_x - \bar{\sigma}_y = \frac{2n_1 f}{t} \tag{24}$$

we obtain

$$\bar{\sigma}_y = \frac{1-\nu}{1-2\nu}\frac{2n_1 f}{t} \tag{25}$$

and

$$\bar{\sigma}_x = \bar{\sigma}_z = \frac{\nu}{1-2\nu}\frac{2n_1 f}{t} \tag{26}$$

where n_1 is the fringe order at the interface along the y-axis, f is the fringe value, and t is the specimen thickness.

The maximum stress at the interface was calculated from the measured fringe order using Equation (25). The stress concentration factor of interest is defined as

$$k_\sigma = \frac{\bar{\sigma}_y}{\sigma_0} \tag{27}$$

where σ_0 is the average applied stress in the y-direction. This factor is plotted as a function of inclusion spacing in Fig. 12. Two-dimensional photoelastic results are compared with theoretical results by FOYE (1967), ADAMS and DONER (1967), and with three-dimensional photoelastic results by MARLOFF and DANIEL (1969). The plane strain analysis of the photoelastic data, which is perfectly valid at the interface, gives results in good agreement with theoretical predictions for fiber volume ratios higher than 0.50 (inclusion spacing ratios $\Delta/R < 0.5$). The experimental values for the stress concentration are higher than theoretical values for lower fiber volume ratios because a smaller portion of the matrix is under nearly plane strain conditions in the case. The stress concentration above is very sensitive to the value of Poisson's ratio for the material because of the $(1-2\nu)$ factor in the denominator of Equation (25).

The quantity of importance in transversely loaded composites is the strain concentration factor in the matrix since many failures originate in the matrix.

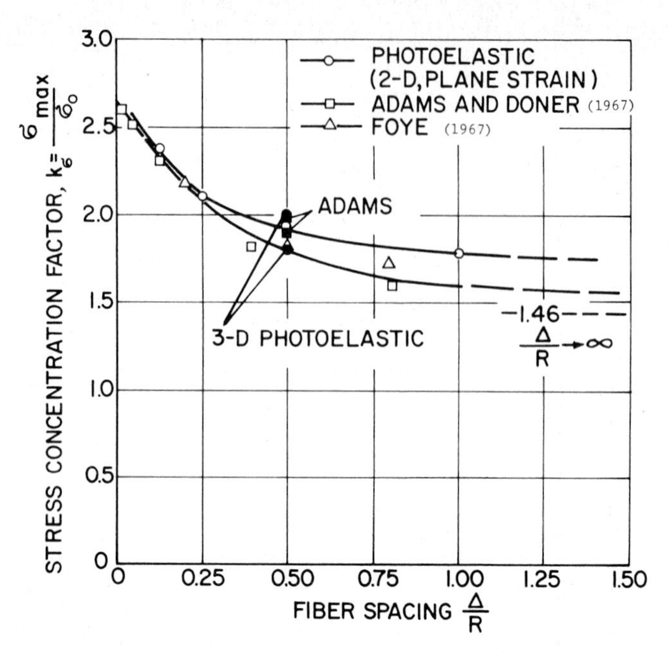

FIG. 12. Stress concentration factor as a function of inclusion spacing for square array under transverse loading.

This factor is defined as the ratio of the maximum interface radial strain to the average strain and it can be computed as follows:

$$k_\varepsilon = \frac{\bar{\varepsilon}_y}{\varepsilon_0} = k_\sigma \left(\frac{E_c}{E_m}\right) \frac{(1+\nu)(1-2\nu)}{1-\nu} \tag{28}$$

where ε_0 is the average strain, E_c the transverse composite modulus, E_m the matrix modulus, and ν the matrix Poisson's ratio. Of the quantities above, k_σ is obtained from photoelastic data only, E_m and ν are obtained from tests of the matrix material, and E_c is either calculated from the constituent properties or measured directly by experiment. The strain concentration factor is plotted as a function of inclusion spacing in Fig. 13. It is seen that this factor rises sharply with increasing fiber volume ratio. Results from a series of two-dimensional photoelastic tests are tabulated in Table 1. The extreme cases of a single inclusion and continuous inclusions are also listed.

The transverse tensile strength of a composite can be predicted by using appropriate failure criteria. The simplest one in this case is the maximum tensile strain criterion. The maximum strain at the interface of a fiber due to combined transverse tensile loading and curing shrinkage is given by

$$(\varepsilon_y)_{\max} = \frac{(1+\nu)(1-2\nu)}{E_m(1-\nu)} k_\sigma \sigma_0 + \varepsilon_R \tag{29}$$

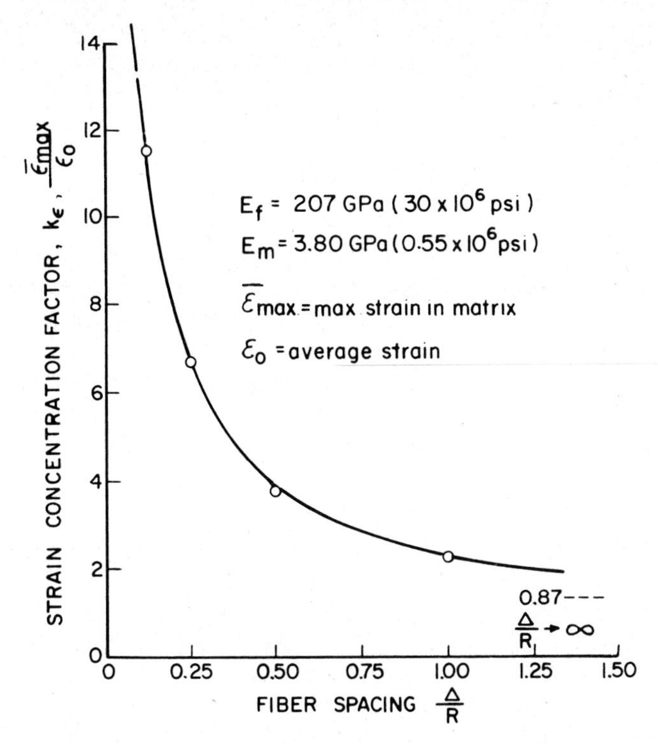

FIG. 13. Strain concentration factor as a function of inclusion spacing for square array under transverse loading.

where ε_R is the radial residual strain at the interface, obtained from the same photoelastic model at no load.

Equating this strain to the ultimate tensile strain in the resin and assuming linear elastic behavior to failure, we obtain:

$$\varepsilon^u = \frac{S_T}{E_m} = (\varepsilon_y)_{max} \tag{30}$$

TABLE 1. Stress and strain concentration factors for two-dimensional photoelastic models.

Inclusion spacing Δ/R	Ratio of composite to matrix modulus E_c/E_m	Stress concentration factor k_σ	Strain concentration factor k_ε
∞ (Single inclusion)	1.00	1.46	0.87
1	2.11	1.78	2.23
0.500	3.31	1.91	3.76
0.250	5.36	2.11	6.73
0.125	8.18	2.37	11.53
0	14.5	2.4	21

where S_T is the tensile strength of the matrix. Solving for σ_0 in the equations above and equating it to the transverse tensile strength of the unidirectional composite we obtain:

$$\sigma_0 = S_{22T} = \frac{(1-\nu)}{k_\sigma(1+\nu)(1-2\nu)}[S_T - \varepsilon_R E_m] . \quad (31)$$

This value is very sensitive to the value of Poisson's ratio because of the $(1-2\nu)$ factor in the denominator. For a typical glass/epoxy composite of 0.50 fiber volume ratio the predicted strength would be $S_{22T} = 46.9$ MPa (6790 psi), whereas the actually measured strength of prototype specimens is $S_{22T} =$ 47.3 MPa (6850 psi).

The prediction above is based on a local failure criterion at a point. Actual fracture of a composite should be based on global failure criteria taking into consideration the statistical distribution of localized failures.

3.3. Microphotoelastic studies

Microphotoelastic experiments utilize models with prototype reinforcing filaments to closer simulate the three-dimensional state of stress and bond characteristics of the prototype composite. Specimens usually consist of a thin layer of epoxy matrix with one or more fibers embedded in it. The method, by its nature, yields normally only average through the thickness results.

SCHUSTER and SCALA (1964) determined stresses around a Sapphire whisker. They outlined a procedure whereby the average through the thickness value of the principal stress difference was correlated to the principal stress difference on the plane through the axis of the whisker. The highest stress concentrations causing matrix cracking were observed at points of whisker fracture occurring after whisker embedment. HAENER (1966) conducted microphotoelastic tests to obtain qualitative correlations with theoretical stress distributions. Qualitative studies of fracture mode and propagation were conducted by ROSEN (1964), ALEXANDER et al. (1965), and CAPUTO and HILZINGER (1969). The latter used one- and two-ply microphotoelastic glass/epoxy and graphite/epoxy models with widely spaced fibers. They studied the influence on fracture of the fiber type, fiber orientation, type of flaw (cut fiber, unbonded fiber, matrix void, exposed edges), and loading condition (static tension, biaxial tension-compression, cyclic tension, and impact). Broken fibers in the 0-deg direction affected strength and failure drastically as was shown by ROSEN (1964). Fracture of $[0/90]_{ns}$ specimens initiated at transverse fiber bundles and was accompanied by stacked cracks and debonding. Specimens with ±45-deg fiber orientation failed in shear along the fibers with failure initiation at crossover points and near the edges of the specimen.

Microphotoelastic studies conducted by the author dealt with shrinkage stresses and effects on matrix stress concentration of fiber orientation, fiber debonding, and matrix cracking at the interface. Specimens were prepared by casting epoxy around boron fibers. Shrinkage stresses at the end of a 0.1 mm (0.004 in.) diameter boron fiber are illustrated in Fig. 14. High stress concen-

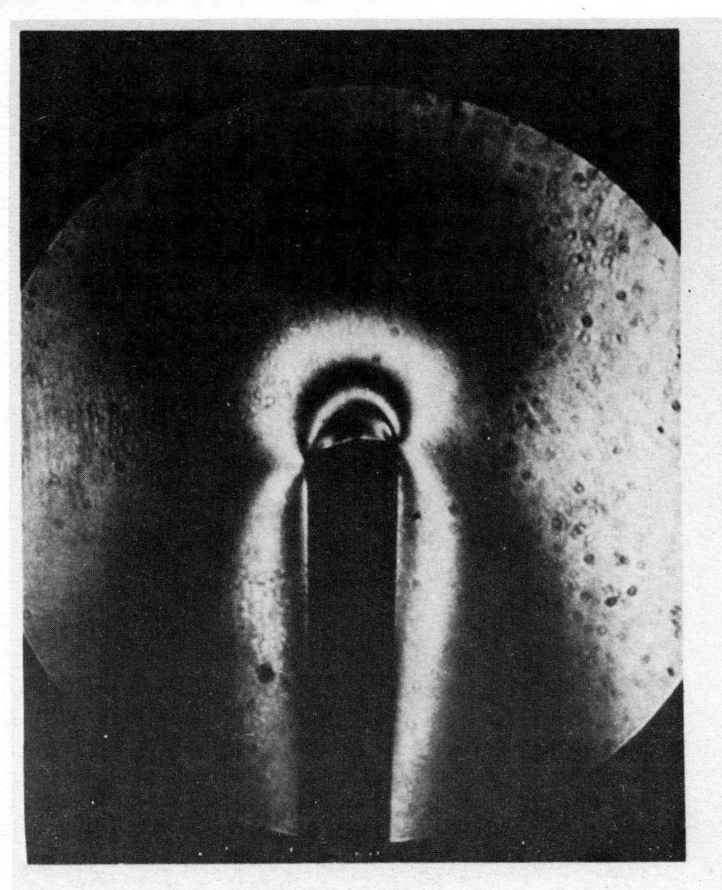

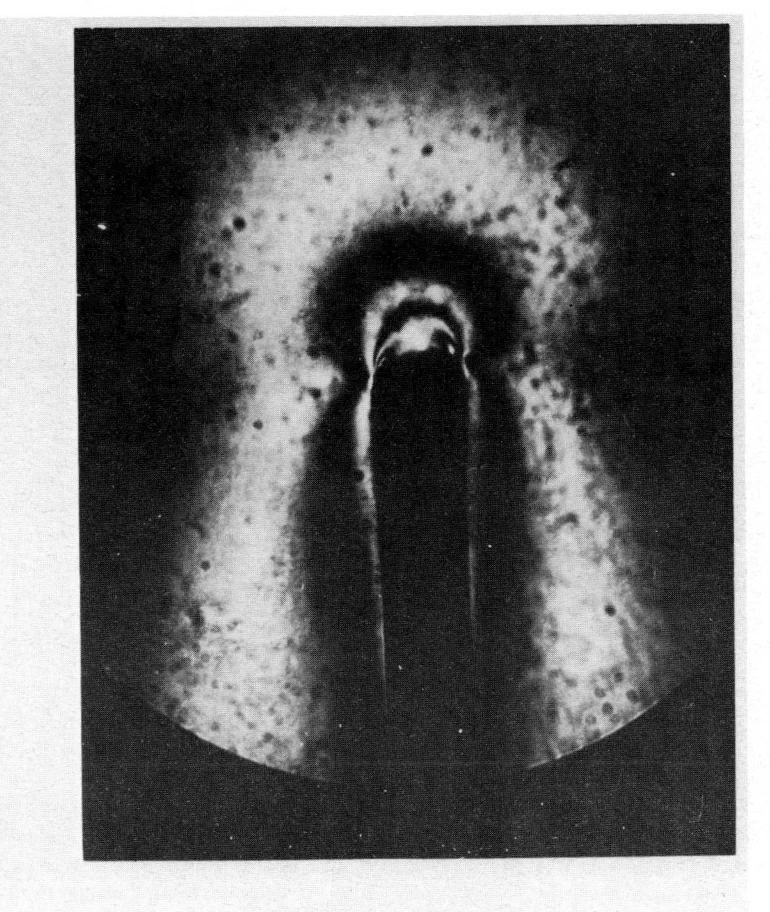
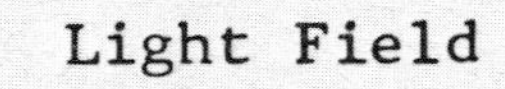

FIG. 14. Isochromatic fringe patterns due to matrix shrinkage around a 112 μm (0.0044 in.) diameter boron fiber.

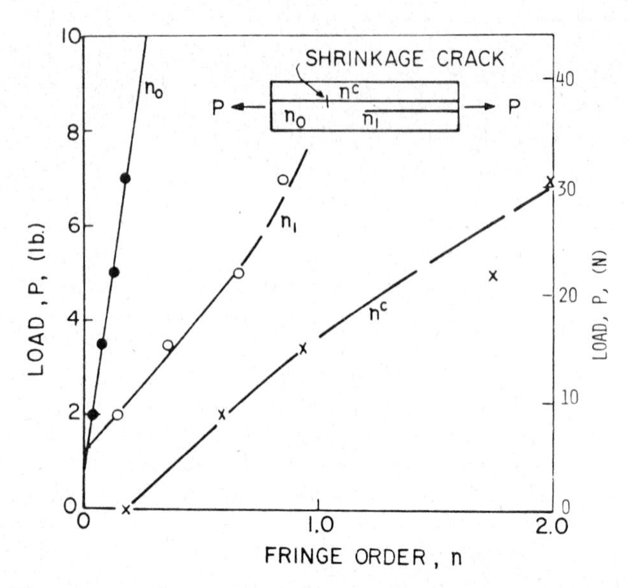

FIG. 15. Birefringence variation with load at fiber end, shrinkage crack and in the far field of microphotoelastic specimen.

trations at the corners of the fiber end are manifested. Phenomena such as fiber debonding and matrix cracking are revealed by the shape of the load-birefringence curves such as those of Fig. 15. Matrix cracking is accompanied by a rapid increase in birefringence with load, whereas fiber debonding produces a different type of nonlinearity with the birefringence increasing at a reduced rate with load. The 'average' maximum shear concentration factor at the fiber end decreases from 2.5 for axial fibers, to 2.1 for 45-deg fibers, to 1.8 for 90-deg fibers. These results are in agreement with those of two-dimensional photoelastic tests.

3.4. *Three-dimensional model studies*

The state of stress in the matrix of a composite is a three-dimensional one and can only be determined accurately by three-dimensional studies. DURELLI et al. (1970), and PARKS et al. (1970) have studied stress distributions around inserts of various geometries in matrices subjected to shrinkage and mechanical loading. They used standard 'stress-freezing' and slicing techniques and assumed the material to be incompressible at the critical temperature. Some experiments with single bar inclusions using the scattered-light technique were described by JENKINS (1968), and SUTLIFF and PIH (1973).

MARLOFF and DANIEL (1969) described an extensive stress-analysis of a realistic three-dimensional composite model. They used standard 'stress-freezing' and slicing techniques to determine stress distributions in the matrix of a unidirectional composite model subjected to matrix shrinkage and transverse

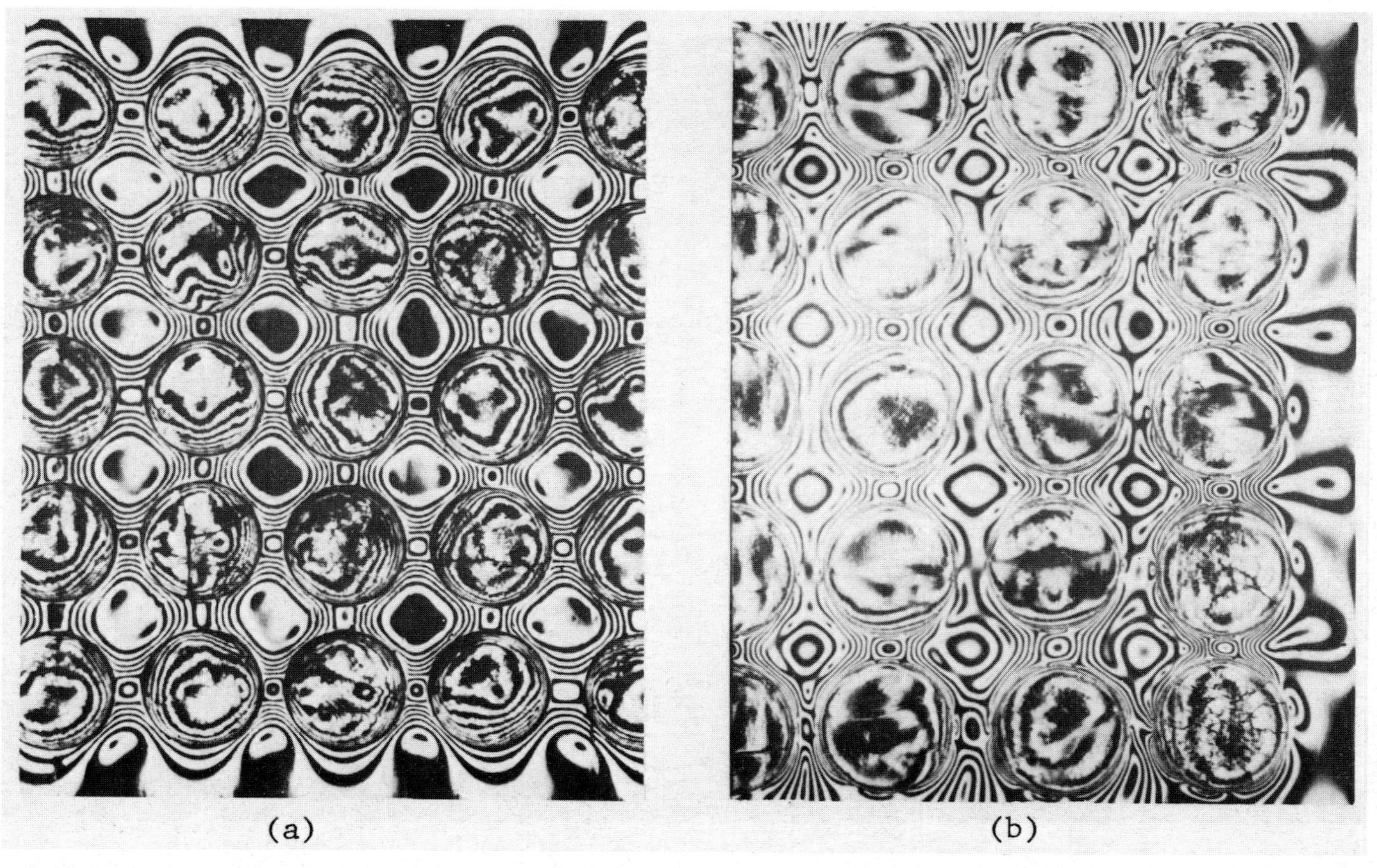

FIG. 16. Isochromatic fringe patterns in slice from model subjected to (a) matrix shrinkage, and (b) combined matrix shrinkage and external loading.

loading. The model was made by casting a low-critical temperature epoxy around polycarbonate (Lexan) rods with a clear spacing of half the rod radius. This combination resulted in a fiber to matrix modulus ratio of 175. Two identical models were prepared and tested to separate the effects of shrinkage from those of loading. One specimen was subjected to uniform uniaxial compression normal to the fibers and the other was left unloaded during the 'stress-freezing' cycle.

Isochromatic fringe patterns of slices taken from these models are shown in Fig. 16. The principle of superposition was used to separate the effects of shrinkage from those of combined shrinkage and loading. Complete stress separation was obtained in both cases and corresponding stress components were subtracted from each other. A numerical integration procedure was used based on Filon's transformation of the Lamé–Maxwell equilibrium equations for an axis of symmetry. The stress component σ_z parallel to the fiber axis was determined by removing subslices in that direction.

The distribution of shrinkage stresses in the matrix along two axes of symmetry is shown in Figs. 17 and 18. These stresses are expressed in

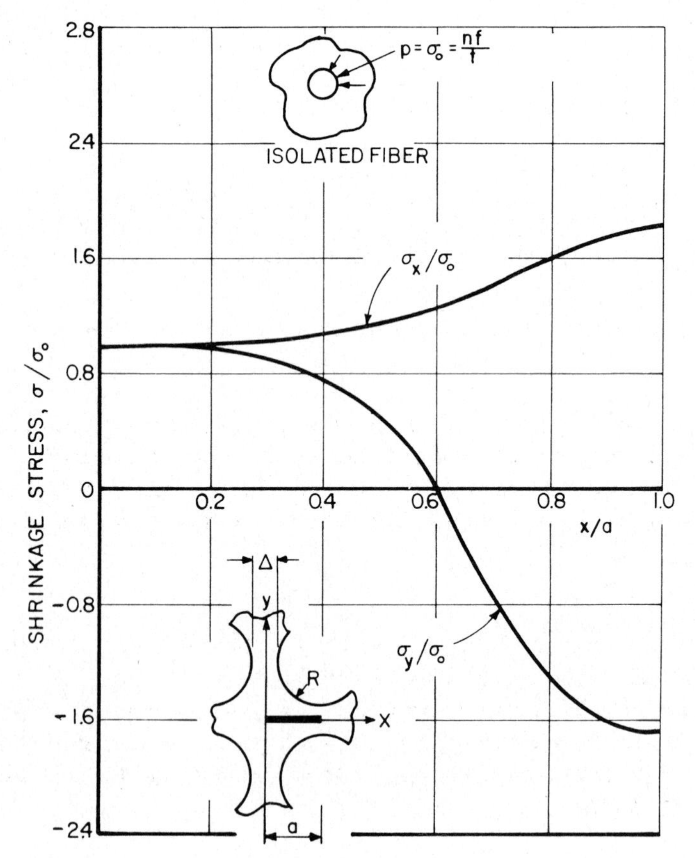

FIG. 17. Distribution of shrinkage stresses along interstitial centerline, $\Delta/R = 0.50$.

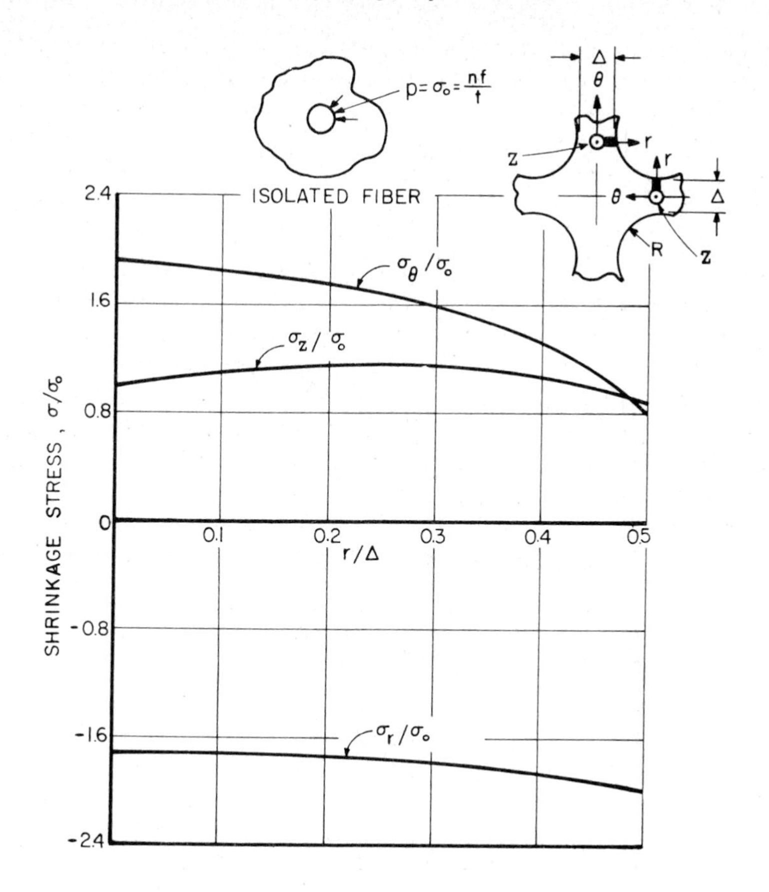

FIG. 18. Distribution of shrinkage stresses across section between fibers, $\Delta/R = 0.50$.

dimensionless form by dividing them by the shrinkage pressure at the interface of an isolated fiber. The circumferential stress σ_θ reaches a maximum value at the center of the minimum section between fibers and it decreases toward the interface (Fig. 18). Conditions of perfect bond and fiber rigidity require that

$$\varepsilon_\theta = \varepsilon_z = \alpha \tag{32}$$

at the interface, where α is the differential free linear shrinkage of matrix and fiber. This leads to the equality $\sigma_\theta = \sigma_z$, which is shown in Fig. 18.

Stress distributions due to external load are given in Figs. 19 and 20. These are presented in dimensionless form by dividing the stress components by the applied average stress σ_0. Conditions of perfect bond and fiber rigidity at the interface require that

$$\varepsilon_\theta = \varepsilon_z = 0 \tag{33}$$

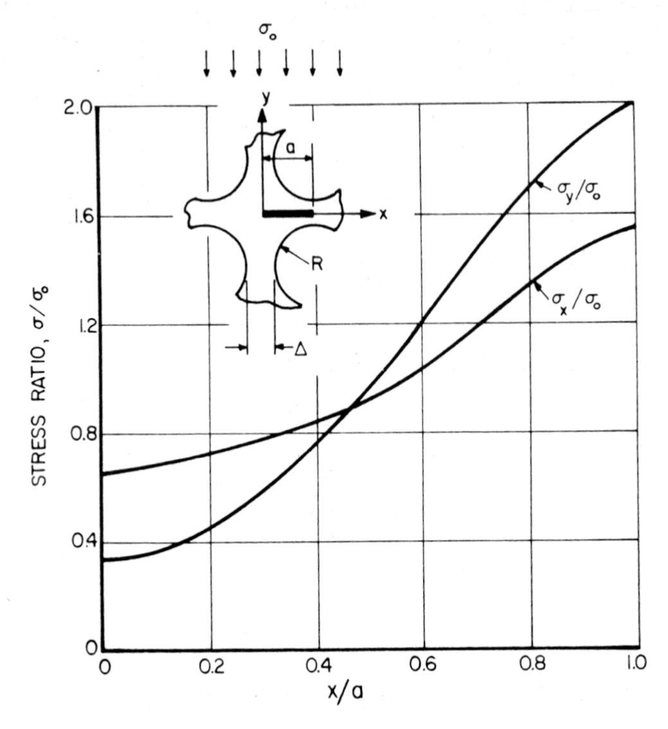

FIG. 19. Stress distributions due to external load along interstitial centerline normal to load direction, $\Delta/R = 0.50$.

which leads for the case of plane strain to

$$\sigma_\theta = \frac{\nu}{1-\nu}\sigma_r , \tag{34}$$

$$\sigma_z = \sigma_\theta . \tag{35}$$

Thus, for an incompressible matrix ($\nu = 0.5$), all principal stresses are equal at the interface at points where a principal direction is normal to the interface. This is evidenced in Fig. 20.

The stress ratio in the direction of loading reaches a maximum value of 2.0 at the midpoint of the matrix section between the fibers as seen in Fig. 20. A slight reduction to 1.80 occurs at the interface. At 90-deg from this point the stress becomes tensile with a ratio equal to −0.5.

Theoretical results by ADAMS and DONER (1967) for a boron/epoxy composite of 0.55 fiber volume ratio give stress ratios of 1.86 and −0.5 at the two points on the interface discussed above. The theory does not predict an increase of stress at the midpoint away from the interface. As pointed out by GOREE (1967) the location and magnitude of the maximum stress may be strongly influenced by fiber spacing and matrix Poisson's ratio.

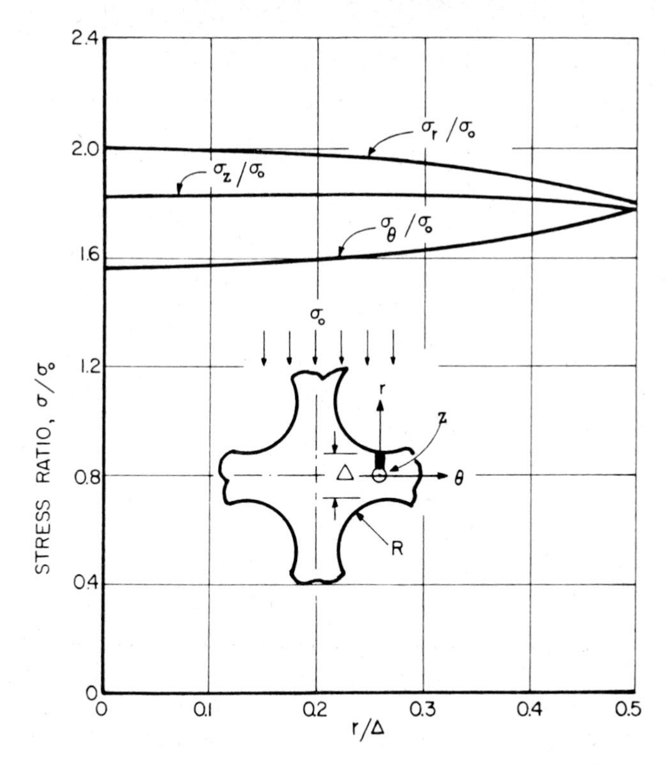

FIG. 20. Stress distributions due to external load across section between fibers parallel to load direction, $\Delta/R = 0.50$.

The three-dimensional photoelastic method is probably the only experimental means for complete micromechanical analysis, however, it has its undesirable features. The separation of residual stresses from those due to external loading requires two identical models and essentially two complete stress analyses. The 'stress-freezing' technique is destructive and allows only a one-time use of expensive models. Poisson's ratio of the matrix at the critical temperature is near 0.5, appreciably different from that of the prototype (0.35). Several attempts to overcome these difficulties have been made. EDELMAN (1969) described a procedure for producing shrinkage-free photoelastic models. JENKINS (1968), SUTLIFF and PIH (1973), and DANIEL (1974) have applied scattered-light techniques, which are nondestructive and allow testing at room temperature where the matrix Poisson's ratio is the same as that of the prototype.

3.5. *Two-dimensional dynamic model studies*

One of the dynamic phenomena in fracture of composites is the crack propagation induced by the energy released when interior fibers break. A dynamic photoelastic study of crack propagation in a composite model was

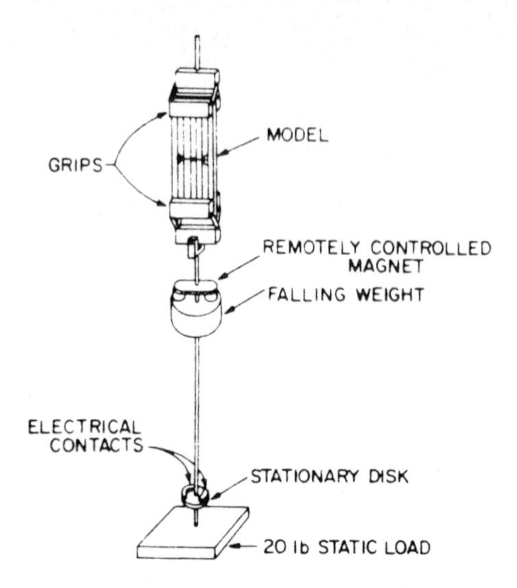

FIG. 21. Fixture for dynamic tensile loading of composite models.

reported by DANIEL (1969, 1970). The models consisted of alternating polyester and glass strips cemented together. A stress raiser was machined in the central glass bar to insure and localize crack initiation. The models were subjected to dynamic longitudinal tension by means of a falling weight system (Fig. 21). The dynamic fringe patterns and the propagating crack were recorded with a multiple spark Cranz–Schardin camera operating at a rate of 200 000 frames per second.

A sequence of isochromatic fringe patterns is shown in Fig. 22. The measured crack propagation velocity was far below the theoretical limiting velocity for the matrix material. The slow velocity and near arrest of the crack can be attributed to 'crack blunting' due to inelastic effects and to reflected stress waves. After passage of the growth inhibiting stress wave the crack was able to resume its propagation to complete failure. Another inhibiting factor may be the type of stress singularity present at the crack tip. Generally, the stress singularity near the crack is of the form r^{-n} with $n=\frac{1}{2}$ for a homogeneous material, $n>\frac{1}{2}$ when the crack enters from a hard into a soft medium, and $n<\frac{1}{2}$ for the reverse case (ZAK and WILLIAMS (1962)). The failure pattern of the model is shown in Fig. 23. An interesting phenomenon observed is the bifurcation occurring when the crack reaches the interface with the glass bar despite the fact that the propagation velocity at this point was far below the limiting value.

In the investigation discussed here, a single crack or fiber break was discussed. In prototype composites containing a random distribution of weak points, cracking starts at one or more of these points as shown by ROSEN (1964). Each fiber break results in a short crack which is arrested by adjacent fibers as suggested in the photoelastic study above. The flaw distribution and severity is

FIG. 22. Transient isochromatic fringe patterns in a glass-plastic composite model under dynamic tension (camera speed: 200 000 frames per second).

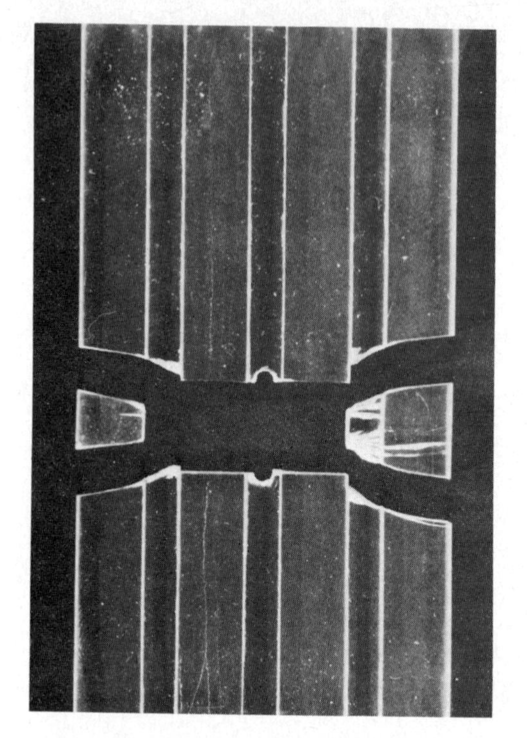

FIG. 23. Failure pattern in model of Fig. 22.

such that fiber breaks occur at other points before the stress concentration at the arrested crack can cause crack extension. Total failure occurs when a sufficient number of these fiber breaks or short cracks line up across the loaded section of the specimen.

4. Composite material characterization

4.1. Material qualification

Composite materials are normally available from the manufacturer in the form of a tape consisting of unidirectional fibers impregnated with the matrix resin partially advanced (B-staged). The prepregs are supplied to a variety of specifications such as fiber volume ratio, ply thickness, and degree of B-staging. Prior to fabrication of larger components and extensive material characterization some preliminary quality-control tests are conducted. Flexural and interlaminar shear strengths are determined from unidirectional coupons and compared with published or expected values.

Qualification testing is done by means of beams subjected to three-point bending. The test specimens are normally 15-ply thick unidirectional coupons. Flexural strength coupons are 10.2 cm (4 in.) long, 1.27 cm (0.50 in.) wide, with a 6.3 cm (2.5 in.) span length. Shear strength coupons are typically 1.5 cm (0.6 in.) long, 0.64 cm (0.25 in.) wide, with a 1 cm (0.4 in.) span length. The standard beam formulas below are used to determine strength values:

$$S_{11}=\frac{3Pl}{2wt^2}, \tag{36}$$

$$S_{12}=\frac{3P}{4wt} \tag{37}$$

where S_{11}, S_{12}=flexural and shear strengths, respectively; P=load; w=beam width; t=thickness.

Laminate fabrication and more extensive material characterization follow the satisfactory completion of the qualification tests.

4.2. Density

The density of composite laminates is usually determined by the displacment method (ASTM D792-66). The sample is first weighed in air, then it is suspended by a wire on the weighing scale and weighed while totally immersed in a liquid of known density and good wettability, such as alcohol. Finally, the suspended wire is weighed immersed in the liquid to the same depth as during the sample weighing. The composite density is then obtained from the relationship

$$\rho_c=\frac{\rho_l W_c}{W_c-W+W_w} \tag{38}$$

where ρ_c=density of composite; ρ_l=density of immersion liquid; W_c=weight of composite sample in air; W=weight of sample and wire while immersed in liquid; W_w=weight of wire while immersed in liquid.

4.3. Fiber volume ratio

The fiber volume ratio (FVR) can be determined by two basic methods. One method involves the dissolution of the matrix in a liquid solvent or, in the case of non-burning fibers, the burning of the matrix, with the weight of the remaining fibers and the fiber and composite densities used in the determination of the percentage volume of fiber. The second method, the gravimetric method, uses the densities of the constituent and composite materials as follows:

$$\text{FVR}=V_f=\frac{\rho_c-\rho_m}{\rho_f-\rho_m} \tag{39}$$

where ρ_c, ρ_m, ρ_f=densities of composite, matrix, and fiber, respectively.

The gravimetric method is valid if the composite is free of voids. In most cases the void content is less than 1%, thus the assumption of a void-free composite is reasonable. The void content can be checked in photomicrographs of composite cross sections.

4.4. *Coefficient of thermal expansion*

Thermal expansion in unidirectional and angle-ply laminates has been measured experimentally for various materials and over wide temperature ranges. FREEMAN and CAMPBELL (1972) used a Leitz dilatometer and strain gages to measure thermal expansion in graphite fiber composites over a temperature range of 78 to 561°K (−320 to 550°F). Thermal expansion has also been measured by WANG et al. (1975) and the author (DANIEL et al. (1975), DANIEL and LIBER (1975, 1976a), DANIEL (1978a)) for a variety of composite materials using strain gages. Commercial strain gages have been found suitable for measuring small thermal strains in composites. Temperature compensation of strain gage readings is done according to the procedure outlined in Section 2.4, Equation (14). Strain gages record local deformations averaged over their gage length, therefore they tend to reflect local material irregularities, inhomogeneities and flaws in a realistic composite. For this reason some variations may be seen between gage readings from different gages at different locations. These strain variations can be of the order of $\pm 100\,\mu\varepsilon$. Readings from surface gages may be different from those of embedded gages because of laminate bending. Realistic results are obtained by averaging readings from a number of embedded gages.

In one application of strain gages to measurement of thermal deformations (DANIEL and LIBER (1976a)), encapsulated gages (WK-00-125TM-350, Option B-157) were embedded between the plies during laminate assembly. The attached ribbon leads were sandwiched between thin (0.013 mm; 0.0005 in.) polyimide strips. A thermocouple was also embedded in each laminate. The instrumented specimens, including a reference titanium silicate specimen, were subjected to the curing and postcuring cycles in the autoclave. Strain gage and thermocouple readings were taken throughout. Subsequently, the same specimens were subjected to a thermal cycle from room temperature to 444°K (340°F) and down to room temperature. Strain gages and thermocouples were recorded at 5.5°K (10°F) intervals. The true thermal strains were obtained from recorded apparent strains as discussed before.

Thermal strains measured on eight-ply unidirectional graphite/epoxy, Kevlar 49/epoxy and S-glass/epoxy specimens are shown in Fig. 24. Both Kevlar 49/epoxy and graphite/epoxy exhibit negative thermal strains in the longitudinal (fiber) direction. The Kevlar 49/epoxy exhibits the largest positive transverse and negative longitudinal strains. The S-glass/epoxy undergoes the lowest thermal deformation in the transverse direction and the highest (positive) in the longitudinal direction. Coefficients of thermal expansion computed from such data are tabulated in Table 2 for eight composite materials.

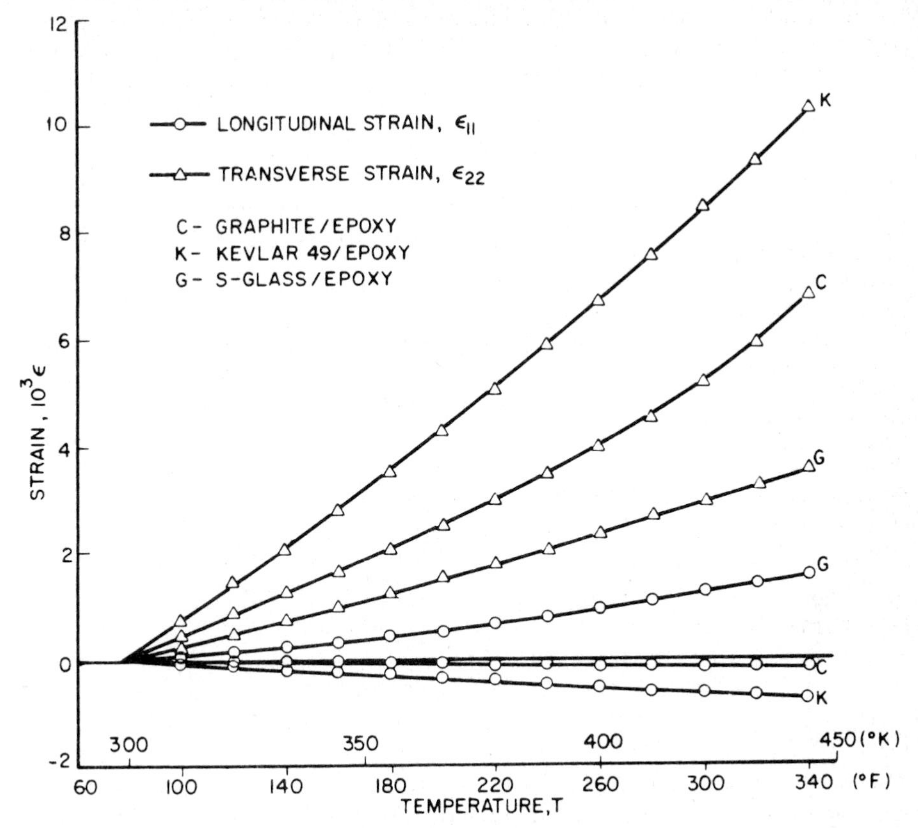

FIG. 24. Thermal strains in unidirectional composites as a function of temperature.

Coefficients are listed at room temperature, 297°K (75°F), and at the elevated temperature of 450°K (350°F). All graphite fiber composites exhibit negative thermal expansion in the fiber direction. The polyimide matrix composites do not show any variation of thermal coefficients with temperature. This is true at least up to the postcuring temperature of 589°K (600°F).

The expansion coefficients along an arbitrary coordinate system x–y rotated by an angle θ with respect to the fiber direction can be obtained by a linear coordinate transformation as

$$\alpha_x = \alpha_{11} m^2 + \alpha_{22} n^2 ,$$
$$\alpha_y = \alpha_{11} n^2 + \alpha_{22} m^2 , \qquad (40)$$
$$\alpha_{xy} = (\alpha_{22} - \alpha_{11}) mn$$

where α_{11}, α_{22} = longitudinal and transverse coefficients of thermal expansion; $m = \cos\theta$; $n = \sin\theta$.

The effective coefficients for multidirectional laminates can be evaluated from the properties of the single lamina by means of lamination theory.

TABLE 2. Thermal expansion coefficients of unidirectional composite materials.

Material	Longitudinal coefficient of thermal expansion, α_{11}, 10^{-6} K^{-1} ($\mu\varepsilon$/°F)		Transverse coefficient of thermal expansion, α_{22}, 10^{-6} K^{-1} ($\mu\varepsilon$/°F)	
	297°K (75°F)	450°K (350°F)	297°K (75°F)	450°K (350°F)
Boron/Epoxy (Boron/AVCO 5505)	6.1 (3.4)	6.1 (3.4)	30.3 (16.9)	37.8 (21.0)
Boron/Polyimide (Boron/WRD 9371)	4.9 (2.7)	4.9 (2.7)	28.4 (15.8)	28.4 (15.8)
Graphite/Epoxy (Modmor I/ERLA 4289)	−1.1 (−0.6)	3.2 (1.3)	31.5 (17.5)	27.0 (15.0)
Graphite/Epoxy (Modmor I/ERLA 4617)	−1.3 (−0.7)	−1.3 (−0.7)	33.9 (18.8)	83.7 (46.5)
Graphite/Polyimide (Modmor I/WRD 9371)	−0.4 (−0.2)	−0.4 (−0.2)	25.3 (14.1)	25.3 (14.1)
S-Glass/Epoxy (Scotchply 1009-26-5901)	3.8 (2.1)	3.8 (2.1)	16.7 (9.3)	54.9 (30.5)
S-Glass/Epoxy (S-Glass/ERLA 4617)	6.6 (3.7)	14.1 (7.9)	19.7 (10.9)	26.5 (14.7)
Kevlar/Epoxy (Kevlar 49/ERLA 4617)	−4.0 (−2.2)	−5.7 (−3.2)	57.6 (32.0)	82.8 (46.0)

4.5. Coefficient of moisture expansion

The coefficients of moisture expansion are determined by measuring specimen deformations over a range of moisture weight gains (WHITNEY et al. (1982)). This type of measurement is difficult because of moisture effects on conventional strain gage adhesives. HAHN and KIM (1978) used a micrometer and a caliper to measure moisture swelling strains in graphite/epoxy composites. In the case of unidirectional composites, most of the swelling occurs in the transverse to the fiber direction, whereas in the case of multidirectional laminates most of the swelling occurs in the thickness direction.

4.6. Static tensile properties of unidirectional composites

Uniaxial tensile tests are conducted on unidirectional laminae to determine the following properties:

E_{11}=longitudinal Young's modulus;
E_{22}=transverse Young's modulus;
ν_{12}=major Poisson's ratio;
ν_{21}=minor Poisson's ratio;
S_{11T}=longitudinal tensile strength;
S_{22T}=transverse tensile length;
ε^{u}_{11T}=ultimate longitudinal tensile strain;
ε^{u}_{22T}=ultimate transverse tensile strain.

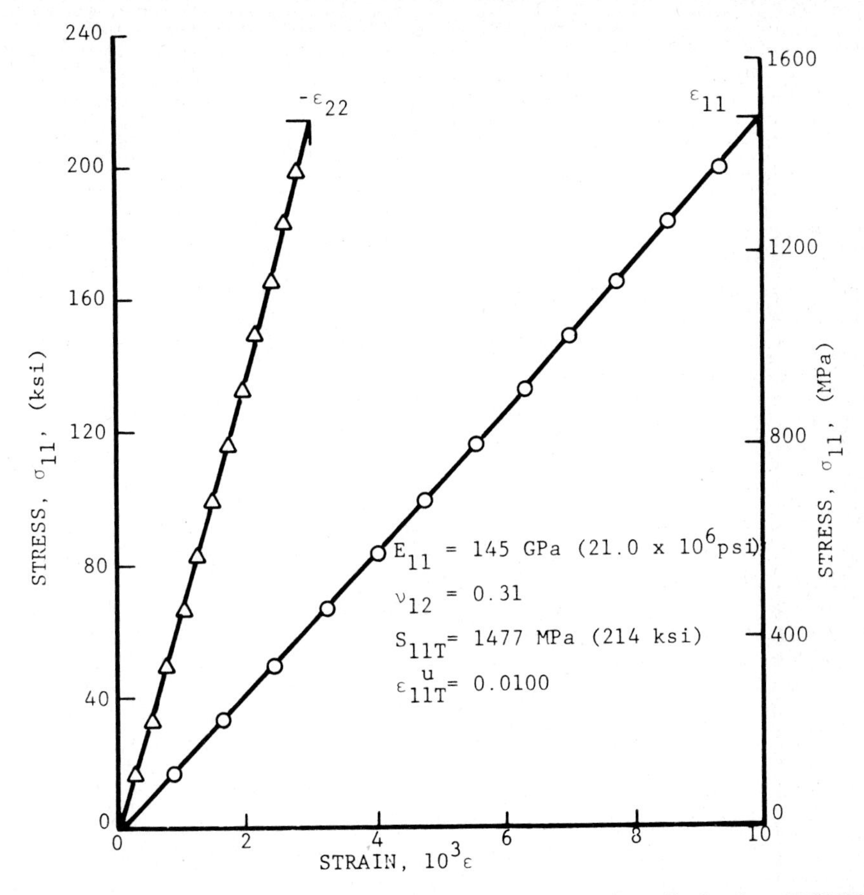

FIG. 25. Strains in 0-deg unidirectional specimen under uniaxial tensile loading (SP286T300 graphite/epoxy).

Tensile specimens are straight-sided coupons of constant cross-section with adhesively bonded beveled glass/epoxy tabs. The longitudinal (0°) coupon is 6 plies thick and 1.27 cm (0.50 in.) wide, while the transvere (90°) coupon is 8 plies thick and 2.54 cm (1.00 in.) wide. Both specimens have an overall length of 23 cm (9 in.) and a 15.2 cm (6 in.) gage length. The specimens are instrumented with a 2-gage 90° rosette bonded at the center of the test section on each side. The specimens are incrementally loaded to failure under uniaxial tensile loading with recording of strains at every step. Typical stress-strain curves for 0° and 90° graphite/epoxy specimens are shown in Figs. 25 and 26. The moduli and Poisson's ratios are determined from the initial slopes of the curves fitted through the data points. Typical fractures of unidirectional tensile coupons are shown in Figs. 27 and 28. In the case of 0° specimens the failure mode consists of fiber breakage, matrix splitting, and fiber pullout. The latter mechanism is much more pronounced in the brooming failure pattern of

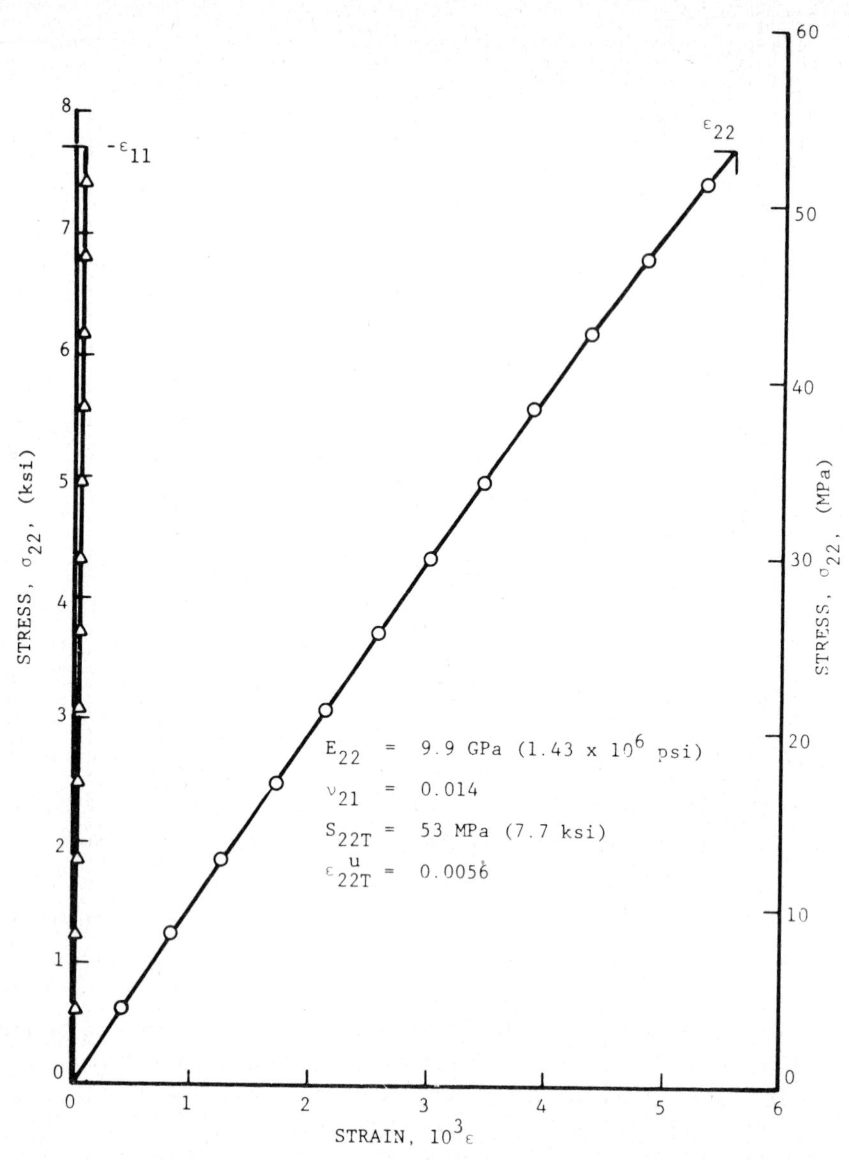

FIG. 26. Strains in 90-deg unidirectional specimen under uniaxial tensile loading (SP286T300 graphite/epoxy).

glass/epoxy. Transverse (90°) specimens fail in a brittle manner by matrix tensile failure between fibers.

4.7. Static compressive properties

Compression testing of composites is one of the most difficult types of testing because of the tendency for premature failure due to crushing or buckling.

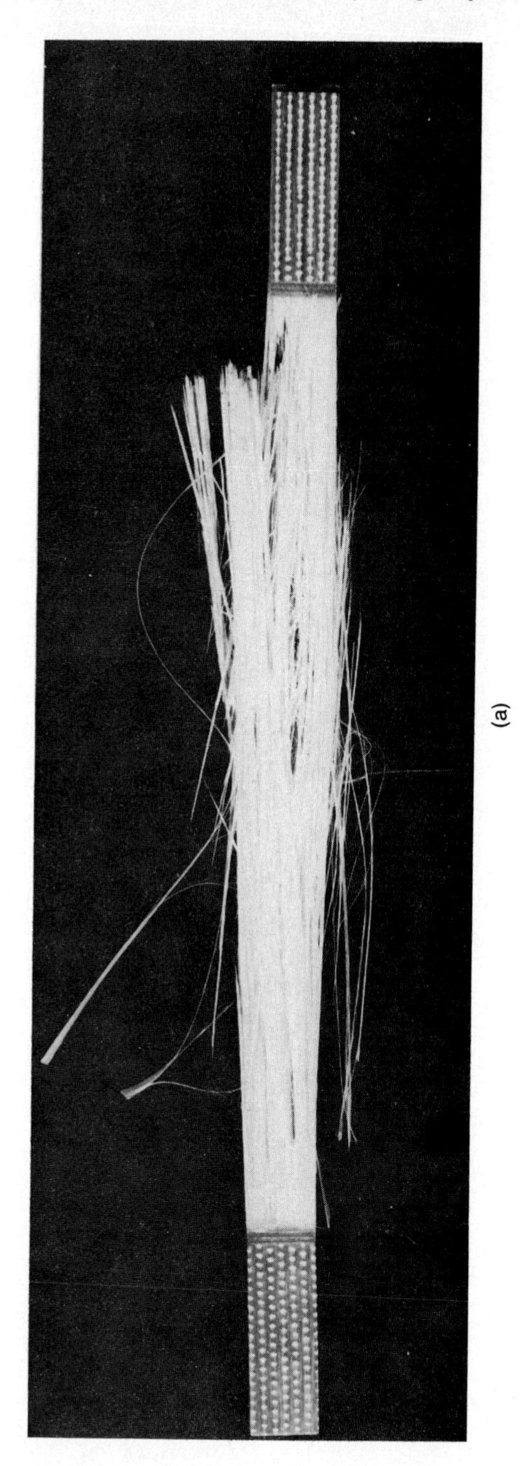

(a)

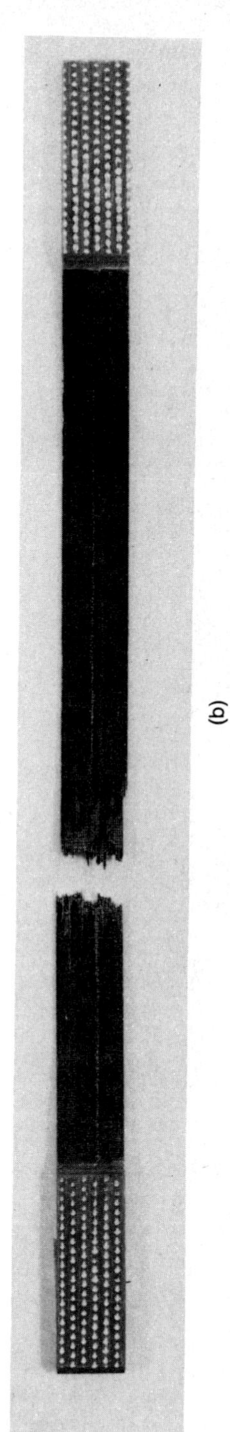

(b)

FIG. 27. Typical fractures of unidirectional 0-deg tensile specimens. (a) s-glass/epoxy; (b) boron/epoxy.

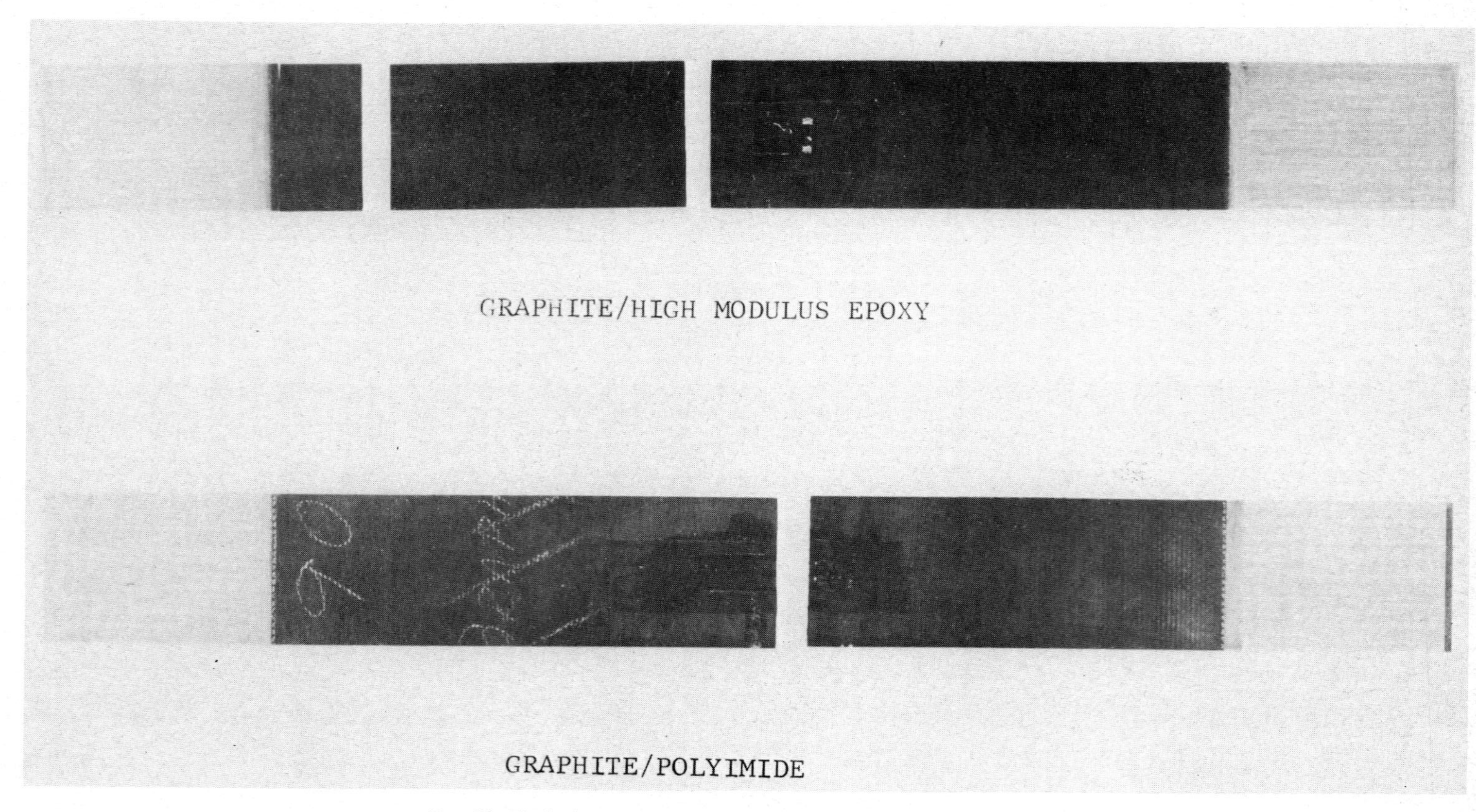

FIG. 28. Typical fractures of unidirectional 90-deg tensile specimens.

Over the years, many test methods have been developed and used, incorporating a variety of specimen designs and loading fixtures. A review of these methods is found in the monograph by WHITNEY et al. (1982).

Compression test methods can be classified into three broad categories. In the first one (Type I) specimens with a very short but unsupported gage length are used. One of these, the so-called Celanese test, makes use of coupon specimens, 14.1 cm (5.5 in.) long, 15–20 plies thick, and 0.64 cm (0.25 in.) wide

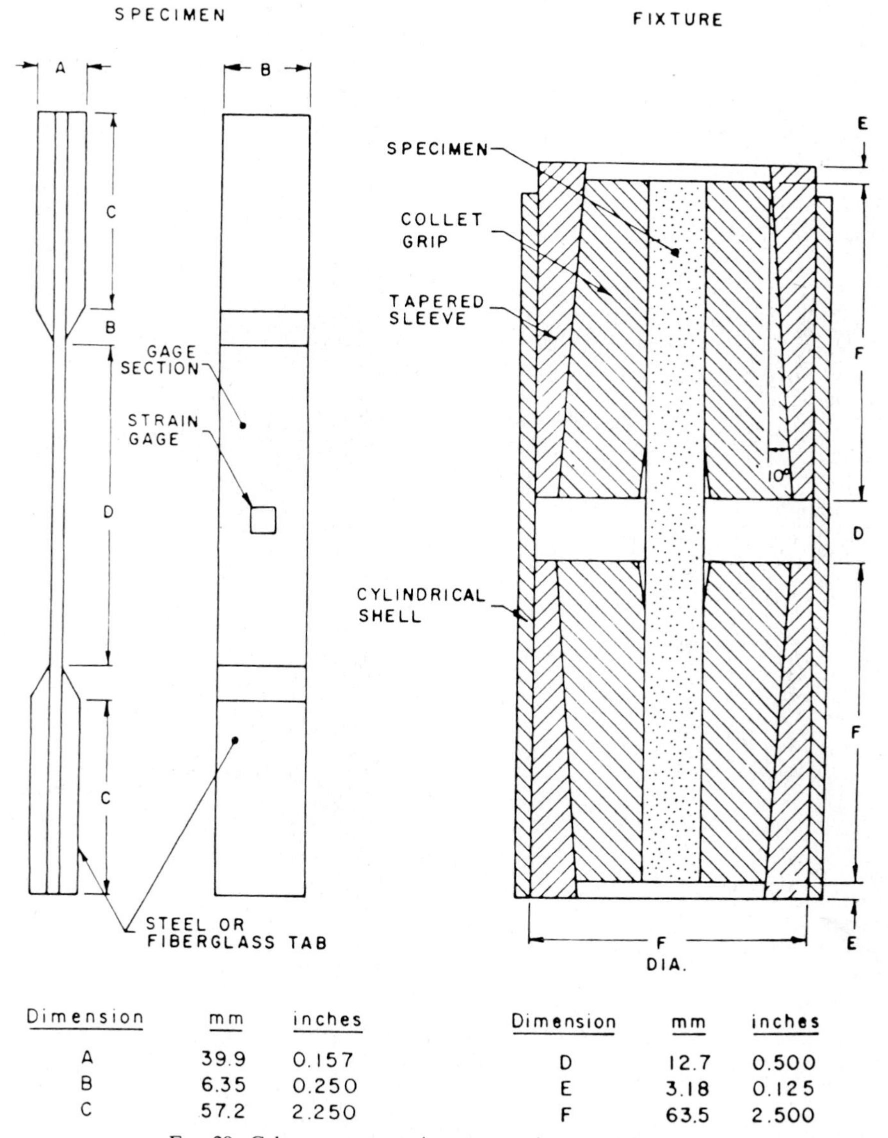

Dimension	mm	inches
A	39.9	0.157
B	6.35	0.250
C	57.2	2.250

Dimension	mm	inches
D	12.7	0.500
E	3.18	0.125
F	63.5	2.500

FIG. 29. Celanese compression test specimen and fixture.

(ASTM D-3410-75). The coupons are tabbed with long tapered glass/epoxy tabs leaving a gage section 1.27 cm (0.5 in.) long. Load is introduced through friction by means of split conical collet grips which fit into matching sleeves which in turn fit into a snugly fitting cylindrical shell (Fig. 29). One major disadvantage of this fixture is that it requires a perfect cone-to cone contact. This contact is not normally achieved due to small variations in tab thickness. Instead, contact is limited to two lines on opposite sides of the specimen. This unstable condition causes a laterial shift in the grips, which then produce high frictional forces in the enveloping cylinder. This situation has resulted in erroneously high values for the stiffness and compressive strength.

The IITRI test method represents a modification of the method above (HOFER and RAO (1977)). The conical grips were replaced with trapezoidal wedges. This eliminates the problem of line contact, since surface-to-surface contact can be attained at all positions of the wedges. Furthermore, it permits precompression of the specimen tabs to prevent slippage early in the load cycle. The lateral alignment of the fixture top and bottom halves is assured by a

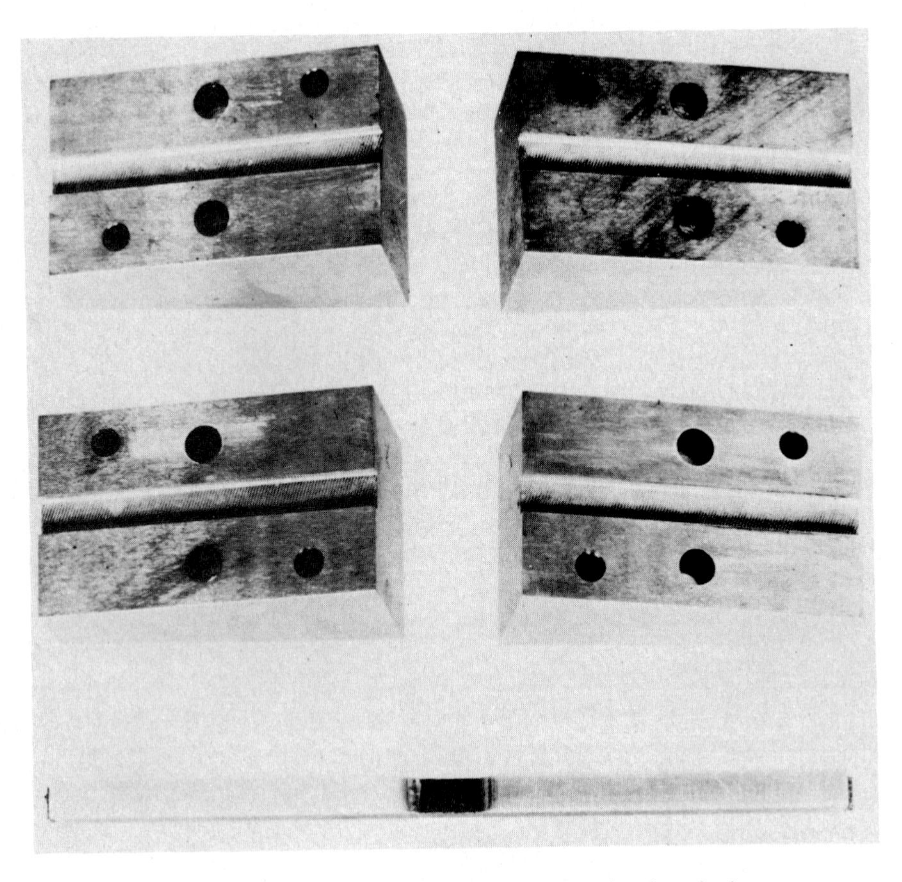

FIG. 30. IITRI compression test specimen and wedge-shaped grips.

guidance system consisting of two parallel roller bushings. The specimen, grips, and fixture assembly are illustrated in Figs. 30, 31, and 32.

Another example of a Type I compression test method is the Northrop method. This method utilizes a small untabbed coupon gripped between grips of unequal length. One elongated grip on each side of the specimen provides lateral support against buckling.

Another Type I test method was developed by the Bureau of Standards. It combines some of the features of the Celanese and IITRI test fixtures, while introducing a feature for introducing the load through tension applied to a pair of harnesses.

In the second category of test methods (Type II), a relatively long, fully supported specimen is used. One example of this type is the SWRI method developed at Southwest Research Institute (GRIMES et al. (1972)). The test specimen is similar to the tensile coupon discussed before but slightly shorter

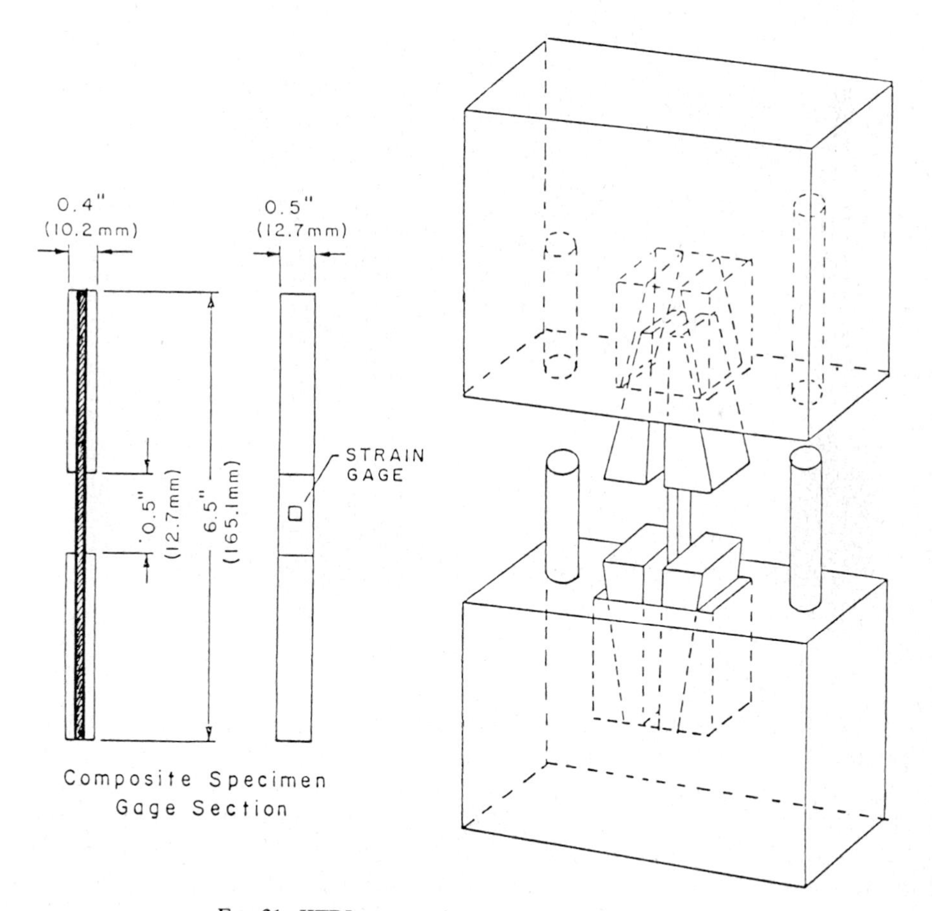

FIG. 31. IITRI compression test specimen and fixture.

FIG. 32. Alignment of IITRI compression test fixture in testing machine.

and with longer tabs. The fixture provides contact support over the entire gage length of the specimen. A second example of the Type II test method is the one developed by Lockheed-California Company (RYDER and BLACK (1977)). This fixture also provides lateral support over the gage length of the specimen. The Type II tests yield data similar to that obtained by the Type I tests, except in the case of 0° specimens when they give consistently lower values. This may be due to some premature buckling despite and lateral support.

FIG. 33. Composite sandwich beam test.

In the third category of compression test methods, the composite laminate is bonded to a honeycomb core which provides the required lateral support. In one case, two composite coupons are bonded to a honeycomb block and the sandwich specimen is loaded edgewise in direct compression. In the second case, a sandwich beam is made by bonding the composite laminate to one side of the honeycomb core and a balancing metal face sheet on the other side. The beam is loaded in four-point bending, as shown in Fig. 33, to produce nearly uniform compression in the test section of the composite sheet. Results obtained from sandwich beam tests tend to be higher than those obtained by the other methods discussed, probably because of the restraint offered by the honeycomb and the biaxial state of stress induced in the composite face sheet.

4.8. In-plane shear properties

Full characterization of a unidirectional composite requires the determination of lamina properties under in-plane shear parallel to the fibers, i.e., shear modulus, G_{12}, shear strength, S_{12}, and ultimate shear strain, ε_{12}^{u}. There are four generally accepted test methods for determination of these properties: (1) the $[\pm 45]_{ns}$ coupon test, (2) the 10° off-axis test, (3) the rail shear test, and (4) the torsion test.

The first test method utilizes an 8-ply $[\pm 45]_{2s}$ coupon of the same dimensions as the 90° unidirectional tensile coupon discussed before. When this coupon is subjected to a uniaxial tensile stress σ_x, the state of stress in each individual lamina, referred to the fiber coordinate system, is shown in Fig. 34 and is expressed as

$$\sigma_{11} = \frac{\sigma_x + \sigma_{xy}}{2}, \qquad \sigma_{22} = \frac{\sigma_x - \sigma_{xy}}{2}, \qquad \sigma_{12} = \frac{\sigma_x}{2} \tag{41}$$

where σ_{xy} is the shear stress in the lamina introduced by shear coupling.

The corresponding strains are

$$\varepsilon_{11} = \varepsilon_{22} = \frac{\varepsilon_x + \varepsilon_y}{2}, \qquad \varepsilon_{12} = \frac{\varepsilon_x - \varepsilon_y}{2} \tag{42}$$

where ε_x, ε_y are the axial and transverse strains in the laminate, measured by using two-gage rosettes on the test coupon. The in-plane (or intralaminar) shear modulus of the unidirectional lamina is obtained from the slope of the σ_{12} vs. ε_{12} curve as

$$G_{12} = \frac{\sigma_x}{2(\varepsilon_x - \varepsilon_y)}, \tag{43}$$

This test method was proposed by ROSEN (1972). The approximation in-

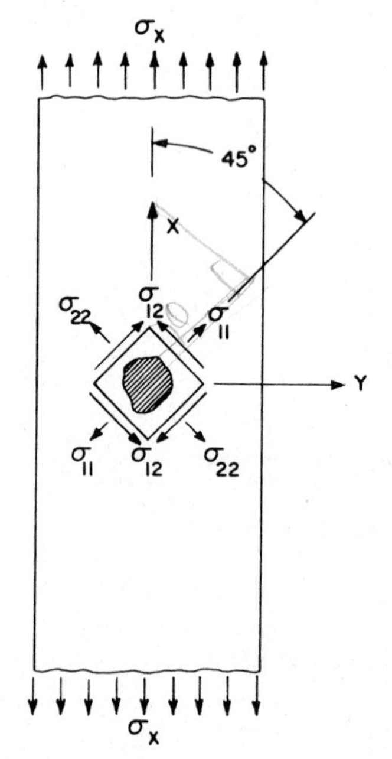

FIG. 34. State of stress in lamina of uniaxially loaded $[\pm 45]_{ns}$ coupon.

volved in this procedure does not take into account edge effects or the influence of the other stress components σ_{11} and σ_{22}. These influences would be especially noticeable in the determination of the shear strength and ultimate shear strain values. A typical shear stress versus shear strain curve obtained from a $[\pm 45]_{2s}$ graphite/epoxy coupon is shown in Fig. 35.

The second test method is the 10° off-axis test (DANIEL and LIBER (1975), CHAMIS and SINCLAIR (1977)). The 10° angle was chosen to minimize the effects of longitudinal and transverse stress components σ_{11} and σ_{22} on the shear response. The specimen is a 6-ply unidirectional coupon with the fibers oriented at 10° with the loading axis, 1.27 cm (0.5 in.) wide and 33 cm (13 in.) long. They are tabbed with tapered tabs and instrumented with 3-gage rosettes on each side of the test section. The 3-gage rosette normally has one gage element oriented axially (ε_x), one at 45° with the loading axis or 55° with the fiber direction (ε_{45}), and one transversely to the loading axis (ε_y). The specimen is subjected to a uniaxial tensile stress σ_x in increments up to failure. The intralaminar shear stress referred to the fiber coordinate system is given by:

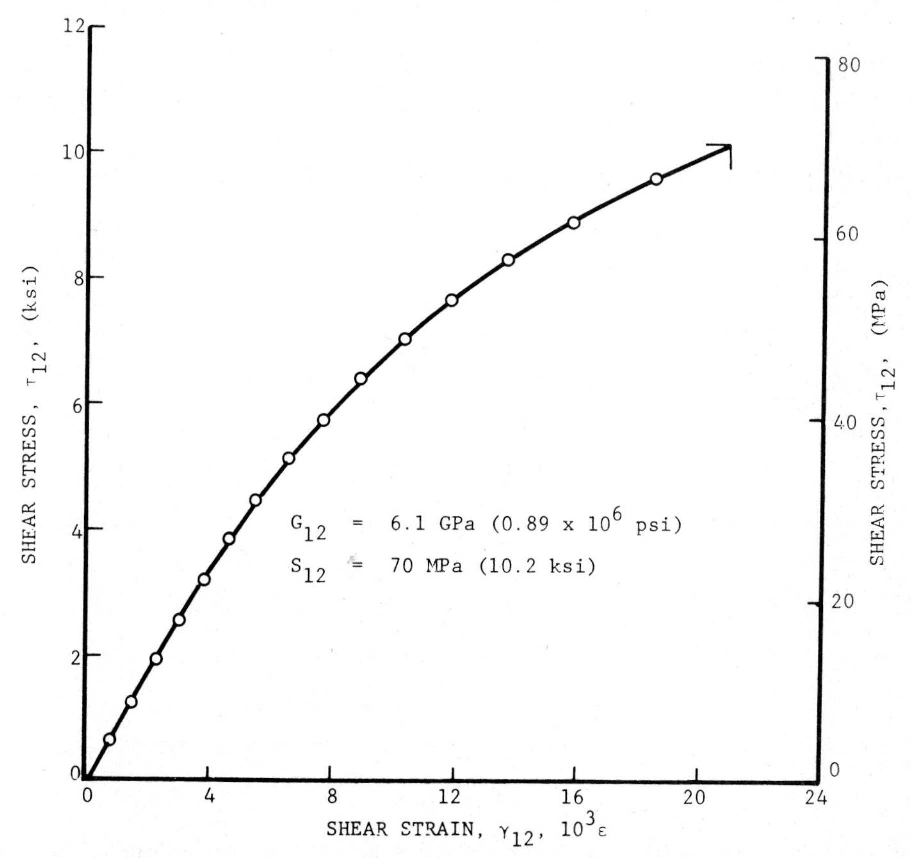

FIG. 35. Shear stress versus shear strain in $[\pm 45]_{2s}$ specimen (SP286T300 graphite/epoxy).

$$\sigma_{12} = \sigma_x \sin\theta \cos\theta = 0.171\,\sigma_x\,. \tag{44}$$

The corresponding shear strain can be obtained in terms of the three strain components measured with the 3-gage rosette as

$$\varepsilon_{12} = \frac{\varepsilon_x - \varepsilon_y}{2}\sin 2\theta + \left(\varepsilon_{45} - \frac{\varepsilon_x + \varepsilon_y}{2}\right)\cos 2\theta \tag{45}$$

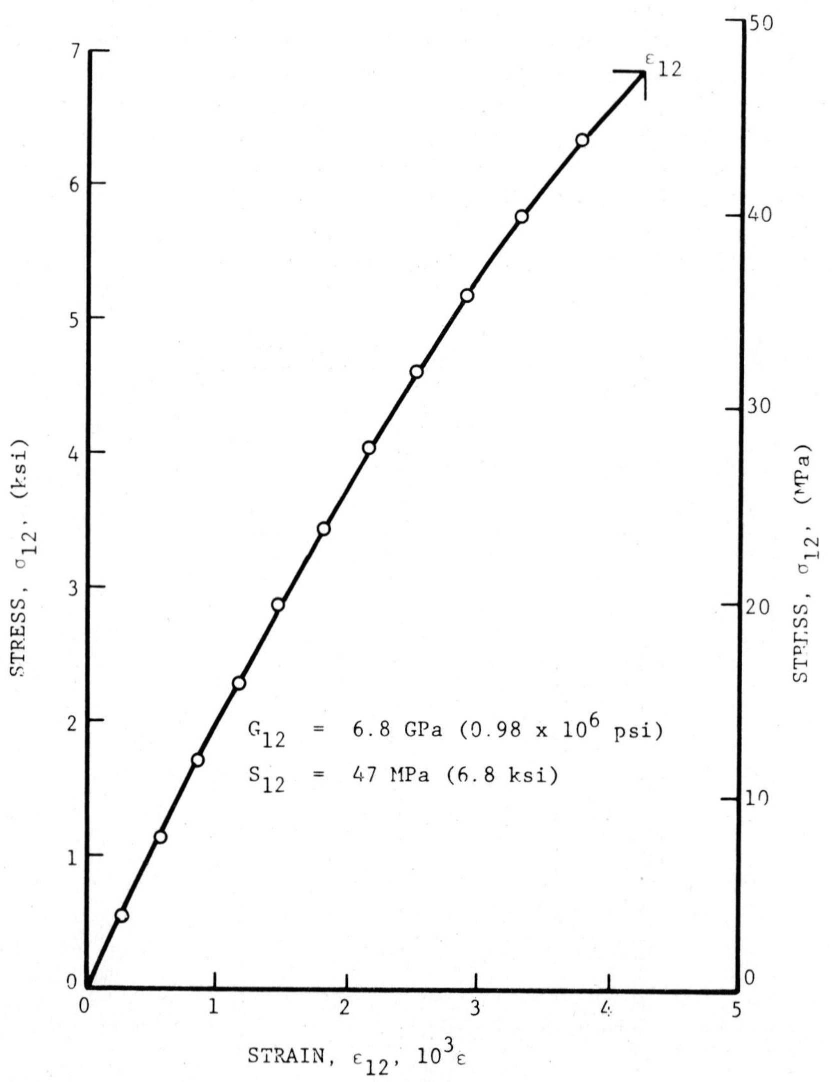

FIG. 36. Shear stress versus shear strain for 10-deg off-axis unidirectional specimen (SP286T300 graphite/epoxy).

where $\theta = 10°$. The in-plane shear modulus is obtained by plotting σ_{12} versus ε_{12} and taking the initial slope of the curve as

$$G_{12} = \frac{\sigma_{12}}{2\varepsilon_{12}}. \tag{45}$$

The ultimate values of σ_{12} and ε_{12} determine the shear strength and ultimate shear strain. A typical shear stress versus shear strain curve obtained from a 10° off-axis graphite/epoxy specimen is shown in Fig. 36. Results obtained from such tests with graphite/epoxy are included in Table 3.

The third method of determining shear properties is the rail shear test. In the two-rail test, a rectangular coupon specimen is bolted and/or bonded to two pairs of rigid steel rails. For determination of in-plane shear properties, the fibers of the specimen are oriented parallel to the axes of the rails. The load is applied at one end of one pair of rails and reacted at the opposite end of the other pair of rails. The loading axis in this case is not parallel to the fibers of the specimen and, thus, it introduces a transverse compressive stress across the fibers in addition to the in-plane shear stress. This situation is alleviated by

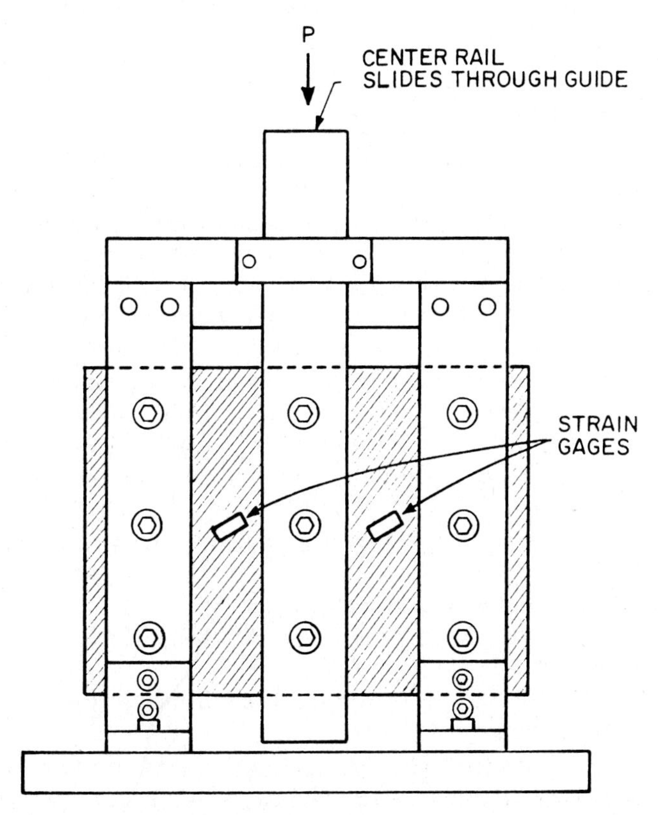

FIG. 37. Three-rail shear fixture.

using the three-rail shear fixture, where the rectangular specimen is clamped between three parallel pairs of rails (Fig. 37). The load is applied to one end of the middle rail and reacted at the opposite ends of the two outer rails.

The average shear stress is obtained by dividing half the applied load by the thickness times the clamped length of the specimen. The shear strain is obtained from a single gage placed at the center of the exposed specimen at 45° with the rail axes. In this type of specimen the state of stress near the ends is not pure shear and may thus result in premature failure due to the other stress components.

The fourth method for determining in-plane shear properties is the torsion method. In this method a solid rod or a hollow tubular specimen is subjected to torque. In the case of the solid rod, the shear stress distribution can only be determined for the initial portion of the load-deformation response. The thin-wall tube is a much better specimen for the torsion test. The main difficulty lies in the fabrication of such a specimen and the load introduction through specially made grips. A shear stress-strain curve can be obtained by plotting torque versus angle of twist or shear strain measured directly with strain gages.

5. Biaxial testing

5.1. Introduction

The response of composite laminates to biaxial states of stress is difficult to predict analytically on the basis of lamina properties because of nonlinear behavior and failure mode interactions. Several theories have been presented to date for predicting biaxial properties of composites, but they all rely to some extent on some limiting assumptions (TSAI and WU (1971), PETIT and WADDOUPS (1969), SANDHU (1974), WU (1974)). To check or verify some of these theories and to generate useful failure envelopes for design purposes, it is necessary to conduct extensive testing of composite laminates under biaxial states of stress. The application of a general in-plane biaxial state of stress, including tension, compression and shear components, poses one of the most difficult problems in composite testing. Some of the basic requirements for a biaxial test specimen are:

(1) A significant volume of the material must be under a homogeneous state of stress.

(2) Primary failure must occur in the test section.

(3) The state of stress must be known without the need for secondary measurements or analysis.

(4) It must be possible to vary the three in-plane stress components (σ_{xx}, σ_{yy}, σ_{xy}) independently.

A variety of specimen types and techniques have been proposed and used for biaxial laminate characterization. They include the off-axis coupon or ring, the crossbean sandwich specimen, bulge plate, rectangular plate loaded in biaxial tension, and the thin-wall tubular specimen.

5.2. *Off-axis specimen*

Uniaxial loading along a direction other than one of the principal material axes produces a biaxial state of stress. This specimen has been used successfully in coupon and ring forms (COLE and PIPES (1972), ROWLANDS (1975)). In the latter case thin-wall rings with the principal material axes at an angle with the circumferential direction are subjected to internal pressure loading. Some of the limitations of this specimen are:

(1) The biaxial normal stresses are always of the same sign.

(2) There is no possibility for independent variation of the three stress components (non-proportional loading).

The off-axis laminate can also be tested in compression by using it as a skin in a honeycomb-core sandwich beam under pure bending. Within the limitations stated before the specimen is simple to use and reliable.

5.3. *Crossbeam sandwich specimen*

This specimen consists of two intersecting beams subjected to pure bending. In one of the many types discussed, the laminate is used as one skin of a honeycomb-core sandwich (Fig. 38). This type of specimen allows, in principle, the application of tension–tension, tension–compression, and compression–compression loading in the central test section. One of the major limitations of this specimen is the disturbing influence of the corners on stress distributions in the test section and on fracture initiation at the points of stress concentration.

Some attempts to alleviate this situation have been made by reducing the thickness in a central portion of the test section. An elliptical reduced-thickness section has been proposed by BERT et al. (1969) to insure a uniform biaxial stress state throughout the test region. The ellipse is selected to satisfy the expression

$$b/a = (\sigma_{22}/\sigma_{11})^{1/2} \tag{46}$$

where a and b are the major and minor axes, and σ_{11} and σ_{22} the principal

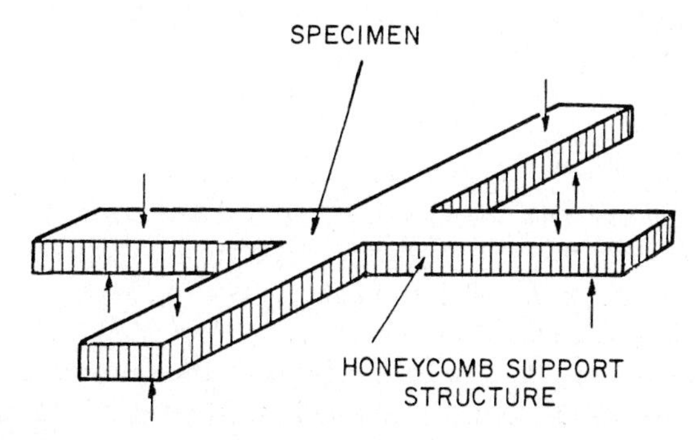

FIG. 38. Crossbeam sandwich specimen for biaxial testing of composite laminates.

stress components. Fabrication of specimens with elliptical reduced-thickness sections is feasible although more expensive. The major drawback of this approach is that the state of stress in the test section is not determinate from the specimen geometry and applied loads.

5.4. *Flat plate specimen*

This specimen is a usually square plate subjected to tension-tension loading on its sides through fiberglass tabs. A variety of biaxial states of stress (in the tension-tension shear space) can be achieved by rotating the principal material axes with respect to the loading directions. Nonproportional loading is possible to some degree. To insure stress homogeneity within a reasonable test section and failure within this region, it is necessary to design the tabs and transition region very carefully.

This specimen is most suitable when the influence of biaxial stress on notches is investigated (DANIEL (1976, 1977, 1980, 1981)). Tensile loading is introduced by means of four whiffle-tree grip linkages designed to insure that four equal tensile loads are applied to each side of the specimen. A photograph of a biaxial specimen with the loading grip linkages is shown in Fig. 39.

Load is applied by means of two pairs of hydraulic jacks attached to the four sides of a reaction frame. The load is transmitted from the hydraulic cylinders to the grip linkages through cylindrical rods going through the bore of these cylinders. The rods are instrumented with strain gages and calibrated in a testing machine to establish the exact relationship between loads and strain gage signals. These strain gage readings are used subsequently both for recording the exact loads applied to the specimen and as feedback signals for controlling the pressures by means of the servo-hydraulic system used. A special fixture is used to help align the specimen in the loading frame.

In the case of unnotched specimens, to achieved a state of uniform stress in the untabbed test section, this section must be shaped in the form of an ellipse with its major and minor axes related to the applied stress biaxiality as expressed by Equation (46).

The major limitations of this type of test are that it is primarily limited to tension-tension loading and that, in the case of unnotched specimens, it is very difficult to induce failure in the test section.

5.5. *Thin-wall tubular specimen*

Of the various biaxial test specimens mentioned, the tubular specimen appears to be the most versatile and offers the greatest potential. It offers the possibility of applying any desired biaxial state of stress with or without proportional loading. The state of generalized plane stress can be achieved by the independent application of axial loads, internal or external pressure, and torque. Tubular specimens have been used successfully with metals because stress concentrations in the load introduction region are relieved by plastic

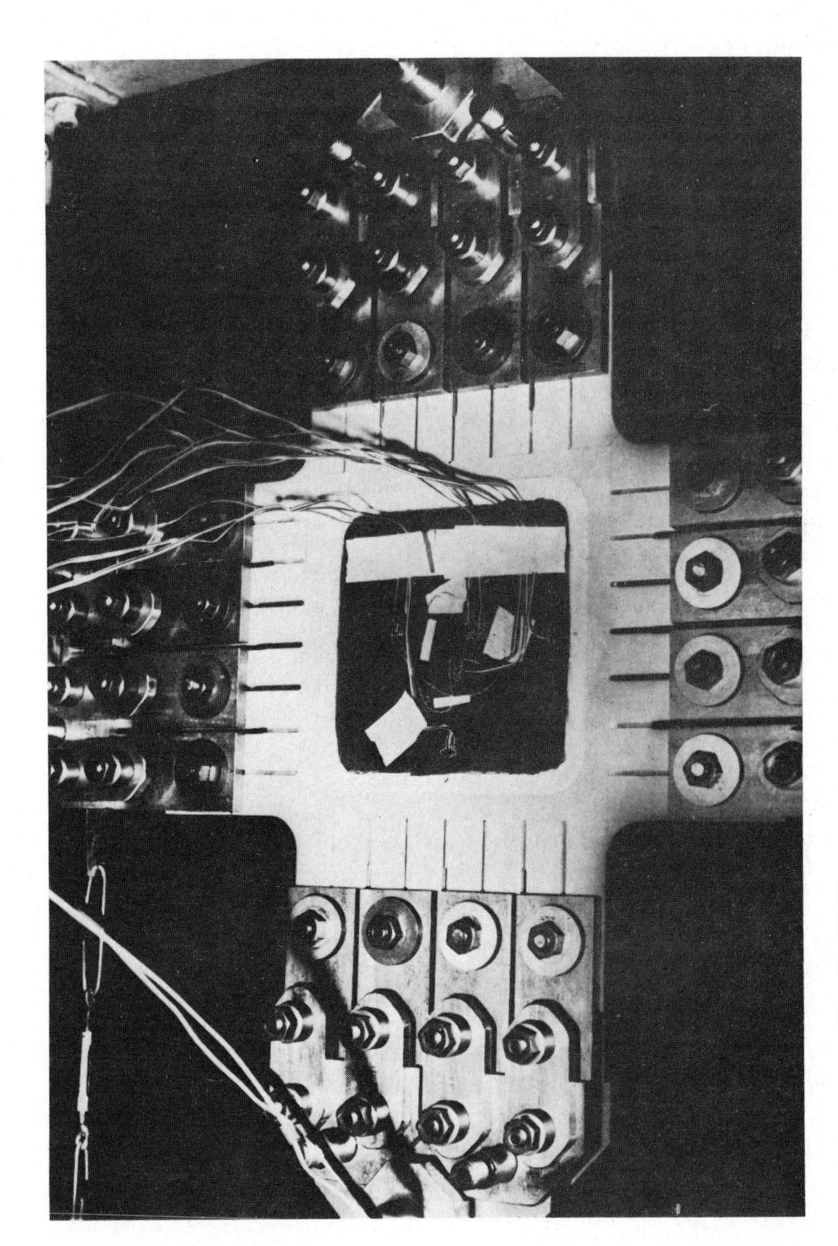

FIG. 39. Whiffle-tree linkage grips for load introduction in biaxial plate specimen.

yielding. This is not the case with brittle-like materials such as composites. For this reason no entirely satisfactory solution has been found to date. In order to achieve the full potential of the thin-wall tubular specimen, the following requirements must be met:

(1) The tube must be loaded without constraints which would produce local extraneous or non-homogeneous stresses.

(2) Surface pressures on the laminate in the test section used for producing circumferential or axial stresses should be minimized to avoid adding a high radial stress component resulting in a triaxial state of stress.

(3) Functional or material failures of the load-introduction tabs must be avoided.

(4) Undesirable buckling prior to material failure must be avoided.

(5) Cost of specimen fabrication, equipment, and testing process must not be prohibitive.

The general biaxial state of stress is produced by means of internal and external pressures p_i and p_o, longitudinal load P_x (which can also be applied by means of p_i and p_o), and torque $T_{x\theta}$ about the longitudinal axis. The axial, circumferential, and shear stresses are obtained as

$$\begin{aligned} \sigma_x &= P_x/2\pi\bar{R}t\,, \\ \sigma_\theta &= (p_i - p_o)\frac{\bar{R}}{t} - \frac{1}{2}(p_i + p_o)\,, \\ \tau_{x\theta} &= T_{x\theta}/2\pi\bar{R}^2 t \end{aligned} \tag{47}$$

where $\bar{R}$ is the mean radius and t the tube thickness.

It is generally assumed that the gradient $\Delta\sigma_\theta$ of the circumferential stress across the thickness is small compared to the average value of σ_θ. However, this is not always the case. Since

$$\Delta\sigma_\theta = (p_i - p_o) \tag{48}$$

the stress gradient ratio is given by

$$\frac{\Delta\sigma_\theta}{\sigma_\theta} = (t/\bar{R})\left[1 - \frac{t}{2\bar{R}}\left(1 + \frac{p_o}{p_i}\right)\Big/\left(1 - \frac{p_o}{p_i}\right)\right]^{-1} \tag{49}$$

For example, for $\bar{R}/t = 10$ and $0 \leq p_o/p_i \leq 0.8$, the stress-gradient ratio varies between 0.10 and 0.18.

The most critical problem and most difficult to solve pertains to the mechanism(s) of load introduction into the specimen. Generally, tubular specimens are provided with end tabs to support seals and to accommodate load introduction. They are finite in length, even short sometimes, to prevent premature buckling failures. The increased stiffness in the tab region, the existence of an unpressurized portion at the end of the tube and the introduction of axial loading (especially if it is done through mechanical gripping) produce discontinuities in the circumferential and axial stresses in the transition between tab and test section. A great deal of effort has been devoted to the load introduction problem in designing systems for biaxial testing of composite tubes.

Testing of composite tubular specimens has been discussed by several investigators who analyzed the various problems arising in this type of specimen (PAGANO and WHITNEY (1970), WIDERA and CHUNG (1972), WHITNEY et al. (1973), SULLIVAN and CHAMIS (1973), GUESS and GERSTLE (1977)). One of the most frequent and most critical problems encountered is that of introducing and maintaining a uniform biaxial state of stress in the specimen test section and inducing failure in the test region. Some testing of tubular specimens has been done without any provisions for relieving end constraints (GUESS and GERSTLE (1977), HÜTTER et al. (1975)). To overcome or minimize end constraints and gripping problems the concepts of tab and grip pressurization have been used (COLE and PIPES (1972, 1974), LINDHOLM et al. (1975)). These concepts, however, have not been implemented with full success because of the inherent difficulty of the problem.

A typical tubular specimen geometry is illustrated in Fig. 40. The specimen is designed to have end tabs for gripping and load introduction. These tabs are made of epoxy or glass/epoxy pre-machined and bonded to the specimen ends. The stiffness of these tabs in the axial and circumferential directions relative to those of the composite tube can be varied by varying the glass/epoxy layup and the tab thickness.

This specimen geometry has been analyzed extensively using finite element methods (DANIEL et al. (1980)). The objective of these analyses was to minimize stress discontinuities in the transition between the test section and tabbed

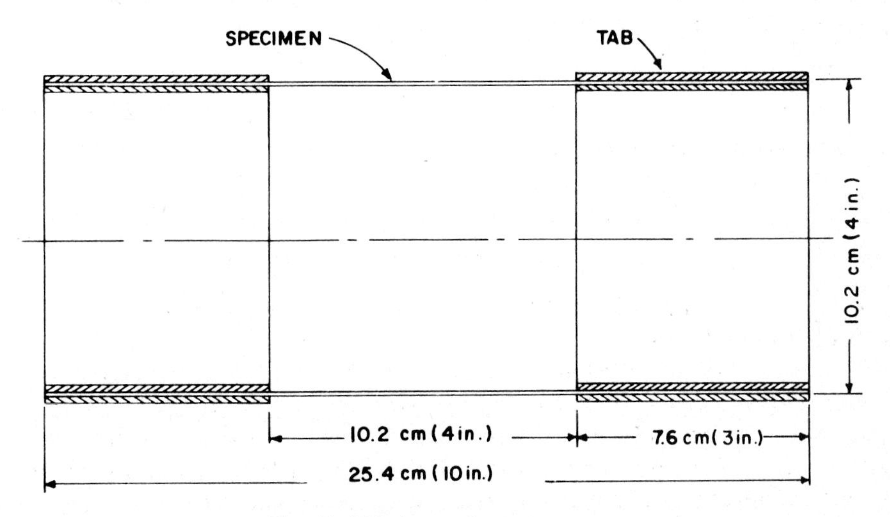

FIG. 40. Tubular specimen geometry.

section of the specimen by varying the tab materials and geometry and the tab compensating pressures. The most critical loading from the point of view of stress discontinuity is pressure loading. The most critical stresses are the axial bending stresses in the transition region.

Fig. 41 shows axial stress distributions along the inner and outer elements of an internally pressurized $[0/\pm45/90]_s$ graphite/epoxy tube with glass/epoxy tabs of rectangular cross section. A peak axial stress of $\sigma_{xx} = 7.5\,p_i$ is reached in the transition region. After a series of iterations it was found that tapered tabs with tapered extensions of a lower stiffness material result is the lowest peak discontinuity stresses as shown in Fig. 42. Here the peak axial stress has been reduced to $\sigma_{xx} = 2.4\,p_i$. In terms of the primary circumferential stress, the peak axial stress above can be expressed as $\sigma_{xx} = 0.048\,\sigma_{\theta\theta}$. The actual deviation from the state of stress in the middle of the test section is even smaller.

A similar analysis for the case of torsional loading showed that no shear stress peaks are produced in the transition region. In the case of axial loading applied through grips at the end of the tab high localized stress peaks are present at the outer surface of the tab (Fig. 43). This localized stress can be alleviated somewhat by a slight rounding off of the grip ends, so that no premature tab failure can result.

The preparation of tubular specimens must be done very carefully to insure that the specimen is of a quality similar to that of flat laminates fabricated and cured in an autoclave. Composite tubes can be fabricated in any desired ply

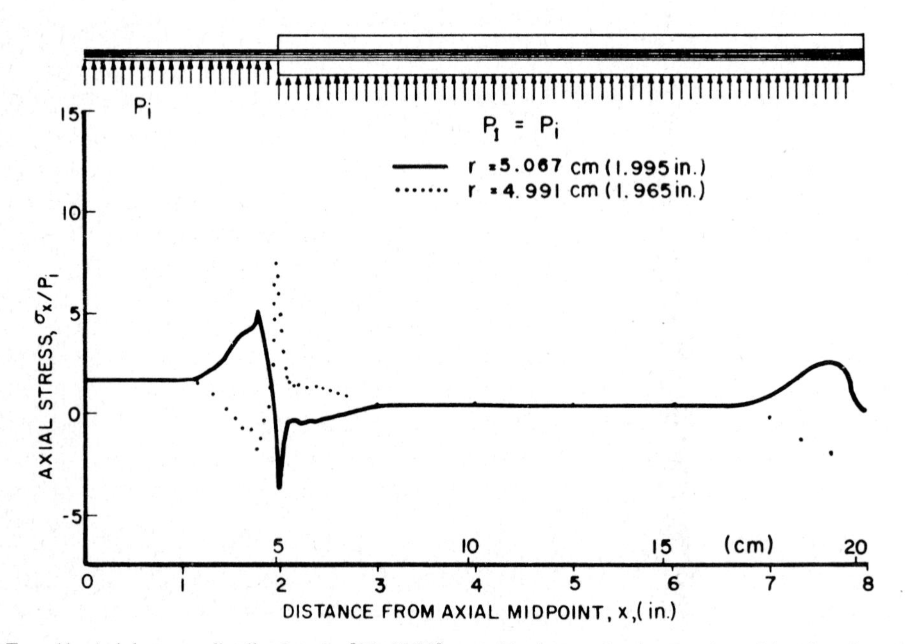

FIG. 41. Axial stress distribution in $[0/\pm45/90]_s$ graphite/epoxy laminate tube with tabs of rectangular cross section for internal pressure loading.

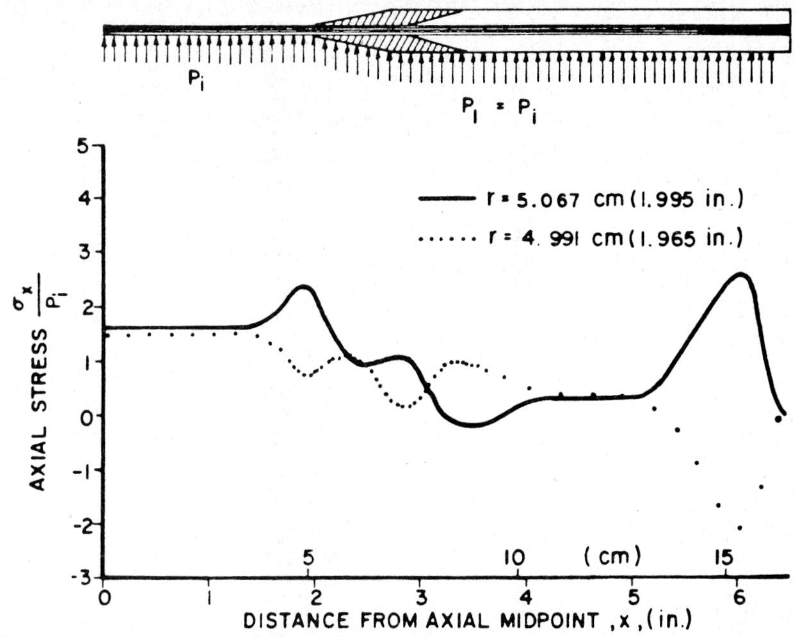

FIG. 42. Axial stress distribution in $[0/\pm45/90]_s$ graphite/epoxy laminate tube with tapered tabs and lower stiffness for internal pressure loading.

orientation and stacking sequence. Fabrication techniques have been developed and described in the technical literature (IIT Research Institute (1970–75), COLE and PIPES (1974), WEED and FRANCIS (1977), LIBER et al. (1977a)). The latter procedure consists of wrapping the prepreg composite tape around a cylindrical perforated mandrel and then, by means of internal pressure, expanding the prepreg tube against the wall of the cylindrical cavity mold tool. Fig. 44 shows the consecutive layers of materials around the steel mandrel required in the fabrication process. The fabrication mold with the composite tube is shown in Fig. 45. The glass/epoxy tabs required for the specimen are fabricated as tubes in a similar manner. They are subsequently machined to size and bonded on the composite tube.

The complete system for biaxial testing of composite tubular specimens requires the introduction and control of internal and external pressures on the specimen, internal and external compensating pressures on the tabs, grip activation pressure, axial load, and torque. The axial loads and torque are introduced through segmented collet grips at the tabs by means of linear and torque actuators. All pressures applied directly to the specimen or to the actuators are applied and controlled independently of each other with an electrohydraulic system. A photograph of a complete biaxial test system is shown in Fig. 46.

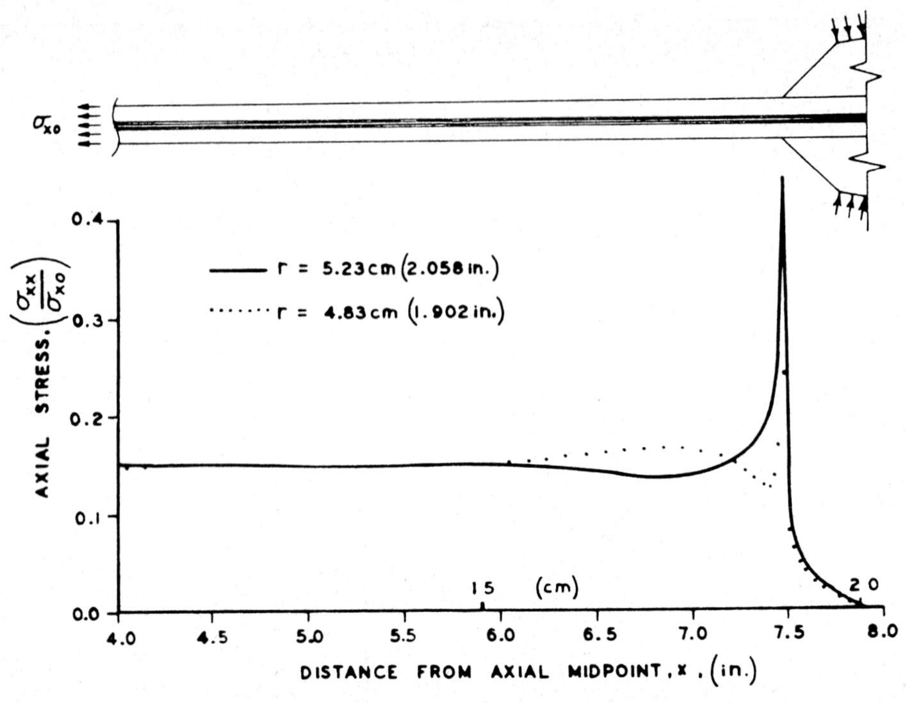

FIG. 43. Axial stress distribution along centroids of inner and outer tab elements for $[0/\pm45/90]_s$ graphite/epoxy specimen and tab layup for axial tensile loading.

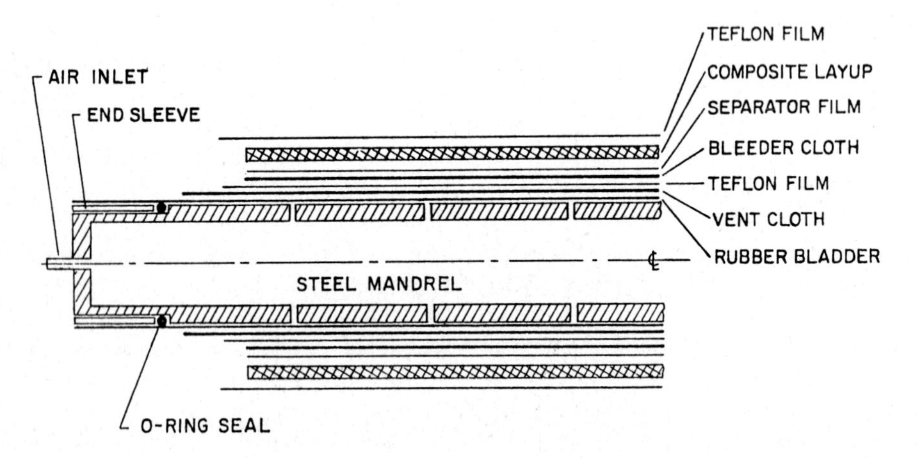

FIG. 44. Sketch showing the center steel mandrel and the sequence of material layers employed in the composite laminate tube fabrication.

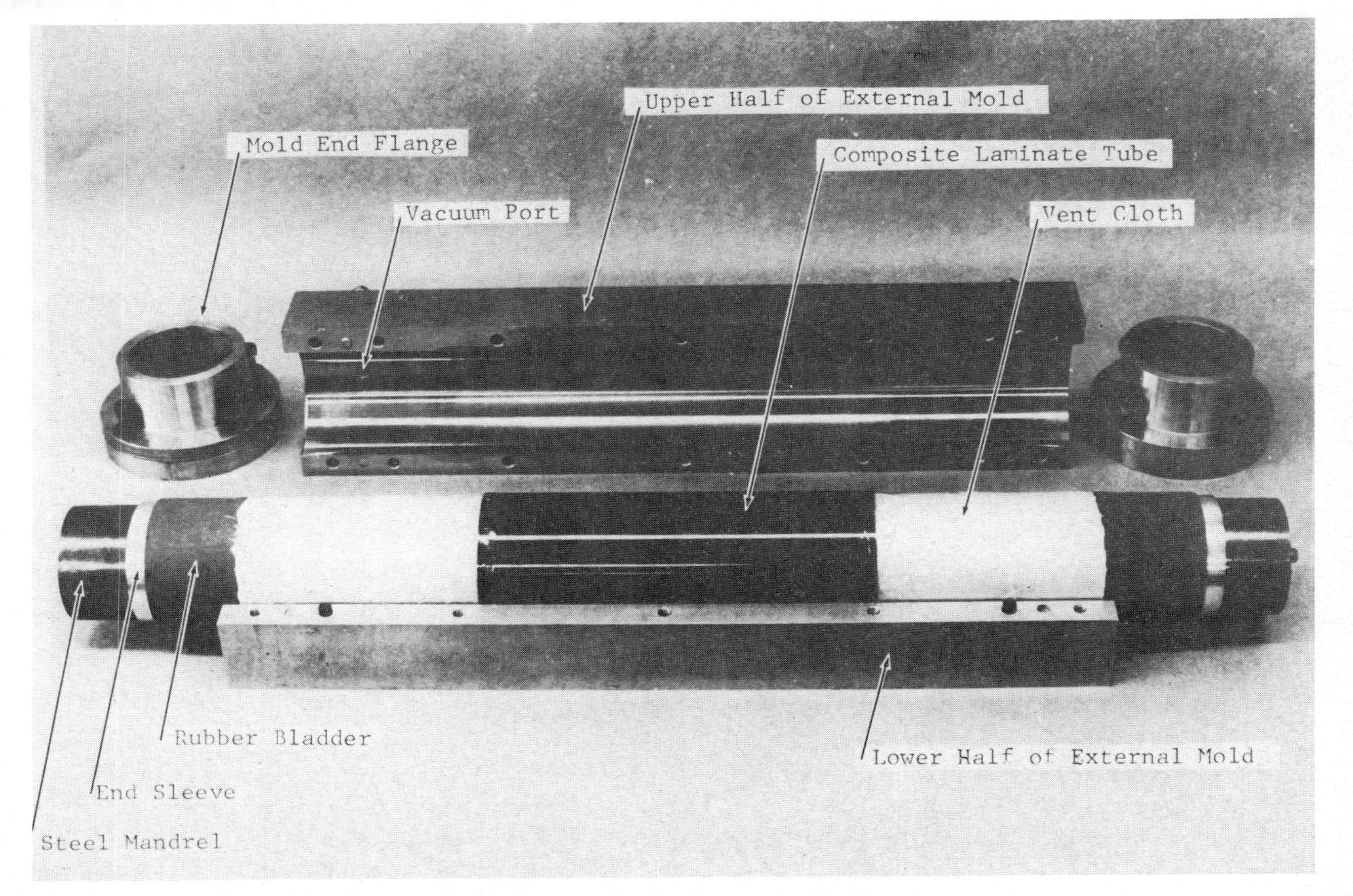

FIG. 45. Components of cavity mold tool for fabricating composite laminate tubes.

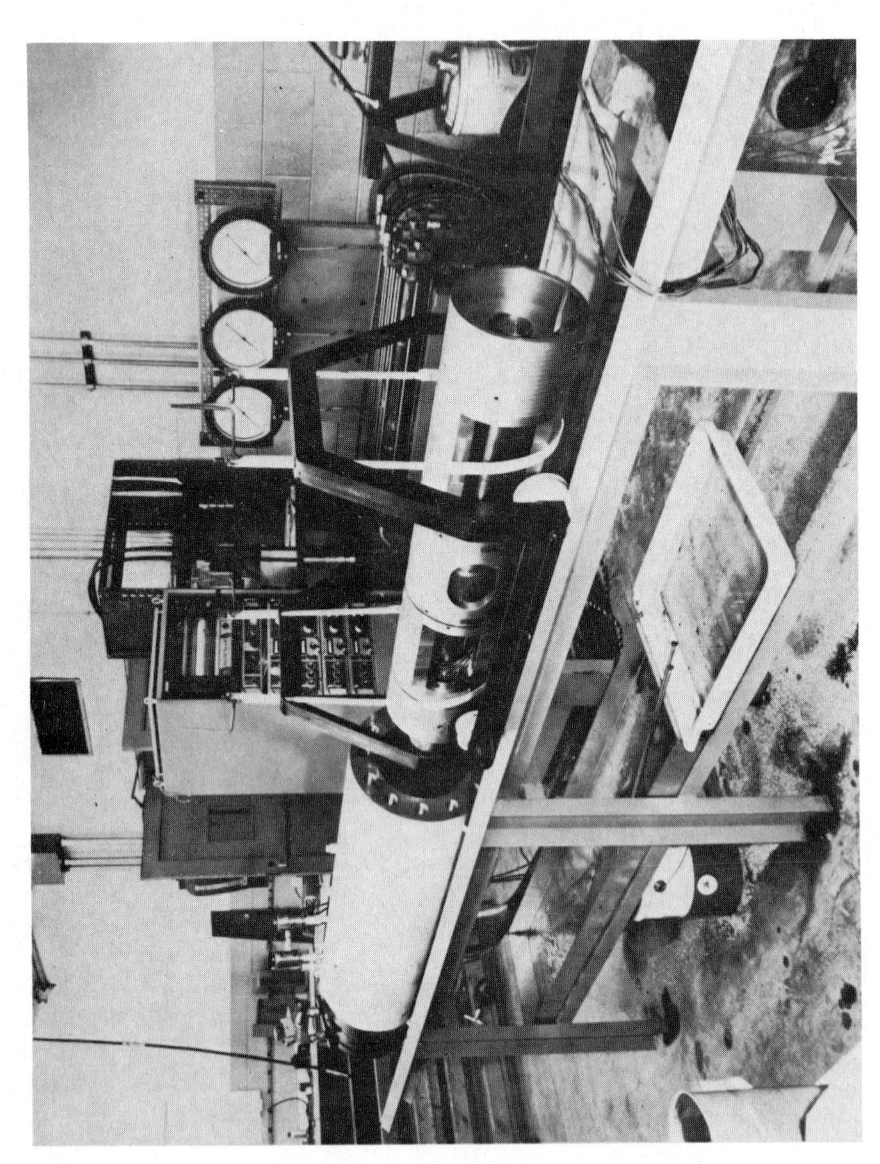

FIG. 46. System for biaxial testing of composite tubes.

6. Applications

6.1. Residual stresses

Lamination residual stresses in angle-ply laminates are produced during curing as a result of the anisotropic thermal deformations of the variously oriented plies. These stresses have been investigated recently both analytically and experimentally (CHAMIS (1971), HAHN and PAGANO (1975), HAHN (1976), DANIEL et al. (1975), DANIEL and LIBER (1975, 1976 a, b, 1977 a, b). They are a function of many parameters, such as type of fiber and matrix, fiber volume ratio, ply orientation, curing temperature, and other variables. They can reach values comparable to the transverse strength of the ply and thus induce cracking of that ply within the laminate. They are equilibrated with interlaminar shear stresses transmitted from adjacent plies and can thus cause delamination.

Residual stresses during curing have been measured in a variety of angle-ply laminates using embedded strain gage techniques (DANIEL and LIBER (1975, 1976a)). Unidirectional and angle-ply specimens were instrumented with surface and embedded gages and thermocouples and the output recorded during curing and postcuring. The unidirectional specimen was used as a reference to determine the unrestrained stress-free thermal expansion of an individual ply.

It was found that apparent strains recorded during the heating stage of the curing cycle are not significant as they correspond to the fluid state of the matrix resin. Residual stress buildup occurs only upon solidification of the matrix at the peak curing temperature and during subsequent cooldown. Strains measured during the cooldown stage of curing as well as those measured during postcuring correspond to the thermal expansion of the laminate.

Thermal strains measured in a unidirectional graphite/epoxy laminate are shown in Fig. 47(a) with room temperature as the reference level. Thermal strains measured in a $[0_2/\pm 45]_s$ graphite/epoxy angle-ply laminate during the cooling stage of postcuring are shown in Fig. 47(b). The residual stresses induced in each ply correspond to the so-called restraint strains, i.e., the difference between the unrestrained thermal expansion of that ply (obtained from the unidirectional specimen) and the restrained expansion of the ply within the laminate (obtained from the angle-ply specimen). Restraint or residual strains in the 0- and 45-deg plies of the $[0_2/\pm 45]_s$ graphite/epoxy laminate are plotted in Figs. 48(a) and (b) as a function of temperature with room temperature taken as the reference level. The stress-free level can be shifted to 444°K (340°F), the temperature at which the matrix solidifies. Other investigators have claimed that the stress-free temperature level might be somewhat lower than the peak curing temperature as indicated by comparing experimental and theoretical results (HAHN (1976)). In the case in question the maximum residual strain at room temperature is 6.43×10^{-3} in the ± 45-deg

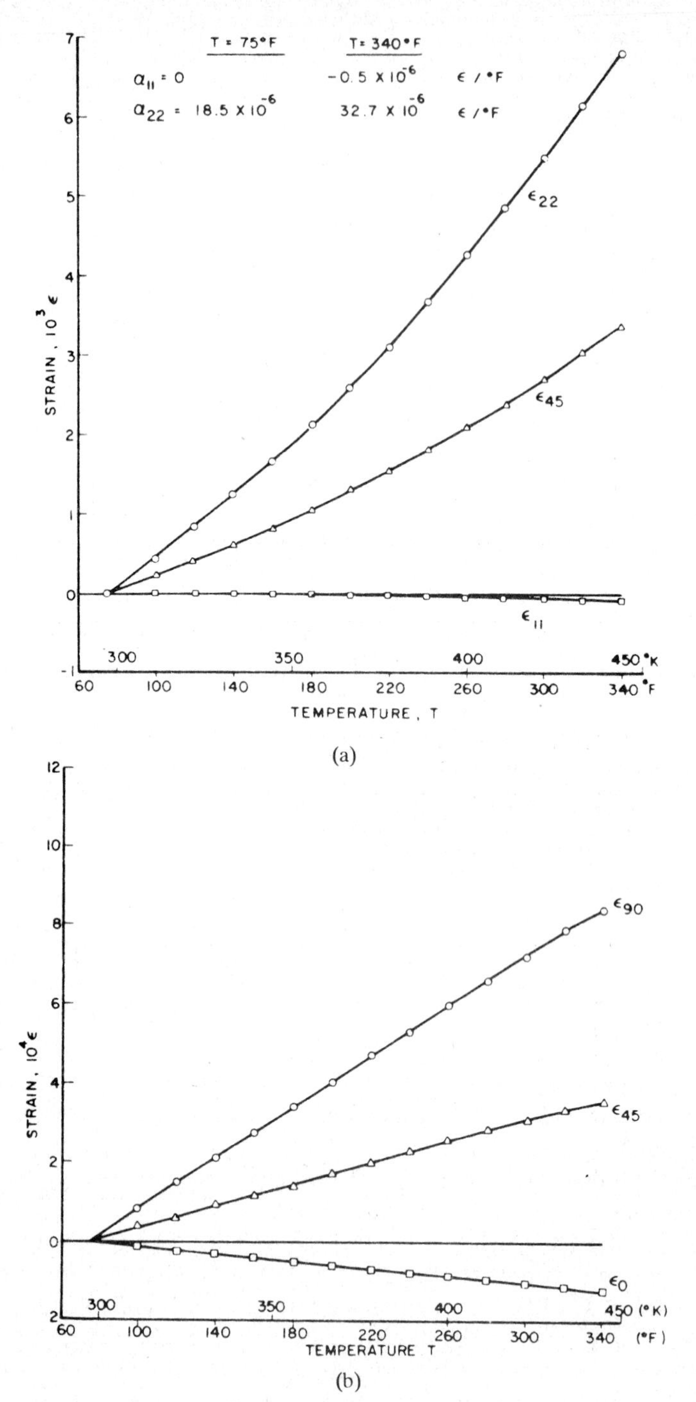

FIG. 47. Thermal strains in graphite/epoxy specimens, (a) $[0_8]$ unidirectional specimen, (b) $[0_2/\pm45]_s$ specimen.

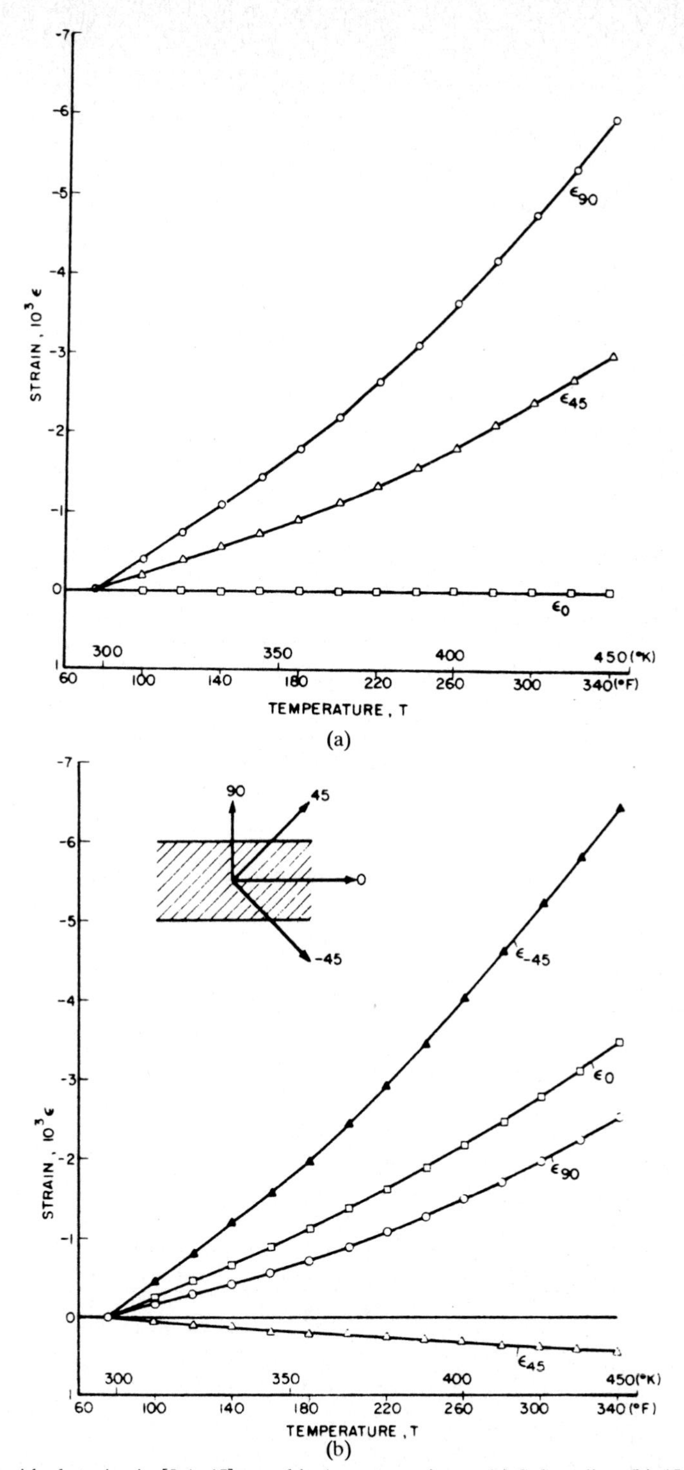

FIG. 48. Residual strains in $[0_2/\pm45]_s$ graphite/epoxy specimen, (a) 0-deg plies, (b) 45-deg plies.

plies in the transverse to the fiber direction. The corresponding maximum residual strain in the 0-deg plies is 5.95×10^{-3}.

Residual stresses in any given ply within the laminate at any given temperature can be computed from the residual strains using the appropriate thermoelastic orthotropic constitutive relations such as

$$\begin{Bmatrix} \sigma_x \\ \sigma_y \\ \sigma_{xy} \end{Bmatrix} = \begin{bmatrix} \bar{Q}_{11} & \bar{Q}_{12} & \bar{Q}_{16} \\ \bar{Q}_{12} & \bar{Q}_{22} & \bar{Q}_{26} \\ \bar{Q}_{16} & \bar{Q}_{26} & \bar{Q}_{66} \end{bmatrix} \begin{Bmatrix} \varepsilon_x - \alpha_x \,\Delta T \\ \varepsilon_y - \alpha_y \,\Delta T \\ 2\varepsilon_{xy} - 2\alpha_{xy} \,\Delta T \end{Bmatrix} \tag{50}$$

where $[\bar{Q}]$ is the stiffness matrix of the lamina referred to the x-y axes of the laminate.

The strain components in the equation above represent the difference between the actual strains in the laminate and the unrestrained thermal expansion in the ply, assuming constant coefficients of thermal expansion.

In a somewhat different approach, residual stresses in a given ply are obtained directly from the measured restraint strains $\varepsilon_{ij}^{\mathrm{r}}$ discussed before. Assuming linear elastic behavior, the residual stresses in any given ply referred to the ply principal directions are given by

$$\begin{aligned} \sigma_{11} &= \frac{E_{11}}{1 - \nu_{12}\nu_{21}} [\varepsilon_{11}^{\mathrm{r}} + \nu_{21}\varepsilon_{22}^{\mathrm{r}}] , \\ \sigma_{22} &= \frac{E_{22}}{1 - \nu_{12}\nu_{21}} [\varepsilon_{22}^{\mathrm{r}} + \nu_{12}\varepsilon_{11}^{\mathrm{r}}] , \\ \sigma_{12} &= 2G_{12}\varepsilon_{12}^{\mathrm{r}} \end{aligned} \tag{51}$$

where subscripts 1 and 2 refer to the fiber and the transverse to the fiber directions.

Residual stresses computed for the 0-deg and ±45-deg plies of the $[0_2/\pm 45]_s$ graphite/epoxy laminate at room temperature are tabulated in Table 3. The transverse to the fibers stress in the ±45-deg plies seems to exceed somewhat the measured transverse tensile strength of the unidirectional material which is 42 MPa (6.1 ksi). This means that these plies are probably damaged in their transverse direction upon completion of curing.

Comparable results for glass/epoxy and boron/epoxy show that residual tensile stresses exhaust a significant portion of the transverse tensile strength of the ply. In the case of Kevlar 49/epoxy computed residual stresses, assuming

TABLE 3. Residual stresses at room temperature in $[0_2/\pm 45]_s$ graphite/epoxy laminate.

Ply (deg)	Stress, MPa (ksi)		
	σ_{11}	σ_{22}	σ_{12}
0	23 (3.3)	42 (6.1)	0
±45	−52 (−7.5)	45 (6.5)	6 (0.9)

linear elastic behavior, far exceed the transverse strength of the ply. It has been shown also that the amount of relaxation of residual stresses is fairly small (DANIEL and LIBER (1976c)).

Further experimental work has been reported on the effects of laminate construction, ply stacking sequence and interply hybridization on residual stresses (DANIEL and LIBER (1977 a, b)).

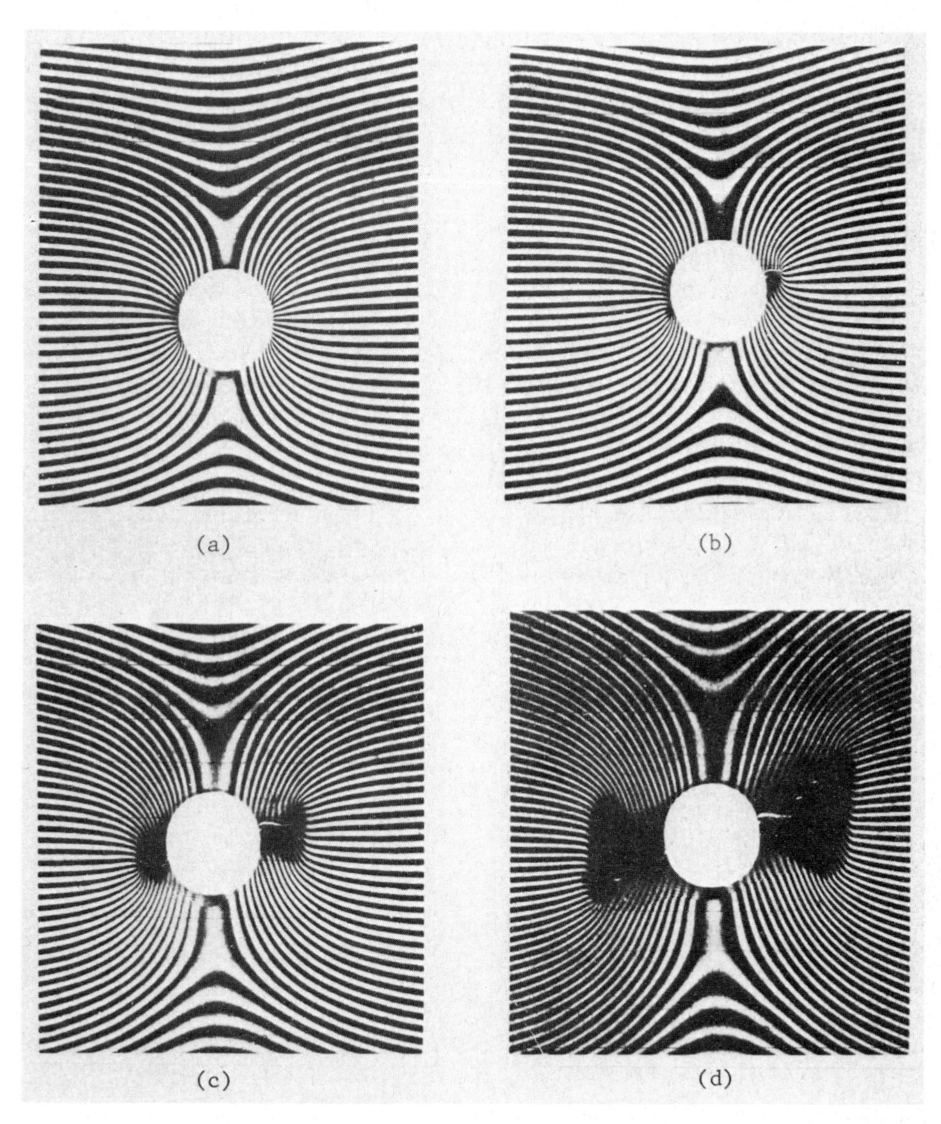

FIG. 49. Sequence of moiré fringe patterns corresponding to vertical displacements in $[0/\pm 45/0/\overline{90}]_s$ glass/epoxy specimen immediately prior to total failure. (a) $\sigma = 198$ MPa (28 700 psi); (b, c) $\sigma = 206$ MPa (29 850 psi); (d) $\sigma = 210$ MPa (30 400 psi).

6.2. Laminates with holes

The behavior of composites with stress concentrations is of great interest in design because of the resulting strength reduction and the damage growth around these stress concentrations. Stress distributions around circular holes in composite plates have been treated analytically using linear anisotropic elasticity (LEKHNITSKII (1963), SAVIN (1961)) and finite element methods (RYBICKI and HOPPER (1973), WHITESIDE et al. (1973)). The latter can be used to account for material inhomogeneity, nonlinearity, and inelasticity. Experimental methods using strain gages, photoelastic coatings, and moiré have proven very useful in verifying theoretical solutions in the linear range and supplementing them in the nonlinear range (DANIEL and ROWLANDS (1971), ROWLANDS et al. (1973, 1974a), DANIEL et al. (1973, 1974), DANIEL (1976, 1977, 1978b, 1980)). Experimental methods are especially useful in studying failure modes.

An application of moiré techniques is illustrated in Fig. 49 (ROWLANDS et al. (1973)). The specimen was a 25.4 cm × 66 cm (10 in. × 26 in.) panel of $[0/\pm45/0/\overline{90}]_s$ layup with a 2.5 cm (1 in.) diameter central hole.[1] A 40 line/mm (1000 lpi) array of horizontal lines was bonded onto the specimen. It was loaded in uniaxial tension to failure. The tensile strain concentration factor determined from these moiré patterns is $(k_\varepsilon)_{\theta=90°} = 3.51$, which compares well with the value of 3.50 determined by strain gages and photoelastic coatings on the same and similar specimens. Crack formation and propagation are evident at the higher stress levels. The cracks start on the boundary of the hole off the horizontal axis where the combination of stress and material strength becomes critical. Delamination, as evidenced by the dark regions in the moiré pattern, seems to follow crack propagation. One interesting phenomenon seen in the fringe pattern at the highest stress level is the seemingly discontinuous pro-

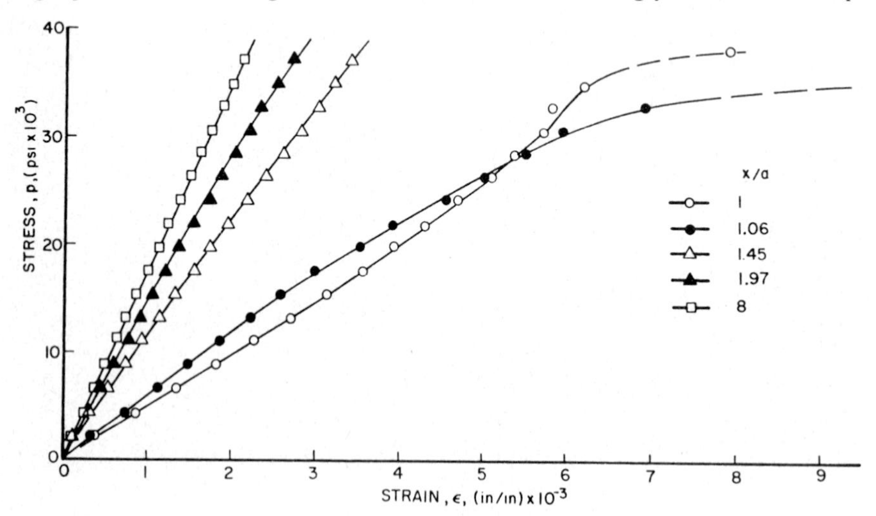

FIG. 50. Vertical strains along horizontal axis of $[0/\pm45/0/\overline{90}]_s$ boron/epoxy plate with a hole under uniaxial tensile loading.

[1] The overbar symbol means that the 90° ply is at the very center of the 9-ply laminate.

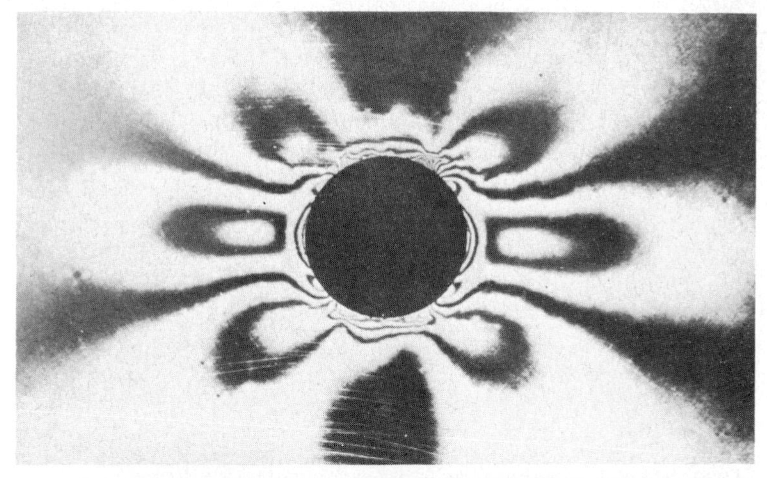

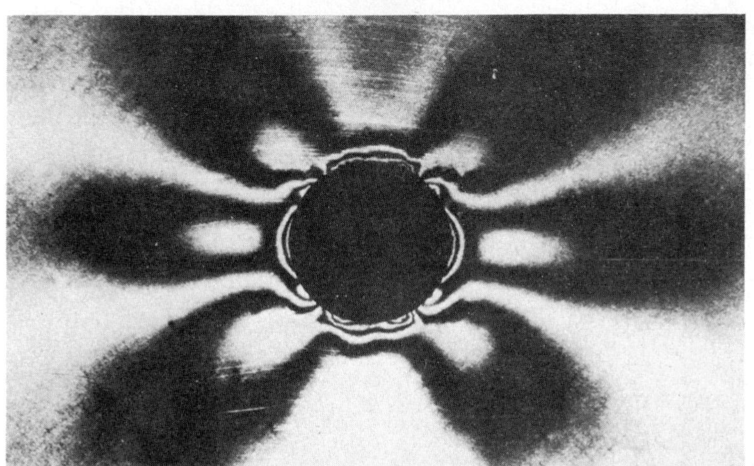

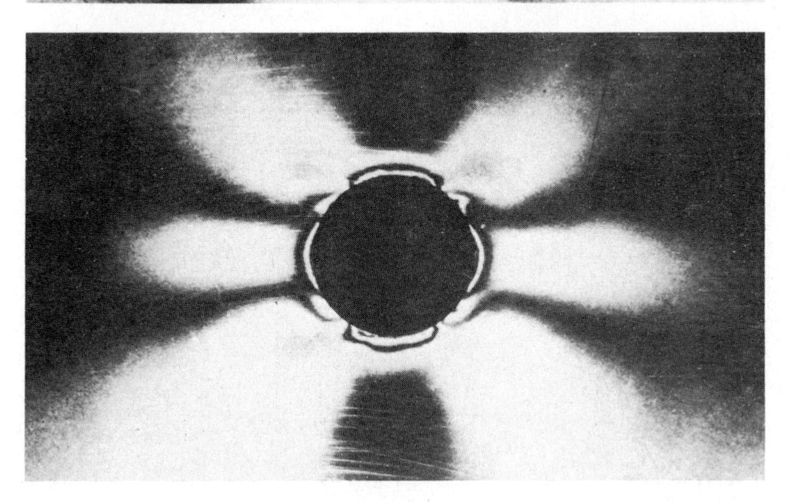

FIG. 51. Isochromatic fringe patterns in photoelastic coating around hole in $[0/\pm 45/0/\overline{90}]_s$ boron/epoxy specimen for applied stresses of 165 MPa (24 000 psi), 225 MPa (32 600 psi), and 292 MPa (42 400 psi).

pagation of the cracks. The cracks seem to appear in discrete, discontinuous horizontal steps with a gross direction of propagation at 45-deg. One possible explanation is the combination of two modes of failure, tensile fracture of the outer vertical fibers and shear delamination in the 45-deg interior plies. The application of moiré fringe multiplication techniques to this problem was discussed by DANIEL et al. (1973).

Strain gages have been used to determine strain distributions to failure in a uniaxially loaded $[0/\pm 45/0/\overline{90}]_s$ boron/epoxy plate with a 2.54 cm (1 in.) diameter hole (Fig. 50). The far-field strains remain fairly linear to failure, whereas the strains on and near the hole boundary become nonlinear at an applied stress of 110 MPa (16 ksi). One interesting phenomenon is the strain redistribution occurring near failure, which makes the strain near the hole boundary increase at a faster rate and overtake the strain on the hole boundary.

This phenomenon of stress (strain) redistribution is better illustrated by the isochromatic fringe patterns on the photoelastic coating applied to the same boron/epoxy specimen (Fig. 51). The birefringence (hence strain) variation around the boundary of the hole for various load levels is shown in Fig. 52. An important and significant result is that the location of maximum birefringence

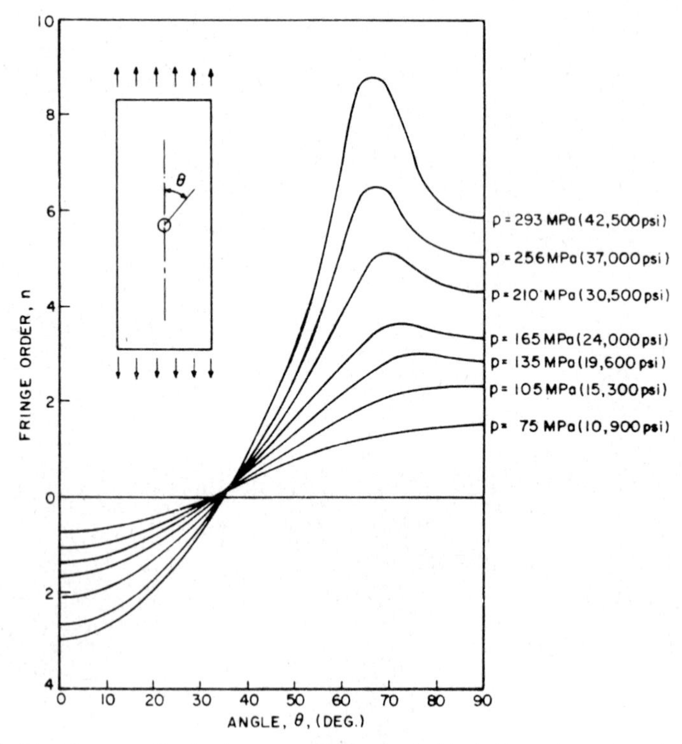

FIG. 52. Birefringence distribution around boundary of hole for various stress levels.

shifts from $\theta = 90°$ (horizontal axis) at low loads to $\theta \cong 67.5°$ near the failure load. This is because the birefringence (tangential normal strain) at $\theta = 67.5°$ increases rapidly in a nonlinear fashion. This is related to the fact that the shear stress-shear strain (σ_{xy} versus ε_{xy}) and circumferential stress-strain ($\sigma_{\theta\theta}$ versus $\varepsilon_{\theta\theta}$) response of the material at $\theta = 67.5$ deg is highly nonlinear. Failure initiated near the points of maximum birefringence. At these points, the elastic membrane and interlaminar shear stresses reach maximum values, and the circumferential normal strain and shear strain increase nonlinearly with load and more rapidly than at any other point. A complex failure combining tensile cracking and shear delamination modes produced a lip on the hole boundary. The crack then progressed parallel to the horizontal axis to complete failure (Fig. 53).

The influence of laminate layup on the behavior of boron/epoxy tensile plates with holes has also been studied with photoelastic coatings (DANIEL et al. (1974)). Some of the laminate constructions studied, in addition to the one discussed above, were $[0/90/0/90]_s$, $[45/90/0/-45]_s$, $[\pm 45/0/\pm 45]_s$, and $[\pm 45/\pm 45]_s$. The first and last of these represent two extremes of behavior illustrated by the fringe patterns of Fig. 54. The $[0/90/0/90]_s$ construction is characterized by a

FIG. 53. Failure around hole in $[0/\pm 45/0/\overline{90}]_s$ boron/epoxy panel under uniaxial tension.

low maximum strain at failure (0.0066) contrasted with very large strains ($>25\times10^{-3}$) at failure for the $[\pm45/\pm45]_s$ construction. Another apparent difference is the sharp strain gradient and high strain concentration (4.82) in the $[0/90]_{2s}$ layup contrasted with a moderate gradient and low strain concentration (2.06) in the $[\pm45]_{2s}$ layup. In the former case the fringes are concentrated near the hole around the horizontal axis and a uniform far-field birefringence is attained at a distance of between half and one radius from the boundary of the hole. In the case of the $[\pm45]_{2s}$ specimen the influence of the hole extends along the 45-deg radii through the entire width of the plate.

The influence of laminate construction on strength is very pronounced. Laminates with a high percentage of 0-deg plies, but with sufficient 45-deg plies to reduce the stress concentration factor, are the strongest. The $[0/90]_{2s}$ construction with 50% 0-deg plies is not strong because of the high stress concentration factor. The $[\pm45]_{2s}$ construction is the weakest because of the absence of 0-deg plies, although the stress concentration factor is the lowest.

Stacking sequence was found to have a noticeable influence on failure patterns and strength of composite laminates with holes (DANIEL et al. (1974)). Stacking sequences resulting in tensile interlaminar normal stresses near the boundary of the hole reduce the strength of the laminate.

The influence of hole geometry was studied experimentally by comparing the

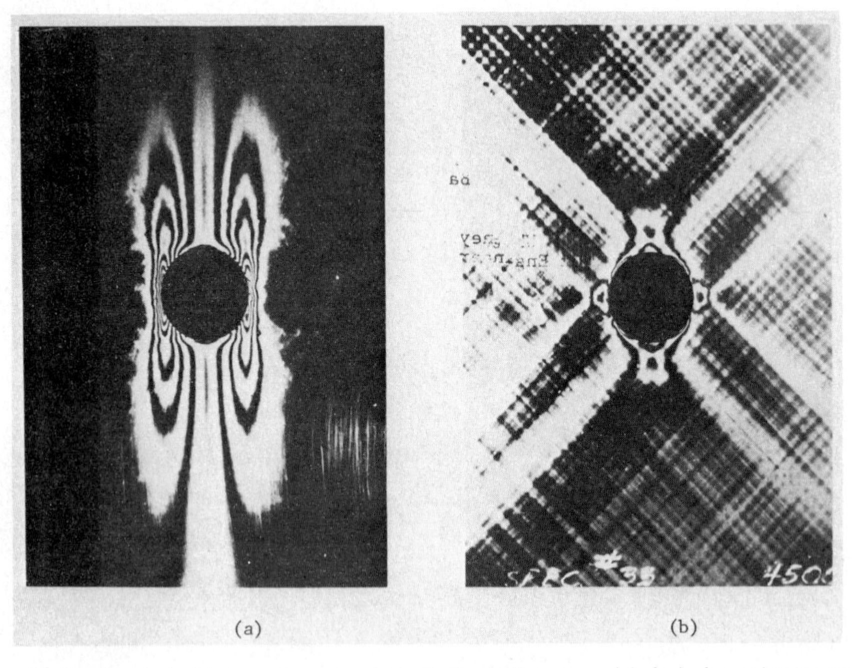

FIG. 54. Isochromatic fringe patterns in photoelastic coating around hole in boron/epoxy specimens. (a) $[0/90/0/90]_s$, $\sigma = 169$ MPa (24 600 psi); (b) $[\pm45/\pm45]_s$, $\sigma = 77$ MPa (11 100 psi).

behavior of boron/epoxy panels with circular, elliptical, and square holes (ROWLANDS et al. (1974)). The strength increases with decreasing stress concentration as expected, but not in a pronounced way. Thus, the strengths ranged between 242 MPa and 369 MPa for stress concentrations of 6.6 and 2.4, respectively.

The influence of hole diameter has been studied experimentally with graphite/epoxy laminates (DANIEL (1976, 1978b, 1980)). The strength of the laminate decreases as the hole diameter increases, although the stress concentration and the maximum stress on the boundary of the hole are the same for all hole diameters. Experimental results for uniaxially loaded quasi-isotropic $[0/\pm 45/90]_s$ graphite/epoxy plates with holes of various diameters are shown in Fig. 55. Two criteria have been used to describe the effect of hole diameter on strength (WHITNEY and NUISMER (1974)). According to the average stress criterion, failure occurs when the axial stress averaged over a characteristic distance a_0 from the hole boundary along the transverse axis equals the strength of the unnotched laminate (Fig. 56). The strength reduction ratio predicted by this criterion is expressed as

$$\frac{S_{yy}}{S_0} = \frac{2}{(1+\xi_2)(2+\xi_2^2)} \tag{52}$$

where $\xi_2 = a/(a + a_0)$ and a_0 = characteristic length dimension.

According to the point stress criterion failure occurs when the axial stress at

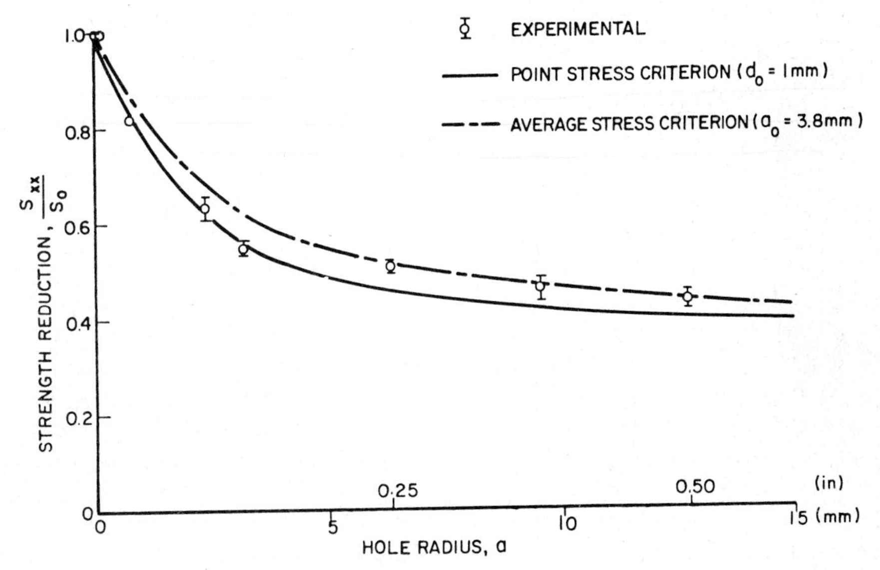

FIG. 55. Strength reduction as a function of hole radius for $[0/\pm 45/90]_s$ graphite/epoxy plates with circular holes under uniaxial tensile loading.

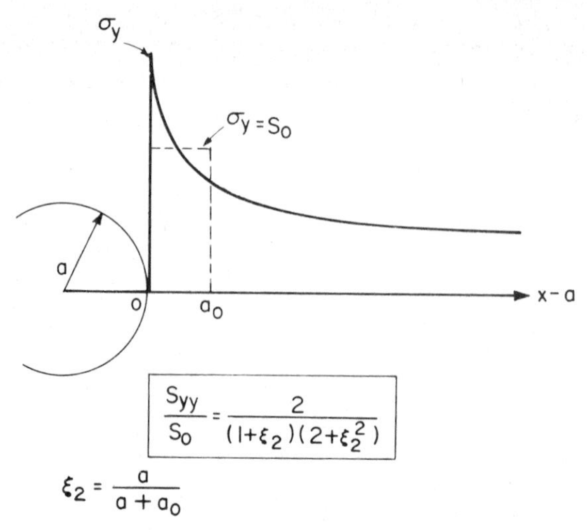

a_0 = Characteristic length dimension (~3.8mm)

S_{yy}, S_0 = Strengths of notched and unnotched laminates, respectively.

FIG. 56. Strength reduction of uniaxially loaded plate with circular hole according to average stress criterion.

some distance d_0 from the hole boundary along the transverse axis equals the strength of the unnotched laminate.

Experimental results were compared with predictions by the two criteria above with satisfactory agreement (Fig. 55). One experimental result which cannot be described by the criteria above is that there is a limiting hole size below which the laminate becomes notch-insensitive, i.e., failure is as likely to occur through the notched as through the unnotched section.

The behavior of graphite/epoxy plates with holes under biaxial loading has been studied experimentally (DANIEL (1976, 1977, 1980)). The specimens in one case were $[0/\pm45/90]_s$ 40 cm × 40 cm (16 in. × 16 in.) square plates with central holes of diameters 2.54 cm (1 in.), 1.91 cm (0.75 in.), 1.27 cm (0.50 in.), and 0.64 cm (0.25 in.). These specimens were tabbed with $[\pm45/+45/\mp45]_s$ glass/epoxy tabs as shown in Fig. 57. Deformations and strains were measured using strain gages and birefringent coatings. Equal biaxial loading was introduced by means of four whiffle-tree grip linkages as shown in Fig. 39, and controlled with a servohydraulic system. The loading frame with the specimen and the associated strain recording instrumentation is shown in Fig. 58. Initially, the circumferential strain is uniform around the boundary of the hole. Subsequently, with increasing load, regions of high strain concentration with nonlinear response develop at eight characteristic locations 22.5° off the fiber axes. Isochromatic fringe patterns in the photoelastic coating around the hole illustrate this phenomenon (Fig. 59). The birefringence variation with load at

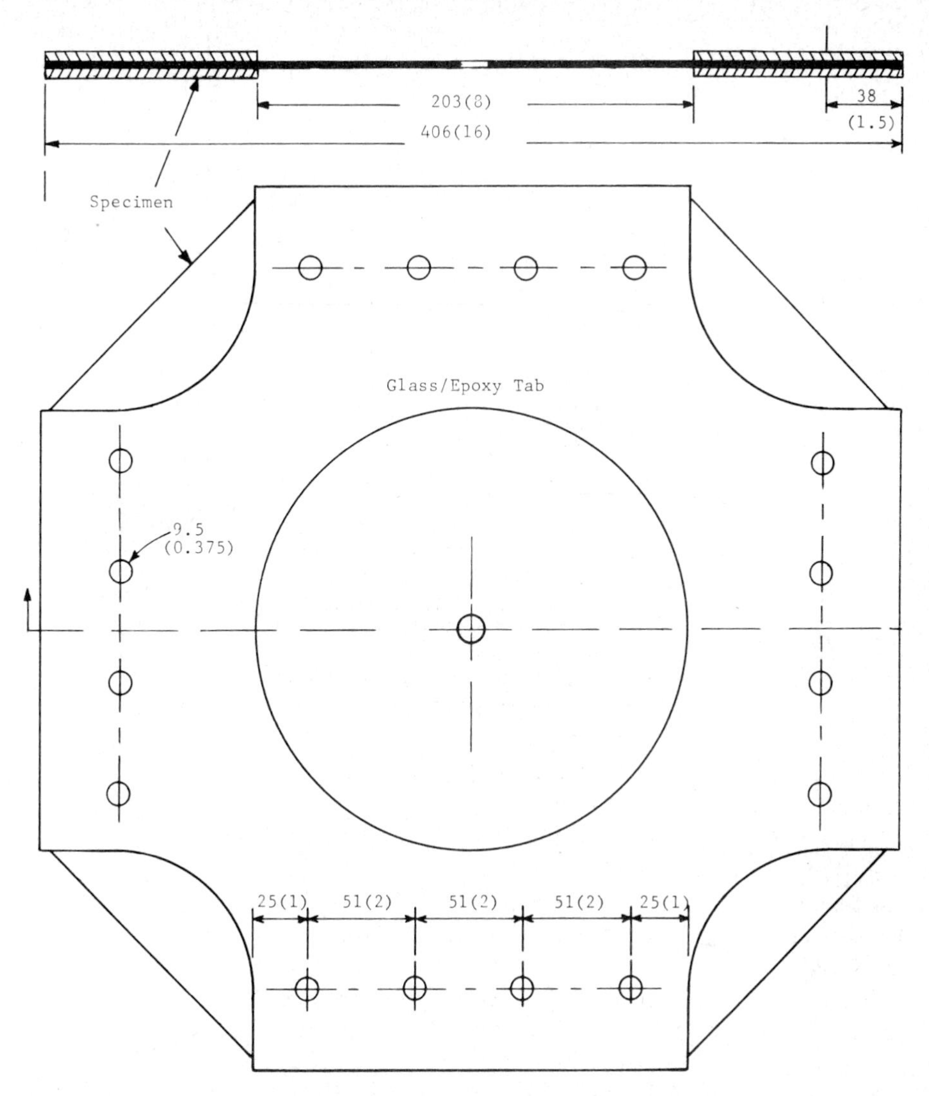

FIG. 57. Sketch of biaxial composite specimen (dimensions are in mm and in.).

two typical characteristic locations on the hole boundary, at 0° and 22.5°, is plotted in Fig. 60. At both locations the fringe order, hence circumferential strain, is the same and varies linearly up to an applied stress of 193 MPa (28 ksi). Thereafter, the strain at the critical 22.5° points increases at an increased rate while at the same time strains at the 0° and 90° locations show an unloading effect.

Failure is initiated at one or more points of high strain concentration (Fig. 61). Maximum strains at failure on the hole boundary reach values up to twice the ultimate strain of the unnotched laminate. The strength reduction ratio,

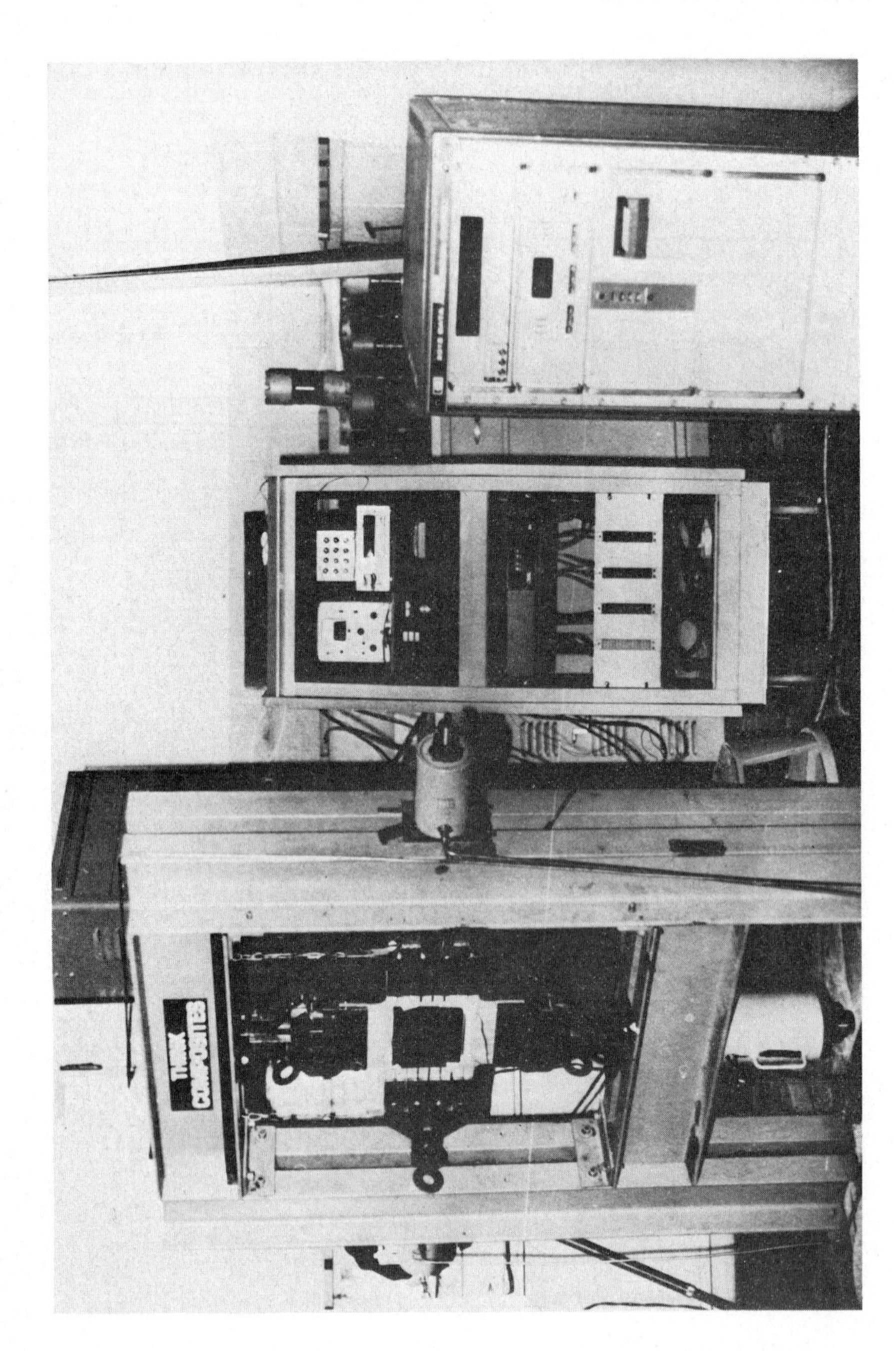

FIG. 58. Loading frame for biaxial testing of flat laminates and associated strain recording instrumentation.

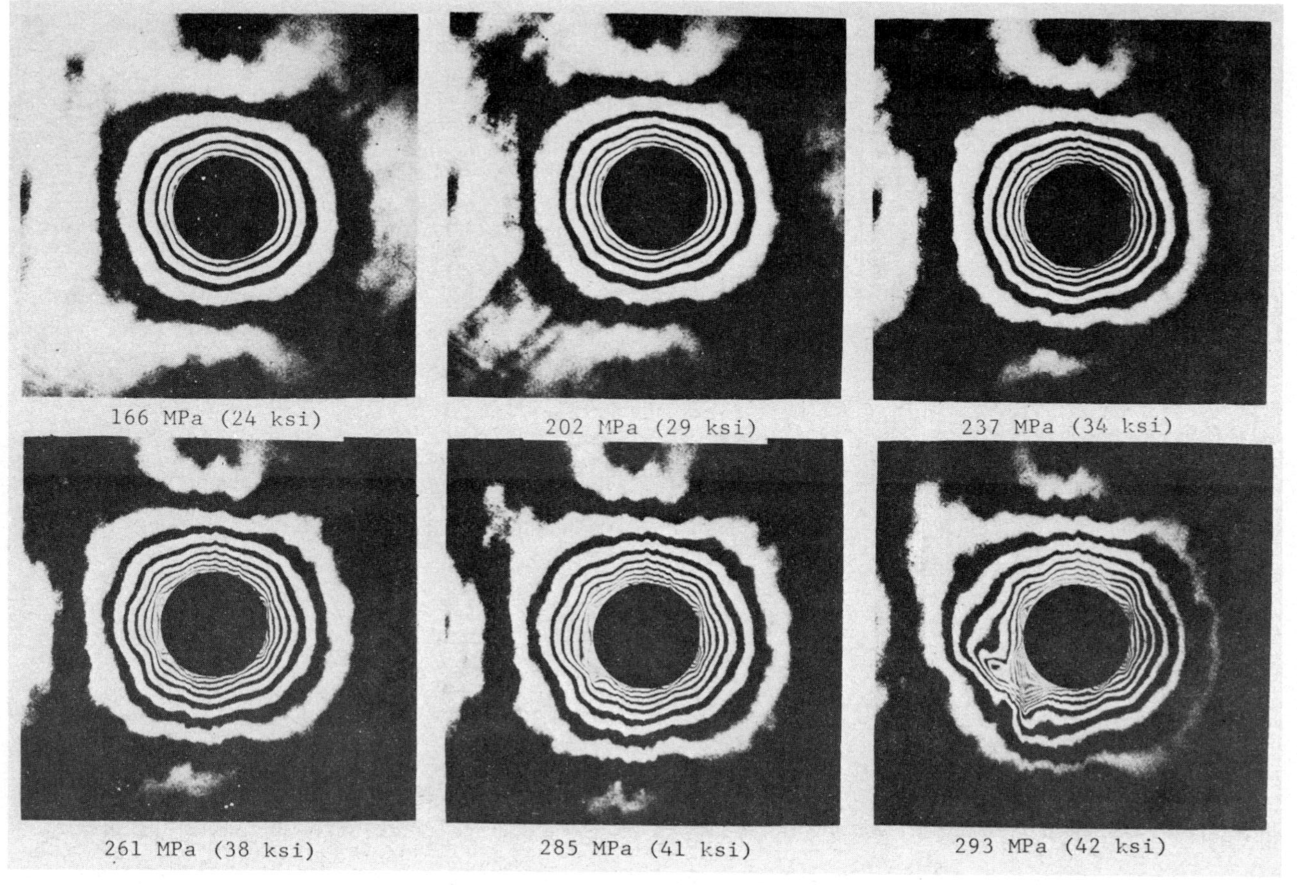

FIG. 59. Isochromatic fringe patterns in photoelastic coating of $[0/\pm45/90]_s$ graphite/epoxy specimen with 2.54 cm (1.00 in.) diameter hole under equal biaxial loading.

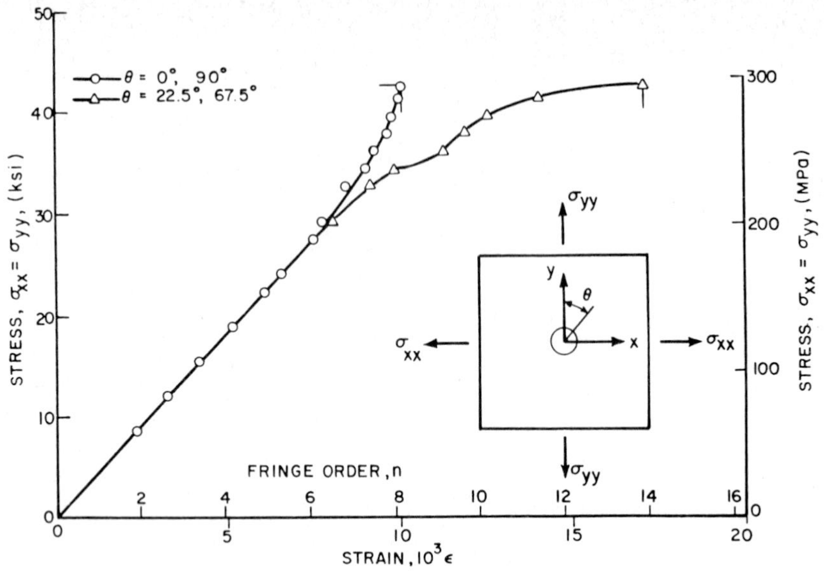

FIG. 60. Fringe order and circumferential strain at two locations on the hole boundary for $[0/\pm45/90]_s$ graphite/epoxy specimen with 2.54 cm (1.00 in.) diameter hole under equal biaxial loading.

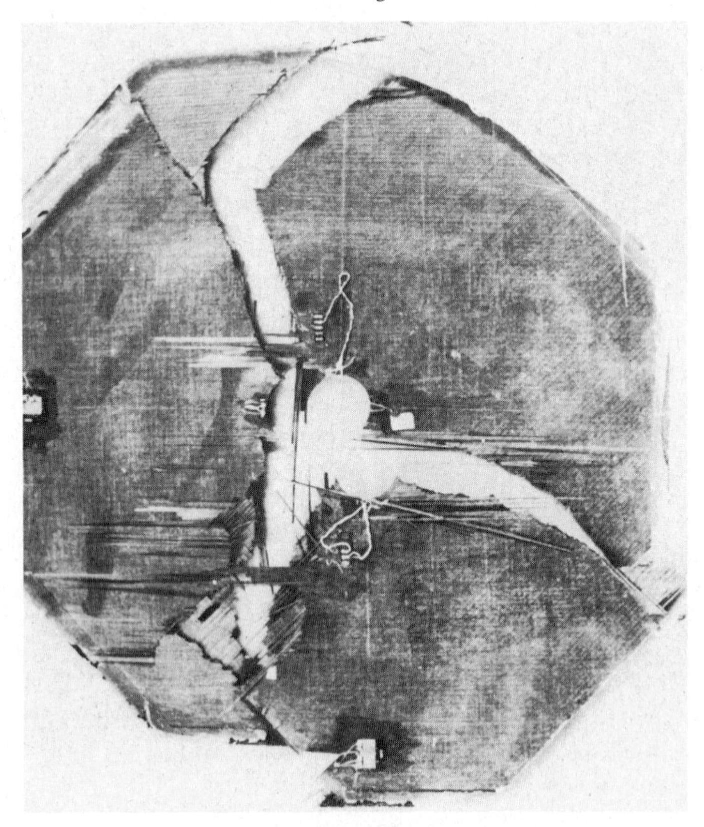

FIG. 61. Failure pattern in $[0/\pm45/90]_s$ graphite/epoxy specimen with 1.91 cm (0.75 in.) diameter hole under equal biaxial loading.

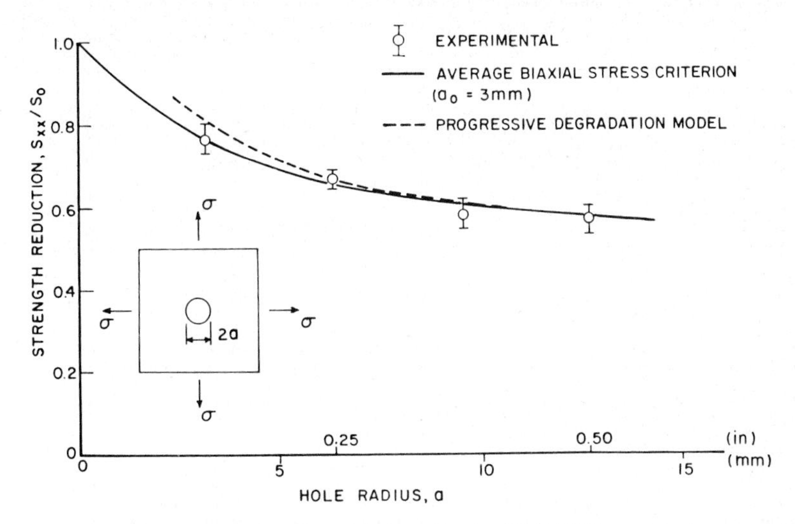

FIG. 62. Strength reduction as a function of hole radius for $[0/\pm45/90]_s$ graphite/epoxy plates with circular holes under 1:1 biaxial tensile loading.

ratio of notched biaxial strength to unnotched uniaxial strength, is plotted versus hole radius in Fig. 62. These ratios are higher than corresponding values for uniaxial loading by approximately 30%. The variation of strength reduction ratio with hole diameter was satisfactorily described by using an average biaxial stress criterion. Radial and circumferential stresses around the hole were averaged over an annulus of 3 mm width and compared with the biaxial strength envelope for the quasi-isotropic unnotched laminate. Results were also in good agreement with predictions based on a tensor polynomial failure criterion for the individual lamina and a progressive degradation model (LO and WU (1978)). The strength reduction ratios for uniaxial and equal biaxial tensile loading represent lower and upper bounds for any tensile loading of this laminate. The strength reduction for any other tensile biaxiality ratio would fall between these two bounds and could be estimated approximately by interpolation.

6.3. Laminates with cracks

Deformations and failure were studied in uniaxially loaded graphite/epoxy plates with cracks and the influence of crack size on failure was determined (DANIEL (1976, 1978c)). The specimens were $[0/\pm45/90]_s$ laminates 12.7 cm (5 in). wide and 56 cm (22 in.) long with transverse through-the-thickness cracks of lengths 2.54 cm (1 in.), 1.91 cm (0.75 in.), 1.27 cm (0.50 in.), and 0.64 cm (0.25 in.). The area around the crack was instrumented with strain gages, photoelastic coatings, and moiré grids. Strain gages were applied along the horizontal (crack) axis with miniature gages in the vicinity of the crack tip. Photoelastic coatings were 0.25 mm (0.01 in.) thick commercial coatings with a reflective backing. For moiré analysis an array of 40 lines/mm (1000 lpi) was applied to the specimen surface around the crack. The specimen surface was made reflective by depositing a thin

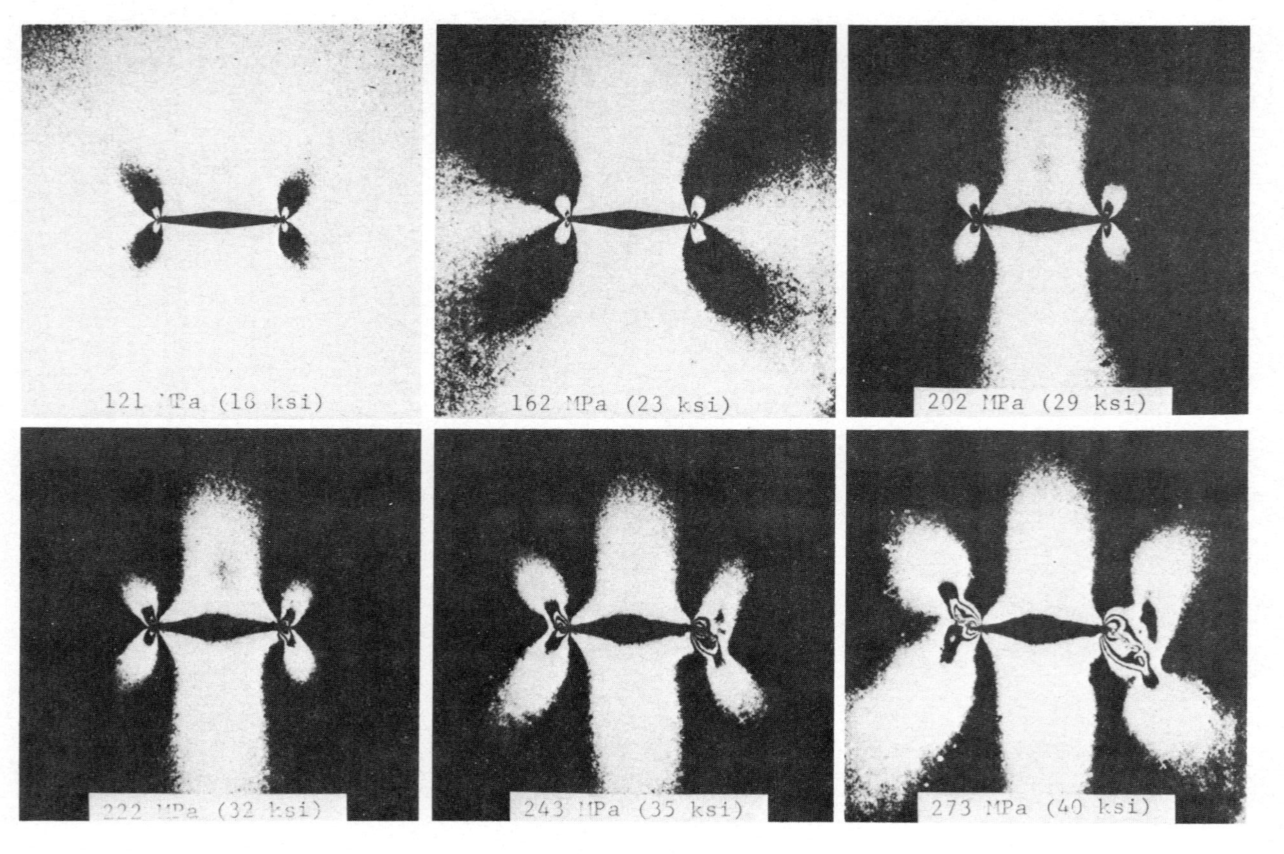

FIG. 63. Isochromatic fringe patterns in photoelastic coating around 1.27 cm (0.50 in.) crack of $[0/\pm 45/90]_s$ graphite/epoxy specimen at various levels of applied stress.

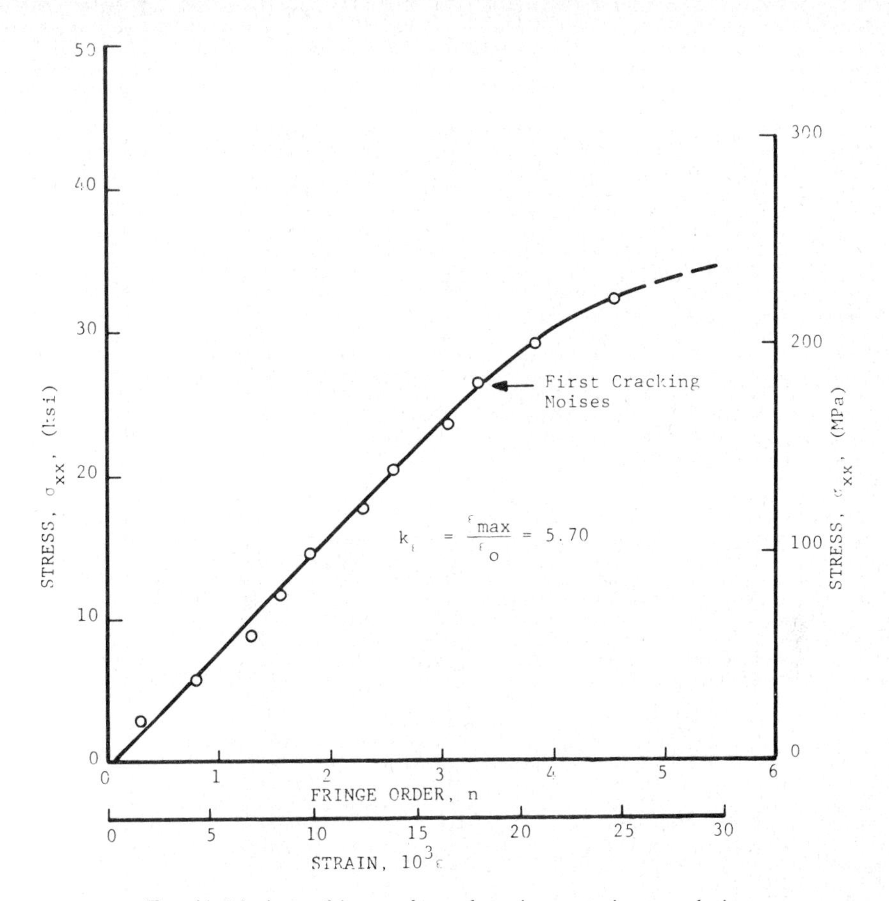

FIG. 64. Maximum fringe order and maximum strain at crack tip.

coating of epoxy dyed white with titanium dioxide. The film with the line array was then bonded on this white reflective surface.

The strain distribution around the crack tip and the phenomenon of damage zone formation and growth are vividly illustrated by the isochromatic fringe patterns in the photoelastic coating (Fig. 63). The maximum fringe order, which occurs at the crack tip off the horizontal axis, and the corresponding maximum strain are plotted in Fig. 64 as a function of applied stress. This curve is nearly linear up to the stress level of 180 MPa (26 ksi) when the first cracking noises were heard. The strain concentration factor computed as the ratio of the maximum strain at the crack tip to the far-field axial strain is 5.70. A noticeable characteristic is the apparent extension of the damage zone at an approximately 45-deg angle with the horizontal. As observed previously by MANDELL et al. (1975), this damage zone consists primarily of subcracks along the fibers of individual plies, local delamination and occasional fiber breakage. The size of this zone increases with applied stress up to some critical value at which point the specimen fails catastrophically.

FIG. 65. Moiré fringe patterns around crack in uniaxially loaded $[0/\pm 45/90]_s$ graphite/epoxy specimen for three levels of applied stress: (a) σ_{xx} = 152 MPa (22 ksi); (b) σ_{xx} = 202 MPa (29 ksi); (c) σ_{xx} = 253 MPa (37 ksi).

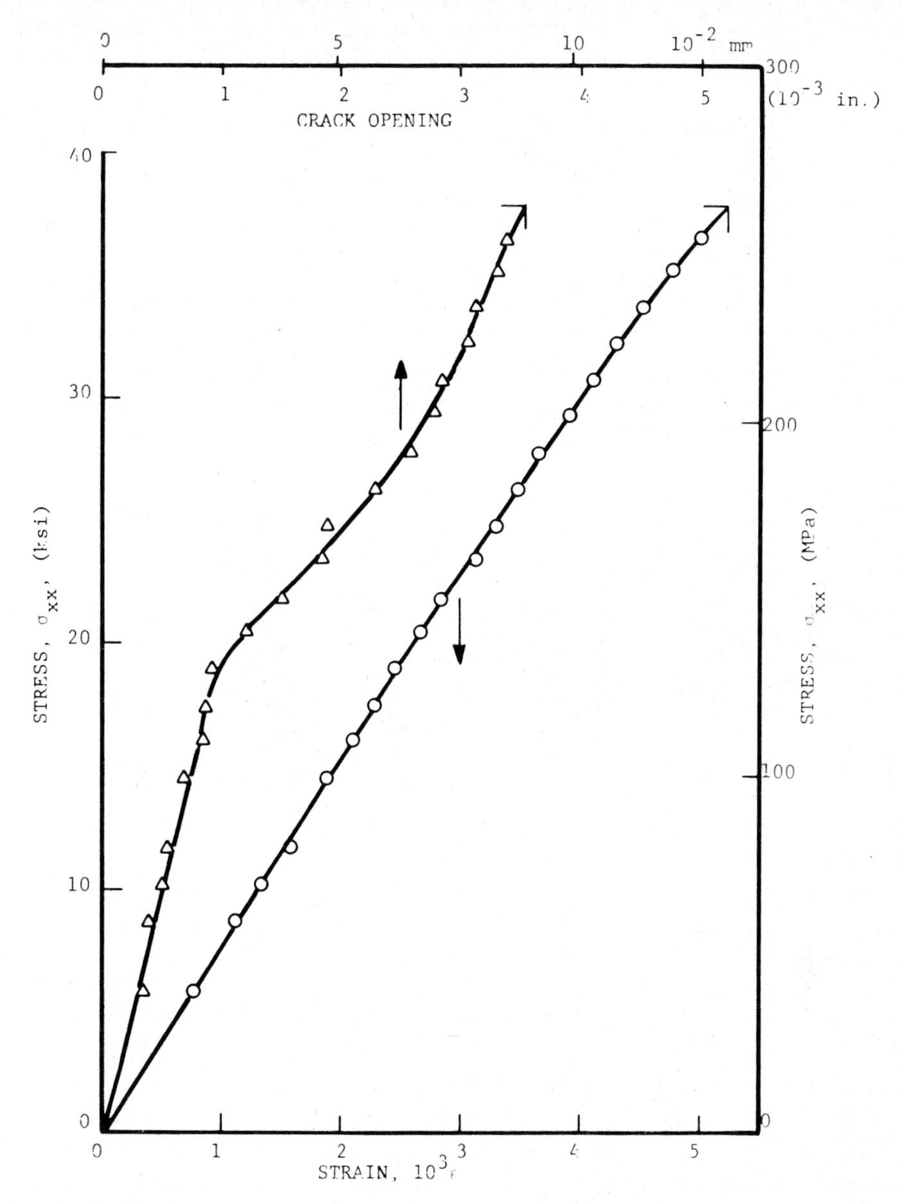

FIG. 66. Crack opening displacement and far-field strain for $[0/\pm45/90]_s$ graphite/epoxy specimen with a 1.27 cm (0.50 in.) horizontal crack.

Moiré fringe patterns corresponding to vertical displacements in the vicinity of the crack are shown in Fig. 65 for three levels of load. Each fringe represents a locus of points of constant vertical displacement of 0.025 mm (0.001 in.) relative to its neighboring fringe. These fringe patterns were analyzed to yield far-field strains at a distance of $y/a \cong 3$ and the crack opening displacment at

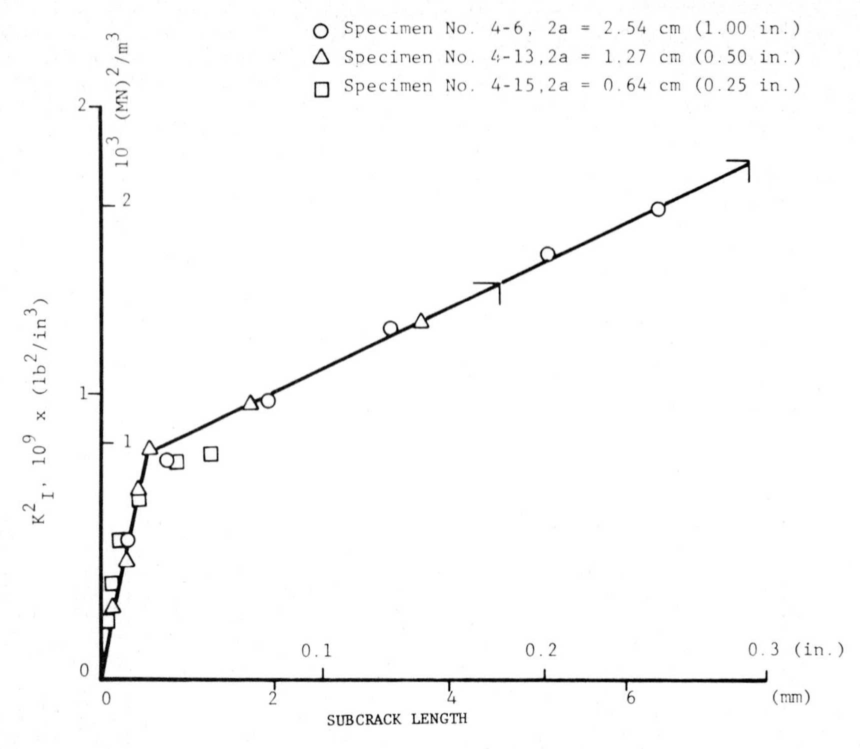

FIG. 67. Variation of length of subcrack with square of stress intensity factor for uniaxially loaded $[0/\pm45/90]_s$ graphite/epoxy plates with cracks.

the center of the crack (Fig. 66). The plot for the latter indicates that the crack opening displacement becomes nonlinear and increases at an increasing rate at an applied stress of approximately 138 MPa (20 ksi) which is near the level of rapid strain increase near the crack tip obtained with strain gages. Final failure was preceded by cracking noises heard at a stress level of 172 MPa (25 ksi).

An attempt was made to measure the damage zone and correlate it with the square of the stress intensity factor as was done by MANDELL et al. (1975). The length of the subcracks producing the damage zone was measured approximately from the photoelastic fringe pattern for three specimens with crack lengths of 2.54 cm (1 in.), 1.27 cm (0.50 in.), and 0.64 cm (0.25 in.). The subcrack length varies linearly with K_I^2 up to a value of $K_I \cong 30\ \text{MPa}\sqrt{\text{m}}$ (27.5 ksi$\sqrt{\text{in.}}$) for all three crack lengths (Fig. 67). Thereafter, the subcrack length increases again linearly with K_I^2 but at a faster rate which is nearly the same for all three crack lengths. This bilinear nature of the curve is characteristic of notch insensitive laminates. Failure patterns for specimens with cracks of various lengths are shown in Fig. 68. They all show extensive delamination near the crack tips and a crack propagation across the width of the specimen that is not too straight.

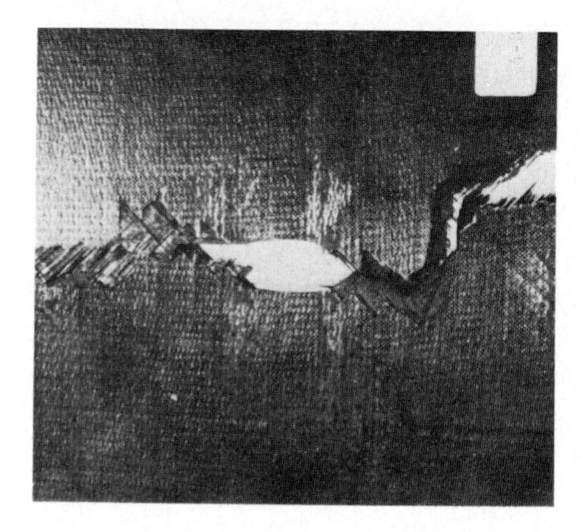
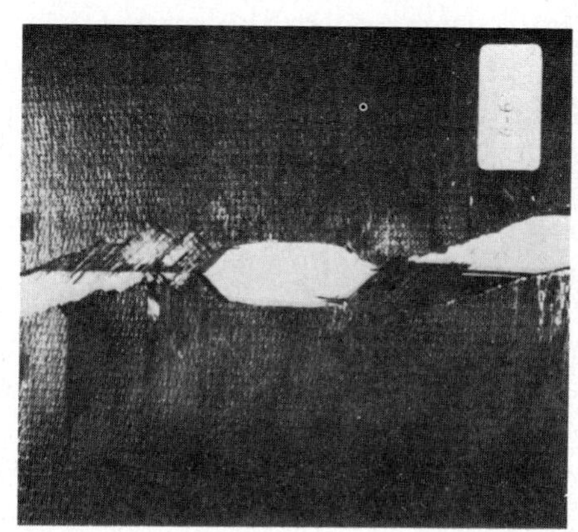

FIG. 68. Failure patterns in uniaxially loaded $[0/\pm45/90]_s$ graphite/epoxy plates with cracks of various lengths (crack lengths are: 0.64 cm (0.25 in.), 1.27 cm (0.50 in.), 1.91 cm (0.75 in.), and 2.54 cm (1.00 in.).

Experimental results agree well with predictions based on the point and average stress criteria, applied in a similar way as in the case of circular holes (Fig. 69). Comparison of results with those from similar specimens with circular holes showed that strength reduction was nearly independent of notch geometry, i.e., specimens with holes and cracks of the same size had nearly the same strength. Combined results for specimens with holes and cracks are shown in Fig. 69 and agree well with the analytical prediction based on the average stress criterion for holes with a characteristic dimension $a_0 = 3.8$ mm.

Similar graphite/epoxy plates with cracks of various lengths were loaded under biaxial tension as shown in Fig. 70 (DANIEL (1981)). The area around the

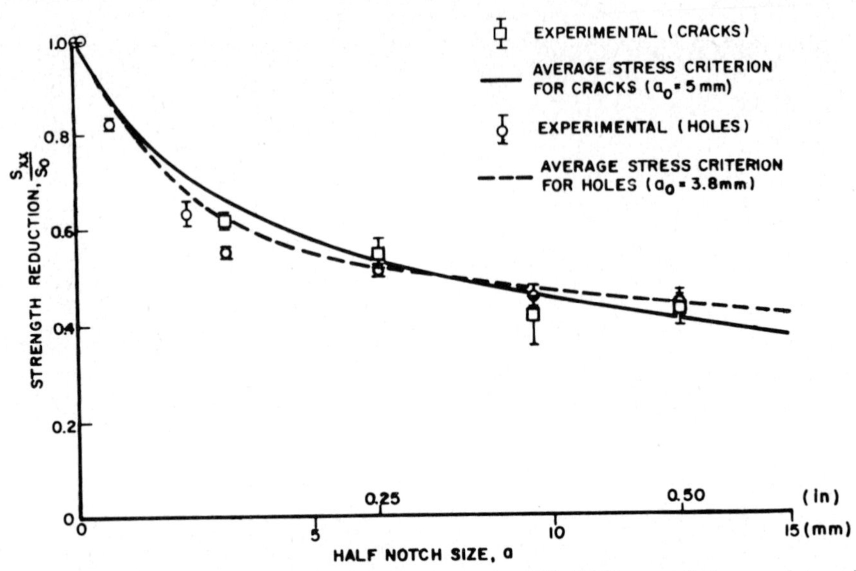

FIG. 69. Strength reduction as a function of notch size for $[0/\pm 45/90]_s$ graphite/epoxy plates with circular holes and horizontal cracks under uniaxial tensile loading.

crack was instrumented with strain gages, moiré grids, and photoelastic coatings.

Isochromatic fringe patterns in the coating around a 1.27 cm (0.50 in.) long crack are shown in Fig. 71 for four load levels. In addition to the primary propagation normal to the initial crack, there is crack extension along the

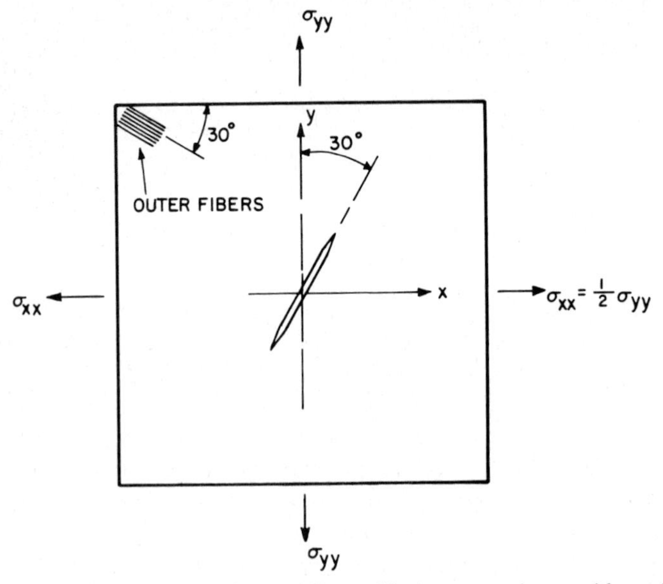

FIG. 70. Biaxial loading of $[0/\pm 45/90]_s$ graphite/epoxy specimens with cracks.

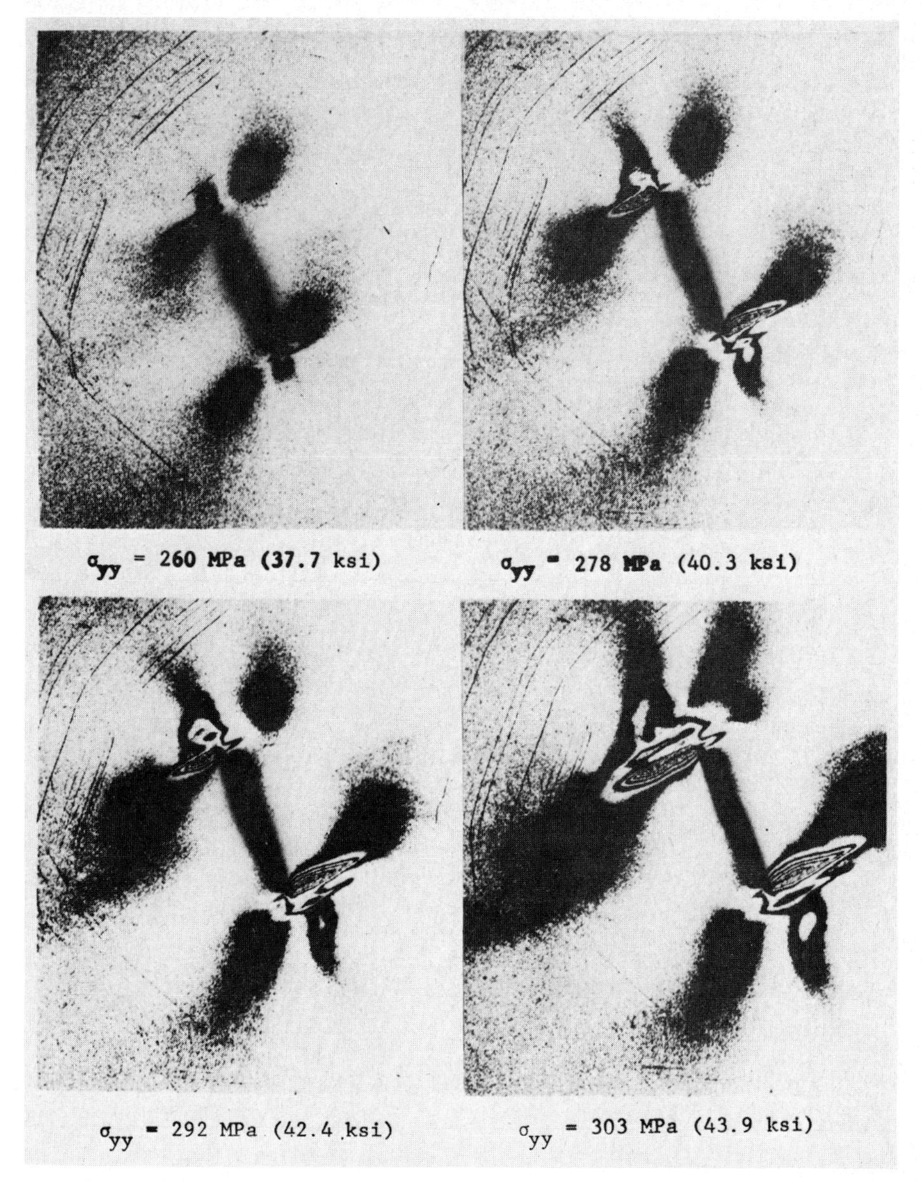

FIG. 71. Isochromatic fringe patterns in photoelastic coating around 1.27 cm (0.5 in.) crack in $[0/\pm45/90]_s$ graphite/epoxy specimen under biaxial loading $\sigma_{yy} = 2\sigma_{xx}$ at 30-deg with crack direction.

original crack direction probably along the fibers of the central plies of the laminate. This is illustrated by the fringe patterns of Fig. 71. There is also evidence of tertiary crack propagation normally to the initial crack direction but initiating at the tip of the subsurface extended crack. Rapid crack propagation occurs in the stress range of 220 to 262 MPa (32 to 38 ksi) correspond-

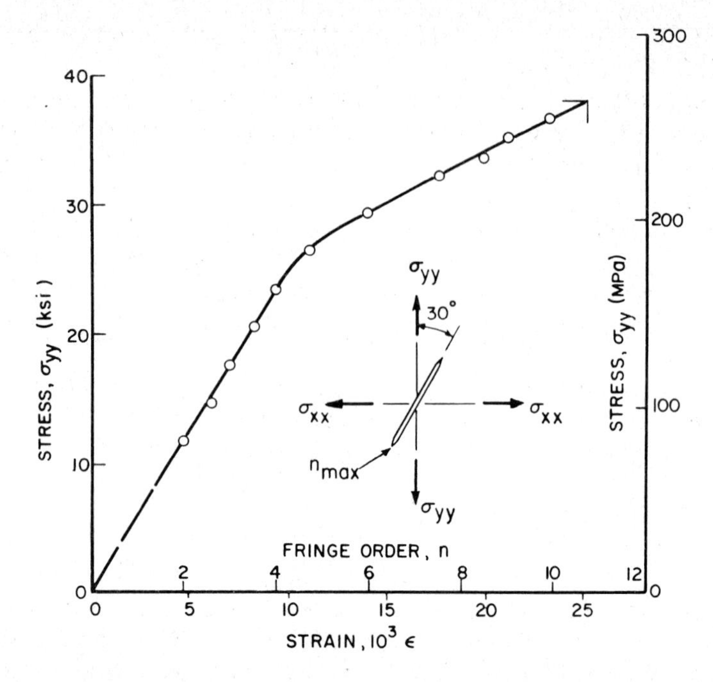

FIG. 72. Maximum fringe order and tangential strain at crack tip of $[0/\pm45/90]_s$ graphite/epoxy specimen with 1.91 cm (0.75 in.) crack under biaxial loading $\sigma_{yy} = 2\sigma_{xx}$.

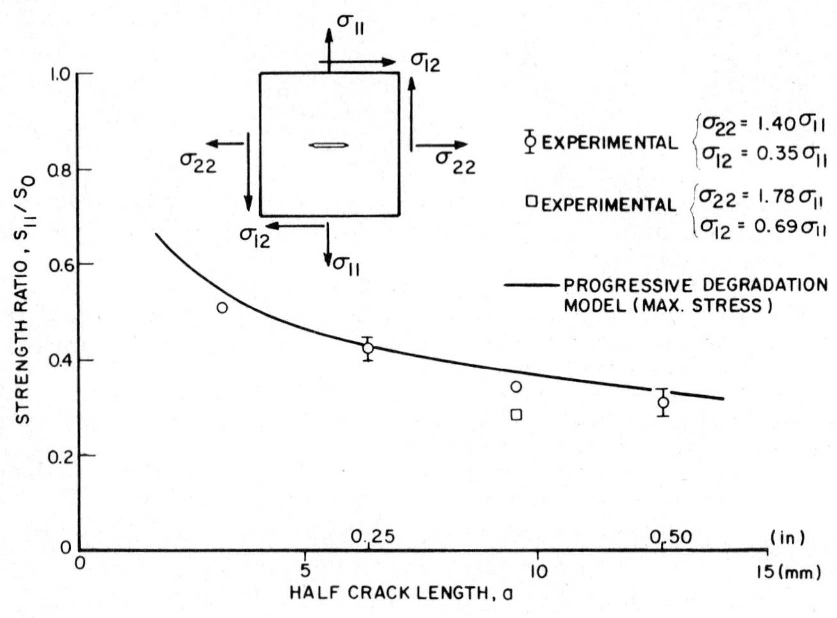

FIG. 73. Comparison of experimental and theoretical results for strength ratio for $[0/\pm45/90]_s$ graphite/epoxy plates with cracks under biaxial loading.

ing to peak strains of the order of 0.030. Total failure occurred at an applied stress of 313 MPa (45.3 ksi).

A measure of the maximum tangential strain at the crack tip is obtained from the maximum fringe order recorded in the coating. The variation of this fringe order and corresponding tangential strain as a function of applied stress is shown in Fig. 72 for a specimen with a 1.91 cm (0.75 in.) long crack. The variation appears bilnear with a knee at a strain of approximately 0.011, which is slightly higher than the ultimate strain of an unnotched coupon of the same layup. The computed stress concentration in the linear range is 3.2. The maximum measured strain at failure, which occurred at $\sigma_{yy} = 263$ MPa (38 ksi) in this case, is over 0.025.

Experimental results were in good agreement with those obtained by an analytical/empirical method using a maximum stress failure criterion for the individual lamina and a progressive degradation model (Fig. 73).

Acknowledgements

This chapter is based to a large extent on previous publications by the author with C.C. Chamis, A.J. Durelli, T. Liber, R.M. Marloff, D. Post, R.E. Rowlands, J.B. Whiteside, and J.M. Whitney. The manuscript was typed proficiently by Mrs. Yolanda Bradley. Chip Cokeing has prepared the drawings.

References

ADAMS, D.F. and D.R. DONER (1967), *J. Composite Materials* **1**, 152.

ALEXANDER, E.L., A.A. CAPUTO, M.E. PRADO and J.E. HILZINGER (1965), Technical Report AFML-TR-65-283.

BERT, C.W. (1972), *Fiber Sci. Technology* **5**, 165.

BERT, C.W. (1974), Experimental Characterization of Composites, in: C.C. CHAMIS, ed., *Structural Design and Analysis*, Part II, Vol. 8, Composite Materials (L.J. Broutman and R.H. Krock, eds.) (Academic Press, New York).

BERT, C.W., B.L. MAYBERRY and J.D. RAY (1969), Behavior of fiber-reinforced plastic laminates under biaxial loading, in: Composite Materials: Testing and Design, ASTM STP **460**, 362–380.

CAPUTO, A.A. and J.E. HILZINGER (1969), Flaw Point and Dynamic Microphotoelasticity Investigation, U.S. Army Materiel Labs, Report 69-42.

CHAMIS, C.C. (1971), Lamination Residual Stresses in Cross-Plied Fiber Composites, in: *Proc. of 26th Annual Conf. of SPI*, Reinforced Plastics/Composites Division, No. 17-D.

CHAMIS, C.C. and J.H. SINCLAIR (1977), *Exper. Mechanics* **17**, 339.

CHANG, F.H., J.C. COUCHMAN, J.R. EISENMANN and B.G.W. YEE (1975), Application of a Special X-Ray Nondestructive Testing Technique for Monitoring Damage Zone Growth in Composite Laminates, in: Composite Reliability, ASTM STP **580**, American Society for Testing and Materials (Philadelphia), 176–190.

CHRISTENSEN, R.M. (1979), *Mechanics of Composite Materials* (Wiley, New York).

COLE, B.W. and R.B. PIPES (1972), Utilization of the Tubular Off-Axis Specimens for Composite Biaxial Characterization, in: *Proc. of Conf. held at Dayton*, OH, AFFDL-TR-72-130.

COLE, B.W. and R.B. PIPES (1974), AFFDL-TR-73-115.

DALLY, J.W. and I. ALFIREVICH (1969), *Exper. Mechanics* **9**, 97.
DALLY, J.W. and R. PRABHAKARAN (1971), *Exper. Mechanics* **11**(8), 346.
DALLY, J.W. and W.F. RILEY (1965), *Experimental Stress Analysis* (McGraw-Hill, New York).
DANIEL, I.M. (1966), Photoelastic Studies of Mechanics of Composites, in: IIT Research Institute Report No. M6132 to General Dynamics, Contract No. AF33(615)–3323.
DANIEL, I.M. (1969), Micromechanics, in: Structural Airframe Application of Advanced Composite Materials **II**, Technical Report AFML-TR-69-101.
DANIEL, I.M. (1970), *J. Composite Materials* **4**, 178.
DANIEL, I.M. (1974), Photoelastic Investigation of Composites, in: G.P. SENDECKYJ, ed., *Mechanics of Composites*, Vol. 2, Composite Materials (L.J. Broutman and R.H. Krock, eds.) (Academic Press, New York).
DANIEL, I.M. (1975a), Optical Methods for Testing Composite Materials, in: Failure Modes of Composite Materials with Organic Matrices and Their Consequences on Design, AGARD-CP-163.
DANIEL, I.M. (1975b), Photoelastic Studies of Mechanics of Composites, in: Progress in Experimental Mechanics, Durelli Anniversary Volume, The Catholic University of America.
DANIEL, I.M. (1976), Part I, AFML-TR-76-244.
DANIEL, I.M. (1977), Part II, AFML-TR-76-244.
DANIEL, I.M. (1978a), Thermal Deformations and Residual Stresses in Fiber Composites, in: IAN D. PEGGS, ed., *Thermal Expansion 6* (Plenum, New York), 203–221.
DANIEL, I.M. (1978b), The Behavior of Uniaxially Loaded Graphite/Epoxy Plates with Holes, in: *Proc. of 2nd International Conf. on Composite Materials*, ICCM/2 (Toronto, Canada), 1019–1034.
DANIEL, I.M. (1978c), *Exper. Mechanics* **18**(7), 246.
DANIEL, I.M. (1980), *Exper. Mechanics* **20**(1), 1.
DANIEL, I.M. (1981), Biaxial Testing of Graphite/Epoxy Laminates with Cracks, in: Test Methods and Design Allowables for Fibrous Composites, ASTM STP **734**, American Society for Testing and Materials, 109–128.
DANIEL, I.M. and A.J. DURELLI (1961), Photoelastic Investigation of Residual Stresses in Glass-Plastic Composites, in: *Proc. of 16th Conf. of Reinforced Plastics Div.*, Society of Plastics Industry, Section 19-A, 1–8.
DANIEL, I.M. and A.J. DURELLI (1962), *Exper. Mechanics* **2**, 240.
DANIEL, I.M. and T. LIBER (1975), IITRI Report D6073-I, NASA CR-134826.
DANIEL, I.M. and T. LIBER (1976a), IITRI Report D6073-II, NASA CR-135085.
DANIEL, I.M. and T. LIBER (1976b), Measurement of Lamination Residual Strains in Graphite Fiber Laminates, in: *Proc. of 2nd International Conf. on Mechanical Behavior of Materials*, ICM-II (Boston, MA).
DANIEL, I.M. and T. LIBER (1976c), Relaxation of residual stresses in angle-ply composite laminates, in: *Composite Materials: The Influence of Mechanics of Failure on Design, Army Symposium on Solid Mechanics* (South Yarmouth, Massachusetts).
DANIEL, I.M. and T. LIBER (1977a), *Exper. Mechanics* **17**(1), 21.
DANIEL, I.M. and T. LIBER (1977b), Lamination Residual Stresses in Hybrid Laminates, in: Composite Materials: Testing and Design (4th Conf.), ASTM STP **617**, American Society for Testing and Materials, 331–343.
DANIEL, I.M. and T. LIBER (1979), Nondestructive Evaluation Techniques for Composite Materials, in: 12th Symposium on Nondestructive Evaluation (San Antonio, TX).
DANIEL, I.M., T. LIBER and C.C. CHAMIS (1975), Measurement of Residual Strains in Boron/Epoxy and Glass/Epoxy Laminates, in: Composite Reliability, ASTM STP **580**, American Society for Testing and Materials, 340–351.
DANIEL, I.M., T. LIBER, R. VANDERBY and G.M. KOLLER (1980), Analysis of Tubular Specimen for Biaxial Testing of Composite Laminates, in: *Proc. of the 3rd International Conf. on Composite Materials, Advances in Composite Materials* (Paris, France), 840–855.
DANIEL, I.M., J.L. MULLINEAUX, F.J. AHIMAZ and T. LIBER (1972), The Embedded Strain Gage Technique for Testing Boron/Epoxy Composites, in: Composite Materials: Testing and Design (2nd Conf.), ASTM STP **497**, American Society for Testing and Materials, 257–272.

DANIEL, I.M., T. NIIRO and G.M. KOLLER (1981), NASA CR-165709.
DANIEL, I.M. and R.E. ROWLANDS (1971), *J. Composite Materials* **5**, 250.
DANIEL, I.M. and R.E. ROWLANDS (1972), Experimental Stress Analysis of Composite Materials, ASME Paper No. 72-DE-6, paper presented at ASME Design Engineering Conference (Chicago, IL).
DANIEL, I.M., R.E. ROWLANDS and D. POST (1973), Exper. Mechanics **13**(6), 246.
DANIEL, I.M., R.E. ROWLANDS and J.B. WHITESIDE (1973), Deformation and Failure of Boron/Epoxy Plate with Circular Hole, in: Analysis of the Test Methods for High Modulus Fibers and Composites, ASTM STP **521**, American Society for Testing and Materials, 143–164.
DANIEL, I.M., R.E. ROWLANDS and J.B. WHITESIDE (1974), *Exper. Mechanics* **14**, 1.
DOMANUS, J.C. and H. LILHOLT (1978), Nondestructive Control of Carbon Fibre Reinforced Composites by Soft X-ray Radiography, in: Proc. of the 1978 International Conf. on Composite Materials, ICCM/2, The American Institute of Mining, Metallurgical, and Petroleum Engineers (Toronto, Canada), 1072–1092.
DURELLI, A.J., V.J. PARKS, H.C. FENG and F. CHIANG (1970), Strains and stresses in matrices with inserts, in: J.W. WENDT, H. LIEBOVITZ and N. PERRONE, eds., *Mechanics of Composite Materials* (Pergamon, New York), 265–335.
EDELMAN, W.E. (1969), *Exper. Mechanics* **9**, 380.
ELSLEY, R.K. (1978), Computer Aided Interpretation of NDE Signals, in: *Proc. of the ARPA/AFML Review of Progress in Quantitative NDE*, AFML-TR-78-55, Air Force Materials Laboratory, 326–330.
ERF, R.K., ed. (1974), *Holographic Nondestructive Testing* (Academic Press, New York).
FOYE, R.L. (1966–1967), Quarterly Report Nos. 1, 2, and 3, Contract No. AF33(615)-5150, North American Aviation, Columbus Division.
FREEMAN, W.T. and M.D. CAMPBELL (1972), Thermal Expansion Characteristics of Graphite Reinforced Composite Materials, in: Composite Materials: Testing and Design (2nd Conf.), ASTM STP **497**, American Society for Testing and Materials, 121–142.
FROCHT, M.M. (1941), *Photoelasticity I* (Wiley, New York), 252.
GOREE, J.G. (1967), *J. Composite Materials* **1**, 404.
GRAHAM, L.J. and R.K. ELSLEY (1978), Characteristics of Acoustic Emission Signals from Composites, in: *Proc. of the ARPA/AFML Review of Progress in Quantitative NDE*, AFML-TR-78-55, Air Force Materials Laboratory, 219–225.
GRIMES, G.C., P.H. FRANCIS, G.E. COMMERFORD and G.K. WOLFE (1972), AFML-TR-72-40, Air Force Materials Laboratory.
GUESS, T.R. and F.P. GERSTLE, Jr. (1977), *J. Composite Materials* **11**, 146.
HAENER, J. (1966), Report 66-62, U.S. Army Aviation Materiel Labs.
HAGEMAIER, D. (1974), *SAE Trans.* **83**, 2767.
HAHN, H.T. (1976), *J. Composite Materials* **10**, 266.
HAHN, H.T. and R.Y. KIM (1978), Swelling of Composite Laminates, in: Environmental Effects on Composite Materials, ASTM STP **658**, American Society for Testing and Materials, 98–120.
HAHN, H.T. and N.J. PAGANO (1975), *J. Composite Materials* **9**, 91.
HARRIGAN, W.C., Jr. (1978), Ultrasonic Inspection of Graphite/Aluminum Composites, in: *Proc. of the 1978 International Conf. on Composite Materials*, ICCM/2 (Toronto, Canada), 1123–1140.
HASHIN, Z. (1972), NASA CR-1974.
HASHIN, Z. and B.W. ROSEN (1964), *J. Appl. Mechanics* **31**, 223.
HILL, R. (1963), *J. Mech. Phys. Solids* **11**, 357.
HILL, R. (1964), *J. Mech. Phys. Solids* **12**, 199.
HOFER, K.E. and P.N. RAO (1977), *J. Testing and Evaluation* **5**(4), 278.
HUNG, Y.Y., I.M. DANIEL and R.E. ROWLANDS (1978), *Exper. Mechanics* **18**(2), 56.
HUNG, Y.Y., R.E. ROWLANDS and I.M. DANIEL (1975), *Applied Optics* **14**(3), 618.
HUNG, Y.Y. and C.E. TAYLOR (1973), Speckle-shearing interferometric camera – A tool for measurement of derivatives of surface displacement, *Proc. of the Society of Photo-Optical Instrumentation Engineers*, Vol. 41 (San Diego, CA), 169–176.
HÜTTER, U., H. SCHELLING, H. KRAUSS (1975), An Experimental Study to Determine Failure

Envelope of Composite Materials with Tubular Specimens Under Combined Loads and Comparison with Several Classical Criteria, presented at AGARD-NATO Meeting, AGARD-CP-163 (Munich, Germany).
IIT Research Institute (1970–1975), IR&D Programs D1039, D1049, D1055, D1061, D1075, and D1096.
JENKINS, D.R. (1968), *Exper. Mechanics* **8**(10), 467.
KIES, J.A. (1962), NRL Report 5752.
KNIGHT, C.E. and H. PIH (1976), *Fibre Sci. Technology* **9**, 297.
KOUFOPOULOS, T. and P.S. THEOCARIS (1969), *J. Composite Materials* **3**, 308.
KULKARNI, S.V., R.B. PIPES, R.L. RAMKUMAR and W.R. SCOTT (1978), The Analytical Experimental and Nondestructive Evaluation of the Criticality of an Interlaminar Defect in a Composite Laminate, in: *Proc. of the 1978 International Conf. on Composite Materials*, ICCM/2 (Toronto, Canada).
LEKHNITSKII, S.G. (1963), Theory of Anisotropic Elastic Body, P. Fern trans., J.J. Brandstatter, ed. (Holden-Day, San Francisco).
LIBER, T., I.M. DANIEL, R.H. LABEDZ and T. NIIRO (1979a), Fabrication and Testing of Composite Ring Specimens, in: 34th Annual Technical Conf., 1979 Reinforced Plastics/Composites Institute, The Society of the Plastics Industry, Inc., Section 22-B (New Orleans, LA).
LIBER, T., I.M. DANIEL and S.W. SCHRAMM (1979b), Ultrasonic techniques for inspecting flat and cylindrical composite specimens, in: R.B. PIPES, ed., Nondestructive Evaluation and Flaw criticality for Composite Materials, ASTM STP **696**, American Society for Testing and Materials (Philadelphia, PA), 5–25.
LINDHOLM, U.S., A. NAGY, L.M. YEAKLEY and W.L. KO (1975), AFFDL-TR-75-83.
LO, K.H. and E.M. WU (1978), Contract No. F33615-76-C-5344, Air Force Materials Laboratory, Quarterly Report No. 7.
MADDUX, G.E. and G.P. SENDECKYJ (1979), Holographic techniques for defect detection in composite materials, in: R.B. PIPES, ed., *Nondestructive Evaluation and Flaw Criticality for Composite Materials*, ASTM STP **696**, American Society for Testing and Materials (Philadelphia, PA), 26–44.
MANDELL, J.F., S. WANG and F.J. MCGARRY (1975), *J. Composite Materials* **9**, 266.
MARLOFF, R.H. and I.M. DANIEL (1969), *Exper. Mechanics* **9**, 156.
MATZKANIN, G.A., G.L. BURKHARDT and C.M. TELLER (1979), Contract No. DLA 900-77-C-3733, Southwest Research Institute for U.S. Army Aviation Research and Development Command (St. Louis, MO).
MOOL, D. and R. STEPHENSON (1971), *Materials Evaluation* **29**(7), 159.
PAGANO, N.J. and J.M. WHITNEY (1970), *J. Composite Materials* **4**, 360.
PARKS, V.J., A.J. DURELLI, K. CHANDRASHEKHARA and T.L. CHEN (1970), *J. Appl. Mechanics* **37**, 578.
PETTIT, D.E. and K.N. LAURAITIS (1977–1980), Quarterly Reports LR 28360-1 to LR 28360-17, Contract No. F33615-77-C-3084, Lockheed California Co. (Burbank, CL).
PETIT, P.H. and M.E. WADDOUPS (1969), *J. Composite Materials* **3**, 2.
PIH, H. and C.E. KNIGHT (1969), *J. Composite Materials* **3**, 94.
PIPES, R.B. and J.L. ROSE (1974), *Exper. Mechanics* **14**(9), 355.
POST, D. and F. ZANDMAN (1961), *Exper. Mechanics* **1**, 21.
RODERICK, G.L. and J.D. WHITCOMB (1976), NASA TM X-73994.
ROSE, J.L., J.M. CARSON and D.J. LEIDEL (1973), Ultrasonic Procedures for Inspecting Composite Tubes, in: Analysis of the Test Methods for High Modulus Fibers and Composites, ASTM STP **521**, American Society for Testing and Materials, 311–325.
ROSEN, B.W. (1964), *AIAA J.* **2**(11), 1982.
ROSEN, B.W. (1972), *J. Composite Materials* **6**(4), 552.
ROTEM, A. (1977), *Fibre Sci.* **10**, 101.
ROWLANDS, R.E. (1975), AFFDL-TR-75-11.
ROWLANDS, R.E. and I.M. DANIEL (1972), *Exper. Mechanics* **12**, 75.
ROWLANDS, R.E., I.M. DANIEL and J.B. WHITESIDE (1973), *Exper. Mechanics* **13**, 31.

ROWLANDS, R.E., I.M. DANIEL and J.B. WHITESIDE (1974a), Geometric and Loading Effects on Strength of Composite Plates with Cutouts, in: Composite Materials: Testing and Design (3rd Conf.), ASTM STP **546**, American Society for Testing and Materials, 361–375.

ROWLANDS, R.E., T. LIBER, I.M. DANIEL and P.G. ROSE (1974b), *AIAA J.* **12**(7), 903.

RYBICKI, E.F. and A.T. HOPPER (1973), AFML-TR-73-100.

RYDER, J.T. and E.D. BLACK (1977), Compression Testing of Large Gage Length Composite Coupons, in: Composite Materials: Testing and Design (4th Conf.), ASTM STP **617**, American Society for Testing and Materials, 170–189.

SAMPSON, R.C. (1962), Report No. 0411-10F, Aerojet-General Corp.

SAMPSON, R.C. (1970), *Exper. Mechanics* **10**, 210.

SANDHU, R.S. (1974), AFFDL-TR-73-137.

SAVIN, G.N. (1961), *Stress Concentration Around Holes* (Pergamon Press, New York).

SENDECKYJ, G.P., G.E. MADDUX and N.A. TRACY (1978), Comparison of Holographic, Radiographic and Ultrasonic Techniques for Damage Detection in Composite Materials, in: *Proc. of the 1978 International Conf. on Composite Materials*, ICCM/2, Metallurgical Section of the American Institute of Mining, Metallurgical, and Petroleum Engineers (Toronto, Canada), 1037–1056.

SCHUSTER, D.M. and E. SCALA (1964), *Trans. Metall. Soc. AIME* **230**, 1635.

SHELDON, W.H. (1978), *Materials Evaluation* **36**, 2, 41.

SULLIVAN, T.L. and C.C. CHAMIS (1973), Some important aspects in testing high modulus fiber composite tube in axial tension, in: Analysis of the Test Methods for High Modulus Fibers and Composites, ASTM STP **521**, 277–292..

SUTLIFF, D.R. and H. PIH (1973), *Exper. Mechanics* **13**, 7, 294.

TSAI, S.W. and E.M. WU (1971), *J. Composite Materials* **5**, 58.

WANG, A.S.D., R.B. PIPES and A. AHMADI (1975), Thermoelastic Expansion of Graphite-Epoxy Unidirectional and Angle-Ply Composites, in: Composite Reliability, ASTM STP **580**, American Society for Testing and Materials, 574–585.

WEED, D.N. and P.H. FRANCIS (1977), *Fibre Sci. Technology* **10**.

WIDERA, O.E. and S.W. CHUNG (1972), *J. Composite Materials* **6**, 14.

WHITESIDE, J.B., I.M. DANIEL and R.E. ROWLANDS (1973), AFFDL-TR-73-48, Air Force Flight Dynamics Laboratory (Dayton, Ohio).

WHITNEY, J.M., I.M. DANIEL and R.B. PIPES (1982), Experimental Mechanics of Fiber Reinforced Composite Materials, Society for Experimental Stress Analysis (Brookfield Center, CT).

WHITNEY, J.M. G.C. GRIMES and P.H. FRANCIS (1873), *Exper. Mechanics* **13**, 5, 185.

WHITNEY, J.M. and R.J. NUISMER (1974), *J. Composite Materials* **8**, 253.

WU, E.M., 1974, Phenomenological anisotropic failure criterion, in: G.P. SENDECKYJ, ed., *Mechanics of Composites*, Vol. 2, Composite Materials (L.J. Broutman and R.H. Krock, eds.) (Academic Press, New York).

ZAK, A.R. and M.L. WILLIAMS (1962), GALCIT SM42-1, California Institute of Technology.

CHAPTER VII

The Effect of Structure Processing Defects on Mechanical Properties of Polymeric Composites

G.M. Gunyaev

All-Union Institute of Aviation Materials
Moscow 107005
USSR

Contents

HANDBOOK OF COMPOSITES, VOL. 3 – Failure Mechanics of Composites
Edited by G.C. SIH and A.M. SKUDRA

List of Symbols

D, d	Diameter
E	Young's modulus
G	Shear modulus
h	Thickness
i, I	Integrity
l	Length
n	Number of fibres
q	Number of fibre breaks
S	Surface area
V	Composite volume fraction
Δ	Distance between fibre ends
ε	Deformation
η	Correspondence coefficient
ϑ	Variation coefficient
ν	Poisson's ratio
σ	Strength
τ	Shear strength
φ	Angle of fibre deflection from the axis
b	Bending
cr	Critical
eff	Effective
f	Fibre
m	Matrix
fm	Interface
o	Original
p	Pores
t	Twisting
u	Surface energy
x, y, z	Coordinate axes
+	(plus) Tension
–	(minus) Compression

1. Processing defects of fibrous composites structure

1.1. Classification of defects

Methods of calculation and prediction of fibrous composites properties are based on the idealized model of integral unidirectional composite, where straight-line elastic reinforcing elements (fibres) with stable geometric dimensions are regularly placed in solid yielding matrix (binder); and there is good adhesion between them at the interface, providing joint deformation of fibres and matrix under various types of loading.

Actually macrostructure of fibrous composites is quite different from idealized model. Current engineering level predetermines the presence of regular (distributed in the total volume) and local structure defects in actual composites.

The appearance of local defects-folds, indents, delaminations, breaks and distortions of layers, scratches, resin extraction from some areas, etc., is, as a rule, the result of deviation from the given parameters or violation of technology, and is of casual character. On the contrary, regular structure defects, the majority of which, as noted by TARNOPOLSKY and ROSE (1969) are 'specially imbedded' in the material at the current technology level of making the components, and that is why should be considered and estimated in calculation and prediction of the composites properties. Carriers of regular structure defects are reinforcing filler, polymeric matrix and interface in the composite. The main defects, as it was shown by TUMANOV, GUNYAEV and LUTSAU (1975) are the violation of fibres continuity, disorientation and distortion, their difference in size and non-uniformity of their volume distribution in the composite, pores and cracks in the matrix, cracks, delaminations and incompleteness of phase contact on fibre-matrix interface. One of the possible structure defects classification in unidirectional polymeric composites is given in Table 1. Some technological factors, contributing to defects formation are also listed there.

1.2. Consideration of structure defects in calculation and prediction of composites properties

The effect of different structure defects in unidirectional composites may be considered by introducing coefficients, which represent the ratio of corresponding indexes of composites properties with the given structure defect type R^* to the composite properties calculated for idealized model R. In this case

$$R^* = RK; \qquad K = f(K_l, K_\varphi, K_v, K_p, K_i, K_s) \tag{1}$$

where $K_l, K_\varphi, K_v, K_p, K_i, K_s$ are coefficients, taking into account the effect of structure processing defects (see Table 1) – such as the violation of fibres

TABLE 1. Processing defects of unidirectional composites structure.

Defect carrier	Type of defect	Conditions of defect formation	Designation of structure defectness coefficient
Reinforcing filler	Violation of fibre continuity-filaments, threads, bundles breaks	Filaments damage during production and manufacturing of reinforcing filler; filaments and threads rupture on tension due to fibres different strength and length in bundles; the use of discrete fibre yarn for reinforcement	K_1
–	Disorientation of fibres in the layer plane	Winding by the tape with the given step; laying-out of layers on tapered, ogival and other double-curvature surfaces	$K_{\varphi 1}$
–	Distortion of fibres in the layer	The use of fabrics and other reinforcing fillers with controllable interweavings of threads and bundles	K_{φ}
–	Space distortion of fibres-twisting	The use twisted threads, bundles as reinforcing filler; fibre distortion due to stability loss in shrinkage	$K_{\varphi 2}$
	Non-uniformity of fibre distribution in the composite volume	Winding or laying-out of fibres with overlapping or discontinuities; fibre agglomeration due to using fibre bundles for reinforcement, in particular twisted ones; filaments different calibre and variation of filaments number along the thread length	K_v
Matrix and phase boundary	Pores	Extraction of solvents, chemical reactions products and sorbed easily volatile substances	K_p
	Cracks and delaminations	Matrix rupture in the volume and along phase boundary, due to stressing as a result of composites chemical and thermal shrinkage	K_i
	Imperfection of phase contact	Incomplete wetting of fibre surface by the binder; pores on the phase boundary; local separation of matrix from the fibre due to breaks in chemical and physical bonds debonding	K_s

continuity, their deviation from straight line, deviation from uniform packing along cross-section, matrix and interface non-integrity, incompleteness of phase contact. The value of R (strength, elasticity modulus) can be calculated either by the known ratios, or it can be determined experimentally, using model defect-free specimen.

2. Reinforcing filler defects

2.1. Violation of fibre continuity

Violation of fiber continuity is the result of damaging filaments due to abrasion, multiple bends and mechanical damages in the production of reinforcing filler: twisting, weaving, passing through filament guides of surface treating units, combining with the binder, winding or lay-up. The possibility of damaging filaments and bundles in tensioning as a result of their different stressing due to different strength (ϑ_σ), modulus (ϑ_E) and length of the fibres in the thread, bundle, strand or layer.

Integrity violation of continuous fibres results in the decrease of their area in the composite cross-section and in the increase (redistribution) of adjacent fibres stressing, the danger of damage increases with the increase of reinforcing element cross-section (diameter d_f). In this case the coefficient, taking into account the violation of continious fibres integrity (K_{l_1}), at the expense of tears number (q) in the cross-section unit of the composite, which height is equal to non-effective fibre length ($l_{f\,cr}$) is determined from the expression, suggested by GUNYAEV (1981).

$$K_{l_1} = \frac{q \cdot V_{f\,max} d_f^2}{V_f} \qquad (2)$$

where $V_{f\,max}$ the marginal fibre content in the composite in the given lay-up.

As follows from Equation (2) and data, given in Fig. 1, the danger of this type defects increases with the increase of the diameter and the decrease of fibre content in the composite.

Reinforcement by large diameter fibres predetermines higher composite sensitivity to the violation of fibre integrity. As it is shown by TUMANOV, GUNYAEV and YARTSEV (1976) the number of reinforcing elements per area unit of the composite cross-section is reduced by a factor of more than 100, if the fibre diameter increases from 7–10 μm to 100–150 μm, and is about 100 fibers per mm^2 for boron fibre reinforced plastics and about 20 000 fibres per mm^2 for carbon and glass-fibre reinforced plastics. On the contary, tear strength of filaments is increased from 0.2–0.4 N (carbon and glass fibres) to 25–45 N (boron fibres). Local weakening of boron fibre reinforced plastic in the break of one fibre is identical to breaking 150–200 fibres in glass or carbon fibre reinforced plastic.

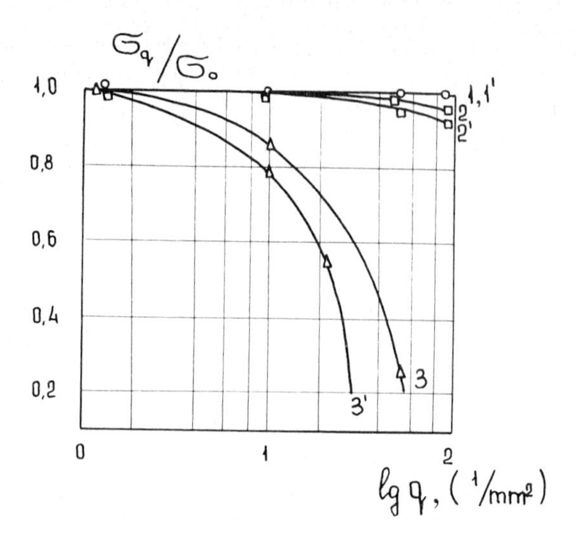

FIG. 1. Effect of fibre breaks accounting for the area unit of carbon (1 and 1′), glass (2 and 2′) and boron (3 and 3′) fibres reinforced plastics cross-section on compression strength. ○ Carbon fibres, diameter 9 μm; □ Glass fibres, diameter 18 μm; △ Boron fibres, diameter 95 μm. Reinforcement extent of composites 1, 2 and 3: 0.6, Composites 1′, 2′ and 3′: 0.4.

The use of yarn, composed of discrete fibres, for composite reinforcing makes it necessary to take into account not only the number of 'breaks' in the cross-section, which may be characterized by the index of fibres discreteness level – the ratio of their length (l_f) to the critical length of the fibre in the composite ($l_{f\,cr}$) – but also by the distance between fibre ends Δ.

It is the existence of this gap, which leads to the decrease of filler cross-section in the composites reinforced by discrete fibre yarn. With the increase of the gap between the fibre ends and the fibre discreteness level the composites strength and elasticity modulus decrease in the reinforcement direction.

2.2. Disorientation and distortion of fibres in layers

The term disorientation signifies the deviation of straight-line fibres in the composite layers from the given reinforcement direction, corresponding to the direction of applying the load. Such structure defects occur in the process of layer winding of the articles over the mandrel with the given step, equal, for example, to the tape width, layup of unidirectional layers of reinforcing filler, tapes, fabrics, etc., on tapered, ogival and other complex shape surfaces. Thus, for instance, on winding of tape single layers having width (h) over the mandrel with diameter (D), adjacent layers turn out to be disoriented through the angle $2\varphi_1 = \mathrm{arctg}(h/\pi D)$. While in the first layer the fibres are deviated through the angle $+\varphi_1$, for the other layer the deviation is through the angle $-\varphi_1$. Such disorientation is called regular antiphase disorientation. If the fibres in each layer are deviated through the same angle, to one side from the given reinforcement direction, uniform disorientation takes place. In most cases the latter type of structure defects is not regular and is of casual character.

Other regular type of structure processing defects is fibre distortion, that is their deviation from a straight line. Regular fibre distortions are inherent in the woven filler reinforced composites. In this case, depending on interweaving type in the fabric: sateen, serge, linen, regular antiphase distortions are imbedded in the composite structure, which are characterized by the extent of fiber bending in the layer. Thus, for the case of sinusoidal distortion $\varphi = A\pi K/l$, where K is the number of l based halfwaves, and A the distortion amplitude. As was shown by TARNOPOLSKY and ROSE (1969) composites with sinusoidal fibre distortion are similar to the material with broken line-shaped fiber distortion as far as their deformation is concerned, that is why considering $\varphi_1 = \varphi = \operatorname{arctg} AK/l$.

Therefore, if texture parameters-diameters and the number of fibres in the weft and warp threads of the fabric d_{fx}, n_{fx} and d_{fy}, n_{fy}, density of fibre lay-up in the fabric threads V_f and l, the interweaving step, are known, for small angle values, according to GUNYAEV (1976), when replacing distortion by disorientation, angle $\varphi = \varphi_1$ is equal to $\varphi = [d_{fx}n_{fx}^{0.5} + d_{fy}n_{fy}^{0.5}]V_f^{0.5}/l$. Coefficients, taking into account the effect of regular antiphase distortions and disorientation on the composite strength and elasticity modulus, considering that $\varphi \leqslant 10°$, $E_x \gg E_y$, may be obtained in using the following equations, suggested by TARNOPOLSKY, ROSE, ZHIGUN and others (1971):

$$K^E = 1 - 2(1 - \nu_{xy})[1 - 2(1 + \nu_{xy})E_x/G_{xz}]\varphi^2\,, \tag{3}$$

$$K^\sigma = \{1 + [E_x/G_{xz} - 2(1 - \nu_{xy})]\varphi^2\}^{-1}\,. \tag{4}$$

Experimental data concerning the effect of regular distortions on elastic strength properties of carbon fibre tapes reinforced composites are given in Table 2. With the reduction of warp and weft threads thickness the extent of fibre distortion decreases and the composite elasticity modulus, compression and tensile strength increases. As the elasticity modulus and the strength of reinforcing fibres are increasing, the composite sensitivity to distortion becomes higher (Fig. 2). The most dramatic decrease of mechanical properties values with the fibre deviation from the direction of testing or straight line is inherent in the composites with maximum anisotropy, characterized by the ratio of elasticity moduli E_x/G_{xz}, as it was shown by GUNYAEV, ZHIGUN and SORINA (1978) and GUNYAEV and SORINA (1980). The reduction of anisotropy level, due to matrix strengthening (increasing G_{xz}) by introducing single crystals into interfibre space of the composites, using whiskerization, reduces composites sensitivity to such structure defects as distortion and disorientation.

2.3. *Space distortion of fibres*

This type of structure defects appear in composites when using twisting of threads, bundles, etc., reinforcing fillers, and also in case of fibre distortion due

TABLE 2. Effect of processing structure defects, caused by fibres distortion in carbon tape, on mechanical properties of epoxy carbon fibre reinforced plastics.

Tape texture[a]		Filament properties				Composite composition and properties				
Distortion angle of warp fibres, (degrees)	Number of warp thread folds	σ_f (GPa)	ϑ_σ (%)	E_f (GPa)	ϑ_E (%)	V_f (%)	V_p (%)	σ_x^+ (GPa)	σ_x^- (GPa)	E_x (GPa)
1	2	3	4	5	6	7	8	9	10	11
2.0	1	2.32	23	241	13	49.3	0.4	0.92	0.48	117
5.0	2	2.26	24	238	15	50.2	0.3	0.91	0.47	112
8.0	4	2.30	22	229	12	47.9	0.5	0.86	0.44	103
8.5	5	2.22	18	232	12	51.3	0.2	0.85	0.45	108
10	6	2.21	26	232	14	50.5	0.5	0.81	0.38	103
12	8	2.28	23	228	9	48.6	0.3	0.77	0.37	92

[a] Warp – number of filaments in primary thread 200, filaments diameter 7.2 μm. Weft – number of filaments in primary thread 200, filaments diameter 7.8 μm, number of weft thread folds: 4.

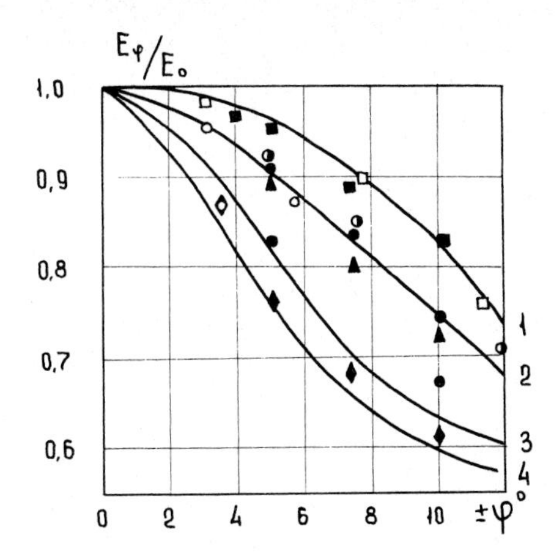

FIG. 2. Effect of regular fibre antiphase distortions and disorientation in composite layers with E_x/G_{xz} ratio 10(1), 20(2), 40(3) and 60(4). Carbon fibres with $E_f = 230$ GPa ○, ●; whiskerized fibres with $E_f = 230$ Gpa ◑; with $E_f = 450$ GPa ◇, ◆; Glass fibres with $E_f = 92$ GPa □, ■; Boron fibres with $E_f = 380$ GPa △, ▲ (○, □, ◇, △ distortions; ●, ■, ◆, ▲ disorientation).

to the loss of stability to chemical and thermal shrinkage (the latter are, as a rule, of a local character).

Twisting is widely used as a means of increasing processability of threads, yarn bundles, eliminating fluffing and breaking of filaments. Space distortion of fibres in twisting may be characterized by two parameters: (φ_1), the fibres disorientation angle relative to yarn axis, equal to the angle of fibre lifting along the helical line, and (φ) the angle, describing the distortion of the same fibre due to its rounding the yarn. The relation between the (φ_1), (φ) parameters and the yarn texture parameters (bundle, thread) – such as twisting (K_t) (the number of twistings per linear metre), diameter (D) and density (V_f) of the yarn, diameter (d_f), the number (n) of filaments in the yarn – is expressed by the equations

$$\varphi_1 = \operatorname{arctg} D \cdot 2K_t/1000 = \operatorname{arctg} d_f(n/V_f)^{0.5} \cdot 2K_t \cdot 10^{-3}\,,$$

$$\varphi_2 = \operatorname{arctg} D \Big/ \left[D^2 + \left(\frac{1000}{2K_t}\right)^2\right]^{0.5} = \operatorname{arctg}(\sin \varphi_1)\,.$$

At sparse twistings $\sin\varphi \simeq \operatorname{tg}\varphi \simeq \varphi$, therefore $\varphi = 2\varphi_1$. The effect of regular fibre space distortions on strength and elasticity modulus of polymeric composites was studied by GUNYAEV (1981). It can be taken into account by corresponding coefficients $K^{\sigma}_{\varphi 3}$ and $K^{E}_{\varphi 3}$ equal to

$$K^{\sigma}_{\varphi 3} = \{1 + [E_x/G_{xz} - 2(1-\nu_{xy})]2\varphi_1^2\}^{-1}\,, \tag{5}$$

$$K^{E}_{\varphi 3} = 1 - 2(1-\nu_{xy})[1 - 2(1+\nu_{xy})E_x/G_{xz}]2\varphi^2\,. \tag{6}$$

Fig. 3 shows the data, illustrating the effect of twisting on strength and elasticity

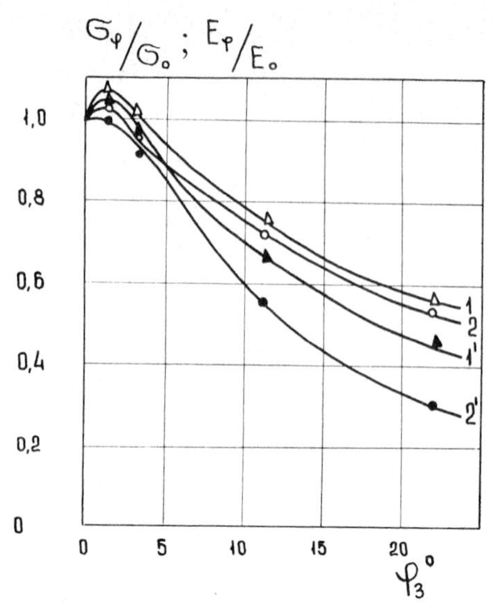

FIG. 3. Effect of fibre distortions during yarn twisting on the strength (1, 1′) and elasticity modulus (2, 2′) of epoxy carbon fibre reinforced plastics. ○, ● carbon fibres with $E_f = 440$ GPa, $\sigma_f =$ 1.97 GPa; △, ▲ carbon fibres with $E_f = 287$ GPa, $\sigma_f = 2.94$ GPa.

modulus of unidirectional epoxy composites, reinforced the yarn, consisting of 10 000 filaments of 9.5 μm diameter and elasticity modulus of 280 and 440 GPa.

The analysis of the data given shows that the composite strength and elasticity modulus depend on the fibre distortion; the optimum is the twisting that has an angle (φ_1) of 1.5–2°. At sparse twisting of the yarn, consisting of a great number of filaments, the values of strength and elasticity modulus may exceed the values obtained with untwisted fillers, as in this case the possibility of fibre disorientation is eliminated, and difference in length and stressing is reduced. The extent of strength and elasticity modulus decrease of the composite with space distortions increases with the increase of the fibre elasticity modulus.

2.4. Non-uniformity of fibre volume distribution

As the type of regular defect can be the result of winding or laying layers with regular overlapping or gapping, agglomeration of fibres when using twisted threads or bundles, woven fillers, lengthwise variation of the bundle metric number and the dispersion of the fibre diameter or cross-section area. As a result, the density of fibre packing varies along the composite cross-section, difference is observed in the distance between individual filaments in the structural elements thread and yarn, and between structural elements, threads and strands. Local dispersion, fibre volume fraction in the composite, results in different stressing of matrix volumes, microinhomogeneities of its properties, in particular those, which are sensitive to variations of reinforcing extent, for example, shear modulus, coefficient of deformation concentration. In the areas of the composite volume, where the content or reinforcing fillers is higher, the concentration coefficient is higher according to SKUDRA and BULAVS (1978), which, combined with technological stresses, due to chemical and thermal shrinkage on curing, leads to the decrease of the composite deformation in transverse direction. The coefficient, accounting for the decrease of tear relative elongation value equals

$$K_v^{\varepsilon} = V_f / V_{f\,max}\,. \tag{7}$$

It should be taken into consideration, that the change in interface size corresponds with the local variation of reinforcing filler volume fraction in the composite. In some cases this results in the change of composition, structure and properties of polymeric matrix in boundary layers, surrounding reinforcing fibres compared to matrix properties at a definite distance from the fibre.

3. Matrix and interface defects

3.1. *Pores*

Pores are formed due to extraction of volatile products of chemical reactions of binder curing, solvents, air and moisture adsorbed by the binder and reinforcing filler. These products, expanding when heated, form a system of open and closed pores, varying in shape and size. The effect of composite porosity is different, depending on the type of stress condition and strain direction (Table 3). It has the greatest effect on the resistance to shear loads and transverse loading, to a lesser extent it effects the resistance to loading in the reinforcement direction. This is due to the fact that pores reduce the matrix' effective

TABLE 3. Porosity effect of epoxy carbon fibre reinforced plastics mechanical properties.

Composite composition		Composite properties						
V_p (%)	V_f (%)	σ_x^+ (GPa)	σ_x^- (GPa)	σ_x^b (GPa)	σ_y^+ (MPa)	τ_{xz} (MPa)	E_x (GPa)	G_{xz} (GPa)
1	2	3	4	5	6	7	8	9
0.2	52.0	1.03	0.82	1.20	22	65	118	3.30
0.8	51.4	1.00	0.81	1.23	22	64	116	3.20
2.5	53.0	1.02	0.79	1.18	21	63	121	3.15
4.0	52.7	1.00	0.76	1.15	20	61	115	3.10
7.0	52.5	0.99	0.74	1.11	19	56	110	3.00
15	50.6	0.93	0.70	1.02	17	50	105	2.90
19	51.8	0.89	0.63	0.99	12	43	98	2.50

cross-section and the area of fibre-matrix contact, and they are also the sources of crack initiation.

Introduction into the equation of a correction which takes into account the porosity effect which describes the interlayer shear strength of unidirectional composite, leads to the following expression, suggested by KOBETS and GUNYAEV (1977):

$$\tau_{xz} = \tau_m S_m K(1-P) + \tau_{fm} K_s^p S_{fm} ,$$

which in its turn makes it possible to obtain the expression for the coefficient, taking into account the effect of pores on interlayer shear strength:

$$K_p = \frac{A(1-P) + K_s^p}{A + K_s} , \qquad A = \tau_m S_m / \tau_{fm} S_{fm} \tag{8}$$

where τ_m is the interlayer shear strength of pore-free matrix; τ_{fm} the interface shear strength; $S_m S_{fm}$ the size of cross-section accounting for matrix and interface in the composite; P the coefficient, taking into account matrix cross-section reduction on uniform distribution of spherical voids, proportional to $(V_p/1-V_f)^{3/2}$; and K_s^p the coefficient, taking into account the reduction of phase contact completeness as a result of pores extension to interface.

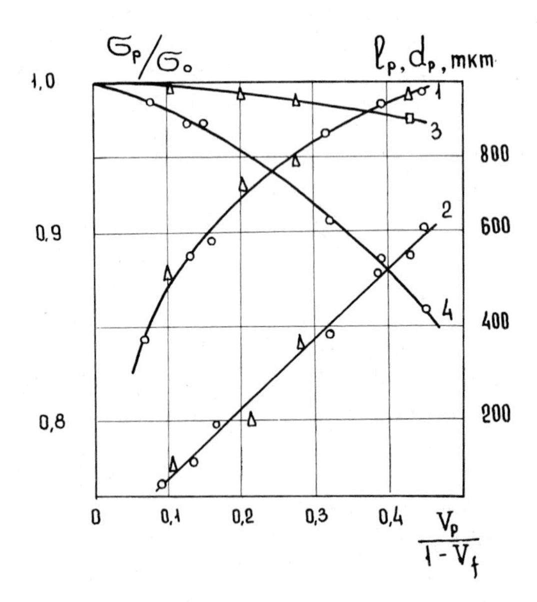

FIG. 4. Effect of composite porosity on pore average size: Length (1) and diameter (2) and tensile strength of boron (3) and carbon (4) fibre reinforced plastics. ○ carbon fibres, $d_f = 7.8$ μm; △ boron fibres, $d_f = 102$ μm.

K is the average coefficient of pore stress concentration in the matrix. A similar expression was obtained by GUNYAEV (1977) for the case of transverse tension. With the increase of porosity the axial compression strength decreases, which is the result of the reduction of the filler working cross-section, due to the loss of stability of some fibres, passing through the pores.

The porosity effect on composites tensile strength was accounted for by TUMANOV, GUNYAEV, MOTDAD and others (1975), when studying a unidirectional composite with the given ratio of (V_f) and (V_p) components, assuming that the fibres in the pore region of (l_p) length behave like unbonded bundles. In this case

$$K_p = 1 - V_p[1 - (l_p/l_{f\,eff})]^{-0.83\vartheta_\sigma}/(1 - V_f) \tag{9}$$

where $l_{f\,eff}$ is the effective fibre length in the composite.

The analysis of the properties of composites with different porosity shows, that the danger of pores increases with the increase of their number and length (Fig. 4). Most dangerous are elongated pores, which length exceeds the critical fibre length in the composite. This fact probably accounts for higher composites sensitivity to porosity increase of fibre thinning. Thus, boron fibre reinforced plastics are less sensitive to porosity than glass and carbon fibre reinforced plastics.

3.2. *Incompleteness of phase contact and cracks*

A decisive effect on polymeric matrix to fibres adhesion along the interface is due to incompleteness of phase contact and cracks. Interface is not only a geometrical boundary between fibres and binders, but also between adjacent areas, influenced by physico-chemical processes, taking place during interaction of fibres and matrix at the stage of composite formulation. The interface condition – its structure and the amount of defects – is effected by physico-chemical and thermomechanical compatibility of composite forming components. The first one defines phase contact completeness, the nature, number and strength of physical and chemical bonds, generated on interaction of polymeric binder with the fibre surface, effects structure formation, the change of matrix and fibre composition and properties in boundary layers due to interdiffusion, selective sorption, the catalytic effect on the binder curing process, etc. The second one defines components intercorrespondence, providing composite integrity and stressing level (which can also be related to structure defects) of interface and components, determined by the difference of their deformation and thermophysical properties.

The completeness of phase contact in the composite can be characterized by coefficient K_s, which is the ratio of the area effected by interaction forces between functional groups and fractions of binder and fibre molecules (S_{fm}) to fibre surface area S_f; thus $K_s = S_{fm}/S_f$, $0 < K_s < 1$.

The interaction surface area accounted for the composite volume unit, depends on reinforcement extent, fibre geometrical sizes, their specific surface and its topology, fibre wettability by the binder, mobility of its molecules or their portions, facilitating sorption and penetration of molecules into pores and cracks on fibre surface, interdiffusion ability. Fibres wettability by the binder, their sorption ability is defined by fibres surface energy, presence of polar reactive groups. Mineral fibres surface energy at the moment of their manufacturing is high enough, but it is quickly reduced as a result of surface contamination due to adsorption of environment products. This leads to the increase of critical wetting angles and the loss of adhesion to binders. To prevent this, reinforcing fillers are subjected to various special treatments, resulting in cleaning, activation, chemical modification and protection of their surface, before making composites.

The effect of phase contact completeness along the interface on the composite properties can be estimated by the results of composites interlayer shear tests, using an approximated equation, as was suggested by KOBETS and GUNYAEV (1977): $\tau_{xz} = K_s^s K_s \tau_{xz}^0$, where K_s is the coefficient of contact completness ($K_s = K_s^p K_s^u$); K_s^p the coefficient taking into account pore extension to interface; K_s^u the coefficient, taking into account surface activity, proportional to the value of surface energy and adsorbability of fibre surface; K_s^s the surface roughness coefficient, proportional to the value of fibre specific surface S_f, and τ_{xz}^0 is the limit shear stress at the interface, defined by the nature of the components combined.

Fig. 5 shows the dependence of interlayer shear strength of epoxy carbon fibre reinforced plastic on the surface roughness coefficient for two kinds of activation treatments. Tangent of straight lines inclination angle, plotted according to the results, obtained by KOBETS, POLYAKOVA, KUZNETSOVA and others (1978), is equal to coefficients of contact completeness along the interface.

In the manufacture of polymeric composites multicomponent binders are used, therefore when combining fibres with these binders, complex processes of fractions and volume redistribution are taking place at the interface. Boundary

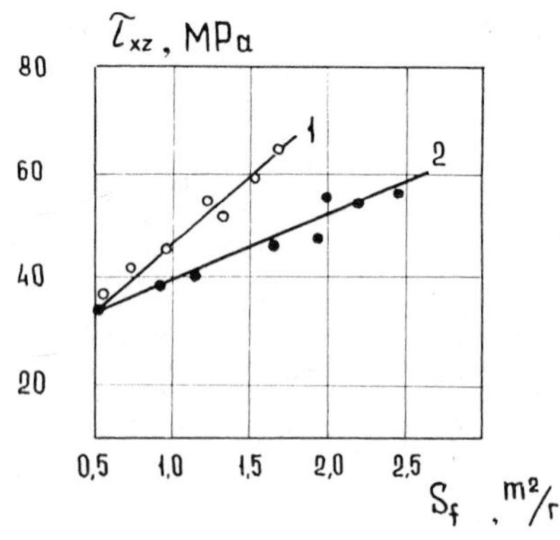

FIG. 5. Dependence of epoxy carbon fibre reinforced plastic interlayer shear strength on roughness coefficient of carbon fibres surface, activated by oxidation in HNO_3 (1) and air (2).

layers are, as a rule, enriched in low-molecular substances, which results in the change of components ratio in binder volume and boundary layers. The above-mentioned factors and the presence of functional groups (having either acidic or basic nature) on the fibre surface can have an inhibiting or catalytic effect on polymer formation in the boundary layer, its curing depth and the structure of polymer network being formed. In some cases this leads to appearance of 'weak' binder boundary layers, characterized by lower density, strength, thermal resistance and higher stressing, compared to the binder in the block. Contrary to mineral fibres, organic fibres are more sensitive to the effect of the binder they are combined with in the composite. As was shown by LIPATOV (1977), diffusion of low-molecular products inside organic fibres may cause their swelling, relaxation and consequently the decrease of fibre and composite strength. This type of defects can be estimated when comparing corresponding properties of the actual composite with corresponding properties of a 'defect-free' composite.

Adhesive interaction of the binder with the fibre prevents free change of their sizes in the process of binder chemical shrinkage in curing and subsequent cooling. Due to the difference in linear thermal expansion coefficients of fibre and matrix on cooling to the temperature lower than glass-transition temperature, some stresses are generated in the composite. Longitudinal, tangential (circumferential) and radial stresses are generated in the fibre and matrix, which character and distribution dependencies are similar to those generated under axial loading. The value of stresses increases with the increase in difference of fibre and matrix thermal expansion coefficients and elasticity moduli. If the stresses generated are commensurable with matrix strength or adhesion strength along the interface, cracks are initiated in the composite, positioned in the direction of fiber orientation.

To estimate the composite quality it is reasonable to use the notion of integrity introduced by RABINOVITCH (1970), which assumes the integrity of all the components and the absence of bonding defects on the composite interface

TABLE 4. Values of the binder elastic-strength properties, providing integrity conditions for glass and organic fibres reinforced composites.

Binder properties	For glass fibre reinforced plastic $\sigma_f = 2.85$ GPa $E_f = 75$ GPa $\varepsilon_f = 3\%$	For organic fibre reinforced plastic $\sigma_f = 4.2$ GPa $E_f = 95$ GPa $\varepsilon_f = 3.8\%$
Tension strength σ_T (MPa)	140	250
Tension modulus E_T (MPa)	4500	5700
Relative tear elongation ε_T (%)	4.5	5.3
Matrix to fibre adhesive strength τ_T (MPa)	94	170

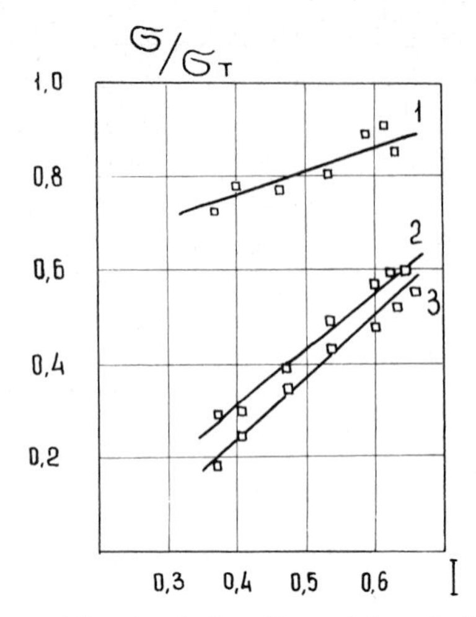

FIG. 6. Interrelation between unidirectional glass fibre reinforced plastic tensile (1), compression (2) and interlayer shear (3), strength and complex integrity (defectness) criterion (I).

not only in as-manufactured condition, but also on deformation of the composite until the fibres are broken, when their strength limit is exhausted. Integrity conditions in the form of a system of inequalities were obtained by RAGINSKY, KONOVITCH and KOLTUNOV (1979), when analyzing composite combined deformation under different loadings. The necessity of fulfilling all and not only one integrity condition should be taken into consideration. Substitution of fibres elastic-strength properties values into the system of inequalities allows to determine the desired elastic-strength properties of the matrix, providing integral structure, Table 4 shows such data for glass and organic fibres reinforced plastics, where reinforcing fibres have different mechanical properties. As follows from Table 4, requirements to matrix properties resulting from integrity conditions, exceed property level of modern binders, that is the production of a 'defective' composite is pre-programmed. The complex integrity factor (I) of the composite is used as the criterion for estimation of the composite integrity (the value inversed to defectness) and strength.

$$I = K_\sigma \sum \eta_i \lambda_i, \quad 0 < I < 1 \tag{10}$$

TABLE 5. Validity coefficients of binder parameters under different types of glass fibre reinforced plastics deformation.

Type of deformation	Validity coefficients			
	λ_ε	λ_E	λ_σ	λ_τ
Tension	0.06	–	0.20	0.74
Compression	–	0.17	0.34	0.49
Shear	–	0.16	0.20	0.64

where $K = K_1 K_\varphi K_v K_p K_i K_s K_{\vartheta_\sigma} K_{\vartheta_E}$ is the coefficient, taking into account composite microstructure defects (K_1, K_φ, K_v, K_p, K_i and K_s), and the deviation of fibres strength and elasticity modulus from average values (K_{ϑ_σ} and K_{ϑ_E}); η_i the correspondence coefficient of binder integrity characteristics: $\eta_E = E/E_T$; $\eta_\sigma = \sigma/\sigma_T$; $\eta_\tau = \tau/\tau_T$ (index T is the theoretical value of the factor determined from integrity conditions). The λ_i are the validity coefficients of binder parameters under various types of deformation.

As an example coefficient values are given in Table 5, and the data illustrating the relation of epoxy glass fibre reinforced plastic strength properties to integrity criterion are given in Fig. 6.

In case of full correspondence of binder properties to integrity conditions $I = 1$ ($K_\sigma = 1$) the deviations of value I from the unit give integral value of the composite defectness extent.

References

GUNYAEV, G.M. (1981), *Struktura i Svoystva Polimernikh Voloknistikh Kompositov* (Khimia, Moscow), 236.

GUNYAEV, G.M. (1977), *Mekhanika Polimerov* **6**, 981.

GUNYAEV, G.M. (1976), Optimizatsiya sostava i makrostrukturi pri konstruirovanii polimernikh kompositsionnikh materialov, in: N.V. AGEEV, A.T. TUMANOV, M.KH. SHORSHOROV, K.I. PORTNOY and V.I. BAKARINOVA, eds., *Voloknistie i Dispersnouprochnennye kompositsionnye Materialy* (Nauka, Moscow), 141.

GUNYAEV, G.M., I.G. ZHIGUN, T.G. SORINA and V.A. YAKUSHIN (1973), *Mekhanika Polimerov* **3**, 492.

GUNYAEV, G.M. and T.G. SORINA (1980), Termoustoichivost' napolnennikh plastikov na osnove fenolo-formal'degidnikh, Poliefirnikh i Epoksidnykh Svyazuyutshchikh, in: E.B. Trostyanskaya, ed., '*Termoustoykhivost' plastikov Konstruktionnogo Naznacheniya* (Khimia, Moscow), 161.

KOBETS, L.P. and G.M. GUNYAEV (1977), *Mekhanika Polimerov* **3**, 445.

KOBETS, L.P., N.B. POLYAKOVA, M.A. KUZNETSOVA, O.B. KOLESINSKAYA, B.C. SAMOYLOV and N.B. BONDARENKO (1978), *Mekhanika Polimerov* **4**, 579.

LIPATOV, U.S. (1977), *Fisicheskaya Khimia Napolnennych Polimerov* (Khimia, Moscow), 196.

RABINOVITCH, A.L. (1970), *Vvedenie v Mekhaniku Armirovannykh Polimerov* (Nauka, Moscow), 287.

RAGINSKY, S.L., M.Z. KANOVITCH and M.A. KOLTUNOV (1979), *Visokoprochnye Stekloplastiki* (Khimia, Moscow), 29.

SKUDRA, A.M. and F.YA. BULAVS (1978), *Stukturnaya Teoria Armirovannykh Plastikov* (Zinatne, Riga), 141.

TARNOPOLSKY, YU.M. and A.V. ROSE (1969), *Osobennosti rascheta Detaley iz Armirovannykh Plastikov* (Zinatne, Riga), 26.

TARNOPOLSKY, YU.M., A.V. ROSE, I.G. ZHIGUN and G.M. GUNYAEV (1971), *Mekhanika Polimerov* **4**, 676.

TUMANOV, A.T., G.M. GUNYAEV, V.G. LUTSAU and E.I. STEPANICHEV (1975), *Mekhanika Polimerov* **2**, 248.

TUMANOV, A.T., G.M. GUNYAEV and V.A. YARTSEV (1976), Svoystva kompositsiy, uprochnennykh voloknami bolshogo diametra, in: N.V. AGEEV, A.T. TUMANOV, M.KH. SHORSHOROV, K.I. PORTNOY and V.I. BAKARINOVA, eds., *Voloknistie i Dispersnouprochnennye kompositsionnye Materialy* (Nauka, Moscow), 156.

CHAPTER VIII

Interaction of Components at Local Failure of Composites

G.A. Vanin

Institute of Mechanics
Academy of Sciences of the Ukrainian SSR
Nesterova 3, Kiev-57,
252057, USSR

Contents

HANDBOOK OF COMPOSITES, VOL. 3 – Failure Mechanics of Composites
Edited by G.C. SIH and A.M. SKUDRA

List of Symbols

$A_0, A_{js}, C_0, C_{js}, D_0, D_{js}$
: coefficients of expansion of complex potentials in series with elliptic functions, see relations (8), (9);

$A_{mn}, A_n, B_{mn}, B_n, C_{mn}, D_{mn}, S_m$
: coefficients of expansion of function of longitudinal displacements in series. They are determined from solution of auxiliary problem of the theory of function of complex variable, see relations (19), (20), (22), (24), (34), (48), (50), (56);

a_j coordinates of the center of jth fiber;

a_n, b_n, q_n, g parameters of expansion of function of longitudinal shear displacement in series with degenerated hypergeometric functions, see relation (18);

a^2, b^2, c^2 parameters of critical stressed state, see relations (31), (32);

$b, b', c, c', d, d', f, f', h$
: parameters of stressed state under transverse extension, see relations (78), (79);

d fiber diameter;

$\mathcal{F}, \mathcal{F}_k$ cell area;

$F(z)$ complex potentials (7), (8), (55);

$F(a_n, b_n, x), U(a_n, b_n, x), U_n(x)$
: degenerated hypergeometric function;

$f_m(\xi), G_a^*, L$ see relations (25), (28), (64), (69);

$G_r, G_{r\varphi}$ shear modulus for fibers with cylindrical anisotropy, see (17);

$G_a, G_0e^{\alpha r}$ variable elasticity modulus;

G_0, g constants that determine non-uniform properties of the fiber with cylindrical anisotropy;

G_{12}, G_{13} first approximation formulas for moduli of material with non-uniform fibers and cracks, see (28);

$l + l_0$ cylindrical fiber surface the regime l_0 of which produces a break of continuity of small width, see Fig. 2;

m_0 number of fibers without imperfections;

m_k number of fibers in k-state;

N number of fibers in specific volume;

n number of fibers nearest to the analyzed fiber;

$P_{n-j}(\cos \theta/2)$ Legendre polynom;

$P(\theta, \Theta)$ function of the distribution of defects of adhesion between fiber and matrix;

$R(z), Q(z)$ see relations (26), (55);

r, φ cylindrical system of coordinates;

r_0 radius of central isotropic part of the fiber;

S_{iksn}, Z_{iksn} parameters of stresses and deformations tensors;
U_j elastic energy of the jth component of the medium;
U_0, V, U_1, U_∞
interaction potentials, see (11);
$\bar{u}$ displacement vector;
u_i components of the elastic displacement vector;
V_n specific volume;
x_i axial coordinates;
$X_m(z)$, $Y_m(z)$, $V_m(z)$, $U_m(z)$
auxiliary piecewise-holomorphic functions, see (23), (24), (34), (41), (51), (52), (53);
α, δ_j, γ_j structure parameters;
B, β, γ, ν^2 see relations (38);
$\Gamma(a_n)$, $\Gamma(2-b_n)$
gamma-functions;
Γ, Γ_0 interphase surface, see Fig. 8;
γ, γ_0 see Fig. 9;
γ surface energy, see (32);
$\delta(\Theta)$ Dirac delta-function;
ε radius of central part of fiber cavity, see (43);
ε_{ik} deformations tensor components;
$\langle\varepsilon_{ik}\rangle$ averaged components of the deformation tensor;
ζ coordinate of point on interphase border;
ζ, ζ_0, ζ' points located on circle of radius λ, on the interphase border, in the zone of coordination for reduced element, respectively, see Fig. 8;
ζ_a, ζ_b see Fig. 2;
$\zeta(z)$, $\eta(z)$ Weierstrass functions, see (8), (9);
$\theta = \dfrac{\varphi_b - \varphi_a}{2}$, $\Theta = \dfrac{\varphi_b - \varphi_a}{2}$
angles specifying values and location of crack center, see Fig. 2, $\kappa = 3 - 4\nu$;
λ radius of crack arc;
$\mu(z)$ see relations (26);
μ_{23}, μ_{32} see relations (28);
ν_n see relations (25);
ξ volumetric content of fibers in CM;
ρ distance from the crack tip;
$\bar{\sigma}_n$ limit value of the stress vector on the platform with normal $\bar{n}$;
σ_{ik} components of stress tensor;
$\langle\sigma_{ik}\rangle$ averaged components of stress tensor;
σ_{ik}^0 stresses in the smoothed field;
$\langle\sigma_{ik}^*\rangle$ critical values of mean stresses;
$\tau = e^{i\psi}$ see relation (19);
$\Phi(z)$, $\Psi(z)$, $\varphi(z)$
complex potentials, see relations (6), (7), (9), (10);

$\psi'(z) = \Psi(z)$
see relations (10);
$\chi(\theta)$, $\pi(\theta)$, $\lambda(\theta)$
see relations (75);
ω_i two-dimensional periods of structure.

Indices:
(a) fiber;
$(+)$, $(-)$ limit values of function along positive and negative direction of normal, respectively;
$(')$ differentiation.

Introduction

The extreme resistance of composites to failure is achieved by a harmonious combination of physico-mechanical and geometrical characteristics of the composites and the parameters determining their interaction. The latter is performed through the intermediate region in the immediate vicinity of the phase contact surface. In composites these regions are more or less uniformly distributed throughout the entire volume, and are of a distinctly marked inhomogeneous structure; they are in a field of high-intensity residual or external stresses being most liable to a diffusive penetration of particles from adjacent phases.

The classical conditions of ideal contact of components, i.e., equality of shifts and stresses at all points of the interphase, do not hold even in faultless combinations of fibres and a matrix of a perfect crystalline structure due to the misfit of the lattice distances, the difference in the interaction potential of the component microelements and other factors. In real systems the local admixtures, boundary structure distortions, barrier or other fibre coatings, as well as the existence of various types of bonds, considerably complicate the formulation of component contact conditions. This leads to a conclusion that the ideal phase contact conditions are acceptable at the initial stage of investigation when the effect of the interphase region state on the material and the bulk properties is not taken into account. A significant effect of the state of the interphase zone is observed in composite materials in which the components differ by an order or more in their physico-mechanical characteristics. Therefore the quantitative relation between the observable properties of materials and the state of their interphase zone, both at the initial stage of loading and in the process of exploitation, are of a special interest. The clarification of the interphase zone effect on the rise of local cracks in the structure clearly observable in the composite materials and the effect on their development of structure continuity, thickness and physico-mechanical properties, is of a special value. The properties of this zone can be effectively controlled by introducing special compounds for coating and fibre surface and by maintaining rational temperature and pressure conditions in the technological processes of material production and by other means.

The analysis of numerous experiments shows that in fibre-glass plastics, for instance, the prevailing appearance of local cracks is observed in the interphase zone or near it at the initial loading stage, or at the initial stage of the material. Therefore, the state of the reinforced medium with the account of the multitude of local cracks is more natural than that without defects. The properties of materials are liable to be mostly effected by the multitude of local cracks oriented equidistantly to the component interfaces when the fibres do not hinder their growth. The growth of crack sizes and concentration alters bulk characteristics of material, and a local denudation of fibres takes place. The latter process is in fact equivalent to the reduction of fibre effective viscosity failure due to surface crack and initiates the development of these cracks into the matrix.

There is a number of causes for arising the defects in composite materials with polymer matrix. The formation of micropores and cracks in the process of hardening because of liberation of by-products of polymeric reactions and the appearance of residual stresses of the first and second order as a result of chemical and thermal shrinkage of the matrix, is a widespread phenomena in this materials.

In metal matrix composites some of the causes of imperfections are the technological stresses and the accumulation of various precipitations in the interphase zone. To describe a more real state of composites, the author (VAN FO FI (1971), VANIN (1977a, b), VANIN (1978), VANIN and BYCHOVETS (1979), VANIN (1979)) proposed a model of materials with a nonideal interaction of components in which the contact surface between the fibres and matrix consists of alternating parts with perfect bonds and regions, where the bond is either totally absent or is partially accomplished by the friction forces or through mutual pressure of the cut edges due to their overlapping during shear or compression stresses. To estimate the dimensions of possible imperfections, the results of water percolation experiments were examined under various types of stressed states and stream gradient directions. In unstressed linearly-reinforced fibreglass plastics the penetrability of liquid along the fibres exceeds the same in the transverse directions by nearly two orders. The comparison of numerical values or water penetrability for the composite and matrix indicates the existence of regions stretched along the fibres with material discontinuity. Under a prolonged stress along the orientation of fibres in a unidirectional fibreglass plastic elongated local cracks appear in the interphase zones due to the rising penetrability. The local cracks are detected by optical and X-ray technique. According to the X-ray diffraction data, the initial transverse and longitudinal sizes of cracks at fibre surfaces are 400–600 Å, and more than 3000 Å respectively. One of the causes of weakened local bonds in fibres is the nonuniform distribution of residual 2nd order stresses in the structure of the material. The theory of residual stresses in linearly reinforced materials of the type of fibreglass plastics has been developed by VAN FO FI (1971) for regular structures with the account of visco-elastic matrix properties and the neglect of chemical shrinkage phenomena at the initial stages of heat cure. In particular, the diagram of shrinkage stress distribution at the interphase boundary in materials with hexagonal fibre packing is shown in Fig. 1. Here, the stress value

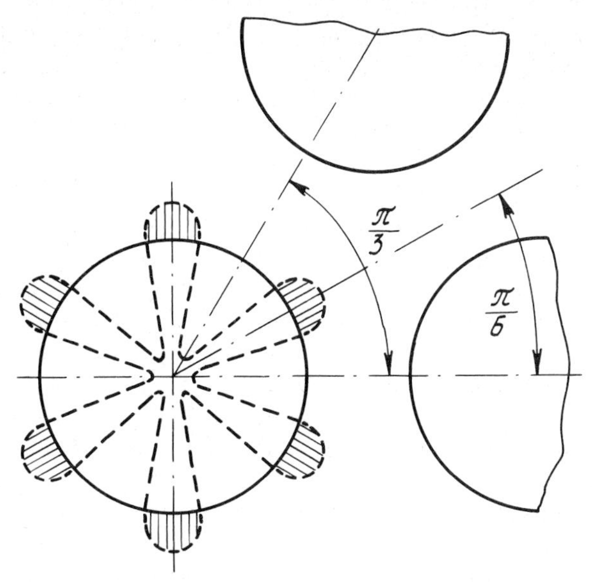

FIG. 1. Distribution of residual stresses at the interphase in a glass-fibre plastic of hexagonal structure.

is counted off from the fibre surface in the direction of the normal and depends on temperature difference $Q = T_g - T_0$ where T_g and T_0 are the cure and room temperatures, respectively, as well as on the cooling conditions. The hatched part of the diagram corresponds to local radial tensile stresses at relative fibre-volume fraction $\xi = 0.7$. The matrix in the interphase zone in this section experienced a two-dimensional tensile stress-along the orientation of the fibre and perpendicular to it. In the region being considered and located between the maximally contiguous fibres the appearance of local cracks is most probable. For fibres with diameter $d = 2 \div 3$ μm the transverse sizes of a region with two-dimensional tensile stress distribution are about 1 μm. Should it be considered that only a part of this region is under maximum stresses, then the width of a local crack calculated theoretically should comprise some tenths of a μm; but experimentally detected cracks are several times smaller. To remove this discrepancy the process of coupling formation between the components should be considered in greater detail with the account of chemical shrinkage phenomena at the initial stage of curing temperature conditions.

The initial longitudinal crack dimensions (3000 Å) are evidently associated with the characteristic dimensions of polymer matrix microstructure inhomogeneity. If the fibre-volume fraction $\xi < 0.5$ and fibres are uniformly distributed, then the local transverse tensile stresses at their boundaries disappear, and the transverse crack size will depend upon characteristic dimensions of the polymer microstructure inhomogeneity.

Additional changes in the states of the composites are contributed by the gradients of the first kind of stresses, i.e., stresses averages by a definite volume. As a result, the total stresses can lead to the rise of high stress concentrations promoting the formation of nonlocal discontinuities between the layers and other structural elements.

1. Starting points

1.1. Elementary states

To simplify the problem, further on linearly-reinforced materials are investigated in which continuous circular cylindrical fibres are ideally oriented along the X_1 axis, whereas the layer edges are indefinitely distant. This is the simplest and most important type of structure, being a component for practically interesting more complex laminates and other structures. It is assumed that in the transverse plane the fibre axes form a common doubly-periodical structure in which each repeated cell contains N fibres of various diameters and distributed at random. The external stress field changes slowly, so that the external field gradient may be neglected within a single cell, i.e., within the parallelogram of periods. In their initial state the crack edges form surfaces equidistant to the interphases, or close to them. In proposed models two-dimensional stress-strained states of a reinforced medium are considered, therefore, the end effects arising at the transverse crack edges are neglected. With the above prerequisites the general case of stressed state of a single cell can be presented in the form of a superposition of three states: longitudinal shear, longitudinal tension-compression and transverse stressed state

$$[\langle\sigma_{ik}\rangle] = \begin{vmatrix} 0 & \langle\sigma_{12}\rangle & \langle\sigma_{13}\rangle \\ \langle\sigma_{21}\rangle & 0 & 0 \\ \langle\sigma_{31}\rangle & 0 & 0 \end{vmatrix} + \begin{vmatrix} \langle\sigma_{11}\rangle & 0 & 0 \\ 0 & 0 & 0 \\ 0 & 0 & 0 \end{vmatrix} + \begin{vmatrix} 0 & 0 & 0 \\ 0 & \langle\sigma_{22}\rangle & \langle\sigma_{23}\rangle \\ 0 & \langle\sigma_{32}\rangle & \langle\sigma_{33}\rangle \end{vmatrix} \tag{1}$$

The angle brackets refer to stresses averaged by the area of a corresponding face of the determining medium volume (Fig. 2). The longitudinal shear state is

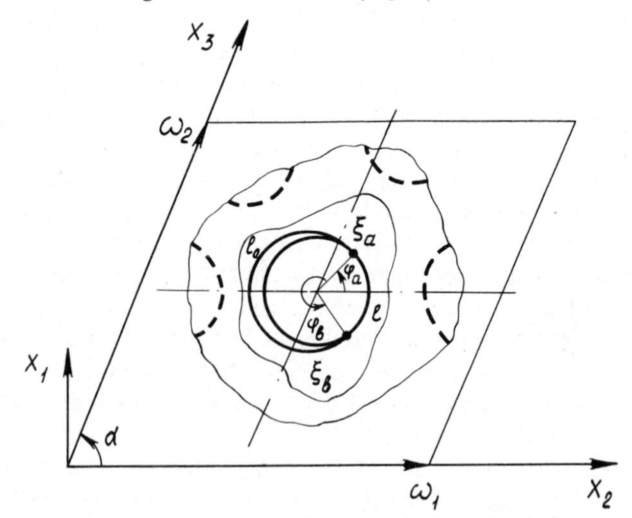

FIG. 2. Determining-volume model of a reduced element.

independent of others due to symmetry plane perpendicular to the fibres. Other states do not commutate with one another, therefore, the solution of the problem at transverse stressed state in a general case can be obtained with a known solution of the problem on longitudinal tension.

1.2. Main problems

The solution of problems for the specified two-dimensional regions and loading cases enables determining the bulk elastic constants of the composition medium by means of which the averaged components of stress tensors $\langle \sigma_{ik} \rangle$ and deformation tensors $\langle \varepsilon_{ik} \rangle$ in the cells are coupled

$$\langle \varepsilon_{ik} \rangle = z_{iksn} \langle \sigma_{sn} \rangle . \tag{2}$$

Here and everywhere below the summation is carried out by the repeated indices. Given the deformation average $\langle \varepsilon_{ik} \rangle$, the equations of the elasticity law will be

$$\langle \sigma_{ik} \rangle = s_{iksn} \langle \varepsilon_{sn} \rangle . \tag{3}$$

Average deformations are determined by difference relations

$$\langle \varepsilon_{ik} \rangle = \frac{1}{2} \left[\frac{\Delta U_i(\omega_k)}{\omega_k} + \frac{\Delta U_k(\omega_i)}{\omega_i} \right], \tag{4}$$

where

$$\Delta U_k(\omega_i) = U_k(x_i + \omega_i) - U_k(\chi_i) \qquad (i, k = 1, 2, 3) .$$

Here ω_i are the two-dimensional structure periods equal to the parallelogram sides (Fig. 2), U_k is the elastic shift vector component. Parameters z_{iksn} (s_{iksn}) depend on the geometry and properties of constitutive phases, as well as on crack localization and size. The relation between the symmetry in mutual fibre arrangement in the cell, the geometry of period parallelogram, the disposition of circular cracks and a number of essentially differing reduced moduli are determined by the structure of the function determining the interaction between the fibres in the cell and the adjacent cells.

In solving the problem about the stressed state of a heterogeneous medium with mixed boundary conditions at the contact surfaces of fibres and the matrix the following steps should be distinguished:

(a) Solution of a mixed boundary-value problem for a separate fibre with a connected matrix, i.e., a reduced element at arbitrary loading;

(b) Account of the interaction between the reduced elements within the cell and between the cells;

(c) Determination of mean stresses;

(d) Defining the integral constants and the elasticity law for the medium;

(e) Evaluation of the ultimate state of the material and investigation of the dependence of the averaged values and local stressed state intensity at the crack tips on the growth of crack sizes and their concentration.

1.3. Methods and solutions

We shall examine separately a reduced element of the medium (Fig. 2). Let ζ be the point coordinate at the interphase boundary, the cylindrical fibre surface $L = l + l_0$ whose part l_0 determines the discontinuity crack of small width, $\bar{U}$ is the shift vector, $\bar{\sigma}_n$ the boundary value of the stress vector on an area with normal $\bar{n}$. There being no overlapping of the cut edges and no friction coupling between them, the mixed conditions of the contact between fibres and matrix will be

$$\begin{aligned} &[\bar{\sigma}_{na} \cdot \bar{n}]^+ = [\bar{\sigma}_n \cdot \bar{n}]^-, \quad \bar{U}_a^+ = \bar{U}^-, \qquad (\zeta \in l), \\ &[\bar{\sigma}_{na} \cdot \bar{n}]^+ = 0, \quad [\bar{\sigma}_n \cdot \bar{n}]^- = 0, \qquad (\zeta \in l_0). \end{aligned} \tag{5}$$

Here superscript plus and minus mean that the ultimate function values are determined along the positive and negative direction of the normal; values pertaining to the fibres are marked by index 'a', and those concerning the matrix or the both components simultaneously have no index.

The boundary-value problem of the elasticity theory (5) for two-dimensional regions and isotropic homogeneous components of the medium can be obtained both by the methods of the complex variable function theory and the method of separation of variables. For composite media with anisotropic heterogeneous fibres the author proposed a new method for the solution of these problems.

In the case of isotropic homogeneous components the complete problem solution reduces to the determination for each region of three complex potentials connected with the stresses by the relations and was shown by VANIN (1977a)

$$\begin{aligned} \sigma_{12} - \mathrm{i}\sigma_{13} &= 2GF'(z), \\ \sigma_{33} - \sigma_{22} - 2\mathrm{i}\sigma_{23} &= 2[\bar{z}\Phi'(z) + \Psi(z)], \\ \sigma_{33} + \sigma_{22} &= 4 \operatorname{Re} \Phi(z), \sigma_{11} = 4\nu \operatorname{Re} \Phi(z), \end{aligned} \tag{6}$$

where $z = x_2 + \mathrm{i}x_3$. The conditions of the stressed state double periodicity put the following restrictions on the matrix potentials

$$\begin{aligned} &F'(z + \omega_j) = F'(z); \qquad \Phi(z + \omega_j) = \Phi(z); \\ &\bar{\omega}_j\Phi'(z) + \Psi(z + \omega_j) = \Psi(z), \qquad (j = 1, 2). \end{aligned} \tag{7}$$

These conditions can be satisfied when the potentials are presented as an expansion in a series of elliptical functions according to GURVITS and KURANT (1968)

$$\left.\begin{matrix} F'(z) \\ \Phi(z) \end{matrix}\right\} = \left\{\begin{matrix} C_0 \\ A_0 \end{matrix}\right\} + \sum_{j=1}^{N} \sum_{s=1}^{\infty} \frac{(-1)^s}{(s-1)!} \left\{\begin{matrix} C_j, s \\ A_j, s \end{matrix}\right\} \zeta^{(s-1)}(z - \alpha_j), \tag{8}$$

where α_j is the coordinate of the jth fibre in the cell, $\zeta^k(z)$ is the derivative of the Weierstrass' zeta function

$$\zeta(z) = \tfrac{1}{2} + \sum_m \sum_n{}' \left(\frac{1}{z-p} + \frac{1}{p} + \frac{z}{p^2}\right), \qquad (m, n = 0 \pm 1 \cdots).$$

Here the prime indicates that the term $m = n = 0$ should be omitted. Then it is accepted that

$$p = m\omega_1 + n\omega_2; \qquad \omega_2 = b\omega_1 \mathrm{e}^{\mathrm{i}\alpha}.$$

One of the subsequently important properties of the Weierstrass' zeta function is its quasi-periodicity

$$\zeta(z + \omega_j) = \zeta(z) + \delta_j,$$

where for the tetragonal ($\alpha = \pi/2$, $b = 1$) and hexagonal ($\alpha = \pi/3$, $b = 1$) cells we should put

$$\delta_1 = \frac{\pi}{\omega_1 \sin \alpha}, \qquad \delta_2 = \frac{\pi \mathrm{e}^{-\mathrm{i}\alpha}}{\omega_1 \sin \alpha}.$$

Function $\Psi(z)$ is constructed with the account of the periodicity condition (1.7) and is found to have the form (VANIN 1977b)

$$\Psi(z) = D_0 + \sum_{j=1}^{N} \sum_{s=1}^{\infty} \frac{(-1)^s}{(s-1)!} [D_{j,s}\zeta^{(s-1)}(z - a_j) - A_{j,s}\eta^{(s)}(z - a_j)], \tag{9}$$

where

$$\eta(z) = \sum_m \sum_n{}' \left(\frac{\bar{p}}{z-p} + z^2 \frac{\bar{p}}{p^3} + z \frac{\bar{p}}{p^2}\right).$$

The derivative of function $\eta(z)$ (2) has the following property

$$\eta'(z + \omega_j) = \eta'(z) + \bar{\omega}_j s'(z) - \gamma_j,$$

where for the tetragonal ($\alpha = \pi/2$, $b = 1$) cell should be accepted that $\gamma_2 = \mathrm{i}\gamma_1$ and for hexagonal ($\alpha = \pi/3$, $b = 1$) $\gamma_1 = 0$, $\gamma_2 = 0$. The elastic displacements in the components are determined by formulae according to MUSKHELISHVILI (1966)

$$U_1 = F(z) + \overline{F(z)},$$
$$2G(U_2 + \mathrm{i}U_3) = \kappa\Phi(z) - z\overline{\Phi(z)} - \overline{\Psi(z)}, \tag{10}$$

where

$$\kappa = 3 - 4\nu, \quad \varphi'(z) = \Phi(z), \quad \Psi'(z) = \Psi(z).$$

More complex methods will be considered below in the investigation of concrete types of fibres anisotropy and inhomogeneity. For periodic structures the interaction between the elements is strictly accounted for by means of the elliptical functions. In more complex cases, when the determining volume contains N inclusions, special methods are to be employed for this problem solution. The author (VANIN 1977a) has proposed a method accounting for the interaction of many inclusions in an elastic medium and reducing the problem to the solution of the sequence of boundary-value problems for complicating regions. In an infinite medium surface S_0 separates the reduced element; the origin of the local system of coordinates is disposed in the centre of gravity of the inclusion. It is assumed that the structure of the remaining part of the medium is obtained by a two- or three-periodical translation of the separated volume. In this approximation the reduced element is in the field with a definite symmetry group. In systems of regular structure the symmetry group is known; in case of an irregular structure it is assumed that the medium outside the volume under consideration has a structure of high-symmetry group. The next approximation considers the inclusions group including the nearest neighbours, etc. The interaction potential is constructed as a series

$$V = U_\infty + U_0 + U_1 + \cdots, \tag{11}$$

where U_∞ is the field potential 'at infinity', there being no interaction between the inclusions; U_0 are the approximations of the smoothed interaction; U_1 is the smoothed field approximation for a complicated-structure volume in the first approximation, etc. The effect of a greater number of adjacent inclusions is accounted for with each approximation owing to which the smoothed field defined more accurately and the oscillating part is separated.

Numerical calculations have shown that the interaction of inclusions in densely packed compositions is divided into two components: smooth and oscillating near a single fibre. The former is determined to a satisfactory accuracy when examining the interaction of a single fibre with the matrix under the effect of a homogeneous internal stress field, making also a major contribution to the value of the reduced elastic constants at relative fibre-volume fraction $\xi < 0.5$. The amplitude of the oscillating interaction is a function of the interfibre distance and their mutual disposition with respect to the inclusion being considered. This interaction determines the increasing concentration of local stresses with the growth of parameter ξ and provides an essential contribution to the value of moduli in heavily filled composites. Therefore, the calculations of material characteristics by the aid of models which do not ensure the account of the oscillating interaction are true only for lightly-field composites. It should be noted that earlier it has been found (VAN FO FI 1971) that in fibrous composites at a certain fibre-volume fraction there exist minimum local stress concentrations in the interphase zone.

1.4. Methods of average

Average stresses in a F_k within volume V_N are determined by the surface integral

$$\langle \sigma_{ik} \rangle = \frac{|\omega_i|}{V_N} \iint_{F_k} \sigma_{ik} \, \mathrm{d}F, \tag{12}$$

which leads to definite state equations of the medium on the basis of thermodynamic relations as was shown (VANIN 1978)

$$\langle \varepsilon_{ik} \rangle = \frac{1}{V_N} \sum_{j=1}^{N+1} \int_{V_j} \left(\frac{\partial U_j}{\partial \langle \sigma_{ik} \rangle} \right)_{T,\ldots} \mathrm{d}v, \tag{13}$$

where U_j is the elastic energy of the jth medium component. In complex fields the elastic energy is substituted by the Helmholtz or Gibbs' potential, depending on the given variables. The bulk value of elastic or other constants is found by repeated differentiation of energy with respect to the appropriate parameter.

Given nothing but bulk stresses and deformations in two-periodic structures at two-dimensional stressed state relations of type (13) and similar to them within may be substituted by equations of type (4) using the quasi-periodicity of special functions in formulas for displacement (10). This determined by the stress (12) and deformation (13) being defined not locally but integrally.

2. Longitudinal shear

The structure of fibres, especially of the high-modulus ones, is characterized by a markedly distinct inhomogeneity due to fibrillar structure in organic fibres, the presence of several preferential-slip planes in carbon fibres, complicated boron fibre structure as well as due to porosity and other imperfections. The investigation of stress fields and integral constants of composition media with fibres of essential anisotropy and inhomogeneity as well as with cracks in the structure of the medium brings about problems the solution of which is considered below. Naturally, the composites of simpler structure can be studied after a series of simplifying assumptions on the basis of known general solutions.

2.1. Common solution

Let the infinite medium with parallel fibres be in the field of average shear stresses $\langle\sigma_{12}\rangle$, $\langle\sigma_{13}\rangle$ or exposed to longitudinal shear deformation $\langle\varepsilon_{12}\rangle$, $\langle\varepsilon_{13}\rangle$. The determining volume V_N contains N fibres and forms a two-periodic structure of the medium. The origin of the local system of cylindrical coordinates r, φ is matched with the fibre axis in a reduced element chosen arbitrarily (Fig. 2). According to formulae (6), (8) and (12), the average shear stresses will be

$$\langle\sigma_{12}\rangle - \mathrm{i}\langle\sigma_{13}\rangle = \sigma_{12}^0 - \mathrm{i}\sigma_{13}^0 + \frac{\mathrm{i}}{F}\sum_{j=1}^{N}\oint[G_aF_a(z) - GF(z)]\,\mathrm{d}\bar{z}\,, \tag{14}$$

where F is the cell area, σ_{12}^0, σ_{13}^0 are the smoothed field stresses; the remaining symbols correspond to the indicated above. For fibres with cylindrical anisotropy the Hooke's law will be

$$\sigma_{1r} = G_r\frac{\partial U_a}{\partial r} + G_{r\varphi}\frac{1}{r}\frac{\partial U_a}{\partial\varphi}\,, \qquad \sigma_{1\varphi} = G_{r\varphi}\frac{\partial U_a}{\partial r} + G_{\varphi}\frac{1}{r}\frac{\partial U_a}{\partial\varphi}\,, \tag{15}$$

where $U_a = U_{1a}$ satisfies the equilibrium equation in translations

$$G\frac{\partial^2 U_a}{\partial r^2} + \left(\frac{G_r}{r} + \frac{\partial G_r}{\partial r} + \frac{1}{r}\frac{\partial G_{r\varphi}}{\partial\varphi}\right)\frac{\partial U_a}{\partial r} + \left(\frac{1}{r^2}\frac{\partial G_{\varphi}}{\partial\varphi} + \frac{1}{r}\frac{\partial G_{r\varphi}}{\partial r}\right)\frac{\partial U_a}{\partial\varphi}$$

$$+ 2\frac{G_{r\varphi}}{r}\frac{\partial^2 U_a}{\partial r\partial\varphi} + \frac{G_{\varphi}}{r^2}\frac{\partial^2 U_a}{\partial\varphi^2} = 0 \tag{16}$$

The alteration of elastic properties of fibres tangentially is unknown, therefore, we put

$$G_r = G_0\left(\frac{r}{r_0}\right)^{2g}\mathrm{e}^{\alpha(r-r_0)}, \quad G_{r\varphi} = \kappa G_r, \quad G_{\varphi} = \delta^2 G_r\,. \tag{17}$$

Constants G_0, g, α, κ, δ^2 determine the inhomogeneous fibre properties in the radial direction; r_0 is the radius of the isotropic fibre core with shear modulus G_0. The solution of Equation (16) is expressed in the form of degenerate hypergeometric function series

$$U_a = 2\,\mathrm{Re}\sum_{n>0}\mathrm{e}^{-x+\mathrm{i}n\varphi}x^{q_n - \mathrm{i}n\kappa}[A_nF(a_n, b_n; x) + B_nU(a_n, b_n; x)]\,, \tag{18}$$

where

$$x = \alpha r, \quad x_0 = \alpha r_0$$

$$F(a_n, b_n; x) = \sum_{m=0}^{\infty}\frac{(a_n)_m}{(b_n)_m}\cdot\frac{x^m}{m!}$$

$$U(a_n, b_n; x) = \frac{\pi}{\sin\pi b_n}\left[\frac{F(a_n, b_n; x)}{\Gamma(1-q_n)\Gamma(b_n)} - x^{1-b_n}\frac{F(1-q_n, 2-b_n; x)}{\Gamma(a_n)\Gamma(2-b_n)}\right]$$

Here, $\Gamma(x)$ is the gamma function

$$q_n = q + \sqrt{q^2 + n^2(\delta^2 - \kappa^2)}\,;$$
$$a_n = 1 + 2q + q_n\,; \qquad b_n = 1 + 2q + 2q_n\,.$$

To avoid unnecessary complications particular cases of solving (18) are examined below, when the isotropic matrix is reinforced by:

(1) isotropic fibres with variable elasticity modulus $G_a = G_0\,e^{\alpha r}$,

(2) fibres with homogeneous cylindrical anisotropic and isotropic core.

The first case at $d > 0$ corresponds to boron fibres fabricated on an isotropic substrate, The core modulus G_0 is, in this case, equal to the tungsten shear modulus.

2.2. *Isotropic inhomogeneous fibres*

For isotropic inhomogeneous fibres with radius d the function of longitudinal displacements will be

$$U_a = \sum_{n>0} \frac{nU_n(x)}{U_n'(x_0)}\, d^n(A_{mn}\tau^n + \bar{A}_{mn}\bar{\tau}^n)\,, \tag{19}$$

where $\tau = e^{i\psi}$, $x_0 = \alpha d$, $U_n(x) = e^{-x}x^n f(1+n, 1+2n; x)$. The explicit form of coefficients A_{mn} is determined by solving the subsidiary problem of the function theory. For the reduced elements local piecewise holomorphic functions of $z = z\,e^{i\psi}$ are introduced

$$X_m(z) = \begin{cases} X_m^+(z) = \sum\limits_{n>0} A_{mn} z^n\,, & (|z| < d) \\ X_m^-(z) = \sum\limits_{n>0} A_{mn}\left(\dfrac{d^2}{z}\right)^n, & (|z| > d) \end{cases} \tag{20}$$

with their ultimate values satisfying boundary condition (5)

$$X_m^+(\zeta) - X_m^-(\zeta) = 0\,, \qquad (\zeta \in l_0) \tag{21}$$

Using condition (5) for solution (19) ($\zeta \in l_0$), we arrive at the equation

$$\sum_{n>0} d^n(A_{mn}\tau^n - \bar{A}_{mn}\bar{\tau}^n) = 0\,,$$

which may be regarded as a limiting relationship for the sequence $X_m(z)$. Local longitudinal shifts in the matrix are determined by the series

$$U = r^m(C_m\tau^m + \bar{C}_m\bar{\tau}^m) + \sum_{n>0}\left(\frac{d^2}{r}\right)^n (B_{mn}\tau^n + \bar{B}_{mn}\bar{\tau}^n) \tag{22}$$

the coefficients of which, according to equalities (5), satisfy the equations

$$d^m(C_m\tau^m - \bar{C}_m\bar{\tau}^m) + \sum_{n>0} B_{mn}\tau^n - \bar{B}_{mn}\bar{\tau}^n = 0\,.$$

The latter is equivalent to functional equations

$$Y_m^+(\zeta) - Y_m^-(\zeta) = 0\,, \qquad (\zeta \in l_0) \tag{23}$$

where

$$Y_m(z) = \begin{cases} Y_m^+(z) = \sum\limits_{n>0} B_{mn}z^n + \bar{C}_m\left(\dfrac{d^2}{z}\right)^m & (|z| < d) \\ Y_m^-(z) = \sum\limits_{n>0} \bar{B}_{mn}\left(\dfrac{d^2}{z}\right)^n + C_m z^m & (|z| > d) \end{cases} \tag{24}$$

At the remaining part of boundary $\zeta \in l$ the ideal-contact conditions are fulfilled, this resulting in functional relationships

$$\begin{gathered} \left[X_m(\zeta) + \frac{G}{x_0 G_a^0}\,Y_m(\zeta)\right]^+ - \left[X_m(\zeta) + \frac{G}{x_0 G_a^0}\,Y_m(\zeta)\right]^- = 0\,, \\ [X_m(\zeta) - Y_m(\zeta)]^+ + [X_m(\zeta) - Y_m(\zeta)]^- = f_m(\zeta)\,. \end{gathered} \tag{25}$$

where $\nu_n = U_n(x_0)/x_0U_n'(x_0)$; $G_a^0 = G_0\mathrm{e}^{x_0}$; $f_m(\zeta) = \Sigma_{n>0}(1 - nx_0\nu_n)(A_{mn}\zeta^n + \bar{A}_{mn}(d^{2n}/\zeta^n))$. The solution of the system of Equations (21), (23) and (25), taking into account the interaction of the reduced elements, which is determined by the nth order polynomial, was found in the form of

$$\begin{gathered} (1 + x_0G_a|G)X_m(z) = Q_m(z) + \mu(z)R_m(z) + \frac{\mu(z)}{2\pi\mathrm{i}}\int \frac{f_m(\zeta)\,\mathrm{d}\zeta}{\mu^+(\zeta)(\zeta - z)} \\ Y_m(z) = Q_m(z) - x_0G_a|GX_m(z)\,.^{e} \end{gathered}$$

Here

$$\begin{gathered} \mu(z) = \sqrt{(z - \zeta_a)(z - \zeta_b)}\,, \\ Q_m(z) = C_m z^m + \bar{C}_m\left(\frac{d^2}{z}\right)^m, \end{gathered} \tag{26}$$

where $R_m(z)$ is the polynomial with arbitrary constants $\zeta_0 = d\,\mathrm{e}^{\mathrm{i}\Theta}$; $\theta = (\varphi_b - \varphi_a)/2$, $\Theta = (\varphi_b + \varphi_a)/2$ are the angles determining the crack (cut) size and location of its centre, the coefficients C_m characterize the interaction between the elements. Expanding $X_m(z)$ and $Y_m(z)$ as series in powers of z, we find the unknown coefficients A_{mn} and B_{mn}. In particular,

$$X_{11}^{+}(z) = A_{11}z + \cdots = \frac{G}{x_0 G_a^0} \cdot \frac{G_1(1+\lambda_1) - \bar{C}_1\lambda_2\,\bar{e}^{2i\Theta}}{1+\nu_1 G/G_a^0} z + \cdots,$$

where $\lambda_1 = -\cos\frac{1}{2}\theta$, $\lambda_2 = \frac{1}{2}\sin^2\frac{1}{2}\theta$, $\nu_1 = (1+(x_0-1)\,e^{x_0})/(e^{x_0} - x_0 - 1)$.

Strict account of the interaction of fibres is accomplished by conforming the stress and shift fields at the outer boundary of the normalized elements and the remaining part of the cell by expanding the elliptical functions and the considered solution as a series in powers of z received by VANIN (1977a). If at the initial state m_0 in the cell have no imperfections in the form of discontinuities at the interphase or near it, then, there being no correlation between the size θ_j and location of the centre Q_j of the crack at jth fibre the distribution of all the defects will be

$$P(\theta, \Theta) = \frac{1}{N}\sum_{j=1}^{N} \delta(\theta-\theta_j)\delta(\Theta-\Theta_j) = \frac{m_0}{N}\delta(\Theta - 2\pi) + \sum_{k=1}^{n}\frac{m_k}{N}\delta(\theta-\theta_k)\delta(\Theta-\Theta_k),$$

where m_k is the number of fibres in kth state, $\delta(\theta)$ is the Dirac delta function. Average stresses, in accordance with formula (14) are related with coefficients C_r

$$\langle\sigma_{12}\rangle - i\langle\sigma_{13}\rangle = 2\eta G C_1 + \frac{2\xi G}{1+\nu_1 G/G_a^0}(C_1\langle 1+\lambda_1\rangle - \bar{C}_1\langle\lambda_2\,e^{2i\Theta}\rangle) + \cdots \qquad (27)$$

where $\eta = 1 - \xi$,

$$\langle\lambda_1\rangle = \frac{m_0}{N} - \sum_{k=1}^{n}\frac{m_k}{N}\cos\frac{\theta_k}{2},$$

$$\langle\lambda_2\,e^{2i\Theta}\rangle = \frac{1}{2N}\sum_{k=1}^{n} m_k\,e^{2i\Theta_k}\sin\frac{\theta_k}{2}.$$

Equations of the averaged law of elasticity will be

$$\langle\varepsilon_{12}\rangle = \frac{1}{F}\iint_F \frac{\partial U}{\partial\langle\sigma_{12}\rangle}\,dv \simeq \frac{\sigma_{12}^0}{G} + \frac{\xi}{d^2 N}\sum_{j=1}^{N}(B_{1j} + \bar{B}_{1j}),$$

where

$$Y_{1j}(z) = \frac{B_{1j}}{z} + \cdots; \qquad C_1 = \frac{\sigma_{12}^0 - i\sigma_{13}^0}{2G}.$$

In the first approximation the main equations will be

$$\langle\varepsilon_{12}\rangle = \frac{1}{G_{12}}\langle\sigma_{12}\rangle + \frac{\mu_{23}}{G_{13}}\langle\sigma_{13}\rangle,$$

$$\langle\varepsilon_{13}\rangle = \frac{\mu_{32}}{G_{12}}\langle\sigma_{12}\rangle + \frac{1}{G_{13}}\langle\sigma_{13}\rangle, \qquad \frac{\mu_{23}}{G_{13}} = \frac{\mu_{32}}{G_{12}},$$

where

$$\left.\begin{matrix} G_{12} \\ G_{13} \end{matrix}\right\} = G\,\frac{(1+\xi\langle\lambda_1\rangle + \eta G/G_a^*)^2 - \xi^2(\langle\lambda_2\cos 2\Theta\rangle^2 + \langle\lambda_2 \sin 2\Theta\rangle^2)}{L \pm 2\xi(1+G/G_a^*)\langle\lambda_2\cos 2\theta\rangle}$$

$$\mu_{23} = \frac{2\xi(1+G/G_a^*)\langle\lambda_2\sin 2\Theta\rangle}{L - 2\xi(1+G/G_a^*)\langle\lambda_2\cos 2\Theta\rangle} + \cdots \tag{28}$$

Here

$$G_a^* = G_a^0\,\frac{e^{x_0} - x_0 - 1}{1 + (x_0 - 1)\,e^{x_0}}$$

$$L = 1 - \xi^2\langle\lambda_1\rangle^2 + 2(1+\xi^2\langle\lambda\rangle)G/G_a^* + (1-\xi^2)(G/G_a^*)^2$$
$$+\ \xi^2(\langle\lambda_2\cos 2\theta\rangle^2 + \langle\lambda_2\sin 2\theta\rangle^2)\,.$$

The first-approximation formulae for moduli (28) of materials with heterogeneous fibres and cracks coincide with those for media with homogeneous fibres obtained by VANIN (1979), when G_a is substituted by G_a^*. Fig. 3 shows

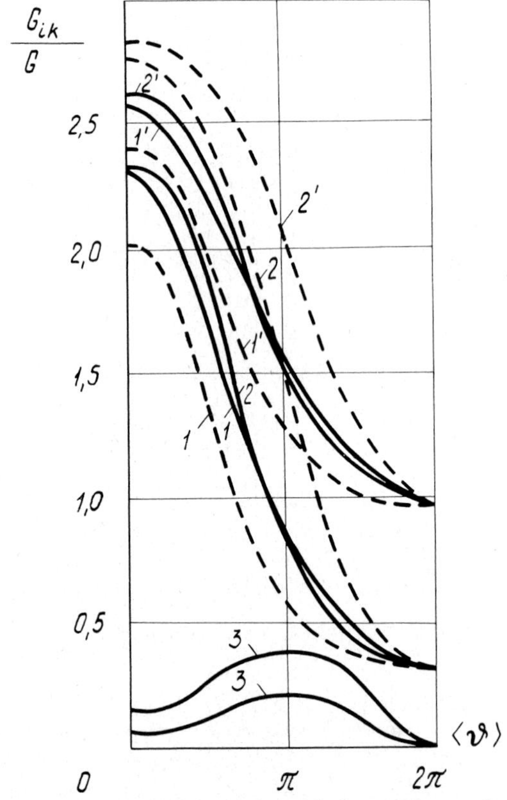

FIG. 3. Change of fibrous materials elastic constants with the growing crack concentration and size.

the curves characterizing the application of elastic constants of a material at $G/G_a^0 = 0.01$, $\xi = 0.5$ with the growth of m_0/N at $X_0 = 1$ when the cracks in the cell are the same. The solid curves were plotted at $\Theta = \pi/4$, the dotted ones at $\Theta = 0$. Curves 1, 2, 3 and 1′, 2′, 3′ correspond to $m_0 = 0$ and $m_0 = 0.5\,N$ and determine the change of G_{12}/G, G_{13}/G and μ_{23} with the growth of θ. The elastic properties are considerably effected by local crack concentration and size. In one of the simplest regular structures, when there is one fibre with a crack in a cell, the calculations were made with complete account of the interaction between the fibres. Curves in Fig. 4 illustrate the changes of the effective shear moduli versus the expansion angle of a symmetrically positioned ($\Theta = 0$) crack at the interphase at $\xi = 0.7$ in a glass fibre plastic. The dashed curves correspond to a tetragonal structure and the solid ones to a hexagonal. Curves 1, 3 determine the changes of G_{13}/G, and curves 2, 4 G_{12}/G. A strict account of the interaction of fibres enables to find more accurately the maximum values of moduli and local variation of curves. The growth of the

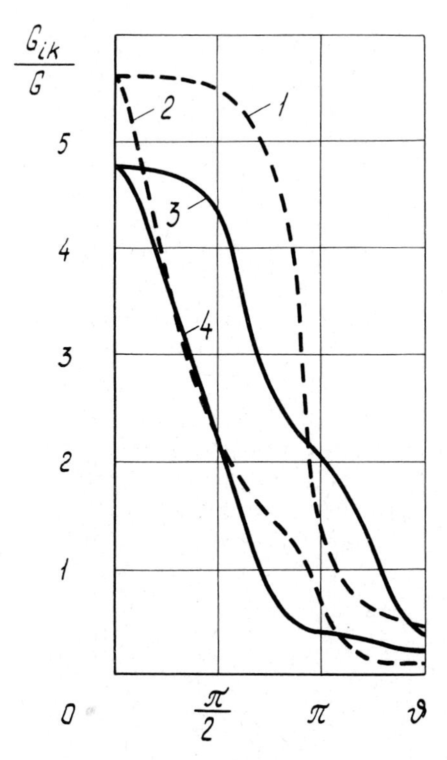

FIG. 4. Change of shear moduli of a heavily-filled glass-fibre plastic with the growth of an interphase crack.

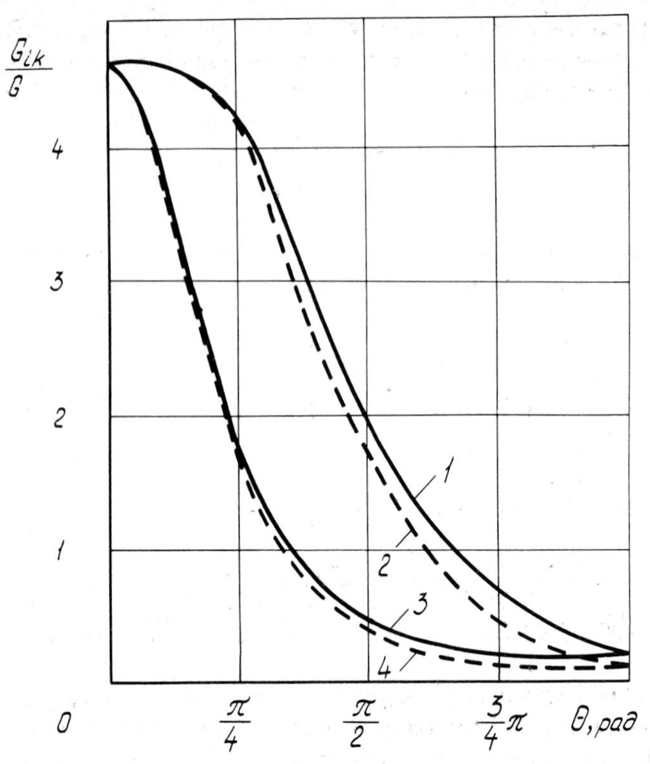

FIG. 5. Approximate change of shear moduli of a glass-fibre plastic with cracks at the interphase (solid-line curves) and in the matrix (dotted curves).

crack, in case of one orientation, leads to a rise of a considerable anisotropy of the material: in tetragonal structure the difference between the initially equal moduli ($G_{13}/G_{12} = 1$) at $\theta = \pi/2$ reaches $G_{13}/G_{12} = 2.8$. For comparison Fig. 5 shows the change of moduli G_{13}/G and G_{12}/G (solid curves 1 and 3) found by an approximating formula (28) at $\xi = 0.7$ for a glass fibre plastic. In this approximation the effect of the mutual fibre arrangement has not been taken into account, but on the whole the curves are close to the abovementioned. Smooth behaviour of the curves indicates the absence of local perturbations playing a great role in the estimation of stress intensities at the crack tips.

2.3. *Critical stresses*

It is interesting to note that the stresses at the crack tips are determined directly through the derivatives of the effective moduli. For the longitudinal shear

$$\sigma_m = -\frac{1}{\sqrt{2\pi\rho}}\sqrt{\frac{\pi d}{\zeta(1+G/G_a^*)}}\langle\sigma_{12}\rangle\sqrt{\frac{\partial}{\partial\theta}\left(\frac{G}{G_{12}}\right)} - \langle\sigma_{13}\rangle\sqrt{\frac{\partial}{\partial\theta}\left(\frac{G}{G_{13}}\right)}, \quad (29)$$

where ρ is the distance from the crack tip. According to the Griffits criterion, the critical stressed state at a random crack orientation will be as was shown BARENBLATT (1959) and CHEREPANOV (1974)

$$\frac{\langle\sigma_{12}^*\rangle^2}{a^2}+\frac{\langle\sigma_{13}^*\rangle^2}{b^2}+\frac{\langle\sigma_{12}^*\rangle\langle\sigma_{13}^*\rangle}{c^2}=1\,. \tag{30}$$

Here the critical values of mean stresses are marked by an asterisk. For the states in the vicinity of point ξ_a it should be assumed that

$$a^2=\frac{8\,\mathrm{d}\gamma}{\left|\left(\frac{\partial}{\partial\theta}-\frac{\partial}{\partial\Theta}\right)\frac{F}{G_{12}}\right|},\quad b^2=\frac{8\,\mathrm{d}\gamma}{\left|\left(\frac{\partial}{\partial\theta}-\frac{\partial}{\partial\Theta}\right)\frac{F}{G_{13}}\right|},\quad c^2=\frac{4\,\mathrm{d}\gamma}{\left|\left(\frac{\partial}{\partial\theta}-\frac{\partial}{\partial\Theta}\right)\frac{\mu_{23}F}{G_{13}}\right|} \tag{31}$$

where γ is the surface energy.

For the states near the tip ζ_b it was assumed that

$$a^2=\frac{b\,\mathrm{d}\gamma}{\left(\frac{\partial}{\partial\theta}+\frac{\partial}{\partial\Theta}\right)\frac{F}{G_{12}}},\quad b^2=\frac{8\,\mathrm{d}\gamma}{\left(\frac{\partial}{\partial\theta}+\frac{\partial}{\partial\Theta}\right)\frac{F}{G_{13}}},\quad c^2=\frac{4\,\mathrm{d}\gamma}{\left(\frac{\partial}{\partial\theta}+\frac{\partial}{\partial\Theta}\right)\frac{\mu_{23}F}{G_{13}}} \tag{32}$$

Equation (30) determines an ellips in the coordinates of mean stresses which fixes the ultimate state of the composite at a longitudinal shear as was shown by VU (1978). For a symmetrical stressed state at $\langle\sigma_{13}\rangle=0$, $\Theta=0$ and provided that the growth of the crack does not violate the symmetry of the medium structure, so that both cracks tips run simultaneously, the critical stress will be

$$\langle\sigma_{12}^*\rangle=\sqrt{\frac{8\,\mathrm{d}\gamma}{\frac{\partial}{\partial\theta}\frac{F}{G_{12}}}}\,.$$

In Fig. 6 curves 1 and 2 represent the change in the critical stresses versus the expansion angle V of an interphase symmetrically growing crack at $\xi=0.7$ for a tetragonal and hexagonal structure of a glass fibre plastic. The existence of a stable crack position at $\theta=\pi$ and $\theta=2\pi/3$ is evident. Additional effects of crack stability are created by local transverse compression stresses in highly-filled composites (Fig. 1). In low-filled composites ($\xi=0.5$), when the fibres are remote from one another, the compression stresses are more or less uniformly distributed around the fibre outline. Calculations indicate that close proximity of the fibres to one another leads to a more pronounced stable crack positioning in the structures considered, so that even barriers occur. The removal of fibres from one another results in the weakening of the stress dependence on the type of packing so that a stable state is maintained only at $\theta=\pi$. The latter is due to the change of the crack tip orientation as regards the stress $\langle\sigma_{12}\rangle$. Therefore, in case of a longitudinal shear, the ultimately possible symmetrical

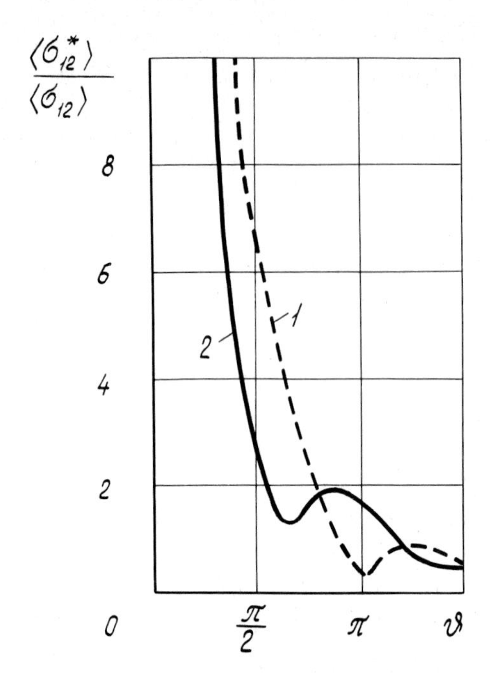

FIG. 6. Dependence of critical shear stresses upon interphase crack size for tetragonal (curve 1) and hexagonal structures.

crack expansion in a tetragonal and hexagonal fibre packings is restricted by angles $\theta = \pi$ and $\theta = 2\pi/3$, respectively (Fig. 7). Hence, the hexagonal fibre packing is the most rational way to enhance the effective rigidity of the composites and its resistance to local fibre delamination.

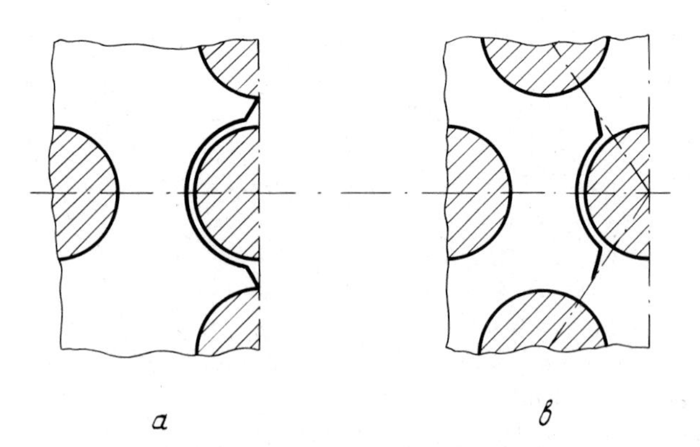

FIG. 7. The ultimate flare angles of cracks for tetragonal and hexagonal fibre packings.

2.4. Cohesive cracks

Crack growth in the vicinity of the interphase is observed at localized cohesive failure of composites. To concretize, a case will be considered, when a crack forms an arch with radius $\lambda > r_0$ and when there are no delaminations at the interphase. The reduced element of the medium includes a homogeneous isotropic fibre and matrix with a crack (Fig. 8). Points located on the circumference of radius λ are designated by ζ, at the interphase by ζ_0 and those in the zone of the reduced element congruence by ζ'. The resultant stressed state consists of a variety of states, i.e., those in the fibre, at the ring between the fibre and surface ζ, and in a region located at the boundary of the reduced element and the remaining part of the cell. Contact conditions between the introduced regions will be:

(1) at the interphase

$$[\bar{\sigma}_{na} \cdot \bar{n}]^{+} = [\bar{\sigma}_{n} \cdot \bar{n}]^{-}, \quad \bar{U}_{a}^{+} = \bar{U}^{-}, \qquad (\zeta_0 \in \Gamma_0)$$

(2) at the circumference $|\zeta| = \lambda$

$$\begin{aligned} &[\bar{\sigma}_{n} \cdot \bar{n}]^{+} = [\bar{\sigma}_{n} \cdot \bar{n}]^{-}, \quad \bar{U}^{+} = \bar{U}^{-}, && (\zeta \in l) \\ &[\bar{\sigma}_{n} \cdot \bar{n}]^{+} = 0, \quad [\bar{\sigma}_{n} \cdot \bar{n}]^{-} = 0, && (\zeta \in l_0) \end{aligned} \tag{33}$$

(3) at the boundary of the reduced element and the remaining part of the cell Γ

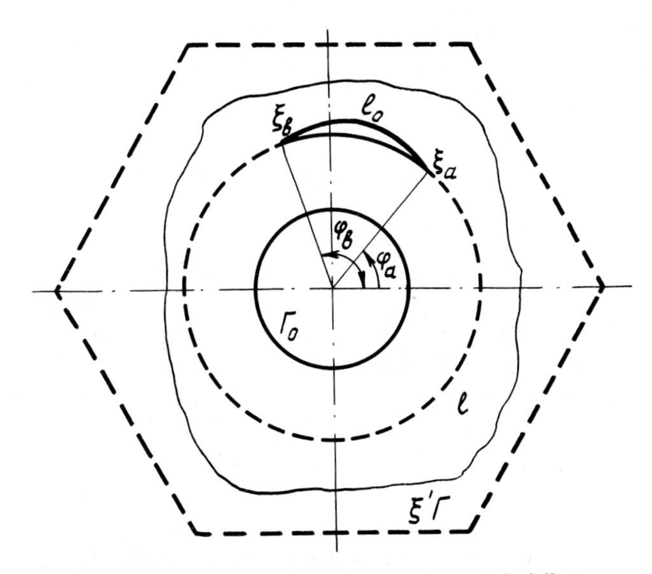

FIG. 8. Reduced element model at chesive matrix failure.

$$[\bar{\sigma}_n \cdot \bar{n}]^+ = [\bar{\sigma}_n \cdot \bar{n}]^- , \quad \bar{U}^+ = \bar{U}^- . \qquad (\zeta' \in \Gamma)$$

The first condition is satisfied automatically if we take

$$U_a = \frac{2}{1 + G_a/G} \sum_{n>0} r^n (A_{mn} \cdot \tau^n + \bar{A}_{mn}\bar{\tau}^n) , \qquad (r \leqslant r_0)$$
$$U = \sum_{n>0} \left(r^n + \frac{G - G_a}{G + G_a} r_0^{2n} r^{-n} \right) (A_{mn}\tau^n + \bar{A}_{mn}\bar{\tau}^n) , \qquad (r_0 \leqslant r \leqslant \lambda) \tag{34}$$
$$U' = r^m (C_m \tau^m + \bar{C}_m \bar{\tau}^m) + \sum_{n>0} r^{-n} (B_{mn}\tau^n + \bar{B}_{mn}\bar{\tau}^n) , \qquad (r > \lambda) .$$

Coefficients A_{mn} and B_{mn} are found from the subsidiary problem by means of functions $X_m(z)$ and $Y_m(z)$, defined ($d = \lambda$) by equalities (20) and (24). The system of functional equations has the form

$$X_m^+(\zeta) - X_m^-(\zeta) = 0 , \quad X_m^+(\zeta) - X_m^-(\zeta) = 0 , \qquad (\zeta \in l_0)$$
$$[X_m(\zeta) - X_m(\zeta)]^+ + [X_m(\zeta) - X_m(\zeta)]^- = 0 , \qquad (\zeta \in l) \tag{35}$$
$$[G/G_a X_m(\zeta) + X_m(\zeta)]^+ - [G/G_a X_m(\zeta) + X_m(\zeta)]^- = 0 , \qquad (\zeta \in l) .$$

The solution of system (35) satisfying all conditions takes the form

$$(1 + G/G_a) X_m(z) = Q_m(z) + \mu(z) R_m(z) ,$$
$$(1 + G/G_a) X_m(z) = Q_m(z) - G/G_a \mu(z) R_m(z) . \tag{36}$$

Here, at $d = \lambda$ the designations coincide with those in Equation (26). The formula for the shear moduli has the form (28) where the substitution should be made

$$G_a^* = G \frac{(1 + G/G_a)(\lambda/r_0)^2 + 1 - G/G_a}{(1 + G/G_a)(\lambda/r_0)^2 - 1 + G/G_a} . \tag{37}$$

Critical shear stresses are determined by Equation (30) with corresponding substitutions of elastic constant by new ones. The stresses around the crack have a singularity of the type $\rho^{-1/2}$, where ρ is the distance from the crack tip. In the case of a simple failure formulae which have been found make it possible to estimate approximately the equality conditions between the cohesive and adhesive strength of composites on the basis of equality of critical stresses for a small crack at the interphase and the region adjoining the fibre. We assume that the interphase zone is homogeneous. If a localized crack is situated symmetrically ($\Theta = 0$) at $\lambda = 1.1\ r_0$, it has been found that for glass fibre plastics at $\xi = 0.7$, $G/G_a = 0.04$ the crack growth resistance at adhesive failure should be 10% higher than that at cohesive failure. The above estimate should be refined for configurations in which separate fibres are arranged rather closely to one another. Undoubtedly that the path of crack running in a matrix along the arch of a constant radius is one of the possible. It is interesting to follow the change of the effective moduli with the crack growth in a glass fibre plastic.

In case of the fibre-volume fraction $\xi = 0.7$ and a growth of symmetric ($\Theta = 0$) crack around all the stressed fibres $\langle\sigma_{12}\rangle$, the dotted curves 2 and 4 in Fig. 5 show the change of G_{13}/G and G_{12}/G. Here $(\lambda/r_0)^2 = 1.21$. With small angle θ, which is of the major interest, the curves characterizing the change of the effective constants practically coincide. Therefore, the investigations of the bulk and effects in high-modulus composites with the account of the interphase cracks suffice to conclude with a satisfactory accuracy that the mentioned effects are influenced by equivalent defects in the matrix.

2.5. *Anisotropic fibres*

The introduction of anisotropic fibres into the composite should result in the increase of the inhomogeneity near the cracks in the stressed state as was shown SIH and LIBOVITS (1975). We assume that for cylindrically anisotropic fibres (15), when $G_2 = \text{const.}$, an isotropic fibre core is existing with a radius ε and shear modulus G_0. The displacement functions of the central fibre part and its jacket are obtained from solution (18) by proceeding to the limit $\alpha = 0$, $g = 0$.

$$U_0 = 2\nu \sum_{n>0} r^n \varepsilon^{n(\nu-1)} \left[\frac{\varepsilon^{i\kappa n}}{B+\nu+i\kappa} A_{mn}\tau^n + \frac{\varepsilon^{i\kappa n}}{B+\nu-i\kappa} \bar{A}_{mn}\bar{\tau}^n \right],$$

$$U_a = \sum_{n>0}{}' \, (r^{n(\nu-i\kappa)} - \gamma\varepsilon^{2n\nu} r^{-n(\nu+i\kappa)}) A_{mn}\tau^n \tag{38}$$

$$+ (r^{n(\nu+i\kappa)} - \bar{\gamma}\varepsilon^{2n\nu} r^{-n(\nu-i\kappa)}) A_{mn}\bar{\tau}^n .$$

Here $\nu^2 = \sigma^2 - \kappa^2$, $B = G_0/G_r$, $\gamma = (B - \nu + i\kappa)/(B + \nu + i\kappa)$. For the isotropic matrix the displacement function in the reduced element is determined by expansion (22). The state in the remaining part of the cell is described by means of elliptical functions (8) and (10). Boundary conditions at the interphase correspond to relations (5). According to the proposed method, the subsidiary problem is formulated for functions defined by the equalities

$$X_m(z) = \begin{cases} X_m^+(z) = \sum_{n>0} (1 + \gamma\varepsilon^{2n\nu}) A_{mn} z^n ; & (|z| < r_0) \\ X_m^-(z) = \sum_{n>0} (1 + \bar{\gamma}\varepsilon^{2n\nu}) \bar{A}_{mn} z^{-n} ; & (|z| > r_0) \end{cases} \tag{39}$$

and functions $Y_m(z)$ according to definition (24). Satisfying the boundary conditions at the interphase, we obtain the system of functional equations

$$X_m^+(\zeta) - X_m^-(\zeta) = 0\,, \quad Y_m^+(\zeta) - Y_m^-(\zeta) = 0\,, \qquad (\zeta \in l_0)$$
$$[\nu X_m(\zeta) + G/G_r Y_m(\zeta)]^+ - [\nu X_m(\zeta) + G/G_r Y_m(\zeta)]^- = 0\,, \qquad (\zeta \in l) \tag{40}$$
$$[X_m(\zeta) - Y_m(\zeta)]^+ + [X_m(\zeta) - Y_m(\zeta)]^- = \varphi_m(\zeta)\,.$$

Here

$$\varphi_m(\zeta) = 2 \sum_{n>0} \varepsilon^{2n\nu}\left(\gamma A_{mn}\zeta^n + \bar{\gamma}\bar{A}_{mn}\frac{r_0^{2n}}{\zeta^n}\right).$$

The solution of this system of functional equations accounting for the mth term of the interaction of the given reduced element with the remaining ones within the cell with be

$$X_m(z) = \left[\mu(z)R_m(z) + \frac{\mu(z)}{2\pi i}\int \frac{\varphi_m(\zeta)\,d\zeta}{\mu+(\zeta)(\zeta - z)} + Q_m(z)\right]\frac{G}{G+\nu G_r} \tag{41}$$

$$Y_m(z) = Q_m(z) - \frac{\nu G_r}{F}X_m(z)\,,$$

where r_0 is the fibre radius; all other symbols correspond to those given in Equations (26). In case of a polynomial load determined by function $G_m z^m$, the explicit form of function $X_m(z)$ will be

$$X_m(z) = \frac{G}{G+\nu G_r}\left\{Q_m(z) + \tfrac{1}{2}\varphi_m(z) + \mu(z)\left[R_m(z)\right.\right.$$
$$\left.\left. - \sum_{n>0}\sum_{j=1}^{n} r_0^{2n\nu}P_{n-j}\left(\cos\frac{\theta}{2}\right)\left(\gamma A_{mn}e^{i(n-j+1)\Theta}a^{m-j}z^{j-1} + \bar{\gamma}\bar{A}_{mn}\frac{a^{m+j-1}}{z_j}e^{-i(n-j)\Theta}\right)\right]\right\}. \tag{42}$$

where $P_n(\cos\theta/2)$ is the Legendre polynomial; for negative n the Legendre polynomials in Equation (42) are omitted. The arbitrary constants included in $R_m(z)$ are chosen by satisfying the limiting condition $Y_m(z) \to C_m z^m$ at $z \to \infty$ and the condition of the absence of singular point of function $X_m(z)$ at $z \to 0$. The unknown coefficients of expansion (40) are determined with the aid of equality

$$X_m^+(\zeta) - X_m^-(\zeta) = \sum_{n>0}{}' a^m[(1+\gamma r_0^{2n\nu})A_{mn}\tau^n - (1+\bar{\gamma}r_0^{2n\nu})\bar{A}_{mn}\bar{\tau}^n]\,,$$

by expanding the left-hand part in Fourier series.

In cases when the fibre core radius z_0 is small as compared to a and it is permissible to neglect the interaction between it and the crack, the terms proportional to $r_0^{2n\nu}$ may be omitted. For these states the coefficients A_{mn} can be found explicitly due to the simplified function $X_m(z)$. This form of $X_m(z)$ plays the role of the first approximation which is successively made more precise by retaining in the next approximation the terms with higher power of $r_0^{2n\nu}$.

In the first approximation the formulae for the bulk shear moduli are identical with relations (26) in which it should be put

$$G_a^* = \sqrt{G_r G_x - G_{rx}^2} = \nu G_r .$$

Parameter νG_r plays the role of a reduced shear modulus of the fibre without the account of the core elasticity effect.

The stress tangents at the crack tips have a classical singularity.

2.6. Hollow fibres

The analysis of formulae for the bulk parameters (28) leads to conclusion that in the first approximation the effective characteristics of the components depend on the averaged elastic constants of the constituents. Therefore, to study in the first approximation the new structures and constituents of the materials with the above indicated cracks, the author proposed a rule of homogeneous transformations according to which the effective characteristics of the composites with other fibres are obtained by transformation

$$G_a \rightarrow G_a^*$$

in formulae (28).

If the medium is reinforced by homogeneous isotropic fibres with a central cavity radius ε, then the effective longitudinal shear modulus in such fibres is easily calculated and determined by formula

$$G_a^* = G_a \frac{r_0^2 - \varepsilon^2}{r_0^2 + \varepsilon^2} . \tag{43}$$

The bulk moduli of media with arch cracks are found through relations (28) with equation (44) taken into account. For materials with cohesive failure cracks and homogeneous isotropic fibres transformation (37) should be employed. In case of hollow fibres, a double transformation should be performed in formula (28) in order to obtain effective constants of the materials.

The final transformation is

$$G_a^* = G \frac{[r_0^2 - \varepsilon^2 + (r_0^2 + \varepsilon^2)G/G_a]\lambda^2/r_0^2 + r_0^2 - \varepsilon^2 - (r_0^2 + \varepsilon^2)G/G_a}{[r_0^2 - \varepsilon^2 + (r_0^2 + \varepsilon^2)G/G_a]\lambda^2/r_0^2 - r_0^2 + \varepsilon^2 + (r_0^2 + \varepsilon^2)G/G_a} . \tag{44}$$

Reduced characteristics of materials with hollow fibres, general-type cylindrical anisotropy and interphase cracks are obtained by transformation

$$G_a^* = \nu G_r \frac{r_0^2 - \varepsilon^2}{r_0^2 + \varepsilon^2} . \tag{45}$$

The transformation found facilitates the assessment of effective parameters of materials with different components.

2.7. *Mixed cracks*

In brittle-fibre and elastic-matrix composites cracks appear simultaneously in all components. A multitude of cracks—at the interphase and in components—results in the appearance of anomalies at the stressed state of the material. Henceforth, to the appearance of two and more adjacent cracks we shall refer as mixed local failure. In large-diameter brittle boron fibres the most commonly encountered are the radial and circular cracks between the jacket and core, therefore, a definite case is considered here: arch cracks in isotropic fibre and at the interphase. The basic designations for a reduced element are given in Fig. 9. The boundary conditions

(a) at the core-jacket boundary

$$\begin{aligned} &[\bar{\sigma}_{an} \cdot \bar{n}]^{+} = [\bar{\sigma}_{an} \cdot \bar{n}]^{-}, \quad U_{0}^{+} = \bar{U}_{a}^{-}, && (\zeta_0 \in l) \\ &[\bar{\sigma}_{an} \cdot \bar{n}]^{+} = 0, \quad [\bar{\sigma}_{an} \cdot \bar{n}]^{-} = 0, && (\zeta_0 \in l_a) \end{aligned} \tag{46}$$

(b) at the fibre-matrix interphase

$$\begin{aligned} &[\bar{\sigma}_{an} \cdot \bar{n}]^{+} = [\bar{\sigma}_{an} \cdot \bar{n}]^{-}, \quad \bar{U}_{a}^{+} = \bar{U}^{-}, && (\zeta \in \gamma) \\ &[\bar{\sigma}_{an} \cdot \bar{n}]^{+} = 0, \quad [\bar{\sigma}_{n} \cdot \bar{n}]^{-} = 0, && (\zeta \in \gamma_0)\,. \end{aligned} \tag{47}$$

Conditions 3 in formula (33) are to be met at the outer boundary of the reduced element. The displacement field in central part of the fibre $r \leqslant \varepsilon$ and in its jacket $\varepsilon \leqslant r \leqslant r_0$ is determined by superposition of suitable elementary functions

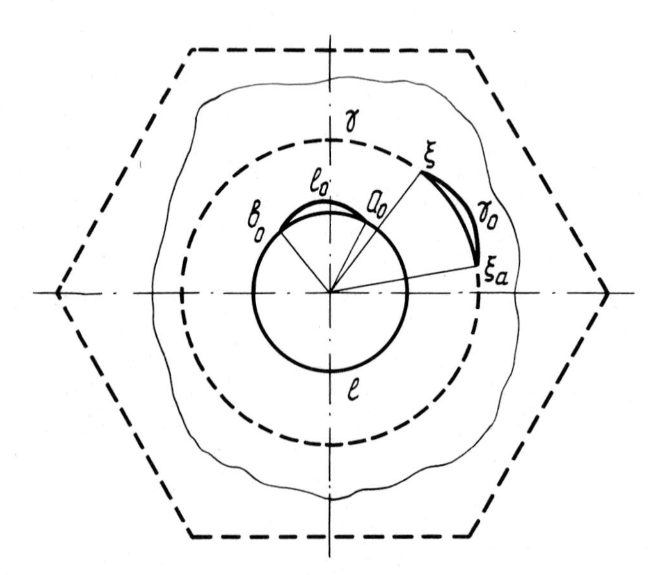

FIG. 9. Reduced element model at mixed failure of fibrous medium.

$$U_0 = \sum_{n>0} r^n (A_{mn}\tau^n + \bar{A}_{mn}\bar{\tau}^n),$$

$$U_a = \sum_{n>0} \left[r^n (C_{mn}\tau^n + \bar{C}_{mn}\bar{\tau}^n) + \frac{\varepsilon^{2n}}{r^n} (B_{mn}\tau^n + \bar{B}_{mn}\bar{\tau}^n) \right]. \tag{48}$$

Subsidiary functions $X_m(z)$ and $Y_m(z)$ introduced by analogy with relations (20) and (24) satisfy the system of functional equations (35) where G is substituted by G_0. Its solution is consistent with functions written down in equation (36). With the help of limiting conditions at an infinitely-remote point and at the centre of the inclusion, the expansion coefficients can be expressed through the remaining arbitrary constants C_{mn}. In particular, for an mth function the above transformations give the explicit form of polynomial $R_m(z)$

$$R_m(z) = \sum_{k=1}^{m} P_{m-k}\left(\cos\frac{\theta}{2}\right)\left[\frac{\bar{Z}_0^{m-k+1}}{\varepsilon^2}\left(\frac{\varepsilon^2}{Z}\right)^k \bar{C}_m - Z_a^{m-k} Z^{k-1} C_m\right] \tag{49}$$

where $Z_0 = \varepsilon\, e^{i\theta}$, $P_{m-k}(\cos\theta/2)$ is the Legendre polynomial.

Constants C_{mn} are chosen to satisfy the boundary conditions at the fibre-matrix interface $r = r_0$. The displacement in the matrix is found in the form

$$U = \sum_{n>0} (D_{mn}\tau^n + \bar{D}_{mn}\bar{\tau}^n)\frac{r_0^{2n}}{\tau^n} + (S_m\tau^m + \bar{S}_m\bar{\tau}^m)r^m, \tag{50}$$

and the corresponding subsidiary piecewise holomorphic function $V_m(z)$ is determined by relations

$$V_m(z) = \begin{cases} V_m^+(z) = \sum_{n>0} D_{mn} z^n + \bar{S}_m \dfrac{r^{2m}}{z^m} & (|z| < r_0) \\ V_m^-(z) = \sum_{n>0} \bar{D}_{mn} \dfrac{r_0^{2n}}{z^n} + S_m z^m & (|z| > r_0). \end{cases} \tag{51}$$

The functions $Y_m(z)$ which have been found are re-expanded power series, this resulting in that the displacements for the fibre jacket take the canonical form

$$U_a = \sum_{n>0} [U_{mn}(z)\tau^n + \bar{U}_{mn}(z)\bar{\tau}^n].$$

Here, each function U_{mn} depends parametrically on the succession C_{mn}. The subsidiary functions $U_m(z)$ are determined by standard relations

$$U_m(z) = \begin{cases} U_m^+(z) = \sum_{n>0} \dfrac{U'_{mn}(r_0)}{n} z^n & (|z| < r_0) \\ U_m^-(z) = \sum \dfrac{\bar{U}'_{mn}(r_0)}{n} \dfrac{r_0^{2n}}{z^n} & (|z| > r_0). \end{cases} \tag{52}$$

To separate the stressed-state singularities at the crack tips and to express the constants C_{mn} through the given coefficients J_m, we construct a second system of functional equations

$$\begin{aligned} &U_m^+(\zeta) - U_m^-(\zeta) = 0\,, \quad V_m^+(\zeta) - V_m^-(\zeta) = 0\,, \qquad (\zeta \in \gamma_0) \\ &[U_m(\zeta) + G/G_a V_m(\zeta)]^+ - [U_m(\zeta) + G/G_a V_m(\zeta)]^- = 0\,, \qquad (\zeta \in \gamma) \\ &[U_m(\zeta) - V_m(\zeta)]^+ + [U_m(\zeta) - V_m(\zeta)]^- = \psi_m(\zeta)\,, \qquad (\zeta \in \gamma) \end{aligned} \tag{53}$$

where

$$\psi_m(\zeta) = \sum \frac{1}{n}\left[(U'_{mn} - nU_{mn})\zeta^n + (\bar{U}'_{mn} - n\bar{U}_{mn})\frac{r_0^{2m}}{\zeta^n}\right].$$

The solution of this system is similar to those given previously (see (25), (35), (40)) and has the form

$$\begin{aligned} &(1 + G_a/G)U_m(z) = Q_m(z) + \mu(z)R_m(z) + \frac{\mu(z)}{2\pi i}\int_l \frac{\psi_m(\zeta)\,dr}{\mu^+(\zeta)(\zeta - z)}, \\ &V_m(z) = Q_m(z) - G_a/GU_m(z) \\ &Q_m(z) = S_m z^m + \bar{S}_m \frac{r_0^{2m}}{z^m}\,. \end{aligned} \tag{54}$$

Using the expansion

$$U_m^+(\zeta) - U_m^-(\zeta) = \sum\left(\frac{U'_{mn}}{n}\zeta^n - \frac{\bar{U}'_{mn}}{n}\bar{\zeta}^n\right).$$

and applying to the left side of the equality the Sokhatski formulae determining the limiting function values found in the form of integrals of the Couchy type, we arrive at a system of algebraic equations

$$\sum_k d_{mk}C_{mk} = S_m$$

where d_{mk} are the known coefficients. To find approximate analytical solutions, the first terms are retained in the left-hand side of the system. The numerical analysis is necessary in order to make the found solution more accurate at concrete values of parameters, with angles Θ and θ determining the crack location and size being included.

The coefficients C_{mk} being known, the integral moduli of cracked fibrous medium are constructed in the same manner as those given in equations (28). The stress intensity at the crack tips has a classical singularity.

2.8. *Influence of packing*

High strength and rigidity properties of composites are to a certain extent attainable by direct increase of the relative volume-content of the reinforcing phase. In high-field materials the fibre interaction effects become essential, so that a correct calculation of a stressed state and bulk characteristics requires for the subsequent terms in function expansions have to be retained. A method used below (VANIN 1977a) is effective for structures close to the regular (see Section 1.3). The distribution of perfect and imperfect fibre combinations is of major importance to the investigations of the role of subsequent approximations of the fibre interaction in the cell. The experiments with optically active glass fibre plastics indicate a more or less uniform cross-sectional appearance of localized cracks in the process of loading, although their subsequent growth is accelerated in the most loaded areas. At the initial stage the cracks, practically, do not interact with one another, so that in a hexagonal structure the simplest distribution model within the limits of an area with identical first-kind residual stresses corresponds to that given in Fig. 10. In a

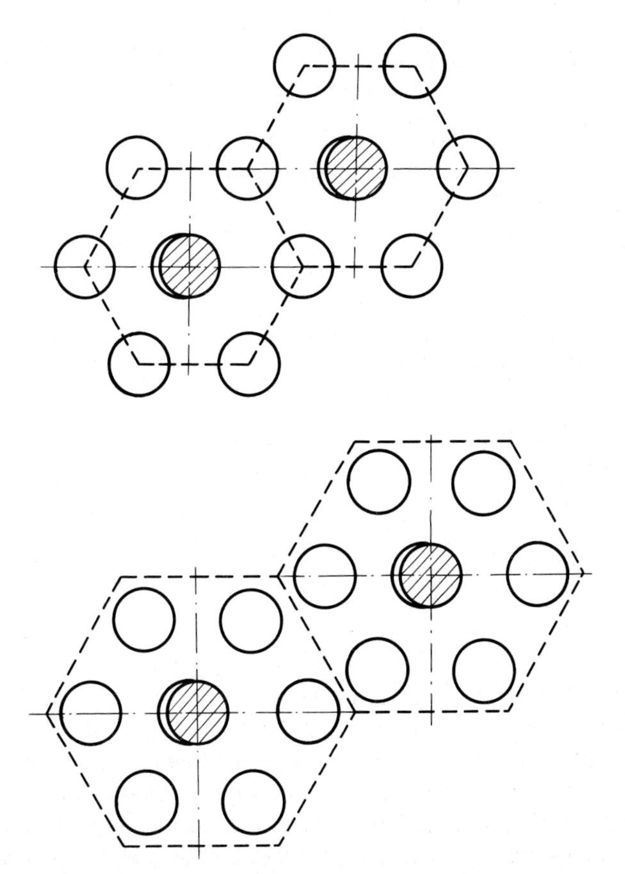

FIG. 10. A model of simple fibre distribution with cracks in hexagonal structure.

cell there is one fibre with a crack per two sound fibres. If the number of rows of the former fibres is increased up to two, then the said number will change by six.

Function $F(z)$ determining the longitudinal shear displacements in homogeneous isotropic components of the medium (see (10) is constructed by taking into account the interaction with the nearest row of the reduced elements. For hexagonal and tetragonal structures, in which the fibre with an interphase crack is enveloped by two rows of perfect elements, the second approximation $F(z)$ has been found in the form

$$F_a(z) = \frac{G}{G+G_a}\left\{Cz + \bar{C}\frac{r_0^2}{z} + \mu(z)\left(\frac{\bar{\zeta}_0}{z}\bar{C} - C\right) + \frac{nr_0^2}{\omega^n}\frac{G_a - G}{G_a + G}\left[\bar{C}z^{n-1} + C\left(\frac{r_0^2}{z}\right)^{n-1} + \mu(z)R_n(z)\right]\right\}.$$

$$F(z) = \frac{1}{G+G_a}\left\{G\left(Cz + \bar{C}\frac{r_0^2}{z}\right) - G_a\mu(z)\left(\frac{\bar{\zeta}}{z}\bar{C} - C\right) + \frac{n^2r_0^2}{\omega^n}\cdot\frac{G_a - G}{G_a + G}\left[G\bar{C}z^{n-1} + GC\left(\frac{r_0^2}{z}\right)^{n-1} - G_a\mu(z)R_n(z)\right]\right\}. \quad (55)$$

Here ω is the lattion constant, n is the number of fibres nearest to the one considered, $n = 6$ for the hexagonal structure and $n = 4$ for the tetragonal one.

$$R_n(z) = \sum_{k=1}^{n-1} P_{n-1-k}\left(\cos\frac{\theta}{2}\right)\left[\frac{\zeta_0^{n-k}}{r_0^2}\left(\frac{r_0^2}{z}\right)C - \zeta_0^{n-k-1}z^{k-1}\bar{C}\right]$$

All other symbols correspond to those given in equations (26). In case of $\langle\sigma_{13}\rangle = 0$ it should be assumed that

$$C = \langle\sigma_{12}\rangle\frac{G+G_a}{2G}\cdot\frac{G_a + \xi G_a\langle\lambda_1 + \lambda_2 e^{2i\Theta}\rangle + \eta G}{(G_a + \xi G_a\langle\lambda_1\rangle + \eta G)^2 - \zeta^2 Ga(\langle\cos 2\Theta_1\lambda_2\rangle^2 + \langle\sin 2\Theta\lambda_2\rangle^2)} \quad (56)$$

where

$$\langle\lambda_1\rangle = \frac{n}{n+1} - \frac{1}{n+1}\cos\frac{\theta}{2},$$

$$\langle\lambda_2 e^{2i\Theta}\rangle = \frac{1}{2(n+1)}e^{2i\Theta}\sin^2\frac{\theta}{2}.$$

Terms in square brackets (55) denote corrections to the first approximation introduced by a more accurate account of the interaction between the adjacent fibres. The investigation of the stress intensities at the cracks shows that corrections enhance the stress concentration in the interphase zone with the growth of the fibre-volume fraction. However, a strict calculation of local stresses at any closer fibre packing is possible when a greater number of the subsequent expansion terms is retained, this leading to cumbersome calculations. The bulk parameters of composites are less sensitive to rapidly changing local stresses and the more accurate consideration of the interaction essentially contributes to the formulae for moduli (28) at a higher filler-volume fraction. To calculate these corrections by formulae (12) and (13), below is the expression for the additional energy in the fibre and the adjoining matrix in the reduced elements with a crack

$$U'_a = 2n^2(n-1)\frac{\xi G^2 G_a}{(G+G_a)^2}\left(\frac{r_0}{\omega}\right)^{2n}\left(\frac{G_a-G}{G_a+G}\right)^2 \times\left(C\bar{C}+\frac{CA_{n-1}}{S_0^{n-1}}+\frac{\bar{C}\bar{A}_{n-1}}{\bar{\zeta}_0^{n-1}}+\sum_{j=1}^{\infty}\frac{jA_{-j}\bar{A}_j}{(n-1)r_0^{2(n-1)}}\right), \tag{57}$$

$$U'_s = 2n^2(n-1)\frac{\xi G}{(G+G_a)^2}\left(\frac{r_0}{\omega}\right)^{2n}\left(\frac{G_a-G}{G_a+G}\right)^2(1-\xi^{n-1})\Big[(G+G_a)^2\frac{C\bar{C}}{\xi^{n-1}} + G^2C\bar{C}-\frac{GG_a}{r_0^{2(n-1)}}(CA_{n-1}\bar{\zeta}_0^{n-1}+\bar{C}\bar{A}_{n-1}\zeta_0^{n-1}) + \sum_{j=1}^{\infty}\frac{1-\xi^j}{1-\xi^{n-1}}\cdot\frac{jB_j\bar{B}_j}{(n-1)r_0^2(n-1)}\Big].$$

Here, $n = 6$ should be taken for the hexagonal and $n = 4$ for the tetragonal structures; parameters being determined by formula (56), $\zeta_0 = r_0\,e^{i\theta}$,

$$A_j = \sum_{k=0}^{n-2} P_k\left(\cos\frac{\theta}{2}\right)(\bar{C}\zeta_0^{n-1}\lambda_{j-n+2+k} - C\bar{\zeta}_0^{n-1}\lambda_{j+n-k-1}),$$

$$\lambda_m = P_m\left(\cos\frac{\theta}{2}\right) - 2\cos\frac{\theta}{2}P_{m-1}\left(\cos\frac{\theta}{2}\right) + P_{m-2}\left(\cos\frac{\theta}{2}\right) \qquad (m \geqslant 2).$$

The remaining symbols correspond to those given above. For elements without cracks $\theta = 2\pi$. In this case we arrive directly at new more accurate dependences of the elasticity moduli upon the properties of the components for two types of fibre packing

$$G_{12} = G_{12}^0\Big\{1 + \frac{n^2(n-1)\xi(G_a-G)}{[(1+\xi)G_a+\eta G][\eta G_a+(1+\xi)G]}\left(\frac{r_0}{\omega}\right)^{2n} \Big[\left(\frac{\xi-\xi^n}{\xi^n}+(1-\xi)^{n-1}\right)\left(\frac{G_a-G}{G_a+G}\right)^2+\frac{4GG_a}{(G+G_a)^2}\Big]\Big\}. \tag{58}$$

Here, for hexagonal and tetragonal packings $(r_0/\omega)^2 = \xi\sqrt{3}/2\pi$ and $(r_0/\omega)^2 = \xi/\pi$

$$G_{12}^0 = G\frac{(1+\xi)G_a+rG}{rG_a+(1+\xi)G}.$$

The latter formula is widely known but is of insufficient accuracy at a dense ($\xi > 0.5$) filling of high-modulus composites. Formula (58) determines the effective moduli of the composites to a high accuracy at a filling reaching up to 0.9 from the ultimately permissible in regular structures. The calculations indicate a possibility of a double rise of the elasticity modulus in high-modulus composites at an ideal component contact due to rational packing. The stress intensity at the crack is sensitive to structure details and to obtain its strict calculation labour-consuming calculations are required at fixed fibre configuration in the material.

3. Longitudinal tension

The longitudinal tension is the most important state of a unidirectional composite and is most frequently encountered in practice. Elastic characteristics of fibres exceed by an order or more those in the matrix which determines a comparative effectiveness of simple notions of ordinary mixture models in the study of the stressed state and the estimation of elastic constants of the composites. In reality, this state, more than any other, depends upon the play of interphase stresses and deformations. The study of stresses in the composites by the aid of simplified models can hardly be correlated with the critical load during fibre or matrix rupture and is of little use in the attempt to clarify the mechanism and peculiarities of material failure. The contact between the fibres and matrix neutralizes, to a certain extent, the growth of surface cracks on them due to the effect of moisture and friction forces, enhancing apparently the critical loads as a result of local redistribution of stresses. To determine the quantitative relations between the crack size and orientation in the fibre, between the mechanical characteristics of the matrix and the critical load the investigation of models accounting for the spatial stressed state is of a considerable interest (VU 1978). In the studies of longitudinal tension of the composites two-dimensional models reveal only some aspects of the general problem.

3.1. Solution of problem

Henceforth the composites with homogeneous and isotropic properties with the interphase cracks accounted for will be considered. In contrast to the designations given in the previous section, angles θ and Θ denote the arch crack size and centre location. In longitudinal tension-contraction the state is represented by a sum of two states:

(1) fibrous medium tension without the account of the interaction between the fibre and matrix due to the action of unknown constant stresses, proceeding on the assumption that transverse to the fibre axes planes $X_1 = \text{const}$ are not distorted in the process of deformation;

(2) flat deformed state, when there are no longitudinal deformations and displacements are given at the interphases.

These displacements are so chosen that, there being no longitudinal deformations, the difference of transverse displacements of the matrix and fibres would ensure continuity of the total displacement at the surface of the ideal contact of the fibres with the matrix. The elementary state of components is determined directly

$$U_1 = x_1\langle\varepsilon_{11}\rangle\,, \qquad U_2 + \mathrm{i}U_3 = -\nu z\langle\varepsilon_{11}\rangle \tag{59}$$

The second state is studied by means of relations (6)–(10). The boundary conditions (5) remain unchanged, except the second one for the displacements, it being formulated with condition of component displacement difference (59)

$$(U_2 + \mathrm{i}U_3)^+ = (U_2 + \mathrm{i}U_3)^- + (\nu_a - \nu)z\langle\varepsilon_{11}\rangle \qquad (\zeta \in l)\,. \tag{60}$$

The mean tension stresses with the account of component interaction will be

$$\langle\sigma_{11}\rangle = (\xi E_0 + \eta E)\langle\varepsilon_{11}\rangle + 2\nu\langle\Phi(z) + \overline{\Phi(z)}\rangle + 2\nu_a\langle\Phi_a(z) + \overline{\Phi_a(z)}\rangle\,. \tag{61}$$

All other components of mean stress tensor are zero, therefore, the cell equilibrium conditions lead to additional relations resultant from the properties of elliptical functions (8) and (9). For simple hexagonal and tetragonal lattices they have the form

$$\begin{aligned}&\frac{\omega_j + \bar{\omega}_j}{2}\langle\sigma_{33}\rangle + \frac{\omega_j - \bar{\omega}_j}{2}\langle\sigma_{22}\rangle - \mathrm{i}\bar{\omega}_j\langle\sigma_{23}\rangle \\ &\quad = \omega_j(A_0 + \bar{A}_0) - \delta_j A_2 - \bar{\gamma}_j\bar{A}_2 + \bar{\omega}_j D_0 - \bar{\delta}_j\bar{D}_2 \qquad (j = 1, 2)\,.\end{aligned} \tag{62}$$

Here ω_1, $\omega_2 = \omega_1\,\mathrm{e}^{\mathrm{i}a}$ are lattice periods. For general two-periodical structures in which the cell contains N fibres equalities (62) are complicated and are, therefore, not given here. The stressed state of the reduced element in the first approximation is determined by piecewise holomorphic functions.

$$\begin{aligned}\Phi_a(z) = b_0 + \frac{P}{1-g} + \frac{P_0}{(1-g)z^2} + \mu(z)\Bigg[c + \frac{P\,\mathrm{e}^{\mathrm{i}\Theta}}{1-g}\left(\cos\frac{\theta}{2} + 2\beta\sin\frac{\theta}{2} - z\,\mathrm{e}^{\mathrm{i}\Theta}\right) \\ - \frac{P_0}{1-g}\,\frac{1 - z\,\mathrm{e}^{-\mathrm{i}\Theta}\left(\cos\frac{\theta}{2} - 2\beta\sin\frac{\theta}{2}\right)}{z^2\mu(0)}\Bigg]\,.\end{aligned} \tag{63}$$

$$\begin{aligned}\Phi(z) = \Gamma + \frac{\Gamma'}{z^2} - \frac{P}{1-g} - \frac{P_0}{(1-g)z^2} + \mu(z)\Bigg[-\frac{P\,\mathrm{e}^{\mathrm{i}\Theta}}{1-g}\left(\cos\frac{\theta}{2} + 2\beta\sin\frac{\theta}{2} - z\,\mathrm{e}^{-\mathrm{i}\Theta}\right) \\ + \frac{P_0}{1-g}\,\frac{1 - z\,\mathrm{e}^{-\mathrm{i}\Theta}\left(\cos\frac{\theta}{2} - 2\beta\sin\frac{\theta}{2}\right)}{z^2\mu(0)}\Bigg]\,.\end{aligned}$$

Here

$$\mu(z) = e^{i\theta}\sqrt{1 - 2z^{-i\theta}\cos\theta + z^2 e^{-2i\theta}}\left(\frac{z\,e^{i\theta} - e^{i\theta}}{z\,e^{-i\theta} - e^{-i\theta}}\right)^{i\beta}.$$

$$g = -\frac{\kappa G_a + G}{\kappa_a G + G_a}, \quad \beta = \frac{1}{2\pi}\ln(-g), \quad \mu(0) = -g\,e^{-i\theta - 2\beta\theta} \tag{64}$$

The constants introduced are related with internal matrix stresses

$$\Gamma' = \frac{\sigma_{33} - \sigma_{22}}{2} + i\sigma_{23}, \qquad \Gamma = \frac{\sigma_{22} + \sigma_{33}}{4} + \frac{2iG\varepsilon_\infty}{1+\kappa}$$

$$P_0 = \frac{1-\kappa}{\kappa_0 G + G_a}\, G_a\bar{\Gamma}',$$

$$P(\kappa_a G + G_a) = 2(\nu_a - \nu)G_a G\langle\varepsilon_{11}\rangle + (1+\kappa)G_a\Gamma - (1+\kappa_a)Gb_0.$$

In addition to the indicated, there are restrictions for the coefficients of function expansion (63) in power series resulting from their holomorphism as was shown by VANIN (1977a).

3.2. *Efficient constants*

The linearly-reinforced material model under consideration has, in a general case, one symmetry plane perpendicular to fibre orientation. Homogeneous media with one plane of symmetry belong to the monoclinic system so that the averaged law of elasticity for this material can be reduced to take the form

$$\begin{aligned}
\langle\varepsilon_{11}\rangle &= z_{11}\langle\sigma_{11}\rangle + z_{12}\langle\sigma_{22}\rangle + z_{13}\langle\sigma_{33}\rangle + z_{16}\langle\sigma_{23}\rangle,\\
\langle\varepsilon_{22}\rangle &= z_{21}\langle\sigma_{11}\rangle + z_{22}\langle\sigma_{22}\rangle + z_{23}\langle\sigma_{33}\rangle + z_{26}\langle\sigma_{23}\rangle,\\
\langle\varepsilon_{33}\rangle &= z_{31}\langle\sigma_{11}\rangle + z_{32}\langle\sigma_{22}\rangle + z_{33}\langle\sigma_{33}\rangle + z_{36}\langle\sigma_{23}\rangle,\\
\langle\varepsilon_{23}\rangle &= z_{61}\langle\sigma_{11}\rangle + z_{62}\langle\sigma_{22}\rangle + z_{63}\langle\sigma_{33}\rangle + z_{66}\langle\sigma_{23}\rangle.
\end{aligned} \tag{65}$$

Here $z_{ik} = z_{ki}$ and the number of essential constants is ten. The relations between the introduced designations and technical constants are as follows

$$\begin{aligned}
z_{12} &= -\frac{\nu_{12}}{E_{22}} = -\frac{\nu_{21}}{E_{11}} = -\nu_{21}z_{11}, & \frac{z_{12}^2}{z_{11}} &= -\nu_{21}z_{12},\\
z_{13} &= -\frac{\nu_{31}}{E_{11}} = -\nu_{31}z_{11}, & \frac{z_{12}z_{13}}{z_{11}} &= -\nu_{31}z_{12},\\
z_{16} &= \frac{\nu_{16}}{G_{23}} = \frac{\nu_{61}}{E_{11}} = \nu_{61}z_{11}, & \frac{z_{12}z_{16}}{z_{11}} &= \nu_{61}z_{12}.
\end{aligned} \tag{66}$$

In a discrete structure medium the symmetry is determined by parameters ω_1,

ω_2 and angle α between them, as well as by the structure symmetry within the cell. The first type corresponds to the translational structure symmetry, the second to the point symmetry. The simplest—hexagonal and tetragonal—structures can be regarded as systems with a lower symmetry on account of the misfit of the cell and crack symmetry axes or due to other imperfections. It should be borne in mind that in the successive approximation method additional corrections to parameters characterizing the symmetry of a medium can appear during a subsequent refinement of the solution found. In case of tension along the fibres, only $\langle\sigma_{11}\rangle \neq 0$ therefore, the increment of displacements (10) on the lattice period will be

$$\Delta(U_2 + \mathrm{i}U_3) = \left(\frac{\omega_j + \bar{\omega}_j}{2} z_{21} + \frac{\omega_j - \bar{\omega}_j}{2} z_{31} + \frac{\mathrm{i}\bar{\omega}_j}{2} z_{16}\right)\langle\sigma_{11}\rangle$$

$$= \frac{\kappa}{2G}(\omega_j A_0 - \delta_j A_2) - \frac{\omega_j}{2G}\bar{A}_0 - \frac{\bar{\omega}_j}{2G}\bar{D}_0 + \frac{\delta_j}{B}\bar{D}_2, \qquad (j = 1, 2) \tag{67}$$

The longitudinal elongation modulus is determined from equality (61), the remaining constants from the above formula. The final result

$$z_{11}^{-1} = E_{11} = \xi E_a + \eta E + \frac{8\xi\eta(\nu_a - \nu)^2 G G_a L}{\xi(1+\kappa)LG_a + 2\eta(G_a + \kappa_a G) - \eta(1+\kappa_a)GL}, \tag{68}$$

where

$$L(1-g) = 1 - g\,\mathrm{e}^{-\beta\theta}\left(\cos\frac{\theta}{2} + 2\beta\sin\frac{\theta}{2}\right) \tag{69}$$

For an ideal contact $\theta \to 0$, $L \to 1$, then the formula coincides completely with that known for the Young's modulus of a reinforced material when there is no coupling between the fibre and matrix, $\theta \to 2\pi$, $L \to 0$ and the last term disappears. Although this term reflects the result of composite components interaction, its contribution is quite negligible as was shown by SKUDRA and BULAVS (1982). The following transformations of relationship (67) lead to dependence

$$z_{16} \simeq 0, \qquad z_{31} \simeq z_{21},$$

$$z_{21} = -z_{11}\left[\nu + \frac{2\xi(\nu_a - \nu)\{(1+\xi-2\nu)G_a L + \eta[G_a + \kappa_a(1-L)G]\}}{\xi(1+\kappa)G_a L + 2\eta(G_a + \kappa_a G) - \eta(1+\kappa_a)GL}\right] = -\nu_{21} z_{11} \tag{70}$$

Here ν_a and ν are the Poisson ratios of the fibres and matrix, $\kappa = 3 - 4\nu$. The expression in square brackets characterizes the transverse composite deformations at longitudinal tension, i.e. the effective Poisson ratio. The distribution of stresses at the interphase in hexagonal fibre packing, there being no imperfections at the interphase, fully complies with the distribution of shrinkage stresses (Fig. 1). For this reason in the longitudinal tension the interphase stresses breaking the matrix off the fibres sum up with the residual stresses, promoting, thus, the appearance of local cracks. The stress intensity at the crack apex has a typical singularity. In particular, for a symmetrically disposed crack ($Q=0$) the first term of the series will be

$$\begin{aligned}\sigma_{rr}+\mathrm{i}\sigma_{r\varphi} &= \frac{P\,\mathrm{e}^{-\beta(\theta/2)}}{1-g}\sqrt{\frac{\sin(\theta/2)}{2\rho}}\,[(2\beta-\mathrm{i})(g\,\mathrm{e}^{\beta(\pi/2)}+\mathrm{i}\,\mathrm{e}^{-\beta(\pi/2)})]\\ &\times\exp\Big[\mathrm{i}\Big(\beta\,ln\,\frac{\rho}{2\sin(\theta/2)}-\frac{\theta}{4}-\frac{\pi}{4}\Big)\Big]+\mathrm{i}g(1+4\beta^2)\\ &\times\exp\Big[\frac{\beta\pi}{2}-\mathrm{i}\Big(\beta\,ln\,\frac{\rho}{2\sin(\theta/2)}-\frac{\theta}{4}-\frac{\pi}{4}\Big)\Big]\end{aligned}\qquad(71)$$

Here ρ is the distance from the crack apex. When $\rho\to 0$, the stresses at the crack tip oscillate. In the limiting case of a homogeneous medium $G_a\to G$, $\kappa_a\to\kappa$, $\beta\to 0$, and the oscillations disappear

$$\sigma_{rr}+\sigma_{r\varphi}=\frac{P}{1-g}\sqrt{\frac{\sin(\theta/2)}{2\rho}}\Big[\mathrm{e}^{-\mathrm{i}(\theta/4+\pi/4)}-2g\sin\Big(\frac{\theta+\pi}{4}\Big)\Big]+\cdots$$

3.3. *Media with several cracks*

If there are several cracks at the interphase, then the first terms of the expansion of the effective parameters into series retain their previous form (68), (70), where, in case of n symmetrically located cracks, we take

$$L(1-g)=1-g\,\mathrm{e}^{-n\beta\theta}\Big(\cos n\,\frac{\theta}{2}+2\beta\sin n\,\frac{\theta}{2}\Big),\qquad 0\leqslant\theta\leqslant 2\,\frac{\pi}{n}.$$

The first terms of the expansion of local stresses at the tips of two symmetrically disposed cracks in a power series of the interaction of the reduced elements take the form

$$\begin{aligned}\sigma_{rr}+\mathrm{i}\sigma_{r\varphi} &= \frac{P\,\mathrm{e}^{\beta(\pi/2-\theta)}}{2(1-g)}\sqrt{\frac{\sin\theta}{\rho}}\Big\{(2\beta-\mathrm{i})(g+\mathrm{i}\,\mathrm{e}^{-\beta\pi})\\ &\times\exp\Big[\mathrm{i}\Big(\beta\,ln\,\frac{\rho}{\sin\theta}-\frac{\theta}{2}-\frac{\pi}{4}\Big)\Big]+\mathrm{i}g(1+4\beta^2)\\ &\times\exp\Big[-\mathrm{i}\Big(\beta\,ln\,\frac{\rho}{\sin\theta}-\frac{\theta}{2}-\frac{\pi}{4}\Big)\Big]\Big\}+\cdots\end{aligned}$$

In this term the oscillations disappear in a medium in which the mechanical characteristics of components do not differ from one another.

3.4. Longitudinal compression

There being no overlapping of crack edges and fibre distortions, the longitudinal compression of a reinforced medium causes opposite-sign stresses which, at a certain loading stage, compensate the second-type interphase shrinkage stresses. In this case the residual compressive stresses in the fibre sum up with the outer stresses, enhancing, thus, the loss of fibre stability. Cracks near the interphase change the inertia moments of the reduced elements and promote their bending in a corresponding plane. The considered stress re-distribution mechanism in the structure of a medium preceeds a general failure of material under compression. In structure elements an essential role belongs to the first-type residual stresses which depend on the geometry of the element and can change the initial mechanisms of its failure.

4. Transverse shear

The transverse shear and tension of linearly reinforced composites amount to problems for two-dimensional regions, provided there are no perturbations liable to cause a transition from two-dimensional to spatial stressed state. This restraint, as in the previous cases, precludes, naturally, the investigation of stresses close to the fibre edges, as well as under the action of fields changing rapidly at distances of the order of characteristic dimensions of composites.

4.1. Initial relations

By excluding stresses $\langle \varepsilon_{11} \rangle = 0$ from equations (64) and using designations from (65), for a body with a monoclinic symmetry which is in a flat deformed state $\langle \sigma_{11} \rangle$ we obtain

$$
\begin{aligned}
\langle \varepsilon_{22} \rangle &= (z_{22} + \nu_{21} z_{12}) \langle \sigma_{22} \rangle + (z_{23} + \nu_{31} z_{12}) \langle \sigma_{33} \rangle + (z_{26} - \nu_{61} z_{12}) \langle \sigma_{23} \rangle \,, \\
\langle \varepsilon_{33} \rangle &= (z_{23} + \nu_{21} z_{13}) \langle \sigma_{22} \rangle + (z_{33} + \nu_{31} z_{13}) \langle \sigma_{33} \rangle + (z_{36} - \nu_{61} z_{13}) \langle \sigma_{23} \rangle \,, \\
\langle \varepsilon_{23} \rangle &= (z_{26} + \nu_{21} z_{16}) \langle \sigma_{22} \rangle + (z_{36} + \nu_{31} z_{16}) \langle \sigma_{33} \rangle + (z_{66} - \nu_{61} z_{16}) \langle \sigma_{23} \rangle \,.
\end{aligned}
\tag{72}
$$

From the above equations follows that the transverse state has to be considered after the longitudinal tension characteristics have been determined. Neglecting the second bracketed terms in (72) results in noticeable errors for some composits and violates all the limiting procedures. Average stresses are given at lateral cell boundary by relations (62), the boundary conditions at the interphases by equalities (5), and the first-approximation functions by formulae (63). At the outer boundary of a reduced element functions (63) are joined with expansions (8) and (9), ensuring the continuity and smoothness of stresses displacements. More accurate results are obtained by means of a computer count with retaining the functions of a subsequent approximations.

4.2. *Efficient constants*

The effective modulus of a transverse shear and the side effects in Hook's law are determined by the equality of an increment of the transverse displacements within the structure period to the corresponding changes in the quasi-periodic functions in expansions (8) and (9)

$$\langle\sigma_{23}\rangle = \left[\frac{\omega_j+\bar{\omega}_j}{2}(z_{26}-\nu_{61}z_{12})+\frac{\omega_j-\bar{\omega}_j}{2}(z_{36}-\nu_{61}z_{13})+\frac{\mathrm{i}\omega_j}{2}(z_{66}-\nu_{61}z_{16})\right]$$

$$=\frac{\kappa}{2G}(\omega_jA_0-\delta_jA_2)-\frac{\bar{\omega}_j}{2G}\bar{D}_0+\frac{\bar{\delta}_j}{2G}\bar{D}_2+\frac{\bar{\omega}_j}{2G}\bar{A}_o\,. \tag{73}$$

In the first approximation the dependence of integral constants of a fibrous medium under a transverse shear upon dimensions and orientation of an interphase crack will be

$$z_{66}=\nu_{61}z_{16}+\frac{1}{G}\frac{(1+\kappa)[1-\xi\kappa\chi(\theta)]G_a+(1+\kappa_a)(1+\xi\kappa)G}{(1+\kappa)[1-\xi\chi(\theta)]G_a+\eta(1+\kappa_a)G},$$

$$z_{26}=\nu_{61}z_{12}+\xi\frac{1+\kappa}{2G}\cdot\frac{G_a\pi(\theta)\sin 2\Theta-G_a\lambda(\theta)\sin 4\Theta}{(1+\kappa)[1+\xi\chi(\theta)]G_a+\eta(1+\kappa_a)G}, \tag{74}$$

$$z_{36}=\nu_{61}z_{13}+\xi\frac{1+\kappa}{2G}\cdot\frac{G_a\pi(\theta)\sin 2\Theta+G_a\lambda(\theta)\sin 4\Theta}{(1+\kappa)[1+\xi\chi(\theta)]G_a+\eta(1+\kappa_a)G},$$

where

$$\chi(\theta)=-\frac{\mathrm{e}^{-\beta\theta}}{g}\left(\cos\frac{\theta}{2}-2\beta\sin\frac{\theta}{2}\right)$$

$$-\frac{\sin^4(\theta/2)}{(1-g)L}\left(\frac{1}{2}+2\beta^2\right)^2\frac{2a\cos^2 2\Theta-(1+\kappa_a)L}{2a-(1+\kappa_a)L},$$

$$\pi(\theta)=\frac{1}{4}+\frac{3}{4}\cos\theta-2\beta\sin\theta+2\beta^2\sin\frac{\theta}{2}+\left(\frac{1}{2}+2\beta^2\right)\frac{(1+\kappa_a)L\sin^2(\theta/2)}{2a-(1+\kappa_a)L}, \tag{75}$$

$$\lambda(\theta)=\left(\frac{1}{2}+2\beta^2\right)^2\frac{\sin^4(\theta/2)}{(1-g)L}\cdot\frac{\kappa_a}{2a-(1+\kappa_a)L},$$

$$a=\kappa_a+G_a/G\,.$$

The remaining designations are introduced by formulae (64) and (69). The side effects z_{26} and z_{36} in this approximation are caused by the break of symmetry in rack orientation as regards the average shear stresses. The transverse shear state can be obtained by superposition of the transverse shear tensions and transverse compression stresses on perpendicular areas. The latter stresses may cause an overlapping of crack edges which would result in a change of the stressed state. Therefore, at the transverse shear the critical states can be attained from tangents and tension stresses on areas inclined with respect to the main ones. At the crack tips the stresses have the singularities of the type (71). The all-over failure of a structure element is decidedly influenced by the first-type residual stresses which, when summed up with the external stresses promote the merging of the local cracks into a macrocrack.

5. Transverse tension

Unidirectional composites are the least resistant to failure at transverse tension due to the concentration of local stresses at the interphase even with an ideal contact of the components. The growth and appearance of a multitude of local cracks under this load favours a sharp decrease of the elasticity moduli of the material, and the contribution of transverse layers to the total resistance to failure of a laminated composite is frequently disregarded. In fact, when at transverse tension the cracks do not fully divide the weakened layer owing to the support of the adjacent layers, the weakened layer plays an active role in taking up the shear stresses and in distributing the stress between the adjoining layers. It is, therefore, of interest to reveal the change of the elastic constants of a composite with the growth of concentration and size of the transverse cracks.

5.1. Parameters of first state

To make it more definite, a state of material is considered when only normal stresses $\langle\sigma_{22}\rangle$ are acting. The problem is absolutely equivalent to the previous one: the average stresses at the side surfaces of a separated volume are given by formulae (62), the boundary conditions by (5), and the functions of first approximation by (63). The field in the reduced volume is made consistent with the field in the remaining part of the cell at its external surface. The effective elastic constants are determined by two equations of type (73)

$$\langle\sigma_{22}\rangle\left[\frac{\omega_j+\bar{\omega}_j}{2}(z_{22}+\nu_{21}z_{12})+\frac{\omega_j-\bar{\omega}_j}{2}(z_{23}+\nu_{21}z_{13})+\frac{\mathrm{i}\bar{\omega}_j}{2}(z_{26}+\nu_{21}z_{16})\right]$$

$$=\frac{\kappa}{2G}(\omega_jA_0-\delta_jA_2)-\frac{\omega_j}{2G}\bar{A}_0-\frac{\omega_j}{2G}\bar{D}_0+\frac{\bar{\delta}_j}{2G}\bar{D}_2\,;\qquad (j=1,2)\,. \tag{76}$$

In the first approximation the parameters to be found take the form

$$z_{22}=-\nu_{21}z_{12}+\frac{1+\kappa}{4G}\cdot\frac{\kappa G_a+(1+2\xi)[G_a+(1+\kappa_a)G]-2\xi G_a(\kappa d+b+c)}{(1+\kappa)[1+\eta+\xi(d+b+f)]G_a+2\eta(1+\kappa_a)G}\,,$$

$$z_{23}=-\nu_{21}z_{13}-\frac{1}{2G}\cdot\frac{(1+\kappa)G_a+(1+\xi\kappa)(1+\kappa_a)G-\xi(1+\kappa)(\kappa d+b+h)G_a}{(1+\kappa)[1+\eta+\xi(d+b+f)]G_a+2\eta(1+\kappa_a)G}\,. \tag{77}$$

The other parameters are bulky and of small values. Here

$$d = -\frac{e^{\beta\theta}}{g}\left(\cos\frac{\theta}{2} - 2\beta\sin\frac{\theta}{2}\right) - \frac{\sin^4(\theta/2)}{1-g}\left(\frac{1}{2} + 2\beta^2\right)^2\frac{\sin^2 2\Theta}{L}$$

$$-\frac{1}{2}\cdot\frac{1+\kappa_a}{1-g}\left(\frac{1}{2} + 2\beta^2\right)\sin^2\frac{\theta}{2}\cos 2\Theta\,\frac{(1-g)L - (1+4\beta^2)\sin^2\frac{\theta}{2}\cos 2\Theta}{2a - (1+\kappa_a)L},$$

$$2b = (1-g)L - 2\cos^2\frac{\theta}{2} + 4\beta\sin^2\theta - 8\beta^2\sin^2\frac{\theta}{2}$$

$$+ (1+\kappa_a)L\cdot\frac{(1-g)L - (1+4\beta^2)\sin^2\frac{\theta}{2}\cos 2\Theta}{2a - (1+\kappa_a)L}, \tag{78}$$

$$f = \left(\frac{1}{8} + \frac{7}{8}\cos\theta - 2\beta\sin\theta + \beta^2\sin^2\frac{\theta}{2}\right)\cos 2\Theta,$$

$$h = \frac{\kappa}{2}\left(\frac{1}{2} + 2\beta^2\right)\sin^2\frac{\theta}{2}\cos 2\Theta + \left(\frac{1}{4} + \frac{3}{4}\cos\theta - 2\beta\sin\theta + 2\beta^2\sin^2\frac{\theta}{2}\right)\cdot\cos 2\Theta,$$

$$c = -\frac{\kappa}{2}\left(\frac{1}{2} + 2\beta^2\right)\sin^2\frac{\theta}{2}\cos 2\Theta + \left(\frac{1}{4} + \frac{3}{4}\cos\theta - 2\beta\sin\theta + 2\beta^2\sin^2\frac{\theta}{2}\right)\cdot\cos 2\Theta.$$

The remaining designations correspond to those given in formulae (64) and (69).

5.2. *Second state*

If the transverse tension of a composite is performed by normal stresses $\langle\sigma_{33}\rangle \neq 0$, then, as a result of the boundary problem solution, the remaining constants z_{33} and z_{36}, as well as the symmetry relations between the elastic parameters are determined. In particular, for a fixed system of coordinates

$$z_{33} = -\nu_{31}z_{13} + \frac{1+\kappa}{4G}\cdot\frac{\kappa G_a + (1+2\xi)[G_a + (1+\kappa_a)G] - 2\xi G_a(\kappa d' + b' + c')}{(1+\kappa)[1+\eta+\xi(d'+B'+f')]G_a + 2\eta(1+\kappa_a)G},$$

$$d' = -\frac{e^{\beta\theta}}{g}\left(\cos\frac{\theta}{2} - 2\beta\sin\frac{\theta}{2}\right) - \frac{\sin^4(\theta/2)}{1-g}\left(\frac{1}{2} + 2\beta^2\right)^2\frac{\sin^2 2\Theta}{L}$$

$$+\frac{1}{2}\frac{1+\kappa_a}{1-g}\left(\frac{1}{2} + 2\beta^2\right)\sin^2\frac{\theta}{2}\cos 2\Theta\,\frac{(1-g)L + (1+4\beta^2)\sin^2\theta/2\cos 2\Theta}{2a - (1+\kappa_a)L},$$

$$2b' = (1-g)L + 2\cos^2\frac{\theta}{2} - 4\beta\sin\theta + 8\beta^2\sin^2\frac{\theta}{2} \tag{79}$$

$$+ (1+\kappa_a)L\cdot\frac{(1-g)L + (1+4\beta^2)\sin^2\theta/2\cos 2\Theta}{2a - (1+\kappa_a)L},$$

$$f' = -f, \qquad c' = -c.$$

The change of the elastic moduli with the growth of angle θ in this approximation is monotonic from the maximum value in a composite with an ideal contact of components to a minimum value in a porous matrix. Transverse effects for symmetrically disposed cracks reach the maximum value near $\theta = \pi$. The stress intensity at the crack tips has a typical singularity (71).

6. Conclusion

The present state of the mechanics of composites is characterized by the construction of quantitative relations between parameters determining the effective properties of materials and their resistance to failure and by values determining the properties of components, their interaction and geometry. Mathematical formulation of conditions of a nonideal phase contact results in raising complex mixed boundary problems of the elasticity theory for multiple-connected regions the solution procedure of which has not been developed. In this chapter, on the example of a problem on longitudinal shear in fibrous medium, new results are presented obtained by the author for materials with inhomogeneous and anisotropic components. The latter enables one to take into account the effect of the characteristics of high-modulus materials upon the process of origination and development of local cracks. The investigation of changes in the bulk characteristics of composites with the account of the growing crack concentration, size and orientation in the structure leads to new problems on the changes of the stressed structure state in the process of their long-term failing and to the estimation of their safe durability life. In this aspect the nearest problems of the mechanics of composites are related to the investigation of the spatial stressed state of the composites with the account of the third dimension of defects, especially in the problem of longitudinal tension – compression, as well as to the formulation of criteria for short-term and long-term failing of inhomogeneous bodies.

References

BARENBLATT, G.I. (1959), *Prikl. Mate. Mech.* **23**, 622.
VAN FO FI, G.A. (1971), *The Theory of Reinforced Materials* (Naukova Dumka, Kiev) 199.
VANIN, G.A. (1977a), *Prikl. Mech.* **8**, 35.
VANIN, G.A. (1977b), *Prikl. Mech.* **10**, 14.
VANIN, G.A. (1978), *DAN Ukr. SSR* **7**, 603.
VANIN, G.A. and BYKHOVETS, A.N. (1979), *Durability Problems* **3**, 3.
VANIN, G.A. (1979), *Mech. Composite Materials* **2**, 205.
VU, E. (1978), Durability and failure of composites, in: L.J. BRAUTMAN, ed., *Composites Materials*, Vol. 5 (Mir, Moscow) Ch. 5.
GURVITS, A. and R. KURANT (1968), *The Function Theory* (Nauka, Moscow) 178.
MUSKHELISHVILI, N.I. (1966), *Some Fundamental Problems of Mathematical Elasticity Theory* (Nauka, Moscow) 111.
SIH, J. and J. LIBOVITS (1975), Mathematical theory of brittle failure, in: J. LIBOVITS, ed., *Failure*, Vol. 2 (Mir, Moscow) Ch. 2.
SKUDRA, A.M. and F. YA. BULAVS (1982), *Reinforced Plastics Strength* (Chimia, Moscow) 116.
CHEREPANOV, G.P. (1974), *Brittle Failure Mechanics* (Nauka, Moscow) 145.

Subject Index